EMBEDDED SOPC DESIGN WITH NIOS II PROCESSOR AND VHDL EXAMPLES

EMBEDDED SOPC DESIGN WITH NIOS II PROCESSOR AND VHDL EXAMPLES

Pong P. Chu
Cleveland State University

A JOHN WILEY & SONS, INC., PUBLICATION

Published by John Wiley & Sons, Inc., Hoboken, New Jersey
Published simultaneously in Canada

For general information on our other products and services or for technical support, please contact our Customer Care Department within the United States at (800) 762-2974, outside the United States at (317) 572-3993 or fax (317) 572-4002.

Wiley also publishes its books in a variety of electronic formats. Some content that appears in print may not be available in electronic formats. For more information about Wiley products, visit our web site at www.wiley.com.

Library of Congress Cataloging-in-Publication Data is available.

ISBN 978-1-118-00888-1

Printed in the United States of America.

oBook ISBN: 978-1-118-14653-8
ePDF ISBN: 978-1-118-14650-7

10 9 8 7 6 5 4 3 2 1

To my mother, Chi-Te, my wife, Lee, and my daughter, Patricia

CONTENTS

PART III   CUSTOM I/O PERIPHERAL DEVELOPMENT

13 Custom I/O Peripheral with PIO Cores   **297**

14 Avalon Interconnect and SOPC Component   **305**

PART IV   HARDWARE ACCELERATOR CASE STUDIES

PREFACE

An *SoC* (system on a chip) integrates a processor, memory modules, I/O peripherals, and custom hardware accelerators into a single integrated circuit. As the capacity of *FPGA* (field-programmable gate array) devices continues to grow, the same design methodology can be realized in an FPGA chip and is sometimes known as *SoPC* (system on a programmable chip). In a traditional embedded system, the hardware is constructed around a fixed-sized processor and off-the-shelf peripherals and the software is customized to implement the desired functionalities. The emerging SoPC-based design provides a new alternative. Because of the programmability of FPGA devices, *customized hardware* can be incorporated into the embedded system as well. We can tailor the processor, select only the needed I/O peripherals, create a custom I/O interface, and develop specialized hardware accelerators for computation-intensive tasks.

The current development of *HDL* (hardware description language) synthesis and FPGA devices and the availability of soft-core processors allow designers to quickly develop and simulate custom hardware and software, realize the entire system on a prototyping device, and verify the operation of the physical implementation. We can now use a PC and an inexpensive FPGA prototyping board to construct a sophisticated embedded system. This book uses a "learning by doing" approach and illustrates the hardware and software design and development process by a series of examples. An Altera FPGA prototyping board and its *Nios II soft-core processor* are used for this purpose.

The book is divided into four major parts. Part I covers HDL and synthesis of custom hardware. Part II provides an overview of embedded software development with the emphasis on low-level I/O access and drivers. Part III demonstrates the

design and development of hardware and software for several complex I/O peripherals, including a PS2 keyboard and mouse, a graphic video controller, an audio
codec, and an SD (secure digital) card. Part IV provides several case studies of
the integration of hardware accelerators, including a custom GCD (greatest common divisor) circuit, a Mandelbrot set fractal circuit, and an audio synthesizer
based on DDFS (direct digital frequency synthesis) methodology. All the hardware
and software examples can be synthesized, compiled, and physically tested on the
prototyping board.

Focus and audience

Focus   The embedded system is studied extensively and many books cover this
subject. The coverage is mostly on the software development, usually around a
specific processor. The new "hardware programmability" of the SoPC platform
provides a new dimension on the embedded system development. This book mainly
focuses on this aspect and the relevant design issues, including the derivation of a
soft-core processor and *IP* (*intellectual property*) core based system, the partition
and integration of software and hardware, and the development of custom I/O
peripherals and hardware accelerators.

Audience and prerequisites   The intended audience is students in an advanced digital
design, embedded system, or software–hardware codesign course as well as prac-
ticing engineers who wish to learn FPGA-, HDL-, and SoPC-based development.
Readers need to have a basic knowledge of digital systems, usually a required course
in electrical engineering and computer engineering curricula, and a working knowledge of the C language. Prior exposure to computer architecture, microcontroller,
and operating system is not necessary but will be helpful.

Logistics

FPGA prototyping board   This book is prepared to be used with an Altera *DE1*
board (also known as *Cyclone II FPGA Starter Development Kit*) and *DE2* board.
All HDL and C codes and discussions can be applied to the two boards directly.
Most peripherals discussed in this book are de facto industrial standards, and the
corresponding codes can be used as long as a board contains an Altera FPGA device
and provides proper analog interface circuits and connectors.

PC accessories   The design examples include interfaces to several PC peripheral
devices. A PS2 keyboard, a PS2 mouse, and a VGA compatible monitor, a pair of
earphones or powered speakers, and an SD card are required for the respective I/O
peripherals. These accessories are widely available and probably can be obtained
from an old PC.

Software   Three Altera software packages are needed for the Nios II-based system:
Quartus II Web edition, which performs HDL synthesis and simulation, *SOPC
Builder*, which configures and creates a Nios II-based system, and *Nios EDS* (*embedded design suite*), which is the integrated software development platform. All
three software packages can be downloaded from Altera's web site.

Codes and tutorials   The HDL and C codes of the book can be obtained from the companion web site. The codes and tutorials are developed and tested with *Altera Quartus II Web Edition v10 sp1* and *Altera Nios II EDS v10 sp1*. The software packages are running under Windows 7 32-bit with administrator privileges. Minor differences in the procedure may occur for other versions and operating systems.

Book organization

The book consists of four parts plus an introductory chapter. It starts with the "big picture":

- Chapter 1 provides an overview of embedded system and introduces the SoPC concept and development flow.

Part I introduces the basic HDL constructs and synthesis procedure and demonstrates the construction of custom digital circuits. It consists of six chapters:

- Chapter 2 describes the skeleton of an HDL program, basic language syntax, and logical operators. Gate-level combinational circuits are derived with these language constructs.
- Chapter 3 provides an overview of an FPGA device, prototyping board, and development flow. The development process is demonstrated by a tutorial of the Altera Quartus II synthesis software.
- Chapter 4 introduces HDL's relational and arithmetic operators and routing constructs. These correspond to medium-sized components, such as comparators, adders, and multiplexers. Module-level combinational circuits are derived with these language constructs.
- Chapter 5 presents the description of memory elements and the construction of "regular" sequential circuits, such as counters and shift registers, in which the state transitions exhibit a regular pattern, as well as a discussion of the use and inference of Cyclone II device's internal memory modules.
- Chapter 6 discusses the construction of a finite state machine (FSM), which is a sequential circuit whose state transitions do not exhibit a simple, regular pattern.
- Chapter 7 presents the construction of an FSM with data path (FSMD). The FSMD is used to implement register transfer (RT) methodology, in which the system operation is described by data transfers and manipulations among registers.

Part II introduces the construction of a Nios II-based system and the development of embedded software. A simple flashing-LED design is used to illustrate the key concepts of this process. It consists of five chapters:

- Chapter 8 provides an overview of the Nios II soft-core processor and examines its key components.
- Chapter 9 introduces the construction of a Nios II-based system and the basic coding techniques to access low-level I/O peripherals. The derivation of hardware and software is demonstrated by a tutorial of Altera SOPC Builder and Nios II EDS, respectively.
- Chapter 10 examines the structure and use of several IP cores (i.e., pre-designed I/O peripherals) of SOPC Builder and covers the development of ad hoc I/O driver software routines.

- Chapter 11 provides an overview of the Altera *HAL* (*hardware abstraction layer*) run-time environment and illustrates its usage.
- Chapter 12 discusses the interrupt structure, including the operation of Nios II's interrupt controller and the development of software interrupt service routines.

Part III applies the techniques from Parts I and II to design an array of peripheral modules on the prototyping board. Each module consists of custom hardware and a basic software driver. These can be considered as primitive IP cores and incorporated into a larger project. Part III consists of seven chapters:

- Chapter 13 demonstrates the I/O interfacing with PIO IP cores. This scheme can be used for simple I/O peripherals and avoids the overhead of creating a new SOPC component.
- Chapter 14 gives an overview of Altera's *Avalon interface*, which functions as a "bus structure" for a Nios II processor to connect memory and I/O modules, and demonstrates the procedure of creating a customized IP core.
- Chapter 15 covers the interface to the external SRAM (static RAM) and SDRAM (synchronous dynamic RAM) devices and the basic testing procedure.
- Chapter 16 covers the design of the PS2 interface. The hardware portion consists of a PS2 controller to generate and process the PS2 clock and data signals. The software portion is composed of two sets of drivers: one for the PS2 keyboard, which reads and decodes scan codes from a keyboard, and one for the PS2 mouse, which obtains and processes the button and movement information from a mouse.
- Chapter 17 presents the design and implementation of a graphic video controller. The hardware portion covers the generation of video synchronization signals and the construction and interface of a custom SRAM-based video memory module. The software portion covers the basic driver routines to draw pixels and to display and process bitmap images and texts.
- Chapter 18 discusses the design of the audio codec chip interface. The hardware portion consists of an I^2C bus controller for codec configuration and a serial bus controller to transmit and receive digitalized audio data streams. The software portion is composed of routines to set codec parameters and to generate and record the audio data.
- Chapter 19 presents the design of the SD card interface. The hardware portion is done by an SPI bus controller and the software portion consists of driver routines for card initialization and basic file read and write operations.

Part IV presents three case studies of hardware accelerators, which utilize custom hardware to perform computation intensive tasks. It includes three chapters:

- Chapter 20 shows the design of a custom GCD (greatest common divisor) accelerator based on the binary Euclid algorithm. Its performance is compared with software-based implementation.
- Chapter 21 illustrates the construction and integration of a Mandelbrot set fractal accelerator, which can select any portion of the set and displays the fractal on a VGA screen.
- Chapter 22 discusses the implementation of a direct digital frequency synthesis and modulation circuit. The circuit is used for an audio synthesizer with adjustable envelops.

Companion Web Site

An accompanying web site (`http://academic.csuohio.edu/chu_p/rtl`) provides additional information, including the following materials:

- Errata
- Code listing and relevant files
- Links to Altera software
- Links to referenced materials
- Additional project ideas

Errata   The book is self-prepared, which means that the author has produced all aspects of the text, including illustrations, tables, code listings, indexing, and formatting. As errors are always bound to happen, the accompanying web site provides an updated errata sheet and a place to report errors.

P. P. Chu

Cleveland, Ohio
January, 2011

ACKNOWLEDGMENTS

The author wishes to express his thanks for Blair Fort, Ralene Marcoccia, and Stephen Brown of Altera University Program for their help.

The author also thanks Ari Feldman for giving permission to use the sprite page and Altera Corporation for giving permission to use figures from various handbooks and manuals.

Altera is a trademark and service mark of Altera Corporation in the United States and other countries. Altera products are the intellectual property of Altera Corporation and are protected by copyright laws and one or more U.S. and foreign patents and patent applications. All other trademarks used or referred to in this book are the property of their respective owners.

P. P. Chu

CHAPTER 1

OVERVIEW OF EMBEDDED SYSTEM

An embedded system is a special type of computer system. In this chapter, we examine the basic characteristics of an embedded system, highlight its differences from a general-purpose computer system, and introduce the concept and development flow of a "high-end" FPGA-based embedded system, which the focus of this book.

1.1 INTRODUCTION

1.1.1 Definition of an embedded system

An *embedded system* (or *embedded computer system*) can be loosely defined as a computer system designed to perform one or a few specific tasks. The computer system is not the end product but a dedicated "embedded" part of a larger system that often includes additional electronic and mechanical parts. By contrast, a *general-purpose computer system*, such as a PC (personal computer), is a general computing platform and itself is the end product. It is designed to be flexible and to support a variety of end-user needs. Application programs are developed based on the available resource of the general-purpose computer system.

Since an embedded system is dedicated to specific tasks, its design can be optimized to reduce cost. A good design should contain just enough hardware resources to meet the application's required functionalities. On the other hand, a general-purpose computer system is expected to support a variety of needs and thus an ap-

Embedded SOPC Design with Nios II Processor and VHDL Examples. By Pong P. Chu
Copyright © 2011 John Wiley & Sons, Inc.

plication program is provided with a relatively abundant hardware resource. From this perspective, an embedded system can be thought of as *a computer system with severely resource constraint.*

The terms "embedded system" and "general-purpose computer system" are not strictly defined, as most systems have some elements of extensibility or programmability. For example, a cell phone can be treated as an embedded system since it is mainly for wireless communication. However, an advanced phone allows users to load other types of applications, such as simple video games, and thus exhibits the characteristics of a general-purpose computer system.

In our book, we refer to a general-purpose computer system as a "desktop system" since a desktop computer it is the most commonly used general-purpose system.

1.1.2 Example systems

Embedded systems are used in a wide range of applications and each application has its own specific requirements. We examine three example systems to illustrate the basic characteristics of embedded applications:

- Microwave oven.
- Digital camera.
- Vehicle stability control system.

Microwave oven   A microwave oven cooks or heats food with microwave radiation generated by a magnetron. A microwave oven usually has a keypad to select the cooking time and power level and an LCD or LED display that shows the status or time. It contains an embedded computer that processes the keypad input, keeps track of timing, generates the display patterns, and controls the magnetron unit.

The operation of the microwave oven requires no extensive computation and does not involve high-speed data transfer. The tasks can be accomplished by a very simple 8-bit processor (i.e., a processor with 8-bit internal data width) and a small read-only program memory. The entire embedded system can be implemented by a *microcontroller*, which is usually a single IC chip containing the 8-bit core processor, small memory, and simple I/O peripherals.

The microwave oven is a representative "low-end" embedded system.

Digital camera   A digital camera takes photographs by recording images electronically via an image sensor and stores the digitized image in a flash memory card. The image sensor contains millions of pixel sensors. A pixel sensor converts light to an electronic signal. The output of the pixel sensors is digitized and stored as an image file. A typical digital camera contains a set of buttons and knobs to control and adjust camera operation and a small LCD display to preview the stored pictures.

The embedded system in the camera performs two major tasks. The first task involves the general "housekeeping" I/O operations, including processing the button and knob activities, generating the graphic on an LCD display, and writing image files to the storage device. These operations are more involved than those of a microwave oven and the system requires a more capable 16- or 32-bit processor as well as a separate memory chip. The second task is to process the image and perform data compression to reduce the file size. Because of the large number of pixels and the complexity of the compression algorithm, it requires a significant amount of computation. An embedded processor is usually not powerful enough to handle the computation-intensive operation. A custom digital circuit can be designed to perform this particular task and take the load off the processor. This type of circuits is known as *hardware accelerators*.

The digital camera is a representative "high-end" embedded system.

Vehicle electronic stability control system   A vehicle ESC (electronic stability control) system helps to improve a vehicle's maneuverability by detecting and minimizing skids. During driving, it continues comparing the driver's intended direction with the vehicle's actual direction. When the loss of steering control is detected (e.g., due to a wet or iced surface), the ESC system intervenes automatically and applies the brakes to individual wheels to steer the vehicle to the intended direction.

The embedded system obtains the intended direction from the steering wheel angle and obtains the actual direction from the vehicle lateral acceleration and the individual wheel's rotating speed. It determines the occurrence and nature of the skid and then calculates and applies brake forces to individual wheels to offset the skid condition.

The ESC embedded system has two special characteristics. First, the ESC system imposes a *real-time constraint* — an operational deadline from the triggering event (i.e., onset of skid condition) to the system response (i.e., application of the brake forces). The system fails to work if the brake is not applied within a specific amount of time. Second, since the steering concerns the driver's safety, the embedded system is *mission critical* and thus must be robust and reliable.

1.2 SYSTEM DESIGN REQUIREMENTS

When designing a computer system, we must consider a variety of factors:
- Cost
- General computation speed
- Special computation need
- Real-time constraint
- Reliability
- Power consumption

The term *special computing need* means the type of computation task, such as data compression, encryption, pattern recognition, etc., which cannot be easily accomplished by a general-purpose processor.

In general, we wish that every computer system would be inexpensive, fast, reliable, and would use little power. However, these criteria are frequently fighting against each other. For example, a faster processor is more expensive and consumes more power. An embedded system can be used in a wide range of applications and each system has its own unique needs. For each system, we need to identify the key requirements and seek the best trade-off. One way to illustrate these requirements is to use a "radar chart" shown in Figure 1.1. There are six axes in the chart, each indicating the importance of a factor. As a point in an axis moves outward from the center, its importance increases from "not important" to "extremely important."

A desktop PC is for general use and thus does not place weight on a particular factor. Its chart is "well rounded," as shown in Figure 1.1(a). A microwave oven

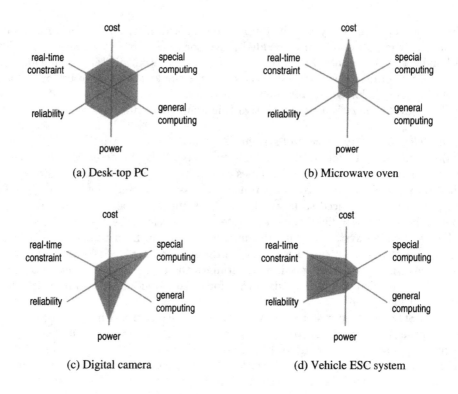

Figure 1.1   Radar charts of various systems.

can be considered as a "commodity" and its profit margin is not very high. Thus, it is extremely sensitive to the part cost. The embedded system for the microwave is very simple and its key requirement is to reduce the cost. Its chart is shown in Figure 1.1(b). A digital camera requires special image processing and compression capability. Since it is a handheld device powered by a battery, reducing power usage is important. Thus, the two key requirements of the camera's embedded system are the power and special computation need. Its chart is shown in Figure 1.1(c). A vehicle ESC system imposes a strict operational deadline and is mission critical. The key requirements of the ESC embedded system are the real-time constraint and reliability. Its chart is shown in Figure 1.1(d).

From the requirement's point of view, we can treat an embedded system as a computer system with extreme design requirements.

1.3   EMBEDDED SOPC SYSTEMS

The main focus of this book is on the "high-end" embedded systems similar to the digital camera. This type of system usually has a processor and simple I/O peripherals to perform general user interface and housekeeping tasks and special hardware accelerators to handle computation-intensive operations. These components can be integrated into a single integrated circuit, commonly referred to as *SoC* (*system on a chip*). As the capacity of *FPGA* (*field-programmable gate ar-*

ray) devices continues to grow, the same design methodology can be realized in an FPGA chip and is sometimes known as *SoPC* (*system on a programmable chip*) or *PSoC* (*programmable system on a chip*). We use the term SoPC in the book.

While designing a system based on a conventional embedded processor, we examine the required functionalities and then select a processor, external I/O peripherals, and ASSP (application specific standard product) devices to construct the hardware platform. Because of the fixed-sized processor architecture, a limited choice of ASSP devices, and the cost of manufacturing printed circuit boards, the hardware configuration is usually rather "rigid" and the desired system functionalities are usually done by *customized software*.

An FPGA device contains logic cells and interconnects that can be configured (i.e., "programmed") to perform a specific function. The desired hardware functionalities are usually described in *HDL* (*hardware description language*) code, which is then synthesized and implemented by the FPGA device. Because of the programmability of FPGA devices, *customized hardware* can be incorporated into the embedded system as well. We can tailor the processor, select only the needed I/O peripherals, create a custom I/O interface, and develop specialized hardware accelerators for computation-intensive tasks. The SoPC-based embedded system provides a new dimension of flexibility because both the hardware and software can be customized to match specific needs.

1.3.1 Basic development flow

The embedded SoPC system development consists of the following parts:
- Partition the tasks to software and hardware accelerators.
- Develop the hardware, including the hardware accelerators and I/O peripherals, and integrate it with the processor.
- Develop the software.
- Implement the hardware and software and perform testing.

Since the design examples in this book are targeted for Altera prototyping boards, our discussion uses the Altera development platform and its *Nios II* processor. Note that Nios II is a *soft-core* processor, which means the processor is described in HDL code and synthesized later by using FPGA's generic logic cells.

The basic Nios II-based development flow is shown in Figure 1.2. The four basic parts are elaborated in the following subsections.

Software–hardware partition   Step 1 (labeled 1 in the diagram) is to determine the software–hardware partition. An embedded application usually performs a collection of tasks. In an SoPC-based design, a task can be implemented by hardware, software, or both. Based on the performance requirement, complexity, and hardware core availability, we can decide the type of implementation accordingly.

Hardware development flow   The left branch represents the hardware design flow. Step 2 derives the basic hardware architecture. The custom hardware can be divided into three categories:
- *Nios II processor and standard I/O peripherals* (labeled "Nios configuration" in the diagram). Altera provides the soft cores of the processor and a collection of frequently used I/O peripherals. A third-party vendor supplies

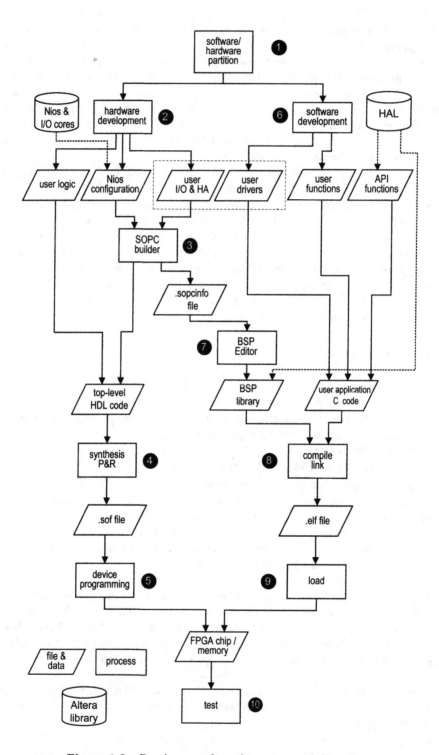

Figure 1.2   Development flow of a system with Nios II.

additional I/O cores as well. We can select the needed I/O peripherals and configure the basic Nios II system.

- *User I/O peripherals and hardware accelerators* (labeled "User I/O & HA" in the diagram). For certain specialized I/O functions or computation-intensive tasks, a pre-designed core may not exist or cannot satisfy the performance requirement. We must design the hardware from scratch and integrate it into the Nios II system as a custom I/O peripheral.
- *User logic.* Some portion of the hardware may be separated from the Nios II system. It is not attached to the Nios interconnect structure and does not interact directly with the processor.

Step 3 generates the HDL code from the customized Nios II system. It is done by using Altera's *SOPC Builder* software package. In this software, we can configure the processor, select the desired standard I/O cores, and incorporate the user-designed I/O peripherals. SOPC Builder then generates the HDL codes for the customized Nios II system and also generates the .sopcinfo file that contains system configuration information. We can combine this code with the other use logic codes to form the final top-level HDL description.

The top-level HDL code contains the description of the complete hardware. Step 4 performs synthesis and placement and routing and eventually generates the FPGA configuration file (i.e., the .sof file).

Software development flow   The right branch represents the software design flow. Step 6 derives the basic software structure. Altera provides a software library, which is integrated into its *HAL* (*hardware abstraction layer*) platform, for the Nios II system. It consists of *I/O device drivers*, which are low-level routines to access I/O peripherals, and a collection of high-level functions in an *application programming interface* (*API*). From the hardware–software interface's point of view, we can divide the software code into three categories:

- *API functions.* These are the functions from the Altera HAL platform.
- *User I/O drivers.* When designing a custom I/O peripheral or hardware accelerator, we also need to develop software I/O routines to control its operation and to exchange its data with the processor.
- *User functions.* These implement the needed functionalities for the embedded application.

We can utilize these drivers and functions to construct the application program.

When a Nios II system is created, the processor and I/O configuration is recorded in the .sopcinfo file. In Step 7, the *BSP Editor* software program examines this file, extracts the needed device drivers from the HAL library, and builds up a *BSP* (*board support package*) library to support the system.

Step 8 compiles and links the software routines and BSP library and builds the final software image file (i.e., the .elf file).

Physical implementation and test   Physically implementing the system involves two steps. We first download the FPGA configuration file to the FPGA device (i.e., "program" the device), as in Step 5, and then load the software image into Nios II's memory, as in Step 9. The physical system can be tested afterwards, as in Step 10.

The most unique characteristics of an SoPC-based embedded system are that custom I/O peripherals and hardware accelerators can be integrated into the system. The major task involves the development of custom hardware and a software driver, as shown in the dotted box in Figure 1.2. This is the main focus of the book.

1.4   BOOK ORGANIZATION

The remaining book is divided into four parts. Part I introduces the basic HDL constructs and synthesis procedure and discusses the development of custom digital circuits. Part II provides an overview of a Nios II-based system and embedded software development with the emphasis on low-level I/O access and drivers. A simple flashing-LED design is used to illustrate the key concepts. Part III applies the techniques from Parts I and II to design an array of complex I/O peripheral modules on the Altera DE1 prototyping board, including a PS2 keyboard and mouse controller, a graphic video controller, an audio codec controller, and an SD (secure digital) card controller. Part IV presents three case studies of the integration of hardware accelerators, including a custom GCD (greatest common divisor) circuit, a Mandelbrot set fractal circuit, and an audio synthesizer based on DDFS (direct digital frequency synthesis) methodology.

1.5   BIBLIOGRAPHIC NOTES

In this book, a short bibliographic section appears at the end of each chapter to provide the most relevant references for further exploration. A more comprehensive bibliography is included at the end of the book.

Embedded systems encompass a spectrum of design issues. The two books, *Embedded System Design: A Unified Hardware/Software Introduction* by F. Vahid and T. D. Givargis and *Computers as Components: Principles of Embedded Computing System Design, 2nd edition* by W. Wolf, provide a comprehensive discussion. Most processor-oriented embedded system books are around specific low-end microcontrollers. However, *Programming 32-bit Microcontrollers in C: Exploring the PIC32* by L. Di Jasio, as its title indicates, is based on 32-bit PIC processors and covers more advanced design examples.

Software-hardware co-design is an emerging research area. *A Practical Introduction to Hardware/Software Codesign* by P. R. Schaumont addresses the basic concepts and issues of combining hardware and software into a single system design process.

PART I

BASIC DIGITAL CIRCUITS DEVELOPMENT

CHAPTER 2

GATE-LEVEL COMBINATIONAL CIRCUIT

HDL (hardware description language) is used to describe and model digital systems. VHDL is one of the two major HDLs. In this chapter, we use a simple comparator to illustrate the skeleton of a VHDL program. The description uses only logical operators and represents a gate-level combinational circuit, which is composed of simple logic gates.

2.1 OVERVIEW OF VHDL

VHDL stands for "VHSIC (very high-speed integrated circuit) hardware description language." It was originally sponsored by the U.S. Department of Defense and later transferred to the IEEE (Institute of Electrical and Electronics Engineers). The language is formally defined by IEEE Standard 1076. The standard was ratified in 1987 (referred to as VHDL 87), and revised several times. This book mainly follows the revision in 1993 (referred to as VHDL 93).

VHDL is intended for describing and modeling a digital system at various levels and is an extremely complex language. The focus of this book is on hardware design rather than the language. Instead of covering every aspect of VHDL, we introduce the key VHDL synthesis constructs by examining a collection of examples. Detailed VHDL coverage may be explored through the sources listed in the Bibliography.

In this chapter, we introduce the HDL concepts, basic VHDL language constructs, logical operators, and program structure. A simple gate-level combinational

Embedded SOPC Design with Nios II Processor and VHDL Examples. By Pong P. Chu   **11**
Copyright © 2011 John Wiley & Sons, Inc.

Table 2.1   Truth table of a 1-bit equality comparator

input $i0\ i1$	output eq
0 0	1
0 1	0
1 0	0
1 1	1

circuit is used for demonstration. In Chapter 4, we cover the more sophisticated VHDL operators and constructs and examine module-level combinational circuits, which are composed of intermediate-sized components, such as adders, comparators, and multiplexers.

2.2 GENERAL DESCRIPTION

Consider a 1-bit equality comparator with two inputs, i0 and i1, and an output, eq. The eq signal is asserted when i0 and i1 are equal. The truth table of this circuit is shown in Table 2.1.

Assume that we want to use basic logic gates, which include *not*, *and*, *or*, and *xor* cells, to implement the circuit. One way to describe the circuit is to use a sum-of-products format. The logic expression is

$$eq = i0 \cdot i1 + i0' \cdot i1'$$

One possible corresponding VHDL code is shown in Listing 2.1. We examine the language constructs and statements of this code in the following subsections.

Listing 2.1   Gate-level implementation of a 1-bit comparator

```
library ieee;
use ieee.std_logic_1164.all;
entity eq1 is
   port(
5      i0, i1: in std_logic;
       eq: out std_logic
   );
end eq1;

10 architecture sop_arch of eq1 is
   signal p0, p1: std_logic;
   begin
   -- sum of two product terms
   eq <= p0 or p1;
15 -- product terms
   p0 <= (not i0) and (not i1);
   p1 <= i0 and i1;
   end sop_arch;
```

2.2.1   Basic lexical rules

VHDL is case insensitive, which means that upper- and lowercase letters can be used interchangeably, and free formatting, which means that spaces and blank lines can be inserted freely. It is good practice to add proper spaces to make the code clear and to associate special meaning with cases. In this book, we reserve uppercase letters for constants.

An *identifier* is the name of an object and is composed of 26 letters, digits, and the underscore (_), as in `i0`, `i1`, and `data_bus1_enable`. The identifier must start with a letter.

The comments start with `--` and the text after it is ignored. In this book, the VHDL keywords are shown in boldface type, as in **entity**, and the comments are shown in italics type, as in

```
--  this  is  a  comment
```

2.2.2   Library and package

The first two lines,

```
library ieee;
use ieee.std_logic_1164.all;
```

invoke the `std_logic_1164` package from the `ieee` library. The package and library allow us to add additional types, operators, functions, etc., to VHDL. The two statements are needed because a special data type is used in the code.

2.2.3   Entity declaration

The entity declaration

```
entity eq1 is
   port(
      i0, i1: in std_logic;
      eq: out std_logic
   );
end eq1;
```

essentially outlines the I/O signals of the circuit. The first line indicates that the name of the circuit is `eq1`, and the port section specifies the I/O signals. The basic format for an I/O port declaration is

```
signal_name1, signal_name2, ... : mode data_type;
```

The **mode** term can be **in** or **out**, which indicates that the corresponding signals flow "into" or "out of" of the circuit. It can also be **inout**, for bidirectional signals.

2.2.4   Data type and operators

VHDL is a *strongly typed language*, which means that an object must have a data type and only the defined values and operations can be applied to the object. Although VHDL is rich in data types, our discussion is limited to a small set of predefined types that are suitable for synthesis, mainly the `std_logic` type and its variants.

std_logic type The std_logic type is defined in the std_logic_1164 package and consists of nine values. Three of the values, '0', '1', and 'Z', which stand for logical 0, logical 1, and high impedance, can be synthesized. Two values, 'U' and 'X', which stand for "uninitialized" and "unknown" (e.g., when signals with '0' and '1' values are tied together), may be encountered in simulation. The other four values, '-', 'H', 'L', and 'W', are not used in this book.

A signal in a digital circuit frequently contains multiple bits. The std_logic_vector data type, which is defined as an array with elements of std_logic, can be used for this purpose. For example, let a be an 8-bit input port. It can be declared as

```
a: in std_logic_vector(7 downto 0);
```

We can use a term like a(7 downto 4) to specify a desired range and a term like a(1) to access a single element of the array. The array can also be declared in ascending order:

```
a: in std_logic_vector(0 to 7);
```

We generally avoid this format since it is more natural to associate the MSB with the leftmost position.

Logical operators Several logical operators, including **not, and, or**, and **xor**, are defined over the std_logic_vector and std_logic data type. Bit-wise operation is used when an operator is applied to an object with the std_logic_vector data type. Note that the **and, or**, and **xor** operators have the same precedence and we need to use parentheses to specify the desired order of evaluation, as in

```
(a and b) or (c and d)
```

2.2.5 Architecture body

The architecture body,

```
architecture sop_arch of eq1 is
   signal p0, p1: std_logic;
begin
   -- sum of two product terms
   eq <= p0 or p1;
   -- product terms
   p0 <= (not i0) and (not i1);
   p1 <= i0 and i1;
end sop_arch;
```

describes operation of the circuit. VHDL allows multiple bodies associated with an entity, and thus the body is identified by the name sop_arch ("sum-of-products architecture").

The architecture body may include an optional declaration section, which specifies constants, internal signals, and so on. Two internal signals are declared in this program:

```
signal p0, p1: std_logic;
```

The main description, encompassed between **begin** and **end**, contains three *concurrent statements*. Unlike a program in C language, in which the statements

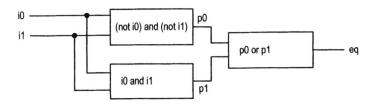

Figure 2.1 Graphical representation of a comparator program.

are executed sequentially, concurrent statements are like circuit parts that operate
in parallel. The signal on the left-hand side of a statement can be considered
as the output of that part, and the expression specifies the circuit function and
corresponding input signals. For example, consider the statement

```
eq <= p0 or p1;
```

It is a circuit that performs the or operation. When p0 or p1 changes its value, this
statement is activated and the expression is evaluated. The new value is assigned
to eq after the default propagation delay.

The graphical representation of this program is shown in Figure 2.1. The three
circuit parts represent the three concurrent statements. The connections among
these parts are implicitly specified by the signal and port names. The order of the
concurrent statements is clearly irrelevant and the statements can be rearranged
arbitrarily.

2.2.6 Code of a 2-bit comparator

We can expand the comparator to 2-bit inputs. Let the input be a and b and the
output be aeqb. The aeqb signal is asserted when both bits of a and b are equal.
The code is shown in Listing 2.2.

Listing 2.2 Gate-level implementation of a 2-bit comparator

```
    library ieee;
    use ieee.std_logic_1164.all;
    entity eq2 is
4      port(
           a, b: in std_logic_vector(1 downto 0);
           aeqb: out std_logic
       );
    end eq2;
9
    architecture sop_arch of eq2 is
       signal p0,p1,p2,p3: std_logic;
    begin
       -- sum of product terms
14     aeqb <= p0 or p1 or p2 or p3;
       -- product terms
       p0 <= ((not a(1)) and (not b(1))) and
             ((not a(0)) and (not b(0)));
       p1 <= ((not a(1)) and (not b(1))) and (a(0) and b(0));
19     p2 <= (a(1) and b(1)) and ((not a(0)) and (not b(0)));
       p3 <= (a(1) and b(1)) and (a(0) and b(0));
    end sop_arch;
```

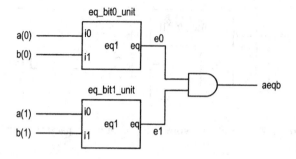

Figure 2.2   Construction of a 2-bit comparator from 1-bit comparators.

The a and b ports are now declared as a two-element std_logic_vector. Derivation of the architecture body is similar to that of a 1-bit comparator. The p0, p1, p2, and p3 signals represent the results of the four product terms, and the final result, aeqb, is the logic expression in sum-of-products format.

2.3   STRUCTURAL DESCRIPTION

A digital system is frequently composed of several smaller subsystems. This allows us to build a large system from simpler or predesigned components. VHDL provides a mechanism, known as *component instantiation*, to perform this task. This type of code is called a *structural description*.

An alternative to the design of the 2-bit comparator of Section 2.2.6 is to utilize the previously constructed 1-bit comparators as the building blocks. The diagram is shown in Figure 2.2, in which two 1-bit comparators are used to check the two individual bits and their results are fed to an and cell. The aeqb signal is asserted only when the two bits are equal.

The corresponding code is shown in Listing 2.3. Note that the entity declaration is the same and thus is not included.

Listing 2.3   Structural description of a 2-bit comparator

```
architecture struc_arch of eq2 is
   signal e0, e1: std_logic;
begin
4    — instantiate two 1-bit comparators
   eq_bit0_unit: entity work.eq1(sop_arch)
      port map(i0=>a(0), i1=>b(0), eq=>e0);
   eq_bit1_unit: entity work.eq1(sop_arch)
      port map(i0=>a(1), i1=>b(1), eq=>e1);
9    — a and b are equal if individual bits are equal
   aeqb <= e0 and e1;
end struc_arch;
```

The code includes two component instantiation statements, whose syntax is

```
unit_label: entity lib_name.entity_name(arch_name)
   port map(
      formal_signal=>actual_signal,
      formal_signal=>actual_signal,
      . . .
```

```
);
```

The first portion of the statement specifies which component is used. The `unit_label` term gives a unique id for an instance, the `lib_name` term indicates where (i.e., which library) the component resides, and the `entity_name` and `arch_name` terms indicate the names of the entity and architecture. The `arch_name` term is optional. If it is omitted, the last compiled architecture body will be used. The second portion is port mapping, which indicates the connection between *formal signals*, which are I/O ports declared in a component's entity declaration, and *actual signals*, which are the signals used in the architecture body.

The first component instantiation statement is

```
eq_bit0_unit: entity work.eq1(sop_arch)
    port map(i0=>a(0), i1=>b(0), eq=>e0);
```

The `work` library is the default library in which the compiled entity and architecture units are stored, and `eq1` and `sop_arch` are the names of the entity and architecture defined in Listing 2.1. The port mapping reflects the connections shown in Figure 2.2. The component instantiation statement is also a concurrent statement and represents a circuit that is encompassed in a "black box" whose function is defined in another module.

This example demonstrates the close relationship between a block diagram and code. The code is essentially a textual description of a schematic. Although it is a clumsy way for humans to comprehend a diagram, it puts all representations into a single HDL framework. The Altera Quartus package includes a simple schematic editor utility that can perform a schematic capture in graphic format and then convert the diagram into an HDL structural description.

The component instantiation statement is added in VHDL 93. Older codes may use the mechanism in VHDL 87, in which a component must first be declared (i.e., made known) and then used. The code in this format is shown in Listing 2.4.

Listing 2.4 Structural description with VHDL-87

```
   architecture vhd_87_arch of eq2 is
     -- component declaration
 3   component eq1
       port(
           i0, i1: in std_logic;
           eq: out std_logic
       );
 8   end component;
     signal e0, e1: std_logic;
   begin
     -- instantiate two 1-bit comparators
     eq_bit0_unit: eq1    -- use the declared name, eq1
13     port map(i0=>a(0), i1=>b(0), eq=>e0);
     eq_bit1_unit: eq1    -- use the declared name, eq1
       port map(i0=>a(1), i1=>b(1), eq=>e1);
     -- a and b are equal if individual bits are equal
     aeqb <= e0 and e1;
18 end vhd_87_arch;
```

Note that the original clause

```
eq_bit0_unit: entity work.eq1(sop_arch)
```

is replaced by a clause with the declared component name

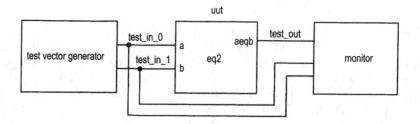

Figure 2.3   Testbench for a 2-bit comparator.

```
eq_bit0_unit: eq1
```

2.4 TESTBENCH

After code is developed, it can be *simulated* in a host computer to verify the correctness of the circuit operation and can be *synthesized* to a physical device. Simulation is usually performed within the same HDL framework. We create a special program, known as a *testbench*, to mimic a physical lab bench. The sketch of a 2-bit comparator testbench program is shown in Figure 2.3. The uut block is the unit under test, the **test vector generator** block generates testing input patterns, and the **monitor** block examines the output responses.

A simple testbench for the 2-bit comparator is shown in Listing 2.5.

Listing 2.5   Testbench for a 2-bit comparator

```
library ieee;
use ieee.std_logic_1164.all;
entity eq2_testbench is
4 end eq2_testbench;

architecture tb_arch of eq2_testbench is
    signal test_in0, test_in1: std_logic_vector(1 downto 0);
    signal test_out: std_logic;
9 begin
    -- instantiate the circuit under test
    uut: entity work.eq2(struc_arch)
        port map(a=>test_in0, b=>test_in1, aeqb=>test_out);
    -- test vector generator
14   process
    begin
        -- test vector 1
        test_in0 <= "00";
        test_in1 <= "00";
19      wait for 200 ns;
        -- test vector 2
        test_in0 <= "01";
        test_in1 <= "00";
        wait for 200 ns;
24      -- test vector 3
        test_in0 <= "01";
        test_in1 <= "11";
        wait for 200 ns;
        -- test vector 4
29      test_in0 <= "10";
```

```
         test_in1 <= "10";
         wait for 200 ns;
         -- test vector 5
         test_in0 <= "10";
34       test_in1 <= "00";
         wait for 200 ns;
         -- test vector 6
         test_in0 <= "11";
         test_in1 <= "11";
39       wait for 200 ns;
         -- test vector 7
         test_in0 <= "11";
         test_in1 <= "01";
         wait for 200 ns;
44     end process;
   end tb_arch;
```

The code consists of a component instantiation statement, which creates an instance of a 2-bit comparator, and a process statement, which generates a sequence of test patterns.

The process statement is a special VHDL construct in which the operations are performed sequentially. Each test pattern is generated by three statements. For example,

```
         -- test vector 2
         test_in0 <= "01";
         test_in1 <= "00";
         wait for 200 ns;
```

The first two statements specify the values for the **test_in0** and **test_in1** signals, and the third indicates that the two values will last for 200 ns.

The code has no monitor. We can observe the input and output waveforms on a simulator's display, which can be treated as a "virtual logic analyzer." The simulated timing diagram of this testbench is shown in Figure 3.23.

Writing code for a comprehensive test vector generator and a monitor requires detailed knowledge of VHDL and is beyond the scope of this book. This listing can serve as a testbench template for other combinational circuits. We can substitute the **uut** instance and modify the test patterns according to the new circuit.

2.5 BIBLIOGRAPHIC NOTES

A short bibliographic section appears at the end of each chapter to provide some of the most relevant references for further exploration. A comprehensive bibliography is included at the end of the book.

VHDL is a complex language. *The Designer's Guide to VHDL* by P. J. Ashenden provides detailed coverage of the language's syntax and constructs. The author's *RTL Hardware Design Using VHDL: Coding for Efficiency, Portability, and Scalability* provides a comprehensive discussion on developing effective, synthesizable codes. The derivation of the testbench for a large digital system is a difficult task. *Writing Testbenches: Functional Verification of HDL Models, 2nd edition*, by J. Bergeron focuses on this topic.

2.6 SUGGESTED EXPERIMENTS

At the end of each chapter, some experiments are suggested as exercises. The experiments help us to better understand the concepts and provide a hands-on opportunity to design and debug actual circuits.

2.6.1 Code for gate-level greater-than circuit

Develop the HDL codes in Experiment 3.8.1. The code can be simulated and synthesized after we complete Chapter 3.

2.6.2 Code for gate-level binary decoder

Develop the HDL codes in Experiment 3.8.2. The code can be simulated and synthesized after we complete Chapter 3.

CHAPTER 3

OVERVIEW OF FPGA AND EDA SOFTWARE

Developing a large FPGA-based system is an involved process that consists of many complex transformations and optimization algorithms. Software tools are needed to automate some of the tasks. We use the Altera *Quartus II Web Edition* package for synthesis, implementation, and device programming, and use the Mentor Graphics *ModelSim Altera Starter Edition* package for HDL simulation. In this chapter, we give a brief overview of the FPGA device and the DE1 prototyping board, and provide short tutorials for the two software packages to "jump-start" the learning process.

3.1 FPGA

3.1.1 Overview of a general FPGA device

A *field-programmable gate array* (FPGA) is a logic device that contains a two-dimensional array of generic logic cells and programmable switches. The conceptual structure of an FPGA device is shown in Figure 3.1. A logic cell can be configured (i.e., *programmed*) to perform a simple function, and a programmable switch can be customized to provide interconnections among the logic cells. A custom design can be implemented by specifying the function of each logic cell and selectively setting the connection of each programmable switch. Once the design and synthesis are completed, we can use a simple adaptor cable to download the desired logic cell and switch configuration to the FPGA device and obtain the custom circuit. Since

Embedded SOPC Design with Nios II Processor and VHDL Examples. By Pong P. Chu   **21**
Copyright © 2011 John Wiley & Sons, Inc.

S | programmable switch

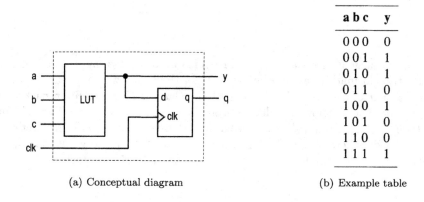

Figure 3.1   Conceptual structure of an FPGA device.

a b c	y
0 0 0	0
0 0 1	1
0 1 0	1
0 1 1	0
1 0 0	1
1 0 1	0
1 1 0	0
1 1 1	1

(a) Conceptual diagram   (b) Example table

Figure 3.2   Three-input LUT-based logic cell.

this process can be done "in the field" rather than "in a fabrication facility (fab),"
the device is known as *field programmable*.

LUT-based logic cell   A logic cell usually contains a small configurable combinational circuit with a D-type flip-flop (D FF). The most common method to implement a configurable combinational circuit is a *lookup table* (LUT). An n-input LUT
can be considered as a small 2^n-by-1 memory. By properly writing the memory content, we can use the LUT to implement any n-input combinational function. The
conceptual diagram of a three-input LUT-based logic cell is shown in Figure 3.2(a).

An example of three-input LUT implementation of $a \oplus b \oplus c$ is shown in Figure 3.2(b). Note that the output of the LUT can be used directly or stored to the D FF. The latter can be used to implement sequential circuits.

Macro cell   Most FPGA devices also embed certain *macro cells* or *macro blocks*. These are designed and fabricated at the transistor level, and their functionalities complement the general logic cells. Commonly used macro cells include memory blocks, combinational multipliers, clock management circuits, and I/O interface circuits. Advanced FPGA devices may even contain one or more prefabricated processor cores.

3.1.2   Overview of the Altera Cyclone II devices

The Altera DE1 prototyping board is used in this book and it contains an FPGA device from Altera's *Cyclone II* family. Although Cyclone II devices are low-cost entry-level FPGA devices, they have all the key features of advanced devices and support the use of soft-core processor. Their basic characteristics are examined in the following subsections.

Logic cell   A logic cell in the Cyclone II device is known as an *LE* (*logic element*). An LE contains a four-input lookup table, which is a function generator that can implement any function of four variables, and a D FF. The lookup table can be split into two three-input lookup tables to implement a carry-chain used in arithmetic circuits. The FF also contains additional enable and loading logic so that it can be configured to support various types of register operations. The block diagram of an LE is shown in Figure 3.3. The lookup table and the D FF are highlighted with thick dotted boxes.

Interconnect structure   As the size of the transistor shrinks, the routing delay becomes a significant portion of a circuit's overall propagation delay. To optimize the performance, routing within a Cyclone II device is performed on several levels. The conceptual diagram is shown in Figure 3.4. On the local level, 16 LEs are grouped together to form a *LAB* (*logic array block*). The LEs within the same LAB are connected via the *local interconnect* and signals can be routed within the LAB directly. There are also *direct link interconnects* for routing signals between the adjacent LABs. On the global level, the LABs and macro cells are connected via a two-dimensional row- and column-based *MultiTrack interconnect* structure. The MultiTrack interconnect is made of dedicated fixed-length horizontal (row) and vertical (column) routing channels. A horizontal channel consists of R4 and R24 interconnects, which traverse a distance of 4 and 24 blocks to the right or left, respectively. A vertical channel consists of C4 and C16 interconnects, which traverse a distance of 4 and 16 blocks in an up or down direction, respectively. An LE's output can be connected to any type of interconnect structure, as shown in Figure 3.3.

Macro cell   The Cyclone II device contains four types of macro blocks: *combinational multiplier, embedded memory block, phase-locked loop* (*PLL*), and *input/output element* (*IOE*). The combinational multiplier accepts two 18-bit numbers as inputs and calculates the product. Its usage is discussed in Section 4.1.2. The embedded memory block, known as *M4K block*, is a 4K-bit synchronous SRAM that

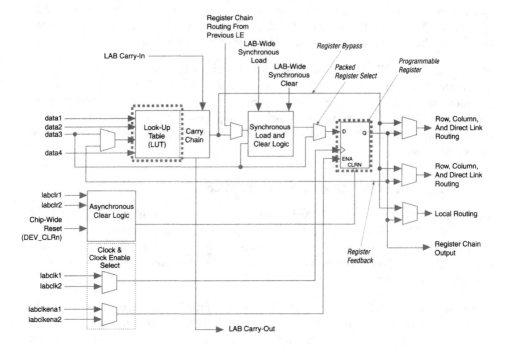

Figure 3.3   Block diagram of a Cyclone II logic element. (Courtesy of Altera Corp.)

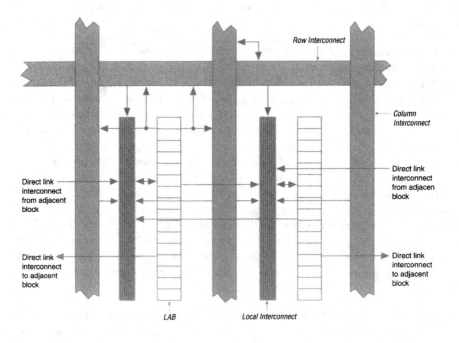

Figure 3.4   Conceptual diagram of Cyclone II interconnect structure. (Courtesy of Altera Corp.)

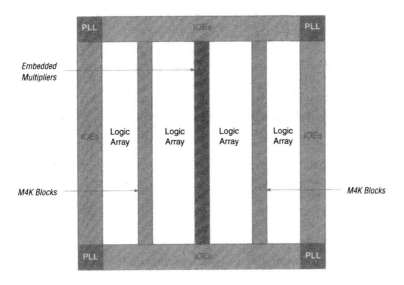

Figure 3.5 Top-level diagram of Cyclone II EP2C20 device. (Courtesy of Altera Corp.)

Table 3.1 Devices in the Cyclone II family

Device	Number of LEs	Number of M4K blocks	Total RAM bits	Number of multipliers	Number of PLLs
EP2C5	4,608	26	120K	13	2
EP2C8	8,256	36	166K	18	2
EP2C15	14,448	52	240K	26	4
EP2C20	18,752	52	240K	26	4
EP2C35	33,216	105	484K	35	4
EP2C50	50,528	129	594K	86	4
EP2C70	68,416	250	1152K	150	4

can be arranged in various types of configurations. Its inference is discussed in Section 5.7. A PLL provides general-purpose clocking with clock synthesis and phase shifting and can be used to reduce clock skew. Its instantiation and configuration are discussed in Section 15.7. An IOE is associated within a physical I/O pin. It is located at the ends of LAB rows and columns around the periphery of the device. The IOE controls the flow of data between the device's I/O pins and the internal logic and can be configured to support a wide variety of I/O signaling standards. The top-level layout of a Cyclone II device EP2C20 is shown in Figure 3.5.

Devices in the Cyclone II family The Cyclone II family contains a collection of devices. The devices have similar types of logic cells and macro cells but their densities differ. The numbers of LEs, M4K RAM blocks, multipliers, and PLLs of these devices are summarized in Table 3.1.

3.2 OVERVIEW OF THE ALTERA DE1 AND DE2 BOARDS

The Altera DE1 board is based on a Cyclone II EP2C20 device and has an array of built-in peripherals. The main parts and connectors are as follows:

- Altera Cyclone II EP2C20 FPGA device
- Onboard USB blaster for device programming
- Altera EPCS4 serial configuration EEPROM
- 512K byte SRAM
- 8M byte SDRAM
- 4M byte Flash memory
- Four pushbutton switches
- Ten slide switches
- Ten red user LEDs
- Eight green user LEDs
- Four seven-segment LED displays
- Audio codec with line-in, line-out, and microphone-in jacks
- VGA port with three 4-bit DACs
- RS-232 transceiver and 9-pin port
- PS2 mouse and keyboard port
- SD Card socket
- Two 40-pin expansion headers
- 50-MHz oscillator, 27-MHz oscillator, and 24-MHz oscillator for clock sources

The layout of the board is shown in Figure 3.6.

The Altera DE2 board is similar to the DE1 board but with larger parts and additional peripherals. It increases the sizes of the following parts:

- Altera Cyclone II EP2C35 FPGA device
- 18 slide switches
- 18 red user LEDs
- Nine green user LEDs
- Eight seven-segment LED displays
- VGA port with three 10-bit DACs

and adds several more advanced peripherals:

- Video-in port
- USB 2.0 port (type A and type B)
- Ethernet port
- 16-by-2 LCD display
- Infrared port

The discussion and design examples in this book are based on the DE1 board and can be adopted for the DE2 board with minor or no modification. The advanced peripherals of the DE2 board are not covered.

3.3 DEVELOPMENT FLOW

Steps 4 and 5 in Figure 1.2 represent the HDL development. The more elaborated flow, which also includes the verification track, is shown in Figure 3.7. To facilitate further reading, we adopt some terms used in the Altera documentation. The left track of the flow is the refinement and programming process, in which a system is

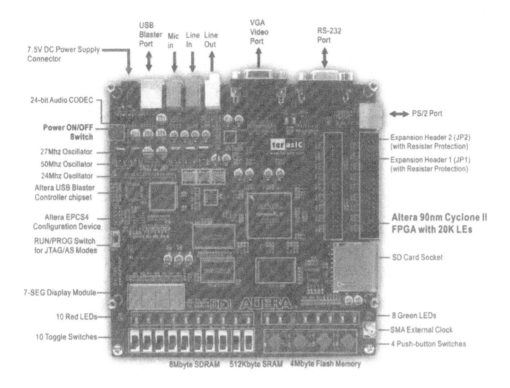

Figure 3.6   Layout of the Altera DE1 board. (Courtesy of Altera Corp.)

transformed from an abstract textual HDL description to a device cell-level config-uration and then downloaded to the FPGA device. The right track is the validation process, which checks whether the system meets the functional specifications and performance goals. The major steps in the flow are:

1. Design the system and derive the HDL file(s). We may need to add a separate constraint file to specify certain implementation constraints.
2. Develop the testbench in HDL and perform *RTL simulation*. The RTL term reflects the fact that the HDL code is done at the register transfer level. The simulation checks whether the initial HDL description meets the specifications and functions properly.
3. Perform *analysis and synthesis*. As the name indicates, this process contains two smaller processes, *analysis* and *synthesis*. The *analysis process* analyzes the file structure, checks the syntax of HDL codes, and performs *elaboration*, which constructs the design hierarchy and establishes signal connectivity. The *synthesis process* transforms the HDL constructs to generic gate-level compo-nents and then maps them to the FPGA's logic elements and IOEs.
4. Perform *placement and routing*. The placement and routing process derives the physical layout inside the FPGA chip. It places the logic elements in physical locations and determines the routes to connect various signals. It is sometimes referred to as *fitting* in Altera documentation.

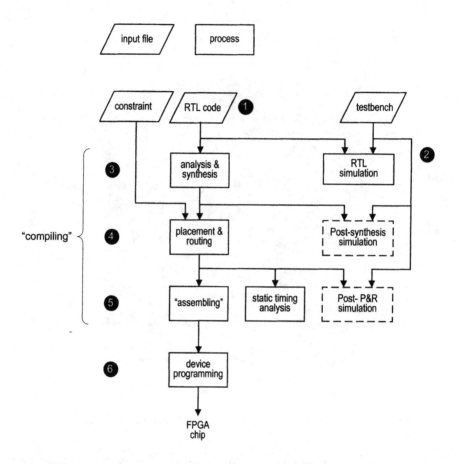

Figure 3.7   Development flow.

5. Generate the configuration file. In this step, the output of the fitter is converted into a "programming image" file for the designated target device. It is referred to as *assembling* in Altera documentation.

6. Program the device. In this process, the configuration file is downloaded into the target device. The action is also referred to as *programming* an FPGA device. The physical circuit can be verified accordingly.

Steps 3, 4, and 5 combined are sometimes referred to as the *compilation* process in Altera documentation.

The optional *post-synthesis simulation* can be performed after synthesis, and the optional *post-placement-and-routing simulation* can be performed after placement and routing. The post-synthesis simulation uses a synthesized netlist to replace the RTL description and checks the correctness of the synthesis process. The post-placement-and-routing simulation uses the final netlist, along with detailed timing data, to perform the simulation. Because of the complexity of the netlist, these types of simulation may require a significant amount of time. If we follow good design and coding practices, the HDL code will be synthesized and implemented correctly. We only need to use RTL simulation to check the correctness of the HDL code and use *static timing analysis* to examine the relevant timing information.

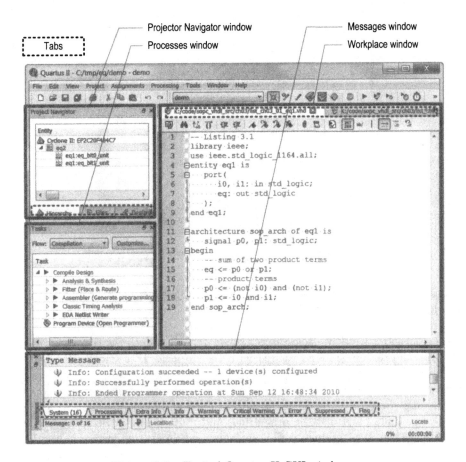

Figure 3.8   Typical Quartus II GUI window.

Both post-synthesis and post-placement-and-routing simulations may be omitted from the development flow.

3.4   OVERVIEW OF QUARTUS II

Altera Quartus II contains software tools for the left track, as depicted in Figure 3.7. It provides a graphic interface for users to access tools and display relevant files. Altera software is updated regularly and the discussion in this section and the tutorial in the next section are based on *Quartus II v10.0*. Some differences may exist in other versions. The default Quartus II GUI window is shown in Figure 3.8. Its menu items and frequently used action icons are displayed on top. The remaining is divided into four smaller windows:

- Project Navigator window (top left)
- Tasks window (middle left)
- Messages window (bottom)
- Workplace area (top right)

Note that a window may contain multiple pages and the tabs at the top or bottom are used to select the desired page. Each window may be resized, moved, docked,

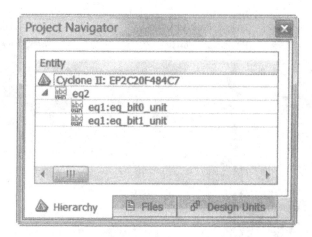

Figure 3.9   Hierarchy display in the Project Navigator window.

or undocked. The default layout can be restored by selecting Tools ≻ Customize...
and then clicking the Reset All button.

Project Navigator window   The Project Navigator window shows the design hierarchy, files, and design units associated with the project. This information becomes
available after we perform the initial analysis and elaboration process. The hierarchy page displays the design in a hierarchical tree, starting with the top-level
module. The lower-level modules can be expanded and hidden as needed. The Files
and Design Units pages list the design files and units in the projects, respectively.
Double-clicking on a module, a file, or a unit will open the corresponding file in
the Workplace area. A sample Project Navigator window is shown in Figure 3.9. It
displays the hierarchy of the 2-bit comparator in Listing 3.2. The top row indicates
the target FPGA device for the design.

Tasks window   The Tasks window allows a user to gather relevant processes (known
as a *flow*) and show them in a flow-based layout. Several predefined flows are
provided and can be selected from the Flow list. The default is the Compilation
flow, which represents the left track of Figure 3.7, and a snapshot flow is shown in
Figure 3.10. It shows the processes as a flow from top to bottom. The left portion
shows the progress of the process and a check mark is placed when a process is
successfully completed. The details of a process can be expanded or hidden as
needed. We only use this flow in our book.

Messages window   The Message window displays status messages, errors, warnings,
etc. We can select the appropriate tab to get the desired information.

Workplace area   The workplace area is the remaining area in the GUI window. It
can contains multiple document windows, such as HDL code, reports, schematics,
and so on. We can view and edit various types of files in this area.

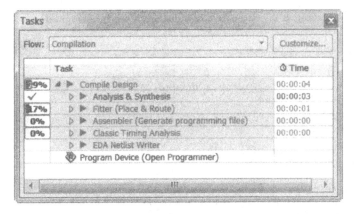

Figure 3.10   Compilation flow in the Tasks window.

3.5   SHORT TUTORIAL OF QUARTUS II

Altera Quartus II consists of an array of software tools, but a detailed discussion of their use is beyond the scope of this book. We present a short tutorial in this section to illustrate the basic development process. A separate HDL simulation tutorial is provided in Section 3.6.

The process is oriented to the DE1 board and the designs in this book. The major steps are:

1. Create the design project with HDL codes and constraints.
2. Create a testbench and perform RTL simulation.
3. Compile the project.
4. Perform a timing analysis.
5. Program the FPGA device.

These steps follow the general development flow discussed in Section 3.3.

We use the 2-bit comparator discussed in Chapter 2 in the tutorial. To physically test the circuit, we connect the four input signals to four slide switches and the output to an LED. The codes are repeated in Listings 3.1 and 3.2.

Listing 3.1   Gate-level implementation of a 1-bit comparator

```
1 library ieee;
  use ieee.std_logic_1164.all;
  entity eq1 is
     port(
        i0, i1: in std_logic;
6       eq: out std_logic
     );
  end eq1;

  architecture sop_arch of eq1 is
11    signal p0, p1: std_logic;
  begin
     eq <= p0 or p1;
     p0 <= (not i0) and (not i1);
     p1 <= i0 and i1;
16 end sop_arch;
```

Listing 3.2   Structural description of a 2-bit comparator

```
library ieee;
use ieee.std_logic_1164.all;
entity eq2 is
    port(
        a, b: in std_logic_vector(1 downto 0);
        aeqb: out std_logic
    );
end eq2;

architecture struc_arch of eq2 is
    signal e0, e1: std_logic;
begin
    -- instantiate two 1-bit comparators
    eq_bit0_unit: entity work.eq1(sop_arch)
        port map(i0=>a(0), i1=>b(0), eq=>e0);
    eq_bit1_unit: entity work.eq1(sop_arch)
        port map(i0=>a(1), i1=>b(1), eq=>e1);
    -- a and b are equal if individual bits are equal
    aeqb <= e0 and e1;
end struc_arch;
```

In the Quartus II GUI, the same action can be invoked by multiple ways. For example, we can start the compiling process by selecting the proper menu item, clicking the icon on the top, or clicking the process in the Tasks window. We use the menu in the tutorial.

3.5.1   Create the design project

A Quartus II project contains basic information of a design, which includes the location of the working directory, the top-level entity, the source files, the target device, the constraints, and tool settings. There are several tasks associated with this step:

- Create a project.
- Assign a device.
- Create new HDL files.
- Check the code syntax.
- Add existing HDL files.
- Import a pin-assignment constraint file.

Create a project   During project creation, Quartus II's New Project Wizard is invoked and it guides users through five pages to set up a new project. The initial project settings and relevant information are specified in this process. For flexibility, we manually update the setting later and only use the wizard to specify the location of the working directory and project name. A new project can be created as follows:

1. Launch the Quartus II program.
2. In the Quartus II GUI, select File ≻ New Project Wizard.... The New Project Wizard dialog appears. Click the Next> button to go to the next page. Enter the working directory name and enter the name of the project as demo.
3. Click the Finish button to exit the dialog.

If all the relevant information and files are known in advance, we can continue with the wizard and skip some of the subsequent tasks.

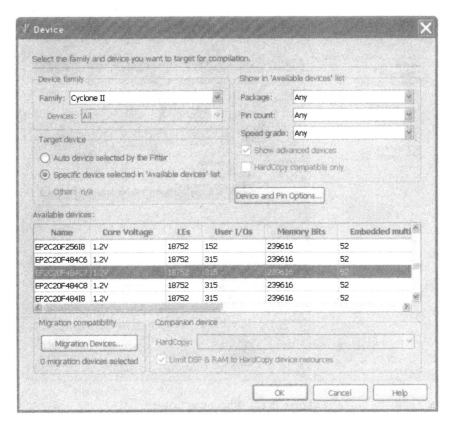

Figure 3.11   Device selection dialog.

Assign a device   Since we plan to implement the design on the prototyping board, we must specify the board's FPGA chip as the target device. On the DE1 board, it is a Cyclone II EP2C20F484C7 device. For other boards, the device name can be found in the board manual or by checking the marking on the actual FPGA chip. For a DE1 board, we can specify the target device as follows:

1. In the Quartus II menu, select Assignments ≻ Device.... The Device dialog appears, as shown in Figure 3.11
2. Select Cyclone II in the Family field and click the Specific device selected in 'Available devices' list button.
3. Scroll down the device list and select EP2C20F484C7.
4. Click the OK button to complete the selection.

The EP2C20F484C7 device has a special dual-purposed pin, known as *nCEO*, which can be configured as a special "programming pin" or a regular I/O pin. The pin is set up as a programming pin in Quartus II by default. On the DE1 board, the nCEO pin is configured as a normal I/O pin and used for the SD card interface. If the SD card is used in a design, we need to properly configure this pin to prevent a compiling error. The procedure to configure this pin is:

1. In the previous Device dialog, click the Device and Pin Options... button. The Device and Pin Options dialog appears, as shown in Figure 3.12.
2. In the left Category: panel. select the Dual-Purpose Pins item.

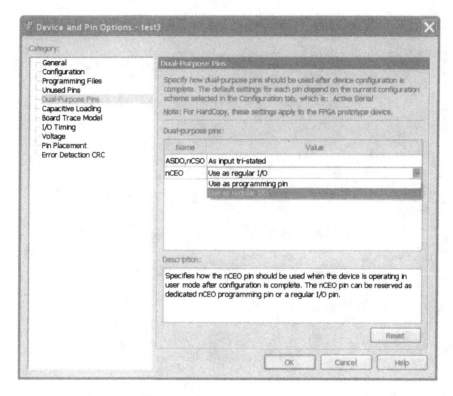

Figure 3.12   Pin option dialog.

3. In the right panel, select the nCEO row.
4. Click the right field of this row and expand the pull-down menu. Select Use as regular I/O, as shown in Figure 3.12.
5. Click the OK button to return to the Device dialog.

Create a new HDL file   A project may contain one or multiple HDL files. If a file does not exist, we must create a new source file. Quartus II contains an HDL editor for this task. For demonstration purposes, we create a file that contains the code in Listings 3.1. The procedure is:

1. Select File ≻ New. The New dialog appears.
2. Select VHDL and then click the OK button. A new text editor window appears in the Workplace area.
3. Select File ≻ Save. The Save as dialog appears. Enter list_ch03_01_eq1.vhd in the File name: field, put a check mark in the Add file to current project box, and click the Save button.
4. Enter the HDL code.
5. Select File ≻ save to save the file.

The Quartus II HDL editor is language sensitive and colors the language constructs for clarity. It also provides a collection of *pre-defined templates* of various language segments to facilitate the code entry. To insert a template into a file, select Edit ≻ Insert Template, expand the VHDL row, and double-click the desired template.

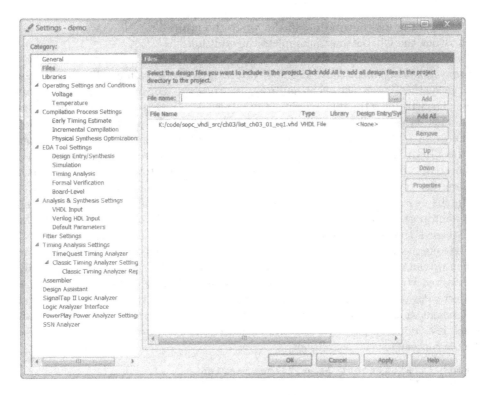

Figure 3.13   Files item in Setting dialog.

Check the code syntax   After completing a new HDL file, we need to check the syntax of the code:

1. Select the desired file window in the **Workplace** area.
2. Select **Processing** ≻ **Analyze Current File**.

The bottom **Messages** window displays the progress of the process and reports errors and warnings. Double-clicking an error message leads to the offending line in the file. We can correct the problem, save the file, and repeat the syntax checking process until all syntax errors are eliminated. The analysis process only checks the syntax of the current file and does not perform elaboration. Other errors may still occur when the entire project is compiled.

Add existing HDL files   A project usually contains multiple files and some files have been developed in previous projects or can be obtained from other sources. We can add existing HDL files to a project. The file containing the codes in Listings 3.2 can be downloaded from the companion web site and its file name is `list_ch03_02_eq2`. The procedure to add the file is:

1. Select **Assignments** ≻ **Settings**.... The **Settings** dialog appears.
2. In the left panel, click on the **Files** item. The **Setting** dialog shows the relevant file information in the right panel, as shown in Figure 3.13. The previously created `list_ch03_01_eq1.vhd` file should be already included in the list.
3. Click the ... button in the **Files Name** row and a **Select File** window appears.
4. Navigate to the proper directory, select the file, and click the **Open** button to return to the **Setting** dialog.

5. Click the **Add** button to add the file to the project.

6. Click the **OK** button to complete the addition.

Import a pin-assignment constraint file   *Constraints* are certain conditions imposed on the synthesis and placement and routing processes. For our purposes, the main type of constraint is the pin assignment of top-level I/O ports and the minimal clock rate. The latter is discussed in Section 5.5.2. During the placement and routing process, an I/O port of the top-level module must be mapped to a physical pin of the FPGA device. Since the peripherals' I/O signals are already permanently connected to the designated FPGA's pins on the DE1 prototyping board, we must ensure that the HDL module's I/O ports are mapped to the corresponding pins.

In the **eq2** circuit, we can connect the **a** and **b** ports to four slide switches and the **aeqb** port to an LED to verify the physical operation of the circuit. For the DE1 board, the corresponding pins are L22, L21, M22, V12, and R20.

The pin assignment can be performed with Quartus II's *Assignment Editor*, in which we can manually assign top-level I/O ports to FPGA's pins as well as specify the desired I/O standards. The process is tedious and error prone, especially for a project with a large number of I/O ports. Quartus II allows the user to save the pin assignment in a file and export and import the file as needed. The .csv file format, which is a comma-separated text format, is used for this purpose. The following is the file content for the **eq2** circuit:

```
From,To,Assignment Name,Value,Enabled
,a[0],Location,PIN_L22,Yes
,a[1],Location,PIN_L21,Yes
,b[0],Location,PIN_M22,Yes
,b[1],Location,PIN_V12,Yes
,aeqb,Location,PIN_R20,Yes
```

The first line is the column header indicating various fields and the remaining lines specify the mappings between I/O ports and pins. We can use the Quartus II built-in editor or any other program to create this file and save it as **eq2_pin.csv**. The procedure to import the pin assignment file is:

1. Select **Assignments** ≻ **Import Assignment....** The **Import Assignment** dialog appears.

2. Click the ... button in the **File Name** row and the **Select File** window appears.

3. Navigate to the proper directory, select the **eq2_pin.csv** file, and click the **OK** button to complete the process.

Since all of our experiments are done in the same prototyping board, the pin assignment is fixed. A pin assignment file that includes all connected I/O peripheral signals of the DE1 board, **chu_de1_pin.csv**, is created for this purpose. We can edit this file according to the I/O port names used in the top-level HDL module and delete the unused pins.

If the top-level module follows the names used in this file, it can be imported to the project without modification. One way to do this is to add a top-level wrapping file with the pre-defined board I/O signal names. The slide switches and red LEDs are named as **sw** and **ledr** in the **chu_de1_pin.csv** file. The wrapping code is shown in Listing 3.3.

Listing 3.3   Top-level wrapping circuit

```
library ieee;
use ieee.std_logic_1164.all;
entity eq_top is
 4    port(
        sw: in std_logic_vector(3 downto 0);    -- 4 switches
        ledr: out std_logic_vector(0 downto 0) -- 1 red LED
      );
 end eq_top;
 9
 architecture struc_arch of eq_top is
 begin
     -- instantiate 2-bit comparator
     eq2_unit: entity work.eq2(struc_arch)
14        port map(a=>sw(3 downto 2), b=>sw(1 downto 0),
                 aeqb=>ledr(0));
 end struc_arch;
```

The code essentially maps the "logical" port names of the comparator to the physical signals on the prototyping board. Note that the output ledr signal is defined as a one-element vector to accommodate future expansion. During compiling, Quartus II may generate warnings for the unused I/O ports.

3.5.2 Create a testbench and perform the RTL simulation

The testbench functions as a virtual lab bench. It consists of the HDL module to be tested and a code segment to generate the stimulus. The RTL simulation verifies the operation of the HDL module in the host computer. We use the *ModelSim* simulator manufactured by Mentor Graphics Corporation for this purpose. Its usage is illustrated in Section 3.6.

3.5.3 Compile the project

There are several tasks in this step:
- Specify the top-level module.
- Compile the project.
- Examine the compilation report.
- Examine the netlists.

The last two tasks provide additional information about the design but can be omitted.

Specify the top-level module   After adding and creating all HDL files, we can specify the top-level module. The procedure is:
1. Select Assignments ≻ Settings.... The Settings dialog appears.
2. In the left panel, click the General item.
3. Enter the top-level entity name, eq2, in the Top-level entity: field.
4. Click the OK button to complete the process.

After compiling, Quartus II analyzes the files based on the top-level module and establishes the hierarchical structure, which is displayed in the Project Navigator window, as shown in Figure 3.9.

Compile the project   Compiling contains the processes to analyze design hierarchy, to perform elaboration, synthesis, and placement and routing, and to generate the

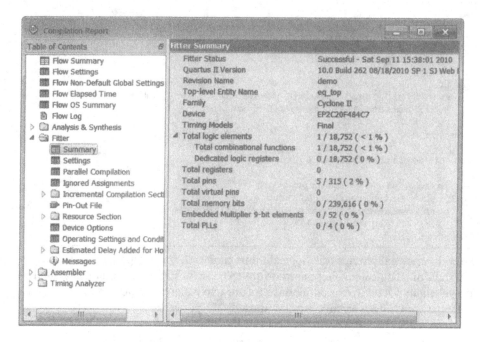

Figure 3.14   Compilation report.

configuration file. It can be invoked by selecting Processing ≻ Start Compilation. The progress of the compiling is displayed on the Tasks window, as shown in Figure 3.10. A green check mark will be placed on the left if the corresponding process is successful.

Although the syntax of individual files is checked earlier, the code may contain constructs that cannot be synthesized or may lead to poor implementation (such as a combinational loop). The error and warning messages are displayed in the Messages window. We must correct the problems and repeat the compiling process if needed.

Examine the compilation report   After successful compilation, a report window is generated and opened automatically on the Workplace area, as shown in Figure 3.14. It can also be invoked later by selecting Processing ≻ Compilation Report.

The window contains information for the overall flow as well as detailed reports for individual processes. A list of more detailed reports can be obtained by clicking on the corresponding directory, as in the Fitter item of Figure 3.14.

The compilation report is quite comprehensive. For our purposes, the following information is of special interest:

- Resource utilization
- Use of I/O pins

The resource utilization basically indicates the size of the resulting circuit in terms of the number of logic elements. It also shows the usage of various macro cells. This information can be found in the Flow summary report or the Fitter's Summary report. The report in Figure 3.14 indicates that one logic element (out of 18,752) is used to synthesize the two-bit comparator.

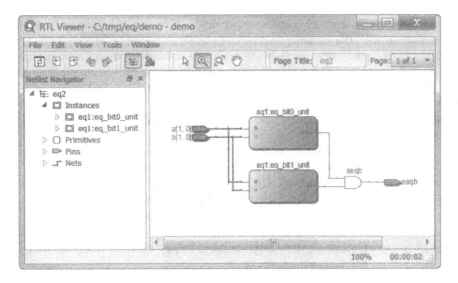

Figure 3.15 Snapshot of RTL Viewer.

Pin assignment is an error-prone task. To verify the pin locations, we can expand Fitter and then Resource Section and check the Input Pins and Output Pins reports for the pin assignment of HDL module's I/O ports.

Examine the netlists Quartus II provides a utility program to show the synthesized netlist in graphic format (i.e., to view the netlist as a schematic) in both the RT (register transfer) level and the logic element level. The former uses generic logic components and maintains the original design hierarchy and thus is more useful. The RT level viewing utility can be invoked by selecting Tools ≻ Netlist Viewers ≻ RTL Viewer. A snapshot of the RTL Viewer window is shown in Figure 3.15. The Netlist Navigator panel shows the original design hierarchy and the right panel displays the schematic of the currently selected eq2_unit, which is similar to that in Figure 2.2.

Because of the simplicity of this circuit, the generated schematic clearly shows the correspondence between the HDL code constructs and the circuit parts. The schematic for complicated HDL codes tends to be complex and involved and is less clear. Nevertheless, it still provides some insights about the HDL description and allows us to check the HDL codes from another perspective.

3.5.4 Perform timing analysis

The timing analysis relates to a system's performance, which is one of the key design criteria. The timing speciation is normally defined around the clock rate of a sequential circuit. We discuss this aspect in Section 5.5.2.

3.5.5 Program the FPGA device

The last step is to download the configuration file to the FPGA device. For the DE1 board, this task can be performed by two methods, known as the *JTAG* (joint test action group) mode and *AS* (active serial) mode. The mode is manually selected by a switch on the DE1 board.

In the JTAG mode, the configuration data is loaded from the host computer into the FPGA device directly. The DE1 board contains a special circuit, known as *USB Blaster*, that accepts a configuration data stream from a USB port and feeds the data into the FPGA device via the device's JTAG port. Since a Cyclone II FPGA is an SRAM-based device, the configuration data is lost when the power is removed.

To overcome this problem, the DE1 board provides an EEPROM (electrically erasable programmable read-only memory) device, which is nonvolatile and thus can maintain data after power-off. On the DE1 board, the configuration data from the EEPROM is automatically loaded into the FPGA chip upon power-up. If a configuration is needed repeatedly, we can load the file into the EEPROM device. This can be done with the AS mode.

Our discussion focuses on the JTAG mode. The key tasks in this step are:

- Set up the DE1 board.
- Download the configuration file.

Set up the DE1 board   The procedure is as follows:

1. Make sure that the programming mode switch is in the RUN position (for the JTAG mode). It is located on the left edge of the DE1 board, labeled sw12.
2. Connect the USB cable.
3. The power of the DE1 board can be provided either by the USB cable or a 7.5V adaptor. Connect the 7.5V adaptor if the host's USB port cannot provide adequate power.
4. Turn on the power by pressing down the red button on the left edge of the DE1 board.

Note that if a configuration file was previously stored in the EEPROM, it will automatically load and lead to some board activities, like flashing LEDs.

Download the configuration file   The procedure is as follows:

1. Select Tools ≻ Programmer. A stand-alone utility program called Programmer is invoked, as shown in Figure 3.16.
2. Check for the following setting:
   (a) Hardware Setup field: USB-Blaster.
   (b) Mode field: JTAG.
   (c) File column: demo.sof (the .sof stands for SRAM object file).
   (d) Device in the bottom: EP2C20F484.
3. The Hardware Setup field specifies the intended programming adapter. It may show No hardware initially. To correct this, click the Hardware Setup button to invoke the Hardware Setup dialog, click the pull-down menu to select the USB-Blaster, and then click the close button. The dialog is shown in Figure 3.17.
4. Click the Start button to start the downloading process.

Now the FPGA device is configured and we can test the circuit with the switches and observe the output LED.

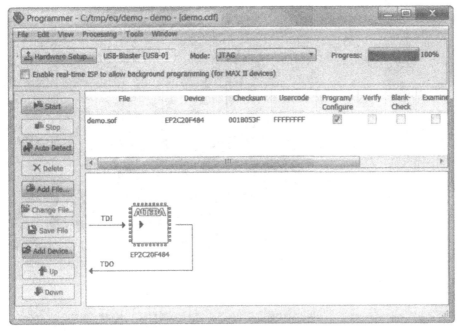

Figure 3.16   Programmer window.

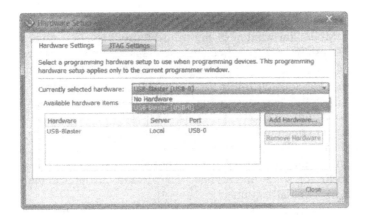

Figure 3.17   Hardware Setup dialog.

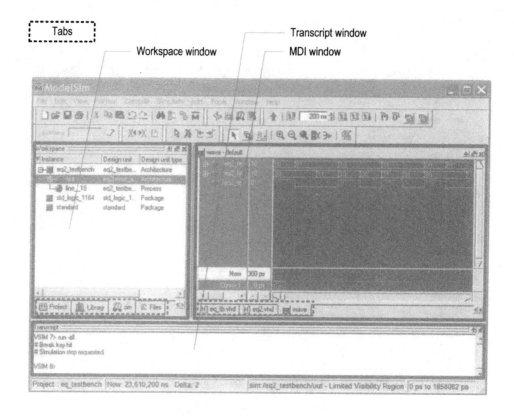

Figure 3.18   Typical ModelSim window.

3.6   SHORT TUTORIAL ON THE MODELSIM HDL SIMULATOR

The ModelSim software is an HDL simulator manufactured by Mentor Graphics Corporation and can run independently without Quartus II.

The default ModelSim window is shown in Figure 3.18. It is divided into three subwindows: Transcript window (bottom), Workspace window, and multiple document interface (MDI) window. The Workspace window displays information on the current process. The bottom tab is used to select the desired process page, which can be Project, Library, Sim, and so on. The Transcript window keeps track of the command history and messages. It can also be used as a command-line interface to enter ModelSim commands. The MDI window is an area to display HDL text, waveform, and so on. The bottom tab selects the desired pages.

Each subwindow may be resized, moved, docked, or undocked. Additional windows may appear for some operations. The default layout can be restored by selecting Window ≻ Initial Layout.

We present a short tutorial in this section to illustrate the basic simulation process. There are three steps:

1. Prepare a simulation project.
2. Compile the HDL codes.
3. Perform a simulation and examine the waveform.

We use the 2-bit comparator testbench discussed in Chapter 2 for the tutorial and the code is repeated in Listing 3.4. An additional assertion statement,

```
assert false
    report "Simulation Completed"
    severity failure;
```

is added to the end of the process. It generates an "artificial failure" and stops the simulation.

Listing 3.4   Testbench of a 2-bit comparator

```
 library ieee; use ieee.std_logic_1164.all;
2 entity eq2_testbench is
 end eq2_testbench;

 architecture tb_arch of eq2_testbench is
    signal test_in0, test_in1: std_logic_vector(1 downto 0);
7    signal test_out: std_logic;
 begin
    -- instantiate the circuit under test
    uut: entity work.eq2(struc_arch)
       port map(a=>test_in0, b=>test_in1, aeqb=>test_out);
12   -- test vector generator
    process
    begin
       -- test vector 1
       test_in0 <= "00";
17      test_in1 <= "00";
       wait for 200 ns;
       -- test vector 2
       test_in0 <= "01";
       test_in1 <= "00";
22      wait for 200 ns;
       -- test vector 3
       test_in0 <= "01";
       test_in1 <= "11";
       wait for 200 ns;
27      -- test vector 4
       test_in0 <= "10";
       test_in1 <= "10";
       wait for 200 ns;
       -- test vector 5
32      test_in0 <= "10";
       test_in1 <= "00";
       wait for 200 ns;
       -- test vector 6
       test_in0 <= "11";
37      test_in1 <= "11";
       wait for 200 ns;
       -- test vector 7
       test_in0 <= "11";
       test_in1 <= "01";
42      wait for 200 ns;
       -- terminate simulation
       assert false
          report "Simulation Completed"
          severity failure;
47   end process;
 end tb_arch;
```

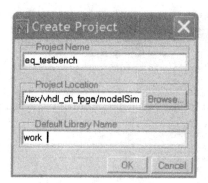

(a) `Create Project` dialog

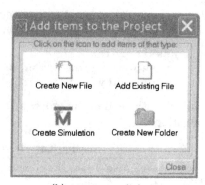

(b) `Add items` dialog

Figure 3.19   New project dialogs.

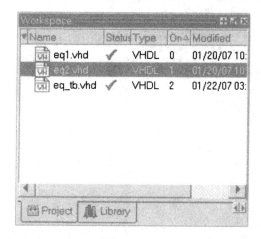

Figure 3.20   Project tab of the workplace panel.

Prepare a simulation project   A ModelSim simulation project consists of the library definition and a collection of HDL files. A testbench is an HDL program and can be created by using the Quartus II text editor, as discussed in Section 3.5.1. Alternatively, ModelSim also has a built-in editor. We assume that all HDL files are already constructed. The procedure to create a project is as follows:

1. Select All Programs ≻ Altera ≻ ModelSim-Altera 6.6c Starter Edition ≻ ModelSim-Altera 6.6c Starter Edition (or wherever ModelSim resides) to launch the ModelSim program.

2. Select File ≻ New ≻ Project and the Create Project dialog appears, as shown in Figure 3.19(a). Enter the project name as `eq_testbench`, select the project location, and set Default Library Name to `work`. Click OK. A blank Project page appears in the main window and the Add items to the project dialog appears, as shown in Figure 3.19(b).

3. In the Add items to the project dialog, click Add Existing File and add the necessary HDL files. Click OK. The project tab appears in the workplace subwindow and displays the selected files, as shown in Figure 3.20.

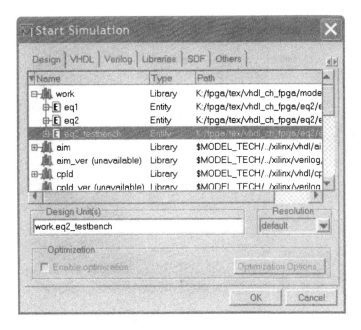

Figure 3.21   Simulate dialog.

Compile the HDL code   The *compile* term here means to convert the HDL code into the ModelSim internal format. In VHDL, the compiling is done on the *design unit* basis. Each entity and architecture is considered as one design unit. The procedure is:

1. Highlight the `eq1` file and right-click the mouse. Select Compile ≻ Compile Selected. Note that the compiling should be started from the modules at the bottom of the design hierarchy. The progress and messages are displayed in the transcript window.

2. If the file contains no syntactical error, a check mark shows up. Otherwise, an X mark shows up. Click the red error line in the transcript window to locate the errors. Correct the problems, save the file, and recompile the file.

3. Repeat the preceding steps to compile the `eq2` file and then the `eq_tb` file.

Perform a simulation and examine the waveform   After compiling the testbench and corresponding files, we can perform the simulation and examine the resulting waveform. This corresponds to running the circuit in a virtual lab bench and checking the waveform in a virtual logic analyzer. The procedure is:

1. Select Simulate ≻ Simulate and the Simulate dialog appears.

2. In the Design tab, find and expand the work library, which is the one defined when we create the project. All compiled units are displayed, as shown in Figure 3.21.

3. Load `eq2_testbench` by double-clicking the corresponding icon. The sim tab appears in the workplace window and the corresponding page displays the structure of the `eq2_testbench` module, as shown in Figure 3.22. An object window, which contains the signals in the selected module, may also appear.

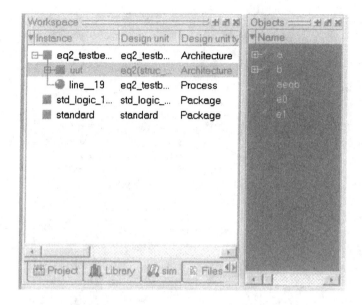

Figure 3.22   Sim panel of the workplace panel.

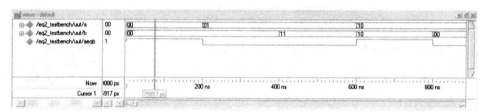

Figure 3.23   Waveform window.

4. Highlight the uut unit and right-click the mouse. Select Add ≻ Add to Wave. This adds all the signals of the uut unit to the waveform page. The waveform page appears in the MDI window.

5. If necessary, rearrange the signals order and set them to the proper formats (decimal, hex, and so on).

6. Select Simulate ≻ Run. There are several commands to control the simulation: Restart (restart the simulation), Run (run the simulation one step), Continue run (resume the run from the interrupt), Run All (run the simulation forever), and Break (break the simulation). These commands are also shown as icons at the top of the window.

7. The waveform window displays the simulated result, shown in Figure 3.23. We can scroll the window, zoom in, or zoom out to check the correctness of the design.

3.7   BIBLIOGRAPHIC NOTES

Both Altera Quartus II and Mentor Graphics ModelSim are complex software packages, and their manuals exceeds several thousand pages. Most documentation can be accessed via the Help menu. Quartus II has a comprehensive 140-page review, titled *Introduction to Quartus II Software*. ModelSim has a detailed tutorial, *ModelSim Tutorial*. These tutorials provide an overview of all features of the software package. Relevant information for the Cyclone II device can be found in its data sheets, *Cyclone II Device Handbook*, which includes a detailed explanation of the logic elements and macro cells. *The Design Warrior's Guide to FPGAs* by Clive Maxfield provides a comprehensive review of FPGA-related issues. The detailed layout and I/O pin assignment of the DE1 board can be found in *DE1 Board User Manual*. Information on other prototyping boards can be found in their manuals.

3.8   SUGGESTED EXPERIMENTS

3.8.1   Gate-level greater-than circuit

The greater-than circuit compares two inputs, a and b, and asserts an output when a is greater than b. We want to create a 4-bit greater-than circuit from the bottom up and use only gate-level logical operators. Design the circuit as follows:

1. Derive the truth table for a 2-bit greater-than circuit and obtain the logic expression in the sum-of-products format. Based on the expression, derive the HDL code using only logical operators.
2. Derive a testbench for the 2-bit greater-than circuit. Perform a simulation and verify the correctness of the design.
3. Use four slide switches as the inputs and one LED as the output. Synthesize the circuit and download the configuration file to the prototyping board. Verify its operation.
4. Use the 2-bit greater-than circuits and 2-bit equality comparators and a minimal number of "glue gates" to construct a 4-bit greater-than circuit. First draw a block diagram and then derive the structural HDL code according to the diagram.
5. Derive a testbench for the 4-bit greater-than circuit. Perform a simulation and verify the correctness of the design.
6. Use eight slide switches as the inputs and one LED as the output. Synthesize the circuit and download the configuration file to the prototyping board. Verify its operation.

3.8.2   Gate-level binary decoder

An n-to-2^n binary decoder asserts one of 2^n bits according to the input combination. The functional table of a 2-to-4 decoder with an enable signal is shown in Table 3.2. We want to create several decoders using only gate-level logical operators. The procedure is as follows:

1. Determine the logic expressions for the 2-to-4 decoder with enable and derive the HDL code using only logical operators.

Table 3.2   Truth table of a 2-to-4 decoder with enable

en	input a(1)	a(0)	output bcode
0	–	–	0000
1	0	0	0001
1	0	1	0010
1	1	0	0100
1	1	1	1000

2. Derive a testbench for the decoder. Perform a simulation and verify the correctness of the design.

3. Use three slide switches as the inputs and four LEDs as the outputs. Synthesize the circuit and download the configuration file to the prototyping board. Verify its operation.

4. Use the 2-to-4 decoders to derive a 3-to-8 decoder. First draw a block diagram and then derive the structural HDL code according to the diagram.

5. Derive a testbench for the 3-to-8 decoder. Perform a simulation and verify the correctness of the design.

6. Use four slide switches as the inputs and eight LEDs as the outputs. Synthesize the circuit and download the configuration file to the prototyping board. Verify its operation.

7. Use the 2-to-4 decoders to derive a 4-to-16 decoder. First draw a block diagram and then derive the structural HDL code according to the diagram.

8. Derive a testbench for the 4-to-16 decoder. Perform a simulation and verify the correctness of the design.

CHAPTER 4

RT-LEVEL COMBINATIONAL CIRCUIT

The gate-level circuits discussed in Chapter 2 utilize simple logical operators to describe gate-level design, which is composed of simple logic cells. In this chapter, we examine the HDL description of module-level circuits, which are composed of intermediate-sized components, such as adders, comparators, and multiplexers. Since these components are the basic building blocks used in *register transfer methodology*, it is sometimes referred to as RT-level design. We first discuss more sophisticated VHDL operators and routing constructs and then demonstrate the RT-level combinational circuit design through a series of examples.

4.1 RT-LEVEL COMPONENTS

In addition to the logical operators, relational operators and several arithmetic operators can also be synthesized automatically. These operators correspond to intermediate-sized module-level components, such as comparators and adders. We examine these operators in this section and also cover miscellaneous synthesis-related VHDL constructs. Tables 4.1 and 4.2 summarize the operators and their applicable data types used in this book.

Embedded SOPC Design with Nios II Processor and VHDL Examples. By Pong P. Chu   **49**
Copyright © 2011 John Wiley & Sons, Inc.

Table 4.1   Operators and data types of VHDL-93 and IEEE std_logic_1164 package

Operator	Description	Data type of operands	Data type of result
a ** b	exponentiation	integer	integer
a * b	multiplication		
a / b	division	*integer type for constants and*	
a + b	addition	*array boundaries, not synthesis*	
a − b	subtraction		
a & b	concatenation	1-D array, element	1-D array
a = b	equal to	any	boolean
a /= b	not equal to		
a < b	less than	scalar or 1-D array	boolean
a <= b	less than or equal to		
a > b	greater than		
a >= b	greater than or equal to		
not a	negation	boolean, std_logic,	same as operand
a and b	and	std_logic_vector	
a or b	or		
a xor b	xor		

Table 4.2   Overloaded operators and data types in the IEEE numeric_std package

Overloaded operator	Description	Data type of operands	Data type of result
a * b	arithmetic	unsigned, natural	unsigned
a + b	operation	signed, integer	signed
a − b			
a = b			
a /= b			
a < b	relational	unsigned, natural	boolean
a <= b	operation	signed, integer	boolean
a > b			
a >= b			

Table 4.3   Type conversions between std_logic_vector and numeric data types

Data type of a	To data type	Conversion function/type casting
unsigned, signed	std_logic_vector	std_logic_vector(a)
signed, std_logic_vector	unsigned	unsigned(a)
unsigned, std_logic_vector	signed	signed(a)
unsigned, signed	integer	to_integer(a)
natural	unsigned	to_unsigned(a, size)
integer	signed	to_signed(a, size)

4.1.1   Relational operators

Six relational operators are defined in the VHDL standard: = (equal to), /= (not equal to), < (less than), <= (less than or equal to), > (greater than), and >= (greater than or equal to). These operators compare operands of the same data type and return a value of the boolean data type. In this book, we don't use the boolean data type directly, but embed it in routing constructs. This is discussed in Sections 4.2 and 4.4. During synthesis, comparators are inferred for these operators.

4.1.2   Arithmetic operators

In the VHDL standard, arithmetic operations are defined for the integer data type and for the natural data type, which is a subtype of integer containing zero and positive integers. We usually prefer to have more control in synthesis and define the exact number of bits and format (i.e., signed or unsigned). The IEEE numeric_std package is developed for this purpose. In this book, we use the integer and natural data types for constants and array boundaries but not for synthesis.

IEEE numeric_std package   The IEEE numeric_std package adds two new data types, unsigned and signed, and defines the relational and arithmetic operators over the new data types (known as *operator overloading*). The unsigned and signed data types are defined as an array with elements of the std_logic data type. The array is interpreted as the binary representation of unsigned or signed integers. We have to add an additional use statement to invoke the package:

```
library ieee;
use ieee.std_logic_1164.all;
use ieee.numric_std.all;    -- invoke numeric_std package
```

The synthesizable overloaded operators are summarized in Table 4.2.

Multiplication is a complicated operation, and synthesis of the multiplication operator * depends on the synthesis software and target device technology. The Altera Cyclone II FPGA family contains prefabricated combinational multiplier blocks. The Quartus II software can infer these blocks during synthesis, and thus the multiplication operator can be used in HDL code. The EP2C20 FPGA device of the DE1 board consists of 26 18-by-18 multiplier blocks. While the synthesis of the multiplication operator is supported, we need to be aware of the limitation on the number and input width of these blocks and use them with care.

Type conversion   Because VHDL is a strongly typed language, std_logic_vector, unsigned, and signed are treated as different data types even when all of them are defined as an array with elements of the std_logic data type. A *conversion function* or *type casting* is needed to convert signals of different data types. The conversion is summarized in Table 4.3. Note that the std_logic_vector data type is not interpreted as a number and thus cannot be converted directly to an integer, and vice versa.

The following examples illustrate the common mistakes and remedies for type conversion. Assume that some signals are declared as follows:

```
library ieee;
use ieee.std_logic_1164.all;
use ieee.numeric_std.all;
. . .
signal s1, s2, s3, s4, s5, s6: std_logic_vector(3 downto 0);
signal u1, u2, u3, u4, u5, u6, u7: unsigned(3 downto 0);
. . .
```

Let us first consider the following assignment statements:

```
u1 <= s1;   -- not ok, type mismatch
u2 <= 5;    -- not ok, type mismatch
s2 <= u3;   -- not ok, type mismatch
s3 <= 5;    -- not ok, type mismatch
```

They are all invalid because of type mismatch. The right-hand-side expression must be converted to the data type of the left-hand-side signal:

```
u1 <= unsigned(s1);      -- ok, type casting
u2 <= to_unsigned(5,4);  -- ok, conversion function
s2 <= std_logic_vector(u3);  -- ok, type casting
s3 <= std_logic_vector(to_unsigned(5,4));  -- ok
```

Note that two type conversions are needed for the last statement.

Let us consider statements that involve arithmetic operations. The following statements are valid since the + operator is defined with the **unsigned** and **natural** types in the IEEE **numeric_std** package.

```
u4 <= u2 + u1;   -- ok, both operands unsigned
u5 <= u2 + 1;    -- ok, operands unsigned and natural
```

On the other hand, the following statements are invalid since no overloaded arithmetic operation is defined for the **std_logic_vector** data type:

```
s5 <= s2 + s1;   -- not ok, + undefined over the types
s6 <= s2 + 1;    -- not ok, + undefined over the types
```

To fix the problem, we must convert the operands to the **unsigned** (or **signed**) data type, perform addition, and then convert the result back to the **std_logic_vector** data type. The revised code becomes

```
s5 <= std_logic_vector(unsigned(s2) + unsigned(s1));  -- ok
s6 <= std_logic_vector(unsigned(s2) + 1);             -- ok
```

Nonstandard arithmetic packages   There are several popular non-IEEE arithmetic packages, which are **std_logic_arith**, **std_logic_unsigned**, and **std_logic_signed**. The **std_logic_arith** package is similar to the **numeric_std** package. The other two packages do not introduce any new data type but define overloaded arithmetic operators over the **std_logic_vector** data type. We do not use these packages in this book.

4.1.3 Other synthesis-related VHDL constructs

Concatenation operator   The concatenation operator, &, combines segments of elements and small arrays to form a large array. The following example illustrates its use:

oe	y
0	Z
1	a_in

Figure 4.1   Symbol and functional table of a tristate buffer.

```
signal a1:  std_logic;
signal a4:  std_logic_vector(3 downto 0);
signal b8, c8, d8: std_logic_vector(7 downto 0);
. . .
b8 <= a4 & a4;
c8 <= a1 & a1 & a4 & "00";
d8 <= b8(3 downto 0) & c8(3 downto 0);
```

Implementation of the concatenation operator involves reconnection of the input and output signals and only requires "wiring."

One major application of the & operator is to perform shifting operations. Although both the VHDL standard and numeric_std package define shift functions, they sometimes cannot be synthesized automatically. The & operator can be used for shifting a signal for a fixed amount, as shown in the following example:

```
signal a: std_logic_vector(7 downto 0);
signal rot, shl, sha: std_logic_vector(7 downto 0);
. . .
-- rotate a to right 3 bits
rot <= a(2 downto 0) & a(7 downto 3);
-- shift a to right 3 bits and insert 0 (logic shift)
shl <= "000" & a(7 downto 3);
-- shift a to right 3 bits and insert MSB
-- (arithmetic shift)
sha <= a(7) & a(7) & a(7) & a(7 downto 3);
```

An additional routing circuit is needed if the amount of shifting is not fixed. The design of a barrel shifter is discussed in Section 4.6.3.

'Z' value of std_logic   The std_logic data type has a value of 'Z', which implies *high impedance* or an open circuit. It is not a normal logic value and can only be synthesized by a *tristate buffer*. The symbol and function table of a tristate buffer are shown in Figure 4.1. The operation of the buffer is controlled by an enable signal, oe ("output enable"). When it is '1', the input is passed to output. On the other hand, when it is '0', the y output appears to be an open circuit. The code of the tristate buffer is

```
y <= a_in when oe='1' else 'Z';
```

The most common application for a tristate buffer is to implement a *bidirectional port* to better utilize a physical I/O pin. A simple example is shown in Figure 4.2. The dir signal controls the direction of signal flow of the bi pin. When it is 0, the tristate buffer is in the high-impedance state and the sig_out signal is blocked. The pin is used as an input port and the input signal is routed to the sig_in signal. When the dir signal is 1, the pin is used as an output port and the sig_out signal

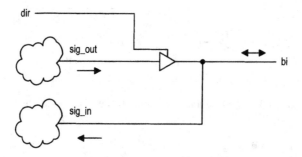

Figure 4.2 Single-buffer bidirectional I/O port.

is routed to an external circuit. The HDL code can be derived according to the diagram:

```
entity bi_demo is
   port (
      bi: inout std_logic;
      . . .
   )
begin
   sig_out <= output_expression;
   . . .
   some_signal <= expression_with_sig_in;
   . . .
   bi <= sig_out when dir='1' else 'Z';
   sig_in <= bi;
   . . .
```

Note that the mode of the `bi` port must be declared as **inout** for bidirectional operation.

For a Cyclone II device, a tristate buffer exists only in the I/O buffer of a physical pin. Thus, the tristate buffer can only be used for I/O ports that are mapped to the physical pins of an FPGA device.

4.1.4 Summary

Because of the nature of a strongly typed language, the data type frequently confuses a new VHDL user. Since this book is focused on synthesis, only a small set of data types and operators are needed. Their uses can be summarized as follows:

- Use the `std_logic` and `std_logic_vector` data types in entity port declaration and for the internal signals that involve no arithmetic operations.
- Use the `'Z'` value only to infer a tristate buffer.
- Use the IEEE `numeric_std` package and its **unsigned** or **signed** data types for the internal signals that involve arithmetic operation.
- Use the data type casting or conversion functions in Table 4.3 to convert signals and expressions among the `std_logic_vector` and various numerical data types.

- Use VHDL's built-in **integer** data type and arithmetic operators for constant and array boundary expressions, but not for synthesis (i.e., not used as a data type for a signal).
- Embed the result of a relational operation, which is in the **boolean** data type, in routing constructs (discussed in Section 4.2).
- Use a user-defined two-dimensional data type for a two-dimensional storage array (discussed in Section 5.2.3).
- Use a user-defined *enumerate data type* for the symbolic states of a finite state machine (discussed in Chapter 6).

4.2 ROUTING CIRCUIT WITH CONCURRENT ASSIGNMENT STATEMENTS

The *conditional signal assignment* and *selected signal assignment* statements are concurrent statements. Their behaviors are somewhat like the if and case statements of a conventional programming language. Instead of being executed sequentially, these statements are mapped to a routing network during synthesis.

4.2.1 Conditional signal assignment statement

Syntax and conceptual implementation   The simplified syntax of a conditional signal assignment statement is

```
signal_name <= value_expr_1 when boolean_expr_1 else
               value_expr_2 when boolean_expr_2 else
                 . . .
               value_expr_n;
```

The Boolean expressions are evaluated successively in turn until one is found to be **true** and the corresponding value expression is assigned to the signal. The **value_expr_n** is assigned if all Boolean expressions are evaluated to be **false**.

The conditional signal assignment statement implies a cascading priority routing network. Consider the following statement:

```
r <= a + b + c when m = n else
     a - b       when m > n else
     c + 1;
```

The routing is done by a sequence of 2-to-1 multiplexers. The diagram and truth table of a 2-to-1 multiplexer are shown in Figure 4.3(a), and the conceptual diagram of the statement is shown in Figure 4.3(b). If the first Boolean condition (i.e., **m=n**) is **true**, the result of **a+b+c** is routed to **r**. Otherwise, the data connected to the 0 port is passed to **r**. We need to trace the path along the 0 port and check the next Boolean condition (i.e., **m>n**) to determine whether the result of **a-b** or **c+1** is routed to the output.

Note that all the Boolean expressions and value expressions are evaluated concurrently. The values from the Boolean circuits set the selection signals of the multiplexers to route the desired value to the output. The number of cascading stages increases proportionally to the number of when-else clauses. A large number of when-else clauses will lead to a long cascading chain and introduce a large propagation delay.

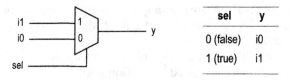

(a) Diagram of a 2-to-1 multiplexer

sel	y
0 (false)	i0
1 (true)	i1

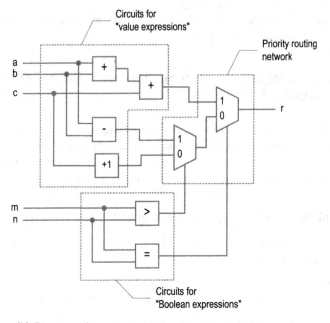

(b) Diagram of a conditional signal assignment statement

Figure 4.3   Implementation of a conditional signal assignment statement.

Table 4.4   Function table of a four-request priority encoder

input r	output pcode
1 – – –	100
0 1 – –	011
0 0 1 –	010
0 0 0 1	001
0 0 0 0	000

Examples   We use two simple examples to demonstrate the use of the conditional signal assignment statement. The first example is a priority encoder. The priority encoder has four requests, r(4), r(3), r(2), and r(1), which are grouped as a single 4-bit r input, and r(4) has the highest priority. The output is the binary code of the highest-order request. The function table is shown in Table 4.4. The HDL code is shown in Listing 4.1.

Table 4.5   Truth table of a 2-to-4 decoder with enable

	input		output
en	a(1)	a(0)	y
0	–	–	0000
1	0	0	0001
1	0	1	0010
1	1	0	0100
1	1	1	1000

Listing 4.1   Priority encoder using a conditional signal assignment statement

```
library ieee;
2 use ieee.std_logic_1164.all;
  entity prio_encoder is
     port(
        r: in std_logic_vector(4 downto 1);
        pcode: out std_logic_vector(2 downto 0)
7 );
  end prio_encoder;

  architecture cond_arch of prio_encoder is
  begin
12    pcode <= "100" when (r(4)='1') else
             "011" when (r(3)='1') else
             "010" when (r(2)='1') else
             "001" when (r(1)='1') else
             "000";
17 end cond_arch;
```

The code first checks the r(4) request and assigns "100" to pcode if it is asserted. It continues to check the r(3) request if r(4) is not asserted and repeats the process until all requests are examined.

The second example is a binary decoder. An n-to-2^n binary decoder asserts 1 bit of the 2^n-bit output according to the input combination. The functional table of a 2-to-4 decoder is shown in Table 4.5. The circuit also has a control signal, **en**, which enables the decoding function when asserted. The HDL code is shown in Listing 4.2.

Listing 4.2   Binary decoder using a conditional signal assignment statement

```
library ieee;
  use ieee.std_logic_1164.all;
3 entity decoder_2_4 is
     port(
        a: in std_logic_vector(1 downto 0);
        en: in std_logic;
        y: out std_logic_vector(3 downto 0)
8 );
  end decoder_2_4;

  architecture cond_arch of decoder_2_4 is
  begin
13    y <= "0000" when (en='0') else
            "0001" when (a="00") else
            "0010" when (a="01") else
```

```
            "0100" when (a="10") else
            "1000";    — a="11"
18 end cond_arch;
```

The code first checks whether **en** is not asserted. If the condition is **false** (i.e., **en** is '1'), it tests the four binary combinations in sequence.

4.2.2   Selected signal assignment statement

Syntax and conceptual implementation   The simplified syntax of a selected signal assignment statement is

```
with sel select
   sig <= value_expr_1 when choice_1 ,
          value_expr_2 when choice_2 ,
          value_expr_3 when choice_3 ,
          . . .
          value_expr_n when others ;
```

The selected signal assignment statement is somewhat like a case statement in a traditional programming language. It assigns an expression to a signal according to the value of the **sel** signal. A choice (i.e., **choice_i**) must be a valid value or a set of valid values of **sel**. The choices have to be *mutually exclusive* (i.e., no value can be used more than once) and *all inclusive* (i.e., all values must be used). In other words, all possible values of **sel** must be covered by one and only one choice. The reserved word, **others**, is used in the end to cover unused values. Since the **sel** signal usually has the **std_logic_vector** data type, the **others** term is always needed to cover the unsynthesizable values ('X', 'U', etc.).

The selected signal assignment statement implies a multiplexing structure. Consider the following statement:

```
signal sel: std_logic_vector (1 downto 0);
. . .
with sel select
   r <= a + b + c   when "00",
        a - b        when "10",
        c + 1        when others ;
```

For synthesis purposes, the **sel** signal can assume four possible values: "00", "01", "10", and "11". It implies a 2^2-to-1 multiplexer with **sel** as the selection signal. The diagram and functional table of the 2^2-to-1 multiplexer are shown in Figure 4.4(a), and the conceptual diagram of the statement is shown in Figure 4.4(b). The evaluated result of **a+b+c** is routed to **r** when **sel** is "00", the result of **a-b** is routed when **sel** is "10", and the result of **c+1** is routed when **sel** is "01" or "11".

Again, note that all value expressions are evaluated concurrently. The **sel** signal is used as the selection signal to route the desired value to the output. The width (i.e., number of input ports) of the multiplexer increases geometrically with the number of bits of the **sel** signal.

Example   We use the same encoder and decoder circuits to illustrate the use of the selected signal assignment statement. The code for the priority encoder is shown in Listing 4.3. The entity declaration is identical to that in Listing 4.1 and is omitted.

(a) Diagram and functional table of a 4-to-1 multiplexer

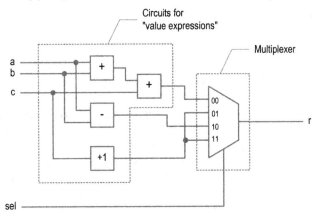

(b) Diagram of a selected signal assignment statement

Figure 4.4   Implementation of a selected signal assignment statement.

Listing 4.3   Priority encoder using a selected signal assignment statement

```
architecture sel_arch of prio_encoder is
begin
   with r select
      pcode <= "100" when "1000"|"1001"|"1010"|"1011"|
                          "1100"|"1101"|"1110"|"1111",
               "011" when "0100"|"0101"|"0110"|"0111",
               "010" when "0010"|"0011",
               "001" when "0001",
               "000" when others;    — r="0000"
end sel_arch;
```

The code exhaustively lists all possible combinations of the r signal and the corresponding output values. Note that the | symbol is used if the choice is more than one value.

The code for the 2-to-4 decoder is shown in Listing 4.4.

Listing 4.4   Binary decoder using a selected signal assignment statement

```
architecture sel_arch of decoder_2_4 is
   signal s: std_logic_vector(2 downto 0);
begin
   s <= en & a;
   with s select
      y <= "0000" when "000"|"001"|"010"|"011",
           "0001" when "100",
           "0010" when "101",
```

```
                "0100" when "110",
  10            "1000" when others;    — s="111"
      end sel_arch;
```

We concatenate **en** and **a** to form a 3-bit signal, **s**, and use it as the selection signal. The remaining code again exhaustively lists all possible combinations and the corresponding output values.

4.3   MODELING WITH A PROCESS

4.3.1   Process

To facilitate system modeling, VHDL contains a number of *sequential statements*, which are executed in sequence. Since their behavior is different from that of a normal concurrent circuit model, these statements are encapsulated inside a *process*. A process itself is a concurrent statement. It can be thought of as a black box whose behavior is described by sequential statements.

Sequential statements include a rich variety of constructs, but many of them don't have clear hardware counterparts. A poorly coded process frequently leads to unnecessarily complex implementation or cannot be synthesized at all. A detailed discussion of sequential statements and processes is beyond the scope of this book. For synthesis, we restrict the use of the process to two purposes:

- Describe routing structures with *if* and *case* statements.
- Construct templates for memory elements (discussed in Chapter 5).

The simplified syntax of a process with a sensitivity list is

```
process(sensitivity_list)
begin
   sequential statement;
   sequential statement;
    . . .
   end process;
```

The **sensitivity_list** is a list of signals to which the process responds (i.e., is "sensitive to"). For a combinational circuit, all the input signals should be included in this list. The body of a process is composed of any number of sequential statements.

4.3.2   Sequential signal assignment statement

The simplest sequential statement is a *sequential* signal assignment statement. The simplified syntax is

```
    sig <= value_expression;
```

The statement must be encapsulated inside a process.

Although its syntax is similar to that of a simple *concurrent* signal assignment statement, the semantics are different. When a signal is assigned multiple times inside a process, only the last assignment takes effect. For example, the code segment

```
process(a,b)
begin
   c <= a and b;
   c <= a or b;
end process;
```

is the same as

```
process(a,b)
begin
   c <= a or b;
end process;
```

On the other hand, if they are concurrent signal assignment statements, as in

```
-- not within a process
c <= a and b;
c <= a or b;
```

the code infers an and cell and an or cell, whose outputs are tied together. It is not allowed in most device technology and thus is a design error.

The semantics of assigning a signal multiple times inside a process is subtle and can sometimes be error-prone. Detailed explanations can be found in the references cited in the bibliographic section. We use multiple assignments only to avoid unintended memory, as discussed in Section 4.4.4.

4.4 ROUTING CIRCUIT WITH IF AND CASE STATEMENTS

If and *case* statements are two other commonly used sequential statements. In synthesis, they can be used to describe routing structures.

4.4.1 If statement

Syntax and conceptual implementation The simplified syntax of an if statement is

```
if boolean_expr_1 then
   sequential_statements;
elsif boolean_expr_2 then
   sequential_statements;
elsif boolean_expr_3 then
   sequential_statements;
 . . .
else
   sequential_statements;
end if;
```

It has one *then branch*, one or more optional *elsif branches*, and one optional *else branch*. The Boolean expressions are evaluated sequentially until an expression is evaluated as **true** or the else branch is reached, and the statements in the corresponding branch will be executed.

An if statement and a concurrent conditional signal assignment statement are somewhat similar. The two statements are equivalent if each branch of the if statement contains only a single sequential signal assignment statement. For example, the previous statement

```
r <= a + b + c when m = n else
    a - b       when m > 0 else
    c + 1;
```

can be rewritten as

```
process(a,b,c,m,n)
begin
  if m = n then
     r <= a + b + c;
  elsif m > 0 then
     r <= a - b;
  else
     r <= c + 1;
  end if;
end;
```

As in a conditional signal assignment statement, the if statement infers a similar priority routing structure during synthesis.

Example The codes of the same priority encoder and written with an if statement are shown in Listings 4.5 and 4.6. They are similar to those in Listings 4.1 and 4.2. Note that the if statement must be encapsulated inside a process.

Listing 4.5   Priority encoder using an if statement

```
   architecture if_arch of prio_encoder is
   begin
     process(r)
     begin
5       if (r(4)='1') then
            pcode <= "100";
        elsif (r(3)='1')then
            pcode <= "011";
        elsif (r(2)='1')then
10          pcode <= "010";
        elsif (r(1)='1')then
            pcode <= "001";
        else
            pcode <= "000";
15      end if;
     end process;
   end if_arch;
```

Listing 4.6   Binary decoder using an if statement

```
   architecture if_arch of decoder_2_4 is begin
     process(en,a)
3    begin
        if (en='0') then
          y <= "0000";
        elsif (a="00") then
          y <= "0001";
8       elsif (a="01")then
          y <= "0010";
        elsif (a="10")then
          y <= "0100";
        else
13        y <= "1000";
```

```
      end if;
   end process;
end if_arch;
```

4.4.2 Case statement

Syntax and conceptual implementation The simplified syntax of a case statement is

```
case sel is
   when choice_1 =>
      sequential statements;
   when choice_2 =>
      sequential statements;
   . . .
   when others =>
      sequential statements;
end case;
```

A case statement uses the `sel` signal to select a set of sequential statements for execution. As in a selected signal assignment statement, a choice (i.e., `choice_i`) must be a valid value or a set of valid values of `sel`, and the choices have to be mutually exclusive and all inclusive. Note that the **others** term at the end covers the unused values.

A case statement and a concurrent selected signal assignment statement are somewhat similar. The two statements are equivalent if each branch of the case statement contains only a single sequential signal assignment statement. For example, the previous statement

```
with sel select
   r <= a + b + c    when "00",
        a - b         when "10",
        c + 1         when others;
```

can be rewritten as

```
process(a,b,c,sel)
begin
   case sel is
      when "00" =>
         r <= a + b + c;
      when "10" =>
         r <= a - b;
      when others =>
         r <= c + 1;
   end case;
end;
```

As in a selected signal assignment statement, the case statement infers a similar multiplexing structure during synthesis.

Example The codes of the same priority encoder and decoder written with a case statement are shown in Listings 4.7 and 4.8. As in Listings 4.3 and 4.4, the codes exhaustively lists all possible input combinations and the corresponding output values.

Listing 4.7   Priority encoder using a case statement

```
architecture case_arch of prio_encoder is
begin
    process(r)
4   begin
        case r is
            when "1000"|"1001"|"1010"|"1011"|
                 "1100"|"1101"|"1110"|"1111" =>
                pcode <= "100";
9           when "0100"|"0101"|"0110"|"0111" =>
                pcode <= "011";
            when "0010"|"0011" =>
                pcode <= "010";
            when "0001" =>
14              pcode <= "001";
            when others =>
                pcode <= "000";
        end case;
    end process;
19 end case_arch;
```

Listing 4.8   Binary decoder using a case statement

```
1 architecture case_arch of decoder_2_4 is
    signal s: std_logic_vector(2 downto 0);
  begin
    s <= en & a;
    process(s)
6   begin
        case s is
            when "000"|"001"|"010"|"011" =>
                y <= "0000";
            when "100" =>
11              y <= "0001";
            when "101" =>
                y <= "0010";
            when "110" =>
                y <= "0100";
16          when others =>
                y <= "1000";
        end case;
    end process;
  end case_arch;
```

4.4.3   Comparison to concurrent statements

The preceding subsections show that the simple if and case statements are equivalent to the conditional and selected signal assignment statements. However, an if or case statement allows *any number* and *any type* of sequential statements in their branches and thus is more flexible and versatile. Disciplined use can make the code more descriptive and even make a circuit more efficient.

This can be illustrated by two code segments. First, consider a circuit that sorts the values of two input signals and routes them to the **large** and **small** outputs. This can be done by using two conditional signal assignment statements:

```
large <= a when a > b else
          b;
small <= b when a > b else
          a;
```

Since there are two relation operators (i.e., two >) in code, synthesis software may infer two greater-than comparators. The same function can be coded by a single if statement:

```
process(a,b)
begin
   if a > b then
      large <= a;
      small <= b;
   else
      large <= b;
      small <= a;
   end if;
end;
```

The code consists of only a single relational operator.

Second, let us consider a circuit that routes the maximal value of three input signals to the output. This can be clearly described by nested two-level if statements:

```
process(a,b,c)
begin
   if (a > b) then
      if (a > c) then
         max <= a;
      else
         max <= c;
      end if;
   else
      if (b > c) then
         max <= b;
      else
         max <= c;
      end if;
   end if;
end process;
```

We can translate the if statement to a "single-level" conditional signal assignment statement:

```
max <= a when ((a > b) and (a > c)) else
       c when (a > b) else
       b when (b > c) else
       c;
```

Since no nesting is allowed, the code is less intuitive. If concurrent statements must be used, a better alternative is to describe the circuit with three conditional signal assignment statements:

```
signal ac_max, bc_max: std_logic;
   . . .
```

```
ac_max <= a when (a > c) else
          c;
bc_max <= b when (b > c) else
          c;
max <= ac_max when (a > b) else
       bc_max;
```

4.4.4   Unintended memory

Although a process is flexible, a subtle error in code may infer incorrect implementation. One common problem is the inclusion of intended memory in a combinational circuit. The VHDL standard specifies that a signal will *keep its previous value* if it is not assigned in a process. During synthesis, this infers an internal state (via a closed feedback loop) or a memory element (such as a latch).

To prevent unintended memory, we should observe the following rules while developing code for a combinational circuit:

- Include all input signals in the sensitivity list.
- Include the else branch in an if statement.
- Assign a value to every signal in every branch.

For example, the following code segment tries to generate a greater-than (i.e., gt) and an equal-to (i.e., eq) output signal:

```
process(a)              -- b missing from sensitivity list
begin
   if (a > b) then      -- eq not assigned in this branch
      gt <= '1';
   elsif (a = b) then   -- gt not assigned in this branch
      eq <= '1';
   end if;              -- else branch is omitted
end process;
```

Although the syntax is correct, it violates all three rules. For example, gt will keep its previous value when the a>b expression is **false** and a latch will be inferred accordingly. The correct code should be

```
process(a,b)
begin
   if (a > b) then
      gt <= '1';
      eq <= '0';
   elsif (a = b) then
      gt <= '0';
      eq <= '1';
   else
      gt <= '0';
      eq <= '0';
   end if;
end process;
```

Since multiple sequential signal assignment statements are allowed inside a process, we can correct the problem by assigning a default value in the beginning:

```
  process(a,b)
  begin
    gt <= '0';           -- assign default value
    eq <= '0';
    if (a > b) then
        gt <= '1';
    elsif (a = b) then
        eq <= '1';
    end if;
  end process;
```

The **gt** and **eq** signals assume '0' if they are not assigned a value later. As discussed earlier, assigning a signal multiple times inside a process can be error-prone. For synthesis, this should not be used in other contexts and should be considered as shorthand to satisfy the "assigning all signals in all branches" rule.

4.5 CONSTANTS AND GENERICS

4.5.1 Constants

HDL code frequently uses constant values in expressions and array boundaries. One good design practice is to replace the "hard literals" with symbolic constants. It makes the code clear and helps future maintenance and revision. The constant declaration can be included in the architecture's declaration section, and its syntax is

```
constant const_name: data_type := value_expression;
```

For example, we can declare two constants as

```
constant DATA_BIT: integer := 8;
constant DATA_RANGE: integer := 2**DATA_BIT - 1;
```

The constant expression is evaluated during preprocessing and thus requires no physical circuit. In this book, we use capital letters for constants.

The use of a constant can best be explained by an example. Assume that we want to design an adder with the carry-out bit. One way to do it is to extend the input by 1 bit and then perform regular addition. The MSB of the summation becomes the carry-out bit. The code is shown in Listing 4.9.

Listing 4.9 Adder using a hard literal

```
library ieee;
use ieee.std_logic_1164.all;
3 use ieee.numeric_std.all;
entity add_w_carry is
    port(
        a, b: in std_logic_vector(3 downto 0);
        cout: out std_logic;
8       sum: out std_logic_vector(3 downto 0)
    );
end add_w_carry;

architecture hard_arch of add_w_carry is
13   signal a_ext, b_ext, sum_ext: unsigned(4 downto 0);
begin
```

```
      a_ext <= unsigned('0' & a);
      b_ext <= unsigned('0' & b);
      sum_ext <= a_ext + b_ext;
18    sum <= std_logic_vector(sum_ext(3 downto 0));
      cout <= sum_ext(4);
   end hard_arch;
```

The code is for a 4-bit adder. Hard literals, such as 3 and 4, are used for the ranges, as in unsigned(4 **downto** 0) and sum_ext(3 **downto** 0), and the MSB, as in sum_ext(4). If we want to revise the code for an 8-bit adder, these literals have to be modified manually. This will be a tedious and error-prone process if the code is complex and the literals are referred to in many places.

To improve the readability, we can use a symbolic constant, N, to represent the number of bits of the adder. The revised architecture body is shown in Listing 4.10.

Listing 4.10   Adder using a constant

```
architecture const_arch of add_w_carry is
   constant N: integer := 4;
   signal a_ext, b_ext, sum_ext: unsigned(N downto 0);
begin
5  a_ext <= unsigned('0' & a);
   b_ext <= unsigned('0' & b);
   sum_ext <= a_ext + b_ext;
   sum <= std_logic_vector(sum_ext(N-1 downto 0));
   cout <= sum_ext(N);
10 end const_arch;
```

The constant makes the code easier to understand and maintain.

4.5.2 Generics

VHDL provides a construct, known as a *generic*, to pass information into an entity and component. Since a generic cannot be modified inside the architecture, it functions somewhat like a constant. A generic is declared inside an entity declaration, just before the port declaration:

```
entity entity_name is
   generic(
      generic_name: data_type := default_values;
      generic_name: data_type := default_values;
          . . .
      generic_name: data_type := default_values
   );
   port(
      port_name: mode data_type;
          . . .
   );
   end entity_name;
```

For example, the previous adder code can be modified to use the adder width as a generic, as shown in Listing 4.11.

Listing 4.11   Adder using a generic

```
library ieee;
use ieee.std_logic_1164.all;
```

```
 3 use ieee.numeric_std.all;
   entity gen_add_w_carry is
      generic(N: integer:=4);
      port(
         a, b: in std_logic_vector(N-1 downto 0);
 8       cout: out std_logic;
         sum: out std_logic_vector(N-1 downto 0)
      );
   end gen_add_w_carry;

13 architecture arch of gen_add_w_carry is
      signal a_ext, b_ext, sum_ext: unsigned(N downto 0);
   begin
      a_ext <= unsigned('0' & a);
      b_ext <= unsigned('0' & b);
18    sum_ext <= a_ext + b_ext;
      sum <= std_logic_vector(sum_ext(N-1 downto 0));
      cout <= sum_ext(N);
   end arch;
```

The N generic is declared in line 5 with a default value of 4. After N is declared, it can be used in the port declaration and architecture body, just like a constant.

If the adder is later used as a component in other code, we can assign the desired value to the generic in component instantiation. This is known as *generic mapping*. The default value will be used if generic mapping is omitted. Use of the generic in component instantiation is shown below.

```
signal a4, b4, sum4: unsigned(3 downto 0);
signal a8, b8, sum8: unsigned(7 downto 0);
signal a16, b16, sum16: unsigned(15 downto 0);
signal c4, c8, c16: std_logic;
. . .
-- instantiate 8-bit adder
adder_8_unit: work.gen_add_w_carry(arch)
   generic map(N=>8)
   port map(a=>a8, b=>b8, cout=>c8, sum=>sum8));
-- instantiate 16-bit adder
adder_16_unit: work.gen_add_w_carry(arch)
   generic map(N=>16)
   port map(a=>a16, b=>b16, cout=>c16, sum=>sum16));
-- instantiate 4-bit adder
-- (generic mapping omitted, default value 4 used)
adder_4_unit: work.gen_add_w_carry(arch)
   port map(a=>a4, b=>b4, cout=>c4, sum=>sum4));
```

A generic provides a mechanism to create *scalable code*, in which the "width" of a circuit can be adjusted to meet a specific need. This makes the code more portable and encourages design reuse.

4.6 DESIGN EXAMPLES

4.6.1 Hexadecimal digit to seven-segment LED decoder

The sketch of a seven-segment LED display is shown in Figure 4.5(a). It consists of seven LED bars and a single round LED decimal point. On the DE1 prototyping

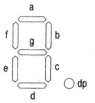

(a) Diagram of a seven-segment LED display

(b) Hexadecimal digit patterns

Figure 4.5   Seven-segment LED display and hexadecimal patterns.

board, the seven-segment LED display devices are active low, which means that an LED segment is lit if the corresponding control signal is 0. The decimal point LED is left unconnected on the board and thus cannot be used.

A hexadecimal digit to seven-segment LED decoder treats a 4-bit binary input as a hexadecimal digit and generates appropriate LED patterns, as shown in Figure 4.5(b). The LED control signals, g, f, e, d, c, b, and a, are grouped together as a single 7-bit signal, sseg. The code is shown in Listing 4.12. It uses one selected signal assignment statement to list all the desired patterns for the sseg signal.

Listing 4.12   Binary code to seven-segment LED decoder

```
library ieee;
use ieee.std_logic_1164.all;
entity bin_to_sseg is
    port(
        bin: in std_logic_vector(3 downto 0);
        sseg: out std_logic_vector(6 downto 0)
    );
end bin_to_sseg;

architecture arch of bin_to_sseg is
begin
    with bin select
        sseg <=
            "1000000" when "0000",
            "1111001" when "0001",
            "0100100" when "0010",
            "0110000" when "0011",
            "0011001" when "0100",
            "0010010" when "0101",
            "0000010" when "0110",
            "1111000" when "0111",
            "0000000" when "1000",
            "0010000" when "1001",
            "0001000" when "1010", --A
            "0000011" when "1011", --B
            "1000110" when "1100", --C
            "0100001" when "1101", --D
```

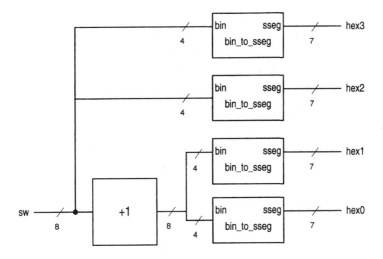

Figure 4.6 Seven-segment LED display testing circuit.

28
```
            "0000110" when "1110",  —E
            "0001110" when others;  —F
    end arch
```

Testing circuit There are four seven-segment LED displays on the DE1 board. We use a simple 8-bit increment circuit to verify operation of the decoder. The sketch is shown in Figure 4.6. The **sw** input is connected to the eight slide switches of the prototyping board. It is fed to an incrementor to obtain **sw+1**. The original and incremented **sw** signals are then passed to four decoders to display the four hexadecimal digits on seven-segment LED displays. The code is shown in Listing 4.13. The four seven-segment LED display devices are named **hex3**, **hex2**, **hex1**, and **hex0**, as labeled on the DE1 board.

Listing 4.13 Seven-segment LED decoder testing circuit

```
library ieee;
use ieee.std_logic_1164.all;
use ieee.numeric_std.all;
entity led_test is
    port(
        sw: in std_logic_vector(7 downto 0);
        hex3, hex2, hex1, hex0: out std_logic_vector(6 downto 0)
    );
end led_test;

architecture arch of led_test is
    signal inc: std_logic_vector(7 downto 0);
begin
    -- increment input
    inc <= std_logic_vector(unsigned(sw) + 1);

    -- instantiate four instances of 7-seg LED decoders
    -- instance for 4 LSBs of input
    sseg_unit_0: entity work.bin_to_sseg
        port map(bin=>sw(3 downto 0), sseg=>hex0);
    -- instance for 4 MSBs of input
```

```
     sseg_unit_1: entity work.bin_to_sseg
        port map(bin=>sw(7 downto 4), sseg=>hex1);
     -- instance for 4 LSBs of incremented value
25   sseg_unit_2: entity work.bin_to_sseg
        port map(bin=>inc(3 downto 0), sseg=>hex2);
     -- instance for 4 MSBs of incremented value
     sseg_unit_3: entity work.bin_to_sseg
        port map(bin=>inc(7 downto 4), sseg=>hex3);
30 end arch;
```

We can follow the procedure in Chapter 3 to synthesize and implement the circuit on the prototyping board and verify its operation.

4.6.2 Sign-magnitude adder

An integer can be represented in *sign-magnitude* format, in which the MSB is the sign and the remaining bits form the magnitude. For example, 3 and -3 become "0011" and "1011" in 4-bit sign-magnitude format.

A sign-magnitude adder performs an addition operation in this format. The operation can be summarized as follows:

- If the two operands have the same sign, add the magnitudes and keep the sign.
- If the two operands have different signs, subtract the smaller magnitude from the larger one and keep the sign of the number that has the larger magnitude.

One possible implementation is to divide the circuit into two stages. The first stage sorts the two input numbers according to their magnitudes and routes them to the `max` and `min` signals. The second stage examines the signs and performs addition or subtraction on the magnitude accordingly. Note that since the two numbers have been sorted, the magnitude of `max` is always larger than that of `min` and the final sign is the sign of `max`.

The code is shown in Listing 4.14, which realizes the two-stage implementation scheme. For clarity, we split the input number internally and use separate sign and magnitude signals. A generic, `N`, is used to represent the width of the adder. Note that the relevant magnitude signals are declared as `unsigned` to facilitate the arithmetic operation, and type conversions are performed at the beginning and end of the code.

Listing 4.14 Sign-magnitude adder

```
   library ieee;
   use ieee.std_logic_1164.all;
   use ieee.numeric_std.all;
   entity sign_mag_add is
5     generic(N: integer:=4);  -- default 4 bits
      port(
         a, b: in std_logic_vector(N-1 downto 0);
         sum: out std_logic_vector(N-1 downto 0)
      );
10 end sign_mag_add;

   architecture arch of sign_mag_add is
      signal mag_a, mag_b: unsigned(N-2 downto 0);
      signal mag_sum, max, min: unsigned(N-2 downto 0);
15    signal sign_a, sign_b, sign_sum: std_logic;
   begin
```

```
        mag_a <= unsigned(a(N-2 downto 0));
        mag_b <= unsigned(b(N-2 downto 0));
        sign_a <= a(N-1);
20      sign_b <= b(N-1);
        -- sort according to magnitude
        process(mag_a,mag_b,sign_a,sign_b)
        begin
            if mag_a > mag_b then
25              max <= mag_a;
                min <= mag_b;
                sign_sum <= sign_a;
            else
                max <= mag_b;
30              min <= mag_a;
                sign_sum <= sign_b;
            end if;
        end process;
        -- add/sub magnitude
35      mag_sum <= max + min when sign_a=sign_b else
                   max - min;
        --form output
        sum <= std_logic_vector(sign_sum & mag_sum);
    end arch;
```

Testing circuit   We use a 4-bit sign-magnitude adder to verify the circuit operation. The testing circuit uses eight slide switches as the two 4-bit inputs and shows the result in two seven-segment LED displays. The rightmost seven-segment LED display shows the 3-bit magnitude and the next LED display shows the sign bit, which is blank for the plus sign and is lit with a middle LED bar for the minus sign. The code is shown in Listing 4.15.

Listing 4.15   Sign-magnitude adder testing circuit

```
1 library ieee;
  use ieee.std_logic_1164.all;
  use ieee.numeric_std.all;
  entity sm_add_test is
      port(
6         sw: in std_logic_vector(7 downto 0);
          hex3, hex2, hex1, hex0: out std_logic_vector(6 downto 0)
      );
  end sm_add_test;

11 architecture arch of sm_add_test is
      signal sum, oct: std_logic_vector(3 downto 0);
  begin
      -- instantiate adder
      sm_adder_unit: entity work.sign_mag_add
16        generic map(N=>4)
          port map(a=>sw(3 downto 0), b=>sw(7 downto 4),
                   sum=>sum);
      -- 3-bit magnitude displayed on rightmost 7-seg LED
      oct <= '0' & sum(2 downto 0);
21    sseg_unit: entity work.bin_to_sseg
          port map(bin=>oct, sseg=>hex0);
      -- sign displayed on 2nd 7-seg LED
      hex1 <= "0111111" when sum(3)='1' else  -- middle bar
              "1111111";                       -- blank
26    -- other two 7-seg LEDs blank
      hex2 <= "1111111";
```

```
    hex3 <= "1111111";
end arch;
```

4.6.3 Barrel shifter

Although VHDL has built-in shift functions, they sometimes cannot be synthesized automatically. In this subsection, we examine an 8-bit barrel shifter that rotates an arbitrary number of bits to the right. The circuit has an 8-bit data input, a, and a 3-bit control signal, amt, which specifies the amount to be rotated. The first design uses a selected signal assignment statement to exhaustively list all combinations of the amt signal and the corresponding rotated results. The code is shown in Listing 4.16.

Listing 4.16   Barrel shifter using a selected signal assignment statement

```
1 library ieee;
  use ieee.std_logic_1164.all;
  entity barrel_shifter is
     port(
        a: in std_logic_vector(7 downto 0);
6       amt: in std_logic_vector(2 downto 0);
        y: out std_logic_vector(7 downto 0)
     );
  end barrel_shifter ;

11 architecture sel_arch of barrel_shifter is
  begin
     with amt select
        y<= a                            when "000",
            a(0) & a(7 downto 1)         when "001",
16          a(1 downto 0) & a(7 downto 2) when "010",
            a(2 downto 0) & a(7 downto 3) when "011",
            a(3 downto 0) & a(7 downto 4) when "100",
            a(4 downto 0) & a(7 downto 5) when "101",
            a(5 downto 0) & a(7 downto 6) when "110",
21          a(6 downto 0) & a(7) when others; -- 111
  end sel_arch;
```

While the code is straightforward, it will become cumbersome when the number of input bits increases. Furthermore, a large number of choices implies a wide multiplexer, which makes synthesis difficult and leads to a large propagation delay. Alternatively, we can construct the circuit by stages. In the nth stage, the input signal is either passed directly to output or rotated right by 2^n positions. The nth stage is controlled by the nth bit of the amt signal. Assume that the 3 bits of amt are $m_2 m_1 m_0$. The total rotated amount after three stages is $m_2 2^2 + m_1 2^1 + m_0 2^0$, which is the desired rotating amount. The code for this scheme is shown in Listing 4.17.

Listing 4.17   Barrel shifter using multi-stage shifts

```
  architecture multi_stage_arch of barrel_shifter is
     signal s0, s1: std_logic_vector(7 downto 0);
3 begin
     -- stage 0, shift 0 or 1 bit
     s0 <= a(0) & a(7 downto 1) when amt(0)='1' else
           a;
     -- stage 1, shift 0 or 2 bits
8    s1 <= s0(1 downto 0) & s0(7 downto 2) when amt(1)='1' else
```

```
              s0;
      -- stage 2, shift 0 or 4 bits
      y <= s1(3 downto 0) & s1(7 downto 4) when amt(2)='1' else
           s1;
13 end multi_stage_arch ;
```

Testing circuit   To test the circuit, we can use eight slide switches for the a signal, three pushbutton switches for the amt signal, and eight discrete LEDs for the
output. The pushbutton switches are labeled key on the DE1 board and a switch
outputs 0 when it is pressed. A top-level wrapping HDL code is create to map the
circuit's I/O port to the prototyping board's signals, as shown in Listing 4.18. Note
that key is inverted for the amt in the code (i.e., a 1 is generated for amt when a
pushbutton switch is pressed).

Listing 4.18   Barrel shifter testing circuit

```
  library ieee;
2 use ieee.std_logic_1164.all;
  use ieee.numeric_std.all;
  entity shifter_test is
     port(
        sw: in std_logic_vector(7 downto 0);
7       key: in std_logic_vector(2 downto 0);
        ledr: out std_logic_vector(7 downto 0)
     );
  end shifter_test;

12 architecture arch of shifter_test is
     signal amt: std_logic_vector(2 downto 0);
  begin
     amt <= not key;
     shift_unit: entity work.barrel_shifter(multi_stage_arch)
17      port map(a=>sw, amt=>amt, y=>ledr);
  end arch;
```

4.6.4 Simplified floating-point adder

Floating point is another format to represent a number. With the same number
of bits, the range in floating-point format is much larger than that in the signed
integer format. Although VHDL has a built-in floating-point data type, it is too
complex to be synthesized automatically.

A detailed discussion of the floating-point representation is beyond the scope of
this book. We use a simplified 13-bit format in this example and ignore the round-
off error. The representation consists of a sign bit, s, which indicates the sign of the
number (1 for negative); a 4-bit exponent field, e, which represents the exponent;
and an 8-bit significand field, f, which represents the significand or the fraction. In
this format, the value of a floating-point number is $(-1)^s * .f * 2^e$. The $.f * 2^e$ is
the magnitude of the number and $(-1)^s$ is just a formal way to state that "s equal
to 1 implies a negative number." Since the sign bit is separated from the rest of
the number, the floating-point representation can be considered as a variation of
the sign-magnitude format.

We also make the following assumptions:

- Both exponent and significand fields are in unsigned format.

		sort	align	add/sub	normalize
eg. 1	+0.54E3	−0.87E4	−0.87E4	−0.87E4	−0.87E4
	−0.87E4	+0.54E3	+0.05E4	+0.05E4	+0.05E4
				−0.82E4	−0.82E4
eg. 2	+0.54E3	−0.55E3	−0.55E3	−0.55E3	−0.55E3
	−0.55E3	+0.54E3	+0.54E3	+0.54E3	+0.54E3
				−0.01E3	−0.10E2
eg. 3	+0.54E0	−0.55E0	−0.55E0	−0.55E0	−0.55E0
	−0.55E0	+0.54E0	+0.54E0	+0.54E0	+0.54E0
				−0.01E0	−0.00E0
eg. 4	+0.56E3	+0.56E3	+0.56E3	+0.56E3	+0.56E3
	+0.52E3	+0.52E3	+0.52E3	+0.52E3	+0.52E3
				+1.08E3	+0.10E4

Figure 4.7   Floating-point addition examples.

- The representation has to be either normalized or zero. *Normalized representation* means that the MSB of the significand field must be '1'. If the magnitude of the computation result is smaller than the smallest normalized nonzero magnitude, $0.10000000 * 2^{0000}$, it must be converted to zero.

Under these assumptions, the largest and smallest nonzero magnitudes are $0.11111111 * 2^{1111}$ and $0.10000000 * 2^{0000}$, and the range is about 2^{16} (i.e., $\frac{0.11111111*2^{1111}}{0.10000000*2^{0000}}$).

Our floating-point adder design follows the process of adding numbers manually in scientific notation. This process can best be explained by examples. We assume that the widths of the exponent and significand are 2 and 1 digits, respectively. The decimal format is used for clarity. The computations of several representative examples are shown in Figure 4.7. The computation is done in four major steps:

1. *Sorting*: puts the number with the larger magnitude on the top and the number with the smaller magnitude on the bottom (we call the sorted numbers "big number" and "small number").
2. *Alignment*: aligns the two numbers so they have the same exponent. This can be done by adjusting the exponent of the small number to match the exponent of the big number. The significand of the small number has to shift to the right according to the difference in exponents.
3. *Addition/subtraction*: adds or subtracts the significands of two aligned numbers.
4. *Normalization*: adjusts the result to normalized format. Three types of normalization procedures may be needed:
   - After a subtraction, the result may contain leading zeros in front, as in example 2.
   - After a subtraction, the result may be too small to be normalized and thus needs to be converted to zero, as in example 3.

- After an addition, the result may generate a carry-out bit, as in example 4.

Our binary floating-point adder design uses a similar algorithm. To simplify the implementation, we ignore the rounding. During alignment and normalization, the lower bits of the significand will be discarded when shifted out. The design is divided into four stages, each corresponding to a step in the foregoing algorithm. The suffixes, b, s, a, r, and n, used in signal names are for "big number," "small number," "aligned number," "result of addition/subtraction," and "normalized number," respectively. The code is developed according to these stages, as shown in Listing 4.19.

Listing 4.19   Simplified floating-point adder

```vhdl
library ieee;
use ieee.std_logic_1164.all;
use ieee.numeric_std.all;
entity fp_adder is
   port (
      sign1, sign2: in  std_logic;
      exp1, exp2: in  std_logic_vector(3 downto 0);
      frac1, frac2: in  std_logic_vector(7 downto 0);
      sign_out: out std_logic;
      exp_out: out std_logic_vector(3 downto 0);
      frac_out: out std_logic_vector(7 downto 0)
   );
end fp_adder ;

architecture arch of fp_adder is
   -- suffix b, s, a, n for
   --         big, small, aligned, normalized number
   signal signb, signs: std_logic;
   signal expb, exps, expn: unsigned(3 downto 0);
   signal fracb, fracs, fraca, fracn: unsigned(7 downto 0);
   signal sum_norm: unsigned(7 downto 0);
   signal exp_diff: unsigned(3 downto 0);
   signal sum: unsigned(8 downto 0); --one extra for carry
   signal lead0: unsigned(2 downto 0);
begin
   -- 1st stage: sort to find the larger number
   process (sign1, sign2, exp1, exp2, frac1, frac2)
   begin
      if (exp1 & frac1) > (exp2 & frac2) then
         signb <= sign1;
         signs <= sign2;
         expb <= unsigned(exp1);
         exps <= unsigned(exp2);
         fracb <= unsigned(frac1);
         fracs <= unsigned(frac2);
      else
         signb <= sign2;
         signs <= sign1;
         expb <= unsigned(exp2);
         exps <= unsigned(exp1);
         fracb <= unsigned(frac2);
         fracs <= unsigned(frac1);
      end if;
   end process;

   -- 2nd stage: align smaller number
```

```
47      exp_diff <= expb - exps;
        with exp_diff select
            fraca <=
                fracs                              when "0000",
                "0"        & fracs(7 downto 1) when "0001",
52              "00"       & fracs(7 downto 2) when "0010",
                "000"      & fracs(7 downto 3) when "0011",
                "0000"     & fracs(7 downto 4) when "0100",
                "00000"    & fracs(7 downto 5) when "0101",
                "000000"   & fracs(7 downto 6) when "0110",
57              "0000000"  & fracs(7)          when "0111",
                "00000000"                     when others;

        -- 3rd stage: add/subtract
        sum <= ('0' & fracb) + ('0' & fraca) when signb=signs else
62             ('0' & fracb) - ('0' & fraca);

        -- 4th stage: normalize
        -- count leading 0s
        lead0 <= "000" when (sum(7)='1') else
67               "001" when (sum(6)='1') else
                 "010" when (sum(5)='1') else
                 "011" when (sum(4)='1') else
                 "100" when (sum(3)='1') else
                 "101" when (sum(2)='1') else
72               "110" when (sum(1)='1') else
                 "111";
        -- shift significand according to leading 0
        with lead0 select
            sum_norm <=
77              sum(7 downto 0)                when "000",
                sum(6 downto 0) & '0'          when "001",
                sum(5 downto 0) & "00"         when "010",
                sum(4 downto 0) & "000"        when "011",
                sum(3 downto 0) & "0000"       when "100",
82              sum(2 downto 0) & "00000"      when "101",
                sum(1 downto 0) & "000000"     when "110",
                sum(0) &          "0000000" when others;

        -- normalize with special conditions
87      process(sum,sum_norm,expb,lead0)
        begin
            if sum(8)='1' then -- w/ carry out; shift frac to right
                expn <= expb + 1;
                fracn <= sum(8 downto 1);
92          elsif (lead0 > expb) then  -- too small to normalize;
                expn <= (others=>'0'); -- set to 0
                fracn <= (others=>'0');
            else
                expn <= expb - lead0;
97              fracn <= sum_norm;
            end if;
        end process;

        -- form output
102     sign_out <= signb;
        exp_out <= std_logic_vector(expn);
        frac_out <= std_logic_vector(fracn);
    end arch;
```

The circuit in the first stage compares the magnitudes and routes the big number to the `signb`, `expb`, and `fracb` signals and the smaller number to the `signs`, `exps`, and `fracs` signals. The comparison is done between `exp1&frac1` and `exp2&frac2`. It implies that the exponents are compared first, and if they are the same, the significands are compared.

The circuit in the second stage performs alignment. It first calculates the difference between the two exponents, which is `expb-exps`, and then shifts the significand, `fracs`, to the right by this amount. The aligned significand is labeled `fraca`. The circuit in the third stage performs a sign-magnitude addition, similar to that in Section 4.6.2. Note that the operands are extended by 1 bit to accommodate the carry-out bit.

The circuit in the fourth stage performs normalization, which adjusts the result to make the final output conform to the normalized format. The normalization circuit is constructed in three segments. The first segment counts the number of leading zeros. It is somewhat like a priority encoder. The second segment shifts the significands to the left by the amount specified by the leading-zero counting circuit. The last segment checks the carry-out and zero conditions and generates the final normalized number.

Testing circuit   The floating-point adder has two 13-bit input operands. Since the prototyping board has only 10 slide switches and four pushbutton switches, it cannot provide enough numbers of physical inputs to test the circuit. To accommodate the 26 bits of the floating-point adder, we must create a testing circuit and assign constants or duplicated switch signals to the adder's input operands. An example is shown in Listing 4.20. It assigns one operand as a constant and uses duplicated switch signals for the other operand. The addition result is passed to the hexadecimal decoders and the sign circuit and is shown on the seven-segment LED displays.

Listing 4.20   Floating-point adder testing circuit

```
library ieee;
use ieee.std_logic_1164.all;
use ieee.numeric_std.all;
entity fp_adder_test is
5    port(
        sw: in std_logic_vector(9 downto 0);
        key: in std_logic_vector(3 downto 0);
        hex3, hex2, hex1, hex0: out std_logic_vector(6 downto 0)
    );
10 end fp_adder_test;

architecture arch of fp_adder_test is
    signal sign1, sign2: std_logic;
    signal exp1, exp2: std_logic_vector(3 downto 0);
15    signal frac1, frac2: std_logic_vector(7 downto 0);
    signal sign_out: std_logic;
    signal exp_out: std_logic_vector(3 downto 0);
    signal frac_out: std_logic_vector(7 downto 0);
begin
20    -- set up the fp adder input signals
    sign1 <= sw(9);
    exp1 <= "1000";
    frac1<= '1' & sw(1) & sw(0) & "10101";
    sign2 <= sw(8);
```

```
25    exp2 <= not key;
      frac2 <= sw(7 downto 0);

      -- instantiate fp adder
      fp_add_unit: entity work.fp_adder
30       port map(
            sign1=>sign1, sign2=>sign2, exp1=>exp1, exp2=>exp2,
            frac1=>frac1, frac2=>frac2,
            sign_out=>sign_out, exp_out=>exp_out,
            frac_out=>frac_out
35       );

      -- instantiate three instances of 7-seg decoders
      -- exponent
      sseg_unit_exp: entity work.bin_to_sseg
40       port map(bin=>exp_out, sseg=>hex0);
      -- 4 LSBs of fraction
      sseg_unit_frac0: entity work.bin_to_sseg
         port map(bin=>frac_out(3 downto 0), sseg=>hex1);
      -- 4 MSBs of fraction
45    sseg_unit_frac1: entity work.bin_to_sseg
         port map(bin=>frac_out(7 downto 4), sseg=>hex2);
      -- sign
      hex3 <= "0111111" when sign_out='1' else  -- middle bar
              "1111111";                          -- blank
50 end arch;
```

4.7 BIBLIOGRAPHIC NOTES

The Designer's Guide to VHDL by P. J. Ashenden provides detailed coverage on the VHDL constructs discussed in this chapter, and the author's *RTL Hardware Design Using VHDL: Coding for Efficiency, Portability, and Scalability* discusses the coding and optimization schemes and gives additional design examples.

4.8 SUGGESTED EXPERIMENTS

4.8.1 Multi-function barrel shifter

Consider an 8-bit shifting circuit that can perform rotating right or rotating left. An additional 1-bit control signal, lr, specifies the desired direction.

1. Design the circuit using one rotate-right circuit, one rotate-left circuit, and one 2-to-1 multiplexer to select the desired result. Derive the code.
2. Derive a testbench and use simulation to verify operation of the code.
3. Synthesize the circuit, program the FPGA, and verify its operation.
4. This circuit can also be implemented by one rotate-right shifter with pre- and post-reversing circuits. The reversing circuit either passes the original input or reverses the input bitwise (for example, if an 8-bit input is $a_7a_6a_5a_4a_3a_2a_1a_0$, the reversed result becomes $a_0a_1a_2a_3a_5a_5a_6a_7$). Repeat steps 2 and 3.
5. Check the report files and compare the number of logic cells and propagation delays of the two designs.
6. Expand the code for a 16-bit circuit and synthesize the code. Repeat steps 1 to 5.

7. Expand the code for a 32-bit circuit and synthesize the code. Repeat steps 1 to 5.

4.8.2 Dual-priority encoder

A dual-priority encoder returns the codes of the highest or second-highest priority requests. The input is a 12-bit `req` signal and the outputs are `first` and `second`, which are the 4-bit binary codes of the highest and second-highest priority requests, respectively.

1. Design the circuit and derive the code.
2. Derive a testbench and use simulation to verify operation of the code.
3. Design a testing circuit that displays the two output codes on the seven-segment LED display of the prototyping board, and derive the code.
4. Synthesize the circuit, program the FPGA, and verify its operation.

4.8.3 BCD incrementor

The binary-coded-decimal (BCD) format uses 4 bits to represent 10 decimal digits. For example, 259_{10} is represented as "0010 0101 1001" in BCD format. A BCD incrementor adds 1 to a number in BCD format. For example, after incrementing, "0010 0101 1001" (i.e., 259_{10}) becomes "0010 0110 0000" (i.e., 260_{10}).

1. Design a three-digit 12-bit incrementor and derive the code.
2. Derive a testbench and use simulation to verify operation of the code.
3. Design a testing circuit that displays three digits on the seven-segment LED display and derive the code.
4. Synthesize the circuit, program the FPGA, and verify its operation.

4.8.4 Floating-point greater-than circuit

A floating-point greater-than circuit compares two floating-point numbers and asserts output, `gt`, when the first number is larger than the second number. Assume that the two numbers are represented in the format discussed in Section 4.6.4.

1. Design the circuit and derive the code.
2. Derive a testbench and use simulation to verify operation of the code.
3. Design a testing circuit and derive the code.
4. Synthesize the circuit, program the FPGA, and verify its operation.

4.8.5 Floating-point and signed integer conversion circuit

A number may need to be converted to different formats in a large system. Assume that we use the 13-bit format in Section 4.6.4 for the floating-point representation and the 8-bit `signed` data type for the integer representation. An integer-to-floating-point conversion circuit converts an 8-bit integer input to a normalized, 13-bit floating-point output. A floating-point-to-integer conversion circuit reverses the operation. Since the range of a floating-point number is much larger, conversion may lead to the underflow condition (i.e., the magnitude of the converted number is smaller than "00000001") or the overflow condition (i.e., the magnitude of the converted number is larger than "01111111").

1. Design an integer-to-floating-point conversion circuit and derive the code.
2. Derive a testbench and use simulation to verify operation of the code.
3. Design a testing circuit and derive the code.
4. Synthesize the circuit, program the FPGA, and verify its operation.
5. Design a floating-point-to-integer conversion circuit. In addition to the 8-bit integer output, the design should include two status signals, uf and of, for the underflow and overflow conditions. Derive the code and repeat steps 2 to 4.

4.8.6 Enhanced floating-point adder

The floating-point adder in Section 4.6.4 discards the lower bits when they are shifted out (it is known as *round to zero*). A more accurate method is to *round to the nearest even*, as defined in the *IEEE Standard for Binary Floating-Point Arithmetic* (IEEE Std 754). Three extra bits, known as the *guard*, *round*, and *sticky bits*, are required to implement this method. If you learned floating-point arithmetic before, modify the floating-point adder in Section 4.6.4 to accommodate the round-to-the-nearest-even method.

CHAPTER 5

REGULAR SEQUENTIAL CIRCUIT

A sequential circuit is a circuit with memory. Modern development follows synchronous design methodology and uses a common clock signal to control storage elements. In this chapter, we describe the HDL codes for basic storage elements, introduce the design and coding of "regular sequential circuits," in which the state transitions in the circuit exhibit a "regular" pattern, as in a counter or shift register, and discuss the use and inference of FPGA's internal memory module.

5.1 INTRODUCTION

A sequential circuit is a circuit with *memory*, which forms the *internal state* of the circuit. Unlike a combinational circuit, in which the output is a function of input only, the output of a sequential circuit is a function of the input and the internal state. The *synchronous design methodology* is the most commonly used practice in designing a sequential circuit. In this methodology, all storage elements are controlled (i.e., synchronized) by a global clock signal and the data is sampled and stored at the rising or falling edge of the clock signal. It allows designers to separate the storage components from the circuit and greatly simplifies the development process. This methodology is the most important principle in developing a large, complex digital system and is the foundation of most synthesis, verification, and testing algorithms. All of the designs in the book follow this methodology.

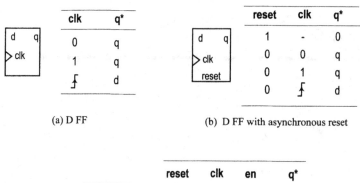

(a) D FF   (b) D FF with asynchronous reset

(c) D FF with synchronous enable

Figure 5.1   Block diagram and functional table of a D FF.

5.1.1   D FF and register

The most basic storage component in a sequential circuit is a D-type flip-flop (D FF). The symbol and function table of a positive edge-triggered D FF are shown in Figure 5.1(a). The value of the d signal is sampled at the rising edge of the clk signal and stored to FF. A D FF may contain an asynchronous reset signal to clear the FF to '0'. Its symbol and function table are shown in Figure 5.1(b). Note that the reset operation is independent of the clock signal.

The three main timing parameters of a D FF are t_{CQ} (clock-to-q delay), t_{SETUP} (setup time), and t_{HOLD} (hold time). t_{CQ} is the time required to propagate the value of d to q at the rising edge of the clock signal. The d signal must be stable around the sampling edge to prevent the FF from entering the metastable state. t_{SETUP} and t_{HOLD} specify the time intervals before or after the sampling edge.

A D FF provides 1-bit storage. A collection of D FFs can be grouped together to store multiple bits and is known as a *register*.

5.1.2   Synchronous system

Block diagram   The block diagram of a synchronous system is shown in Figure 5.2. It consists of the following parts:

- *State register*: a collection of D FFs controlled by the same clock signal
- *Next-state logic*: combinational logic that uses the external input and internal state (i.e., the output of register) to determine the new value of the register
- *Output logic*: combinational logic that generates the output signal

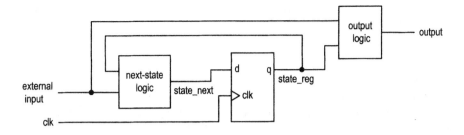

Figure 5.2 Block diagram of a synchronous system.

5.1.3 Code development

Our code development follows the basic block diagram in Figure 5.2. The key is to separate the memory component (i.e., the register) from the system. Once the register is isolated, the remaining portion is a pure combinational circuit, and the coding and analysis schemes discussed in previous chapters can be applied accordingly. While this approach may make the code a little bit more cumbersome at times, it helps us to better visualize the circuit architecture and avoid unintended memory and subtle mistakes.

Based on the characteristics of the next-state logic, we divide sequential circuits into three categories:

- *Regular sequential circuit.* The state transitions in the circuit exhibit a "regular" pattern, as in a counter or shift register. The next-state logic is constructed primarily by a predesigned, "regular" component, such as an incrementor or shifter.
- *FSM.* The state transitions in the circuit do not exhibit a simple, repetitive pattern. The next-state logic is constructed by "random logic" and synthesized from scratch. It should be called a random sequential circuit, but is commonly known as an FSM (*finite state machine*).
- *FSMD.* The circuit consists of a regular sequential circuit and an FSM. The two parts are known as a *data path* and a *control path*, and the complete circuit is known as an FSMD (*FSM with data path*). This type of circuit is used to implement an algorithm represented by *register-transfer* (RT) methodology, which describes system operation by a sequence of data transfers and manipulations among registers.

The three types of circuits are discussed in this and the next two subsequent chapters.

5.2 HDL CODE OF THE BASIC STORAGE ELEMENTS

Describing storage components in HDL is a subtle procedure, and there are many ways to do it. In fact, one common problem encountered by a new HDL user is the inference of unintended latches and buffers. Instead of covering all possible forms of syntactic descriptions, we introduce the code segments for several commonly used memory components. Since our development process separates the register and the combinational circuit, these components are sufficient for all designs in this book. The components are:

- D FF
- Register
- Register file

For demonstration purposes, we also include code for a generic SRAM (static random access memory). Its operation is not controlled by a clock signal and thus it is an *asynchronous* device.

Cyclone II devices contain internal memory modules. These modules are device specific and their use and inferences are discussed in Section 5.7.

5.2.1 D FF

We consider three types of D FFs:

- D FF without asynchronous reset
- D FF with asynchronous reset
- D FF with synchronous enable

The first two are the most basic memory components and can be found in the library of any device technology. The third can be constructed from a simple D FF. We include the code since it is a frequently used memory component and can be mapped to the FF of the Cyclone II device's logic cell.

D FF without asynchronous reset The function table of a D FF is shown in Figure 5.1(a) and the code is shown in Listing 5.1.

Listing 5.1 D FF without asynchronous reset

```
library ieee;
use ieee.std_logic_1164.all;
entity d_ff is
    port(
5       clk: in std_logic;
        d: in std_logic;
        q: out std_logic
    );
end d_ff;
10
architecture arch of d_ff is
begin
    process(clk)
    begin
15      if (clk'event and clk='1') then
            q <= d;
        end if;
    end process;
end arch;
```

The rising edge is checked by the `clk'event and clk='1'` expression, which represents that there is a change in the `clk` signal (i.e., an "event") and the new value is '1'. If this condition is `true`, the value of d is stored to q, and if this condition is `false`, q keeps its previous value (i.e., memorizes the value sampled earlier). Note that only the `clk` signal is included in the sensitive list. This is consistent with the fact that the d signal is sampled only at the rising edge of the `clk` signal, and a change in its value does not trigger any immediate response.

D FF with asynchronous reset   A D FF may contain an asynchronous reset signal, as shown in the function table of Figure 5.1(b). The signal clears the D FF to '0' any time and is not controlled by the clock signal. It actually has a higher priority than the regularly sampled input. Using an asynchronous reset signal violates the synchronous design methodology and thus should be avoided in normal operation. Its major application is to perform system initialization. For example, we can generate a short reset pulse to force a system to an initial state after turning on the power. The code for a D FF with asynchronous reset is shown in Listing 5.2.

Listing 5.2   D FF with asynchronous reset

```
library ieee;
use ieee.std_logic_1164.all;
entity d_ff_reset is
    port(
        clk, reset: in std_logic;
        d: in std_logic;
        q: out std_logic
    );
end d_ff_reset;

architecture arch of d_ff_reset is
begin
    process(clk,reset)
    begin
        if (reset='1') then
            q <='0';
        elsif (clk'event and clk='1') then
            q <= d;
        end if;
    end process;
end arch;
```

Note that the `reset` signal is included in the sensitivity list, and its condition is checked before the rising-edge condition.

D FF with synchronous enable   A D FF may include an additional control signal, `en`, to enable the FF to sample the input value. Its symbol and functional table are shown in Figure 5.1(c). Note that the `en` signal is examined only at the rising edge of the clock and thus is synchronous. If it is not asserted, the FF keeps its previous value. The code is shown in Listing 5.3.

Listing 5.3   One-process coding style for a D FF with synchronous enable

```
library ieee;
use ieee.std_logic_1164.all;
entity d_ff_en is
    port(
        clk, reset: in std_logic;
        en: in std_logic;
        d: in std_logic;
        q: out std_logic
    );
end d_ff_en;

architecture arch of d_ff_en is
begin
    process(clk,reset)
    begin
```

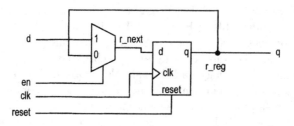

Figure 5.3   D FF with synchronous enable.

```
        if (reset='1') then
           q <='0';
        elsif (clk'event and clk='1') then
19         if (en='1') then
              q <= d;
           end if;
        end if;
     end process;
24 end arch;
```

The enabling feature of this D FF is useful in maintaining synchronism between a fast subsystem and a slow subsystem. For example, assume that the operation rates of a fast and a slow subsystem are 50 MHz and 1 MHz. Instead of using a derived 1-MHz clock to drive the slow subsystem, we can generate a periodic enable tick that is asserted one clock cycle every 50 clock cycles. The slow subsystem is disabled (i.e., keeps the previous state) for the remaining 49 clock cycles. The same scheme can also be applied to eliminate a gated clock signal.

Since the enable signal is synchronous, this circuit can be constructed by a regular D FF and simple next-state logic. The code is shown in Listing 5.4, and its block diagram is shown in Figure 5.3.

Listing 5.4   Two-segment coding style for a D FF with synchronous enable

```
1 architecture two_seg_arch of d_ff_en is
     signal r_reg, r_next: std_logic;
  begin
     -- D FF
     process(clk,reset)
6    begin
        if (reset='1') then
           r_reg <='0';
        elsif (clk'event and clk='1') then
           r_reg <= r_next;
11      end if;
     end process;
     -- next-state logic
     r_next <= d when en ='1' else
               r_reg;
16   -- output logic
     q <= r_reg;
  end two_seg_arch;
```

For clarity, we use suffixes _next and _reg to emphasize the next input value and the registered output of an FF. They are connected to the d and q signals of a

D FF. The earlier one-process code can be considered as shorthand for this more explicit description.

5.2.2 Register

A register is a collection of D FFs that are controlled by the same clock and reset signals. Like a D FF, a register can have an optional asynchronous reset signal and a synchronous enable signal. The code is identical to that of a D FF except that the array data type, std_logic_vector, is needed for the relevant input and output signals. For example, an 8-bit register with an asynchronous reset is shown in Listing 5.5.

Listing 5.5   Register

```
library ieee;
2 use ieee.std_logic_1164.all;
  entity reg_reset is
     port(
        clk, reset: in std_logic;
        d: in std_logic_vector(7 downto 0);
7       q: out std_logic_vector(7 downto 0)
     );
  end reg_reset;

  architecture arch of reg_reset is
12 begin
     process(clk,reset)
     begin
        if (reset='1') then
           q <=(others=>'0');
17      elsif (clk'event and clk='1') then
           q <= d;
        end if;
     end process;
  end arch;
```

Note that the expression (others=>'0') means that all elements are assigned to '0' and it is equivalent to "00000000" in this case.

5.2.3 Register file

A register file is a collection of registers with one input port and one or more output ports. The write address signal, w_addr, specifies where to store data, and the read address signal, r_addr, specifies where to retrieve data. The register file is generally used as fast, temporary storage. The conceptual diagram of a 4-by-8 (i.e., four words and 8 bits per word) register file is shown Figure 5.4. The design consists of four registers with enable signals, a write decoding circuit, and read multiplexing circuits.

The write decoding circuit examines the wr_en signal and decodes the write port address. If the wr_en signal is asserted, the decoding circuit functions as a regular 2-to-2^2 binary decoder that asserts one of the four en signals of the corresponding register. The w_data signal will be sampled and stored into the corresponding register at the rising edge of the clock. The read multiplexing circuit consists of a 4-to-1 multiplexer. It utilizes r_addr as the selection signal to route the desired register output to the read port.

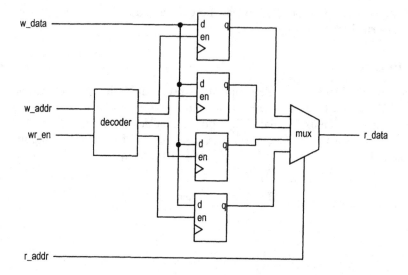

Figure 5.4   Block diagram of a four-word register file.

First, since no built-in two-dimensional array is defined in the **std_logic_1164** package, a user-defined array-of-array data type, **reg_file_type**, is introduced. It is first defined by a type statement and is then used by the **array_reg** signal.

Note that the registers are structured as a two-dimensional 4-by-8 array of D FFs and would best be represented by a two-dimensional data type. There is no pre-defined two-dimensional data type in the IEEE **std_logic_1164** package, and thus we must create a user-defined data type. Assume that there are **ADDR_WIDTH** bits in the address (i.e., 2^{ADDR_WIDTH} words) and there are **DATA_WIDTH** bits per word. The new data type can be defined in a type statement

```
type mem_2d_type is array (0 to 2**ADDR_WIDTH-1) of
    std_logic_vector(DATA_WIDTH-1 downto 0);
```

and then used, as in

```
signal array_reg: mem_2d_type;
```

We can derive the code following the conceptual diagram, as shown in Listing 5.6.

Listing 5.6   Register file with explicit decoding and multiplexing logic

```
library ieee;
use ieee.std_logic_1164.all;
entity reg_file_4x8 is
4    port(
        clk: in std_logic;
        wr_en: in std_logic;
        w_addr, r_addr: in std_logic_vector(1 downto 0);
        w_data: in std_logic_vector(7 downto 0);
9        r_data: out std_logic_vector(7 downto 0)
    );
end reg_file_4x8;

architecture explicit_arch of reg_file_4x8 is
14    constant ADDR_WIDTH: natural:=2; -- bits in address
```

```
        constant DATA_WIDTH: natural:=8;  — bits in data
        type mem_2d_type is array (0 to 2**ADDR_WIDTH-1) of
            std_logic_vector(DATA_WIDTH-1 downto 0);
        signal array_reg: mem_2d_type;
19      signal en: std_logic_vector(2**ADDR_WIDTH-1 downto 0);
    begin
        — 4 registers
        process(clk)
        begin
24          if (clk'event and clk='1') then
                if en(3)='1' then
                    array_reg(3) <= w_data;
                end if;
                if en(2)='1' then
29                  array_reg(2) <= w_data;
                end if;
                if en(1)='1' then
                    array_reg(1) <= w_data;
                end if;
34              if en(0)='1' then
                    array_reg(0) <= w_data;
                end if;
            end if;
        end process;
39      — decoding logic for write address
        process(wr_en,w_addr)
        begin
            if (wr_en='0') then
                en <= (others=>'0');
44          else
                case w_addr is
                    when "00" =>   en <= "0001";
                    when "01" =>   en <= "0010";
                    when "10" =>   en <= "0100";
49                  when others => en <= "1000";
                end case;
            end if;
        end process;
        — read multiplexing
54      with r_addr select
            r_data <=  array_reg(0) when "00",
                       array_reg(1) when "01",
                       array_reg(2) when "10",
                       array_reg(3) when others;
59
    end explicit_arch;
```

It consists of a collection of four registers, a decoding logic to generate the enable signals, and a multiplexer to route the desired data to the read port. We can duplicate the decoding logic and multiplexing logic if additional write ports or read ports are needed.

Although the code is straightforward, the decoding and multiplexing statements become cumbersome as the size of the register file increases. An alternative method is to use *dynamic indexing*, in which a signal is used as an index to access an element in the array. The code for a parameterized register file is shown in Listing 5.7. Two generics are defined in this design. The DATA_WIDTH generic specifies the number of bits in a word and the ADDR_WIDTH generic specifies the number of address bits, which implies that there are 2^{ADDR_WIDTH} words in the file.

Listing 5.7   Parameterized register file

```
library ieee;
library ieee;
use ieee.std_logic_1164.all;
use ieee.numeric_std.all;
5 entity reg_file is
    generic(
        ADDR_WIDTH: integer:=2;
        DATA_WIDTH:integer:=8
    );
10    port(
        clk: in std_logic;
        wr_en: in std_logic;
        w_addr: in std_logic_vector (ADDR_WIDTH-1 downto 0);
        r_addr: in std_logic_vector (ADDR_WIDTH-1 downto 0);
15       w_data: in std_logic_vector (DATA_WIDTH-1 downto 0);
        r_data: out std_logic_vector (DATA_WIDTH-1 downto 0)
    );
  end reg_file;

20 architecture arch of reg_file is
    type mem_2d_type is array (0 to 2**ADDR_WIDTH-1) of
        std_logic_vector(DATA_WIDTH-1 downto 0);
    signal array_reg: mem_2d_type;
  begin
25    process(clk)
    begin
        if (clk'event and clk='1') then
            if wr_en='1' then
                array_reg(to_integer(unsigned(w_addr))) <= w_data;
30          end if;
        end if;
    end process;
    -- read port
    r_data <= array_reg(to_integer(unsigned(r_addr)));
35 end arch;
```

Note that the `array_reg(...w_addr...) <= ...` and `... <= array_reg(...r_addr...)` statements infer decoding and multiplexing logic, respectively. Although the description is more abstract, Altera software recognizes this language construct and can derive the correct implementation accordingly.

Wide decoding and multiplexing circuits have a large number of inputs but are with simple internal logic. Their structure does not match well with the LUT-based cell and leads to poor utilization of FPGA's resources. For example, synthesizing a 2^{12}-by-1 (i.e., 4K-by-1) register file requires about 8700 LEs, which counts about 50% of the LEs in the Cyclone II EP2C20 device. Thus, this method is only feasible for a small register file. The Cyclone II device contains internal synchronous embedded memory modules. These modules can be configured for synchronous operation, and their characteristics are somewhat like a restricted version of the register file. They are a better alternative for larger storage requirements. For example, a Cyclone II EP2C20 device has 52 M4K internal modules and a modified 4K-by-1 register file requires only one module, which counts about 2% of its internal memory resource. The configuration and inference of these modules are discussed in Section 5.7.

5.2.4 SRAM

A register file can be considered as a storage component in which the D FFs constitute the basic memory cells. Since all D FFs are driven by the same clock signal, the operation synchronous. The SRAM (static random access memory) organization is similar to that of a register file except that the D FFs are replaced with the D latches. Since there is no clock, its operation is *asynchronous*. The HDL description for a generic SRAM is similar to that of a register file but without the clock, as shown in Listing 5.8.

Listing 5.8 Generic SRAM

```
library ieee;
use ieee.std_logic_1164.all;
use ieee.numeric_std.all;
entity async_sram is
5    generic(
         ADDR_WIDTH: integer:=2;
         DATA_WIDTH:integer:=8
     );
     port(
10       wr_en: in std_logic;
         w_addr: in std_logic_vector (ADDR_WIDTH-1 downto 0);
         r_addr: in std_logic_vector (ADDR_WIDTH-1 downto 0);
         d: in std_logic_vector (DATA_WIDTH-1 downto 0);
         q: out std_logic_vector (DATA_WIDTH-1 downto 0)
15   );
   end async_sram;

   architecture not_use_arch of async_sram is
     type mem_2d_type is array (2**ADDR_WIDTH-1 downto 0) of
20        std_logic_vector(DATA_WIDTH-1 downto 0);
     signal array_reg: mem_2d_type;
   begin
     process(wr_en, w_addr, w_data)
     begin
25       if wr_en='1' then
            array_reg(to_integer(unsigned(w_addr))) <= d;
         end if;
     end process;
     q <= array_reg(to_integer(unsigned(r_addr)));
30 end not_use_arch;
```

At the transistor level, the area used to construct a latch is much smaller than that of a D FF. However, since there is no inherent structure in the Cyclone II device resembling asynchronous SRAM, it is synthesized from scratch using a feedback circuit with LEs. The implementation is inefficient and frequently leads to difficult timing problems. Thus, using FPGA's internal resource for asynchronous SRAM should be avoided and the code is just for demonstration purposes.

There is an external SRAM chip on the DE1 board. Accessing the device requires a memory controller and the design is discussed in Section 15.3.

5.3 SIMPLE DESIGN EXAMPLES

We illustrate the construction of several simple, representative sequential circuits in this section.

5.3.1 Shift register

Free-running shift register A free-running shift register shifts its content to the left or right by one position in each clock cycle. There is no other control signal. The code for an N-bit free-running shift-right register is shown in Listing 5.9.

Listing 5.9   Free-running shift register

```
library ieee;
use ieee.std_logic_1164.all;
entity free_run_shift_reg is
   generic(N: integer := 8);
5  port(
      clk, reset: in std_logic;
      s_in: in std_logic;
      s_out: out std_logic
   );
10 end free_run_shift_reg;

   architecture arch of free_run_shift_reg is
      signal r_reg: std_logic_vector(N-1 downto 0);
      signal r_next: std_logic_vector(N-1 downto 0);
15 begin
      -- register
      process(clk,reset)
      begin
         if (reset='1') then
20          r_reg <= (others=>'0');
         elsif (clk'event and clk='1') then
            r_reg <= r_next;
         end if;
      end process;
25    -- next-state logic (shift right 1 bit)
      r_next <= s_in & r_reg(N-1 downto 1);
      -- output
      s_out <= r_reg(0);
   end arch;
```

The next-state logic is a 1-bit shifter, which shifts **r_reg** right one position and inserts the serial input, **s_in**, to the MSB. Since the 1-bit shifter involves only reconnection of the input and output signals, no real logic is needed. Its propagation delay represents the smallest possible next-state logic delay, and the corresponding clock rate represents the highest clock rate that can be achieved for a given device technology.

Universal shift register A universal shift register can load parallel data, shift its content left or right, or remain in the same state. It can perform parallel-to-serial operation (first loading parallel input and then shifting) or serial-to-parallel operation (first shifting and then retrieving parallel output). The desired operation is specified by a 2-bit control signal, `ctrl`. The code is shown in Listing 5.10.

Listing 5.10   Universal shift register

```
1 library ieee;
use ieee.std_logic_1164.all;
entity univ_shift_reg is
   generic(N: integer := 8);
   port(
6     clk, reset: in std_logic;
```

```
          ctrl: in std_logic_vector(1 downto 0);
          d: in std_logic_vector(N-1 downto 0);
          q: out std_logic_vector(N-1 downto 0)
       );
11 end univ_shift_reg;

   architecture arch of univ_shift_reg is
       signal r_reg: std_logic_vector(N-1 downto 0);
       signal r_next: std_logic_vector(N-1 downto 0);
16 begin
       -- register
       process(clk,reset)
       begin
          if (reset='1') then
21            r_reg <= (others=>'0');
          elsif (clk'event and clk='1') then
              r_reg <= r_next;
          end if;
       end process;
26     -- next-state logic
       with ctrl select
        r_next <=
          r_reg                              when "00",  --no op
          r_reg(N-2 downto 0) & d(0)         when "01",  --shift left;
31        d(N-1) & r_reg(N-1 downto 1)       when "10",  --shift right;
          d                                  when others; -- load
       -- output
       q <= r_reg;
   end arch;
```

The next-state logic uses a 4-to-1 multiplexer to select the desired next value of the register. Note that the LSB and MSB of d [i.e., d(0) and d(N-1)] are used as the serial input for the shift-left and shift-right operations.

5.3.2 Binary counter and variant

Free-running binary counter A free-running binary counter circulates through a binary sequence repeatedly. For example, a 4-bit binary counter counts from "0000", "0001", ..., to "1111" and wraps around. The code for a parameterized N-bit free-running binary counter is shown in Listing 5.11.

Listing 5.11   Free-running binary counter

```
library ieee;
use ieee.std_logic_1164.all;
use ieee.numeric_std.all;
entity free_run_bin_counter is
5    generic(N: integer := 8);
     port(
        clk, reset: in std_logic;
        max_tick: out std_logic;
        q: out std_logic_vector(N-1 downto 0)
10   );
   end free_run_bin_counter;

   architecture arch of free_run_bin_counter is
       signal r_reg: unsigned(N-1 downto 0);
15     signal r_next: unsigned(N-1 downto 0);
   begin
       -- register
```

Table 5.1   Function table of a universal binary counter

syn_clr	load	en	up	q*	Operation
1	–	–	–	$00\cdots00$	synchronous clear
0	1	–	–	d	parallel load
0	0	1	1	q+1	count up
0	0	1	0	q-1	count down
0	0	0	–	q	pause

```
     process(clk,reset)
     begin
20       if (reset='1') then
             r_reg <= (others=>'0');
         elsif (clk'event and clk='1') then
             r_reg <= r_next;
         end if;
25   end process;
     -- next-state logic
     r_next <= r_reg + 1;
     -- output logic
     q <= std_logic_vector(r_reg);
30   max_tick <= '1' when r_reg=(2**N-1) else '0';
   end arch;
```

The next-state logic is an incrementor, which adds 1 to the register's current value. By the definition of the + operator in the IEEE numeric_std package, the operation implicitly wraps around after the r_reg reaches "1...1". The circuit also consists of an output status signal, max_tick, which is asserted when the counter reaches the maximal value, "1...1" (which is equal to $2^N - 1$).

The max_tick signal represents a special type of signal that is asserted for a single clock cycle. In this book, we call this type of signal a *tick* and use the suffix _tick to indicate a signal with this property. It is commonly used to interface with the enable signal of other sequential circuits.

Universal binary counter   A universal binary counter is more versatile. It can count up or down, pause, be loaded with a specific value, or be synchronously cleared. Its functions are summarized in Table 5.1. Note the difference between the reset and syn_clr signals. The former is asynchronous and should only be used for system initialization. The latter is sampled at the rising edge of the clock and can be used in a normal synchronous design. The code for this counter is shown in Listing 5.12.

Listing 5.12   Universal binary counter

```
library ieee;
use ieee.std_logic_1164.all;
use ieee.numeric_std.all;
4 entity univ_bin_counter is
     generic(N: integer := 8);
     port(
         clk, reset: in std_logic;
         syn_clr, load, en, up: in std_logic;
9        d: in std_logic_vector(N-1 downto 0);
         max_tick, min_tick: out std_logic;
         q: out std_logic_vector(N-1 downto 0)
```

```
    );
  end univ_bin_counter;
14
  architecture arch of univ_bin_counter is
      signal r_reg: unsigned(N-1 downto 0);
      signal r_next: unsigned(N-1 downto 0);
  begin
19    -- register
      process(clk,reset)
      begin
          if (reset='1') then
              r_reg <= (others=>'0');
24        elsif (clk'event and clk='1') then
              r_reg <= r_next;
          end if;
      end process;
      -- next-state logic
29    r_next <= (others=>'0') when syn_clr='1' else
                unsigned(d)    when load='1' else
                r_reg + 1      when en ='1' and up='1' else
                r_reg - 1      when en ='1' and up='0' else
                r_reg;
34    -- output logic
      q <= std_logic_vector(r_reg);
      max_tick <= '1' when r_reg=(2**N-1) else '0';
      min_tick <= '1' when r_reg=0 else '0';
  end arch;
```

The next-state logic follows the function table and uses a conditional signal assignment to prioritize the desired operations.

Mod-m counter   A mod-m counter counts from 0 to $m-1$ and wraps around. A parameterized mod-m counter is shown in Listing 5.13. It has two generics. One is M, which specifies the limit, m, and the other is N, which specifies the number of bits needed and should be equal to $\lceil \log_2 M \rceil$. The code is shown in Listing 5.13, and the default value is for a mod-10 counter.

Listing 5.13   Mod-m counter

```
  library ieee;
2 use ieee.std_logic_1164.all;
  use ieee.numeric_std.all;
  entity mod_m_counter is
      generic(
          N: integer := 4;      -- number of bits
7         M: integer := 10      -- mod-M
      );
      port(
          clk, reset: in std_logic;
          max_tick: out std_logic;
12        q: out std_logic_vector(N-1 downto 0)
      );
  end mod_m_counter;

  architecture arch of mod_m_counter is
17    signal r_reg: unsigned(N-1 downto 0);
      signal r_next: unsigned(N-1 downto 0);
  begin
      -- register
      process(clk,reset)
```

```
22      begin
            if (reset='1') then
                r_reg <= (others=>'0');
            elsif (clk'event and clk='1') then
                r_reg <= r_next;
27          end if;
        end process;
        -- next-state logic
        r_next <= (others=>'0') when r_reg=(M-1) else
                    r_reg + 1;
32      -- output logic
        q <= std_logic_vector(r_reg);
        max_tick <= '1' when r_reg=(M-1) else '0';
    end arch;
```

The next-state logic is constructed by a conditional signal assignment statement. If the counter reaches M-1, the new value is cleared to 0. Otherwise, it is incremented by 1.

Inclusion of the N parameter in the code is somewhat redundant since its value depends on M. A more elegant way is to define a function that calculates N from M automatically. In VHDL, this can be done by creating a user-defined *function* in a *package* and invoking the package before the entity declaration. This is beyond the scope of this book and the details may be found in the references cited in the bibliographic section.

5.4 TESTBENCH FOR SEQUENTIAL CIRCUITS

A testbench is a program that mimics a physical lab bench, as discussed in Section 2.4. Developing a comprehensive testbench is beyond the scope of this book. We discuss a simple testbench for the previous universal binary counter in this section. It can serve as a template for other sequential circuits. The code for the testbench is shown in Listing 5.14.

Listing 5.14 Testbench for a universal binary counter

```
library ieee;
use ieee.std_logic_1164.all;

entity bin_counter_tb is
5 end bin_counter_tb;

architecture arch of bin_counter_tb is
    constant THREE: integer := 3;
    constant T: time := 20 ns; -- clk period
10  signal clk, reset: std_logic;
    signal syn_clr, load, en, up: std_logic;
    signal d: std_logic_vector(THREE-1 downto 0);
    signal max_tick, min_tick: std_logic;
    signal q: std_logic_vector(THREE-1 downto 0);
15 begin
    -- instantiation

    counter_unit: entity work.univ_bin_counter(arch)
20      generic map(N=>THREE)
        port map(clk=>clk, reset=>reset, syn_clr=>syn_clr,
                load=>load, en=>en, up=>up, d=>d,
```

```
                    max_tick=>max_tick, min_tick=>min_tick, q=>q);
```

25
```
   -- clock
```

```
   -- 20 ns clock running forever
   process
30 begin
      clk <= '0';
      wait for T/2;
      clk <= '1';
      wait for T/2;
35 end process;
```

```
   -- reset
```

```
   -- reset asserted for T/2
40 reset <= '1', '0' after T/2;
```

```
   -- other stimulus
```

45 process
   begin

```
      -- initial input
```

50
```
      syn_clr <= '0';
      load <= '0';
      en <= '0';
      up <= '1';   -- count up
      d <= (others=>'0');
55    wait until falling_edge(clk);
      wait until falling_edge(clk);
```

```
      -- test load
```

60
```
      load <= '1';
      d <= "011";
      wait until falling_edge(clk);
      load <= '0';
      -- pause 2 clocks
65    wait until falling_edge(clk);
      wait until falling_edge(clk);
```

```
      -- test syn_clear
```

70
```
      syn_clr <= '1';   -- clear
      wait until falling_edge(clk);
      syn_clr <= '0';
```

```
      -- test up counter and pause
```

75
```
      en <= '1'; -- count
      up <= '1';
      for i in 1 to 10 loop -- count 10 clocks
         wait until falling_edge(clk);
80    end loop;
      en <='0';
      wait until falling_edge(clk);
      wait until falling_edge(clk);
```

```
      en  <='1';
85    wait until falling_edge(clk);
      wait until falling_edge(clk);
```

`— test down counter`

```
90    up  <= '0';
      for i in 1 to 10 loop — run 10 clocks
         wait until falling_edge(clk);
      end loop;
```

`— other wait conditions`

```
      — continue until q=2
      wait until q="010";
      wait until falling_edge(clk);
100   up  <= '1';
      — continue until min_tick changes value
      wait on min_tick;
      wait until falling_edge(clk);
      up  <= '0';
105   wait for 4*T;  — wait for 80 ns
      en  <= '0';
      wait for 4*T;
```

`— terminate simulation`

```
110
      assert false
         report "Simulation Completed"
         severity failure;
      end process ;
115 end arch;
```

The code consists of a component instantiation statement, which creates an instance of a 3-bit counter, and three segments, which generate a stimulus for clock, reset, and regular inputs. Since operation of a synchronous system is synchronized by a clock signal, we define a constant with the built-in data type **time** for the clock period:

```
constant T: time := 20 ns;  — clk period
```

The clock generation is specified by a process:

```
process
begin
   clk <= '0';
   wait for T/2;
   clk <= '1';
   wait for T/2;
end process;
```

The clk signal is assigned between '0' and '1' alternately, and each value lasts for half a period. Note that the process has no sensitivity list and repeats itself forever.

The reset stimulus involves one statement,

```
reset <= '1', '0' after T/2;
```

It indicates that the **reset** signal is set to '1' initially and changed to '0' after half a period. The statement represents the "power-on" condition, in which the **reset** signal is asserted momentarily to clear the system to the initial state. Note that, by default, the 'U' value (for uninitialized), not '0', is assigned to a signal with the std_logic type. Using a short reset pulse is a good mechanism to perform system initialization.

The last process statement generates a stimulus for other input signals. We first test the load and clear operations and then exercise counting in both directions. The final **assert false** statement forces the simulator to terminate simulation, as discussed in Section 3.6.

For a synchronous system with positive edge-triggered FFs, an input signal must be stable around the rising edge of the clock signal to satisfy the setup and hold time constraints. One easy way to achieve this is to change an input signal's value during the '1'-to-'0' transition of the clk signal. The **falling_edge** function of the std_logic_1164 package checks this condition, and we can use it in a wait statement:

```
wait until falling_edge(clk);
```

Note that each statement represents a new falling edge, which corresponds to the advancement of one clock cycle. In our template, we generally use this statement to specify the progress of time. For multiple clock cycles, we can use a loop statement:

```
for i in 1 to 10 loop  -- count 10 clocks
    wait until falling_edge(clk);
end loop;
```

There are other useful forms of wait statements, as shown at the end of the process. We can wait until a special condition, such as "when q is equal to 2",

```
wait until q="010";
```

or wait until a signal changes, such as

```
wait on min_tick;
```

or wait for an absolute time, such as

```
wait for 4*T;  -- wait for 4 clock periods
```

If an input signal is modified after these statements, we need to make sure that the input change does not occur at the rising edge of the clock. An additional

```
wait until falling_edge(clk);
```

statement should be added when needed.

We can compile the code and perform simulation. Part of the simulated waveform is shown in Figure 5.5.

5.5 TIMING ANALYSIS

5.5.1 Timing parameters

In a combinational circuit, the key timing parameter is the propagation delay, which is the longest path within the circuit to propagate a signal from an input port to an output port. The register of a sequential circuit imposes additional the *setup time and hold time constraints* and thus the timing analysis is more involved. The key parameter is f_{max}, *the maximal clock frequency*, which specifies how fast the circuit can operate. Another parameter relevant to the design is t_{CO}, the *clock to output delay*.

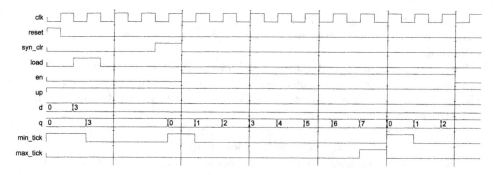

Figure 5.5 Testbench waveform.

Maximal operating frequency One of the most difficult design aspects of a sequential circuit is to ensure that the system timing does not violate the setup and hold time constraints. In a synchronous system, the storage components are grouped together and treated as a single register, as shown in Figure 5.2. We need to perform timing analysis on only one memory component.

The timing of a sequential circuit is characterized by f_{max}. The reciprocal of f_{max} specifies t_{CLOCK}, the minimal clock period, which can be interpreted as the interval between two sampling edges of the clock signal. To ensure correct operation, the next value (i.e., **state_next** in Figure 5.2) must be generated and stabilized within this interval. Assume that the maximal propagation delay of next-state logic is t_{COMB}. The minimal clock period can be obtained by adding the propagation delays and setup time constraint of the closed loop in Figure 5.2:

$$t_{CLOCK} = t_{CQ} + t_{COMB} + t_{SETUP}$$

and the maximal clock rate is the reciprocal:

$$f_{max} = \frac{1}{t_{CLOCK}} = \frac{1}{t_{CQ} + t_{COMB} + t_{SETUP}}$$

For a given FPGA device, t_{CQ} and t_{SETUP} are fixed. The only way to increase f_{max} is to use a faster combinational logic to reduce t_{COMB}. Synthesis software sometimes can identify the slowest path and reduce its delay by adding extra logic (i.e., larger area). A typical area–delay curve is shown in Figure 5.6, in which each point is a possible implementation. The software usually starts with the minimal-area implementation and traverses through the curve to reach a point that satisfies the designated timing constraint. Of course, the trade-off can be achieved only in a limited range. We cannot increase the performance indefinitely.

Clock to output delay A sequential circuit is sometimes used to generate control signals. The timing specification of the output signal plays a key role in this type of application. The main parameter is t_{CO}, the time required to obtain a valid output signal after the rising edge of the clock. The value of t_{CO} is the summation of t_{CQ} and t_{OUTPUT} (the propagation delay of the output logic); that is,

$$t_{CO} = t_{CQ} + t_{OUTPUT}$$

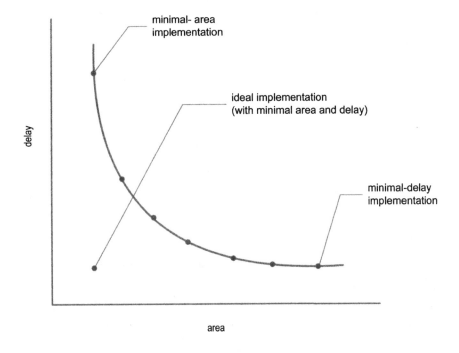

Figure 5.6   Area-delay trade-off curve.

To obtain better performance, we can try to obtain the signal directly from the state register or add an additional output register. After the elimination of the output logic, the t_{CO} becomes

$$t_{CO} = t_{CQ}$$

5.5.2   Timing considerations in Quartus II

Quartus II provides two timing software tools, known as Classic Timing Analyzer and TimeQuest Timing Analyzer. The former is simpler and we use it in this book.

Static timing analysis   *Static timing analysis* is performed to determine various timing parameters of a circuit and can be used to determine whether a system meets the performance goal. During the compiling process, Quartus II automatically starts the Classic Timing Analyzer tool, as shown in the Tasks window in Figure 3.10, and includes a timing analysis report. The Classic Timing Analyzer tool can also be invoked by selecting Processing ≻ Start ≻ Start Classic Timing Analyzer.

The timing analysis report can be examined by expanding the Timing Analyzer item. The summary page of a 32-bit universal binary counter of Listing 5.12 is shown in Figure 5.7. It shows that f_{max} is 209.16 MHz, as indicated in the Clock Setup:'clk' row and the Actual Time column.

Timing constraint in Quartus II   We can specify the desired maximal operating frequency as a synthesis constraint, and the synthesis software will try to obtain a circuit to satisfy this requirement (i.e., a circuit whose f_{max} is equal to or greater than the desired operating frequency). For example, since we use the 50-MHz (i.e.,

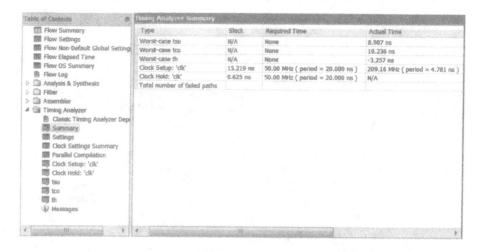

Figure 5.7   Summary page of Classic Timing Analyzer.

20-ns period) oscillator on the DE1 board as the clock source, f_{max} of a sequential circuit must exceed this frequency (i.e., the period must be smaller than 20 ns). The procedure is:

1. Select Assignments ≻ Settings.... The Settings dialog appears.
2. In the left panel, expand the Timing Analysis Settings item and then select Classic Timing Analyzer Settings. The dialog shows the relevant file information in the right panel, as shown in Figure 5.8.
3. Enter 50 MHz in the Default required fmax field.
4. Click the OK button to complete the addition.

After the synthesis and timing analysis are completed, 50 MHz is displayed the Clock Setup:'clk' row and the Required Time column, as shown in Figure 5.7. The actual f_{max} of 209.16 MHz clearly met the imposed 50-MHz constraint.

To find the highest possible f_{max}, we can enter a really high clock rate, such as 300 or 400 MHz, in the Default required fmax field and then check for the actual clock rate that can be achieved.

5.6   CASE STUDY

After examining several simple circuits, we discuss the design of more sophisticated examples in this section.

5.6.1   Stopwatch

We consider the design of a stopwatch in this subsection. The watch displays the time in three decimal digits, and counts from 00.0 to 99.9 seconds and wraps around. It contains a synchronous clear signal, clr, which returns the count to 00.0, and an enable signal, go, which enables and suspends the counting. This design is basically a BCD (binary-coded decimal) counter, which counts in BCD format. In this format, a decimal number is represented by a sequence of 4-bit

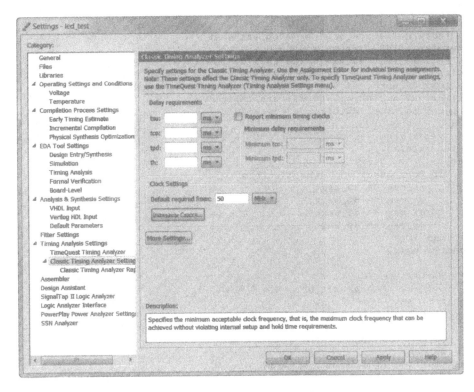

Figure 5.8 Timing Analysis Settings dialog.

BCD digits. For example, 139_{10} is represented as "0001 0011 1001" and the next number in sequence is 140_{10}, which is represented as "0001 0100 0000".

Since the DE1 board has a 50-MHz clock, we first need a mod-5,000,000 counter that generates a one-clock-cycle tick every 0.1 second. The tick is then used to enable counting of the three-digit BCD counter.

Design I Our first design of the BCD counter uses a cascading structure of three decade (i.e., mod-10) counters, representing counts of 0.1, 1, and 10 seconds, respectively. The decade counter has an enable signal and generates a one-clock-cycle tick when it reaches 9. We can use these signals to "hook" the three counters. For example, the 10-second counter is enabled only when the enable tick of the mod-5,000,000 counter is asserted and both the 0.1- and 1-second counters are 9. The code is shown in Listing 5.15.

Listing 5.15 Cascading description for a stopwatch

```
library ieee;
use ieee.std_logic_1164.all;
use ieee.numeric_std.all;
entity stop_watch is
    port(
        clk: in std_logic;
        go, clr: in std_logic;
        d2, d1, d0: out std_logic_vector(3 downto 0)
    );
end stop_watch;
```

```
     architecture cascade_arch of stop_watch is
        constant DVSR: integer:=5000000;
14      signal ms_reg, ms_next: unsigned(22 downto 0);
        signal d2_reg, d1_reg, d0_reg: unsigned(3 downto 0);
        signal d2_next, d1_next, d0_next: unsigned(3 downto 0);
        signal d1_en, d2_en, d0_en: std_logic;
        signal ms_tick, d0_tick, d1_tick: std_logic;
19   begin
        -- register
        process(clk)
        begin
           if (clk'event and clk='1') then
24             ms_reg <= ms_next;
               d2_reg <= d2_next;
               d1_reg <= d1_next;
               d0_reg <= d0_next;
           end if;
29      end process;

        -- next-state logic
        -- 0.1 sec tick generator: mod-5000000
        ms_next <=
34         (others=>'0') when clr='1' or
                              (ms_reg=DVSR and go='1') else
           ms_reg + 1 when go='1' else
           ms_reg;
        ms_tick <= '1' when ms_reg=DVSR else '0';
39      -- 0.1 sec counter
        d0_en <= '1' when ms_tick='1' else '0';
        d0_next <=
           "0000" when (clr='1') or (d0_en='1' and d0_reg=9) else
           d0_reg + 1 when d0_en='1' else
44         d0_reg;
        d0_tick <= '1' when d0_reg=9 else '0';
        -- 1 sec counter
        d1_en <= '1' when ms_tick='1' and d0_tick='1' else '0';
        d1_next <=
49         "0000" when (clr='1') or (d1_en='1' and d1_reg=9) else
           d1_reg + 1 when d1_en='1' else
           d1_reg;
        d1_tick <= '1' when d1_reg=9 else '0';
        -- 10 sec counter
54      d2_en <=
           '1' when ms_tick='1' and d0_tick='1' and d1_tick='1' else
           '0';
        d2_next <=
           "0000" when (clr='1') or (d2_en='1' and d2_reg=9) else
59         d2_reg + 1 when d2_en='1' else
           d2_reg;

        -- output logic
        d0 <= std_logic_vector(d0_reg);
64      d1 <= std_logic_vector(d1_reg);
        d2 <= std_logic_vector(d2_reg);
     end cascade_arch;
```

Note that all registers are controlled by the same clock signal. This example
illustrates how to use a one-clock-cycle enable tick to maintain synchronicity. An
inferior approach is to use the output of the lower counter as the clock signal for

the next stage. Although it may appear to be simpler, it violates the synchronous design principle and is a very poor practice.

Design II   An alternative for the three-digit BCD counter is to describe the entire structure in a nested if statement. The nested conditions indicate that the counter reaches 0.9, 9.9, and 99.9 seconds. The code is shown in Listing 5.16.

Listing 5.16   Nested if-statement description for a stopwatch

```vhdl
architecture if_arch of stop_watch is
    constant DVSR: integer:=5000000;
    signal ms_reg, ms_next: unsigned(22 downto 0);
    signal d2_reg, d1_reg, d0_reg: unsigned(3 downto 0);
    signal d2_next, d1_next, d0_next: unsigned(3 downto 0);
    signal ms_tick: std_logic;
begin
    -- register
    process(clk)
    begin
        if (clk'event and clk='1') then
            ms_reg <= ms_next;
            d2_reg <= d2_next;
            d1_reg <= d1_next;
            d0_reg <= d0_next;
        end if;
    end process;

    -- next-state logic
    -- 0.1 sec tick generator: mod-5000000
    ms_next <=
        (others=>'0') when clr='1' or
                           (ms_reg=DVSR and go='1') else
        ms_reg + 1 when go='1' else
        ms_reg;
    ms_tick <= '1' when ms_reg=DVSR else '0';
    -- 3-digit incrementor
    process(d0_reg,d1_reg,d2_reg,ms_tick,clr)
    begin
        -- default
        d0_next <= d0_reg;
        d1_next <= d1_reg;
        d2_next <= d2_reg;
        if clr='1' then
            d0_next <= "0000";
            d1_next <= "0000";
            d2_next <= "0000";
        elsif ms_tick='1' then
            if (d0_reg/=9) then
                d0_next <= d0_reg + 1;
            else            -- reach XX9
                d0_next <= "0000";
                if (d1_reg/=9) then
                    d1_next <= d1_reg + 1;
                else    -- reach X99
                    d1_next <= "0000";
                    if (d2_reg/=9) then
                        d2_next <= d2_reg + 1;
                    else -- reach 999
                        d2_next <= "0000";
                    end if;
                end if;
```

```
            end if;
54        end if;
      end process;
      -- output logic
      d0 <= std_logic_vector(d0_reg);
      d1 <= std_logic_vector(d1_reg);
59    d2 <= std_logic_vector(d2_reg);
    end if_arch;
```

Verification circuit   To verify operation of the stopwatch, we can use the seven-segment LED displays to show the output of the watch. The code is shown in Listing 5.17. Note that the leftmost LED display is assigned to 0 and the go and clr signals are mapped to two pushbutton switches of the DE1 board.

Listing 5.17   Testing circuit for a stopwatch

```
    library ieee;
    use ieee.std_logic_1164.all;
    entity stop_watch_test is
      port(
5        clk: in std_logic;
         key: in std_logic_vector(1 downto 0);
         hex3, hex2, hex1, hex0: out std_logic_vector(6 downto 0)
      );
    end stop_watch_test;
10
    architecture arch of stop_watch_test is
       signal go, clr: std_logic;
       signal d2, d1, d0: std_logic_vector(3 downto 0);
    begin
15    go <= not key(1);
      clr <= not key(0);
      -- instantiate watch
      watch_unit: entity work.stop_watch(cascade_arch)
        port map(
20          clk=>clk, go=>go, clr=>clr,
            d2 =>d2, d1=>d1, d0=>d0 );
      -- instantiate four instances of 7-seg LED decoders
      sseg_unit_0: entity work.bin_to_sseg
         port map(bin=>d0, sseg=>hex0);
25    sseg_unit_1: entity work.bin_to_sseg
         port map(bin=>d1, sseg=>hex1);
      sseg_unit_2: entity work.bin_to_sseg
         port map(bin=>d2, sseg=>hex2);
      sseg_unit_3: entity work.bin_to_sseg
30       port map(bin=>"0000", sseg=>hex3);
    end arch;
```

5.6.2   FIFO buffer

A FIFO (first-in-first-out) buffer is an "elastic" storage between two subsystems, as shown in the conceptual diagram of Figure 5.9. It has two control signals, wr and rd, for write and read operations. When wr is asserted, the input data is written into the buffer. The read operation is somewhat misleading. The head of the FIFO buffer is normally always available and thus can be read at any time. The rd signal actually acts like a "remove" signal. When it is asserted, the first item

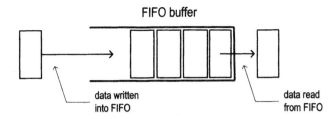

FIFO buffer

data written into FIFO

data read from FIFO

Figure 5.9 Conceptual diagram of a FIFO buffer.

(i.e., head) of the FIFO buffer is removed and the next item becomes available. In this subsection, we introduce a simple, genuine circular-queue-based design.

Circular-queue-based implementation One way to implement a FIFO buffer is to add a control circuit to a register file. The registers in the register file are arranged as a circular queue with two pointers. The *write pointer* points to the head of the queue, and the *read pointer* points to the tail of the queue. The pointer advances one position for each write or read operation. The operation of an eight-word circular queue is shown in Figure 5.10. The top-level block is shown in Figure 5.11.

A FIFO buffer usually contains two status signals, `full` and `empty`, to indicate that the FIFO is full (i.e., cannot be written) and empty (i.e., cannot be read), respectively. One of the two conditions occurs when the read pointer is equal to the write pointer, as shown in Figure 5.10(a), (f), and (i). The most difficult design task of the controller is to derive a mechanism to distinguish the two conditions. One scheme is to use two FFs to keep track of the empty and full statuses. The FFs are set to '1' and '0' during system initialization and then modified in each clock cycle according to the values of the `wr` and `rd` signals. The code of this FIFO controller is shown in Listing 5.18.

Listing 5.18 FIFO controller

```
library ieee;
use ieee.std_logic_1164.all;
use ieee.numeric_std.all;
entity fifo_ctrl is
    generic(ADDR_WIDTH: natural:=4);
    port(
        clk, reset: in std_logic;
        rd, wr: in std_logic;
        empty, full: out std_logic;
        w_addr: out std_logic_vector (ADDR_WIDTH-1 downto 0);
        r_addr, r_addr_next: out std_logic_vector (ADDR_WIDTH-1 downto 0)
    );
end fifo_ctrl;

architecture arch of fifo_ctrl is
    signal w_ptr_reg, w_ptr_next, w_ptr_succ:
        std_logic_vector(ADDR_WIDTH-1 downto 0);
    signal r_ptr_reg, r_ptr_next, r_ptr_succ:
        std_logic_vector(ADDR_WIDTH-1 downto 0);
    signal full_reg, full_next: std_logic;
    signal empty_reg, empty_next: std_logic;
    signal wr_op: std_logic_vector(1 downto 0);
```

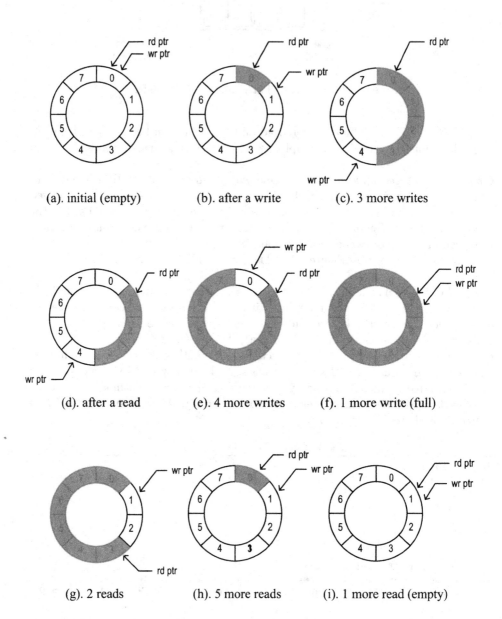

Figure 5.10 FIFO buffer based on a circular queue.

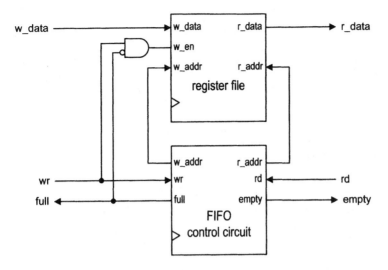

Figure 5.11   Block diagram of a register file based FIFO buffer.

```vhdl
      signal wr_en: std_logic;
24 begin
      -- register for read and write pointers
      process(clk,reset)
      begin
         if (reset='1') then
29          w_ptr_reg <= (others=>'0');
            r_ptr_reg <= (others=>'0');
            full_reg <= '0';
            empty_reg <= '1';
         elsif (clk'event and clk='1') then
34          w_ptr_reg <= w_ptr_next;
            r_ptr_reg <= r_ptr_next;
            full_reg <= full_next;
            empty_reg <= empty_next;
         end if;
39    end process;

      -- successive pointer values
      w_ptr_succ <= std_logic_vector(unsigned(w_ptr_reg)+1);
      r_ptr_succ <= std_logic_vector(unsigned(r_ptr_reg)+1);
44
      -- next-state logic for read and write pointers
      wr_op <= wr & rd;
      process(w_ptr_reg,w_ptr_succ,r_ptr_reg,r_ptr_succ,wr_op,
              empty_reg,full_reg)
49    begin
         w_ptr_next <= w_ptr_reg;
         r_ptr_next <= r_ptr_reg;
         full_next <= full_reg;
         empty_next <= empty_reg;
54       case wr_op is
            when "00" => -- no op
            when "01" => -- read
               if (empty_reg /= '1') then -- not empty
                  r_ptr_next <= r_ptr_succ;
```

```
59              full_next <= '0';
                if (r_ptr_succ=w_ptr_reg) then
                    empty_next <='1';
                end if;
            end if;
64        when "10" => -- write
            if (full_reg /= '1') then -- not full
                w_ptr_next <= w_ptr_succ;
                empty_next <= '0';
                if (w_ptr_succ=r_ptr_reg) then
69                  full_next <='1';
                end if;
            end if;
          when others => -- write/read;
            w_ptr_next <= w_ptr_succ;
74          r_ptr_next <= r_ptr_succ;
        end case;
    end process;
    -- output
    w_addr <= w_ptr_reg;
79  r_addr <= r_ptr_reg;
    r_addr_next <= r_ptr_next;
    full <= full_reg;
    empty <= empty_reg;
end arch;
```

The controller consists of two pointers and two status FFs. Its next-state logic examines the **wr** and **rd** signals and takes actions accordingly. For example, let us consider the "10" case, which implies that only a write operation occurs. The status FF is checked first to ensure that the buffer is not full. If this condition is met, we advance the write pointer by one position and clear the empty status FF. Storing one extra word to the buffer may make it full. This happens if the new write pointer "catches" the read pointer, which is expressed by the **w_ptr_succ=r_ptr_reg** expression.

To increase flexibility, we include an output port, **r_addr_next**, for the unregistered read pointer. It is used in the RAM-based design in Section 5.7.8.

Following the diagram in Figure 5.11, we can combine the controller and the register file to construct the complete FIFO buffer. The is shown in Listing 5.18. Note that the **r_addr_next** port is not used in this design.

Listing 5.19   FIFO buffer

```
library ieee;
use ieee.std_logic_1164.all;
use ieee.numeric_std.all;
entity fifo is
5   generic(
        ADDR_WIDTH: integer:=2;
        DATA_WIDTH:integer:=8
    );
    port(
10      clk, reset: in std_logic;
        rd, wr: in std_logic;
        w_data: in std_logic_vector(DATA_WIDTH-1 downto 0);
        empty, full: out std_logic;
        r_data: out std_logic_vector(DATA_WIDTH-1 downto 0)
15  );
    end fifo;
```

```
    architecture reg_file_arch of fifo is
        signal full_tmp: std_logic;
20      signal wr_en: std_logic;
        signal w_addr: std_logic_vector (ADDR_WIDTH-1 downto 0);
        signal r_addr: std_logic_vector (ADDR_WIDTH-1 downto 0);
    begin
        -- write enabled only when FIFO is not full
25      wr_en <= wr and (not full_tmp);
        full <= full_tmp;
        -- instantiate fifo control unit
        ctrl_unit: entity work.fifo_ctrl(arch)
            generic map(ADDR_WIDTH=>ADDR_WIDTH)
30          port map(clk=>clk, reset=>reset,
                    rd=>rd, wr=>wr,
                    empty=>empty, full=>full_tmp,
                    w_addr=>w_addr,
                    r_addr=>r_addr, r_addr_next=>open);
35      -- instantiate register file
        reg_file_unit: entity work.reg_file(arch)
            generic map(DATA_WIDTH=>DATA_WIDTH,
                    ADDR_WIDTH=>ADDR_WIDTH)
            port map(clk=>clk,
40                  w_addr=>w_addr, r_addr=>r_addr,
                    w_data=>w_data, r_data=>r_data,
                    wr_en=>wr_en);
    end reg_file_arch;
```

The FIFO buffer is a critical component in many applications and its optimized implementation can be quite complex. We discuss an alternative synchronous RAM-based implementation in Section 5.7.5. More efficient, device-specific implementation can be found in the Altera literature.

5.7 CYCLONE II DEVICE EMBEDDED MEMORY MODULE

5.7.1 Overview of memory options of DE1 board

An embedded application may require storage elements for various purposes. No single type of memory can satisfy all criteria. There is usually a trade-off between the size and performance. The Cyclone II device and the DE1 board provide several options for storage elements:

- *EP2C20C's FFs* (for registers): about 20K bits, embedded in logic cells and I/O buffers
- *EP2C20C's on-chip embedded memory modules*: about 212K data bits, configured as 52 4K-bit modules
- *External SRAM*: 4M bits, configured as one 256K-by-16 SRAM device
- *External SDRAM (synchronous dynamic RAM)*: 64M bits, configured as one 4M-by-16 SDRAM device

This helps us to decide which option is most suitable for an application at hand. We examine the use of FPGA's embedded memory module in this section and discuss the access of the external SRAM and SDRAM devices in Chapter 15.

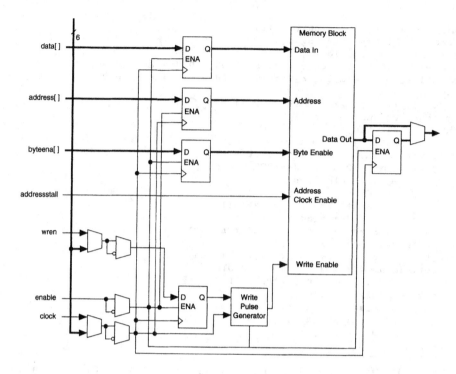

Figure 5.12   Block diagram of an M4K block in single-port mode. (Courtesy of Altera Corp.)

5.7.2   Overview of embedded M4K module

An *M4K block* is a special memory module embedded in a Cyclone II device and is separated from the regular logic cells. Each M4K block consists of 4K (2^{12}) data bits plus optional 512 parity bits. It can be organized in different widths, from 4K by 1 (i.e., 2^{12} by 2^0) to 128 by 32 (i.e., 2^7 by 2^5). Multiple modules can be combined to create larger memory.

An M4K block can be thought of as a fast SRAM with a controller. The controller contains input and output registers and generates control signals for the SRAM. The conceptually diagram of an M4K block in single-port mode is shown in Figure 5.12. The controller essentially creates a synchronous interface wrapping around the SRAM and thus the M4K block is used as a *synchronous SRAM*. The Cyclone II EP2C20C device has 52 M4K blocks, totaling 212K data bits. These blocks can be used for designs requiring intermediate-sized memory storage, such as a FIFO buffer, a large lookup table, or local memory. The m4K module is very flexible and can be configured to perform single- or dual-port access and to support various types of buffering and clocking schemes. We examine several commonly used configurations in following subsections.

5.7.3   Methods to incorporate embedded memory module

Although memory modules have a similar internal structure, there are many subtle differences in their interfaces, such as the numbers of read and write ports, clocking scheme, data and address buffering, enable and reset signals, and initial values. Although it is possible to describe the desired module behaviors in HDL code, the synthesis software may or may not recognize the designer's intention. Therefore, the HDL code cannot always infer the proper memory module and is normally not portable. In Altera Quartus II, there are two methods to incorporate an embedded memory module into a design:

- HDL instantiation via the MegaWizard Plug-in Manager program
- HDL inference with behavioral template

The first one is specific for Altera devices and the second is a semi-device-independent behavioral description.

Instantiation via MegaWizard Plug-in Manager   MegaWizard Plug-in Manager is a utility program to generate Altera-specific components. It can be invoked in the Quartus II GUI by selecting Tool ≻ MegaWizard Plug-in Manager. A dialog appears and the program guides the user through a series of questions and then generates several files. The file with the .qip extension is a text file that contains the information about the core. The file with the .vhd extension contains the instantiated component and wrapping code. The following is the segment of a 4K-by-1 single-port RAM:

```
altsyncram_component: altsyncram
generic map (
    clock_enable_input_a => "BYPASS",
    clock_enable_output_a => "BYPASS",
    intended_device_family => "Cyclone II",
    lpm_hint => "ENABLE_RUNTIME_MOD=NO",
    lpm_type => "altsyncram",
    numwords_a => 4096,
    operation_mode => "SINGLE_PORT",
    outdata_aclr_a => "NONE",
    outdata_reg_a => "UNREGISTERED",
    power_up_uninitialized => "FALSE",
    widthad_a => 12,
    width_a => 1,
    width_byteena_a => 1
)
port map (
    address_a => address,
    clock0 => clock,
    data_a => data,
    wren_a => wren,
    q_a => sub_wire0
);
```

Note that the `altsyncram` component is the Altera's proprietary core for the memory module. Although the code is readily available, we must consult the manual to understand its operation and various configuration parameters.

HDL inference with behavioral template   Although it is not possible to develop a device-independent HDL description, the Quartus II manual suggests a collection of behavioral HDL templates to infer memory modules for Altera FPGA devices. These templates are done by behavioral descriptions and contain no device-specific component instantiation. They are easy to understand and can be simulated without an additional HDL library. However, while the description does not explicitly refer to any Altera component, the code may not be recognized by other third-party synthesis software and the desired memory module cannot always be inferred. Thus, these templates can best be described as "semi-portable" and "semi-device-independent" behavioral descriptions. Because of the clarity of the behavioral description, we use this method in this book.

Note that the Altera embedded memory module can only be configured in few specific ways. If there is no match for the desired description, the module will be synthesized from scratch with the normal LEs. For example, the register file in Listing 5.7 does not use a buffer for a read address or readout data and thus does not match any of the M4K block configuration. The code does not infer an embedded memory module and is synthesized by LEs.

Templates for commonly used memory modules, including a synchronous single-port RAM, a synchronous read-write dual-port RAM, and a ROM, are discussed in the following subsections. It is good practice to separate the memory module code in an individual file and to examine the compiling report to verify the inference of the memory module.

5.7.4 HDL module to infer synchronous single-port RAM

A synchronous single-port memory uses the same address for read and write operation. The code for the single-port memory is shown in Listing 5.20.

Listing 5.20 Altera synchronous single-port RAM (with registered read address)

```
  library ieee;
2 use ieee.std_logic_1164.all;
  use ieee.numeric_std.all;
  entity altera_one_port_ram is
     generic(
        ADDR_WIDTH: integer:=10;
7       DATA_WIDTH: integer:=8
     );
     port(
        clk: in std_logic;
        we: in std_logic;
12      addr: in std_logic_vector(ADDR_WIDTH-1 downto 0);
        d: in std_logic_vector(DATA_WIDTH-1 downto 0);
        q: out std_logic_vector(DATA_WIDTH-1 downto 0)
     );
  end altera_one_port_ram;
17
  architecture beh_arch1 of altera_one_port_ram is
     type mem_2d_type is array (0 to 2**ADDR_WIDTH-1)
        of std_logic_vector (DATA_WIDTH-1 downto 0);
     signal ram: mem_2d_type;
22   signal addr_reg: std_logic_vector(ADDR_WIDTH-1 downto 0);
  begin
     process (clk)
     begin
        if (clk'event and clk = '1') then
27         if (we='1') then
              ram(to_integer(unsigned(addr))) <= d;
           end if;
           addr_reg <= addr;
```

```
              end if;
32      end process;
        q <= ram(to_integer(unsigned(addr_reg)));
      end beh_arch1;
```

In the code, the address is stored into a register and then the registered output is used to retrieve the read data. This reflects that the Cyclone II embedded memory modules are wrapped with a synchronous interface, as shown in Figure 5.12, and the address, input data, and relevant control signals, such as **we** (i.e., write enable), are first sampled and stored into its internal registers.

The existence of the address register implies that the readout data is not immediately available and essentially delayed by one clock cycle. An alternative way to describe the behavior is to register the readout data, as shown in Listing 5.21.

Listing 5.21   Altera synchronous single-port RAM (with registered read data)

```
1 architecture beh_arch2 of altera_one_port_ram is
      type mem_2d_type is array (0 to 2**ADDR_WIDTH-1)
            of std_logic_vector (DATA_WIDTH-1 downto 0);
      signal ram: mem_2d_type;
      signal data_reg: std_logic_vector(DATA_WIDTH-1 downto 0);
6 begin
      process (clk)
      begin
         if (clk'event and clk = '1') then
            if (we='1') then
11              ram(to_integer(unsigned(addr))) <= d;
            end if;
            data_reg <= ram(to_integer(unsigned(addr)));
         end if;
      end process;
16    q <= data_reg;
   end beh_arch2;
```

After synthesis, we can examine the Total memory bits item of the Flow Summary report to verify that embedded memory modules are inferred.

5.7.5   HDL module to infer synchronous simple dual-port RAM

A dual-port RAM includes a second port for memory access and thus has two independent addresses. A commonly used configuration is one port for writing and one port for reading, as in a FIFO buffer. It is called *simple dual-port RAM* in Altera documentation. The coding for a synchronous simple dual-port RAM is similar to that of a synchronous single-port RAM except that a separate address is used for read operation.

The code for the simple dual-port RAM with a read address register is shown in Listing 5.22.

Listing 5.22   Altera synchronous simple dual-port RAM (with new data)

```
library ieee;
use ieee.std_logic_1164.all;
3 use ieee.numeric_std.all;
entity altera_dual_port_ram_simple is
   generic(
      ADDR_WIDTH: integer:=10;
      DATA_WIDTH:integer:=8
```

```
 8      );
        port(
            clk: in std_logic;
            we: in std_logic;
            w_addr: in std_logic_vector(ADDR_WIDTH-1 downto 0);
13          r_addr: in std_logic_vector(ADDR_WIDTH-1 downto 0);
            d: in std_logic_vector(DATA_WIDTH-1 downto 0);
            q: out std_logic_vector(DATA_WIDTH-1 downto 0)
        );
    end altera_dual_port_ram_simple;
18
    architecture new_data_arch of altera_dual_port_ram_simple is
        type mem_2d_type is array (0 to 2**ADDR_WIDTH-1)
            of std_logic_vector (DATA_WIDTH-1 downto 0);
        signal ram: mem_2d_type;
23      signal addr_reg: std_logic_vector(ADDR_WIDTH-1 downto 0);
    begin
        process (clk)
        begin
            if (clk'event and clk = '1') then
28              if (we='1') then
                    ram(to_integer(unsigned(w_addr))) <= d;
                end if;
                addr_reg <= r_addr;
            end if;
33      end process;
        q <= ram(to_integer(unsigned(addr_reg)));
    end new_data_arch;
```

Note that w_addr is used for writing and r_addr is used for reading.

The code for the synchronous simple dual-port RAM is similar to that of the register file in Listing 5.7. The only difference is that the former buffers the read address and thus suffers a delay of one clock cycle. In most applications, we can work around this delay and use the simple dual-port RAM as a large register file.

The code for the synchronous simple dual-port RAM with a readout data register is shown in Listing 5.23.

Listing 5.23   Altera synchronous simple dual-port RAM (with old data)

```
    architecture old_data_arch of altera_dual_port_ram_simple is
        type mem_2d_type is array (0 to 2**ADDR_WIDTH-1)
            of std_logic_vector (DATA_WIDTH-1 downto 0);
        signal ram: mem_2d_type;
 5      signal data_reg: std_logic_vector(DATA_WIDTH-1 downto 0);
    begin
        process (clk)
        begin
            if (clk'event and clk = '1') then
10              if (we='1') then
                    ram(to_integer(unsigned(w_addr))) <= d;
                end if;
                data_reg <= ram(to_integer(unsigned(r_addr)));
            end if;
15      end process;
        q <= data_reg;
    end old_data_arch;
```

Unlike the single-port RAM, there is a minor difference between the two descriptions of the simple dual-port RAM. One new issue for a dual-port RAM is the *read-during-write behavior*, which concerns the memory behavior when the same address is used in write and read operations. In Listing 5.23, writing and reading occur at the clock edge and thus **data_reg** obtained the previous stored data (i.e., old data). In Listing 5.22, writing and storing read address occur at the clock edge. The data is read out one clock later with the registered read address. Thus current data (i.e., new data) is obtained accordingly. For the Cyclone II family, a special "bypass logic" will be added during synthesis to support the "new data" configuration.

5.7.6   HDL module to infer synchronous true dual-port RAM

A full-fledged synchronous dual-port RAM contains two independent access ports and allows two memory operations to be performed simultaneously. It is called *true dual-port RAM* in Altera documentation. Like the simple dual-port RAM, we need to consider memory behavior when the same address is used in write and read operations. An Altera Cyclone II memory module functions as follows:

- When a read operation and a write operation occur on the same port, the new data being written to the memory is read.
- When a read operation and a write operation occur on different ports for the same address, the old data in the memory is read. Additional bypass logic is needed to support to retrieve new data.
- Simultaneous writes to the same location on both ports results in indeterministic behavior.

The code for the synchronous true dual-port RAM without new data bypass logic is shown in Listing 5.24.

Listing 5.24   Altera synchronous true dual-port RAM

```
library ieee;
use ieee.std_logic_1164.all;
3 use ieee.numeric_std.all;
entity altera_dual_port_ram_true is
   generic(
      ADDR_WIDTH: integer:=10;
      DATA_WIDTH:integer:=8
8 );
   port(
      clk: in std_logic;
      we_a, we_b: in std_logic;
      addr_a: in std_logic_vector(ADDR_WIDTH-1 downto 0);
13     addr_b: in std_logic_vector(ADDR_WIDTH-1 downto 0);
      d_a: in std_logic_vector(DATA_WIDTH-1 downto 0);
      d_b: in std_logic_vector(DATA_WIDTH-1 downto 0);
      q_a: out std_logic_vector(DATA_WIDTH-1 downto 0);
      q_b: out std_logic_vector(DATA_WIDTH-1 downto 0)
18 );
 end altera_dual_port_ram_true;

 architecture beh_arch of altera_dual_port_ram_true is
    type ram_type is array (0 to 2**ADDR_WIDTH-1)
23       of std_logic_vector (DATA_WIDTH-1 downto 0);
    signal ram: ram_type;
 begin
    -- port a
    process(clk)
28  begin
       if (clk'event and clk = '1') then
```

```
            if (we_a = '1') then
                ram(to_integer(unsigned(addr_a))) <= d_a;
                q_a <= d_a;
33          else
                q_a <= ram(to_integer(unsigned(addr_a)));
            end if;
        end if;
    end process;
38  -- port b
    process(clk)
    begin
        if (clk'event and clk = '1') then
            if (we_b = '1') then
43              ram(to_integer(unsigned(addr_b))) <= d_b;
                q_b <= d_b;
            else
                q_b <= ram(to_integer(unsigned(addr_b)));
            end if;
48      end if;
    end process;
end beh_arch;
```

5.7.7   HDL module to infer synchronous ROM

Despite its name, a ROM (read-only memory) is a combinational circuit and has no internal state. Its output depends only on its input (i.e., address). There is no real embedded ROM in a Cyclone II device, but it can be emulated by a synchronous single-port RAM with the write operation disabled. The content of the ROM can be expressed as a constant in the HDL code and the values are loaded to the RAM when the device is programmed. A real ROM is a combinational circuit and thus should not have a buffer or a clock signal. However, since the ROM is based on a Cyclone II synchronous RAM module, the emulated ROM needs a clock signal and we call it *synchronous ROM*.

The template of a synchronous ROM is shown by an example in Listing 5.25. The code is to implement the seven-segment LED decoder, similar to that in Listing 4.12. The address of the ROM functions as the 4-bit hexadecimal input and its content is the corresponding LED patterns. The content of the ROM is defined by the HEX2LED_ROM constant and is essentially the truth table of this circuit.

Listing 5.25   Template for a synchronous ROM

```
library ieee;
use ieee.std_logic_1164.all;
use ieee.numeric_std.all;
entity altera_sync_rom_template is
5   port(
        clk: in std_logic;
        addr: in std_logic_vector(3 downto 0);
        q: out std_logic_vector(6 downto 0)
    );
10 end altera_sync_rom_template;

architecture arch of altera_sync_rom_template is
    constant ADDR_WIDTH: integer:=4;
    constant DATA_WIDTH: integer:=7;
15  type mem_2d_type is array (0 to 2**ADDR_WIDTH-1)
        of std_logic_vector (DATA_WIDTH-1 downto 0);
```

```
    — ROM definition
    constant HEX2LED_ROM: mem_2d_type:=(   — 2^4-by-7
       "0000001",  — addr 00
20     "1001111",  — addr 01
       "0010010",  — addr 02
       "0000110",  — addr 03
       "1001100",  — addr 04
       "0100100",  — addr 05
25     "0100000",  — addr 06
       "0001111",  — addr 07
       "0000000",  — addr 08
       "0000100",  — addr 09
       "0001000",  — addr 10
30     "1100000",  — addr 11
       "0110001",  — addr 12
       "1000010",  — addr 13
       "0110000",  — addr 14
       "0111000"   — addr 15
35  );
    signal data_reg: std_logic_vector(DATA_WIDTH-1 downto 0);
  begin
    — registered output to infer embedded RAM
    process (clk)
40  begin
       if (clk'event and clk = '1') then
          data_reg <= HEX2LED_ROM(to_integer(unsigned(addr)));
       end if;
    end process;
45  q <= data_reg;
  end arch;
```

The code is similar to that of the synchronous single-port RAM but with a predefined constant. Note that operation of this ROM depends on the clock signal, and its timing is different from that of a normal ROM. Artificial inclusion of the clock signal and data register is necessary to infer FPGA's internal memory modules for the ROM implementation. During synthesis, the software automatically determines whether to use regular logic elements or M4K modules to realize this circuit.

5.7.8 FIFO buffer revisited

For a FIFO buffer discussed in Section 5.6.2, we can use the synchronous simple dual-port RAM for storage. Because of the registered address signal, direct replacement of the register file will lead to different read behavior. When designing the FIFO controller, we intentionally provide both registered and unregistered read pointers as outputs (i.e., **r_addr** and **r_addr_next**). Note that the registered signal, **r_addr**, is used in Listing 5.19. The unregistered signal, **r_addr_next**, can be used with the synchronous RAM to offset the effect of its internal address register and thus obtain the identical timing behavior for read operation. The revised architecture body is shown in Listing 5.26.

Listing 5.26 FIFO buffer using synchronous SRAM

```
architecture sync_sram_arch of fifo is
   signal full_tmp: std_logic;
   signal wr_en: std_logic;
4  signal w_addr: std_logic_vector (ADDR_WIDTH-1 downto 0);
   signal r_addr_next: std_logic_vector (ADDR_WIDTH-1 downto 0);
```

```
     begin
         —— write enabled only when FIFO is not full
         wr_en <= wr and (not full_tmp);
9        full <= full_tmp;
         —— instantiate fifo control unit
         ctrl_unit: entity work.fifo_ctrl(arch)
             generic map(ADDR_WIDTH=>ADDR_WIDTH)
             port map(clk=>clk, reset=>reset,
14                      rd=>rd, wr=>wr,
                        empty=>empty, full=>full_tmp,
                        w_addr=>w_addr, r_addr=>open,
                        r_addr_next=>r_addr_next);
         —— instantiate synchronous SRAM
19       ssram_unit:
             entity work.altera_dual_port_ram_simple(new_data_arch)
                 generic map(DATA_WIDTH=>DATA_WIDTH,
                             ADDR_WIDTH=>ADDR_WIDTH)
                 port map(clk=>clk,
24                        w_addr=>w_addr, r_addr=>r_addr_next,
                          d=>w_data, q=>r_data,
                          we=>wr_en);
     end sync_sram_arch;
```

5.8 BIBLIOGRAPHIC NOTES

Chapter 8 of the *Cyclone II Device Handbook*, titled *Cyclone II Memory Blocks*, provides detailed information on its internal memory module. Chapter 6 of the *Quartus II Handbook v10.0.1*, titled *Recommended HDL Coding Styles*, discusses the HDL descriptions to infer various memory configurations. Altera's *Embedded Peripherals IP User Guide* provides detailed information of a full-fledged on-chip FIFO buffer.

5.9 SUGGESTED EXPERIMENTS

5.9.1 Programmable square wave generator

A programmable square wave generator is a circuit that can generate a square wave with variable on (i.e., logic 1) and off (i.e., logic 0) intervals. The durations of the intervals are specified by two 4-bit control signals, m and n, which are interpreted as unsigned integers. The on and off intervals are m*100 ns and n*100 ns, respectively (recall that the period of the DE1 onboard oscillator is 20 ns). Design a programmable square wave generator circuit. The circuit should be completely synchronous. We need a logic analyzer or oscilloscope to verify its operation.

5.9.2 Pulse width modulation circuit

The duty cycle of a square wave is defined as the percentage of the on interval (i.e., logic 1) in a period. A PWM (pulse width modulation) circuit can generate an output with variable duty cycles. For a PWM with 4-bit resolution, a 4-bit control signal, w, specifies the duty cycle. The w signal is interpreted as an unsigned integer and the duty cycle is $\frac{w}{16}$. Design a PWM circuit with 4-bit resolution and verify its operation using a logic analyzer or oscilloscope.

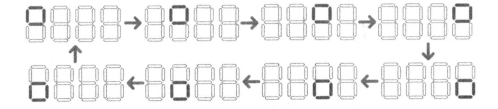

Figure 5.13 Pattern for Experiment 5.9.3.

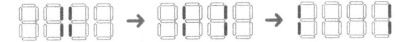

Figure 5.14 Pattern for Experiment 5.9.4.

5.9.3 Rotating square circuit

In a seven-segment LED display, a square pattern can be created by enabling the a, b, f, and g segments or the c, d, e, and g segments. We want to design a circuit that circulates the square patterns on the DE1 board's seven-segment LED displays. The clockwise circulating pattern is shown in Figure 5.13. The circuit should have an input, en, which enables or pauses the circulation, and an input, cw, which specifies the direction (i.e., clockwise or counterclockwise) of the circulation.

Design the circuit and verify its operation on the prototyping board. Make sure that the circulation rate is slow enough for visual inspection.

5.9.4 Heartbeat circuit

We want to create a "heartbeat" for the prototyping board. It repeats the simple pattern in the four-digit seven-segment display, as shown in Figure 5.14, at a rate of 72 Hz. Design the circuit and verify its operation on the prototyping board.

5.9.5 Rotating LED banner circuit

The DE1 prototyping board has four seven-segment LED displays and thus only four symbols can be displayed at a time. We can show more information if the data is rotated and moved continuously. For example, assume that the message is 10 digits (i.e., "0123456789"). The display can show the message as "0123", "1234", "2345", ..., "6789", "7890", ..., "0123". The circuit should have an input, en, which enables or pauses the rotation, and an input, dir, which specifies the direction (i.e., rotate left or right).

Design the circuit and verify its operation on the prototyping board. Make sure that the rotation rate is slow enough for visual inspection.

5.9.6 Enhanced stopwatch

Modify the stopwatch with the following extensions:

- Add an additional signal, up, to control the direction of counting. The stopwatch counts up when the up signal is asserted and counts down otherwise.
- Add a minute digit to the display. The LED display format should be like MSSD, where D represents 0.1 second and its range is between 0 and 9, SS represents seconds and its range is between 00 and 59, and M represents minutes and its range is between 0 and 9.

Design the new stopwatch and verify its operation with a testing circuit.

5.9.7 FIFO with data width conversion

In some applications, the widths of the write port and read port of a FIFO buffer may not be the same. For example, a subsystem may write 16-bit data into the FIFO buffer and another subsystem only reads and removes 8-bit data at a time. Assume that the width of the write port is twice the width of the read port. Redesign the FIFO with a modified controller and register file and verify its operation. The DATA_WIDTH generic should be the width of the read port.

5.9.8 Stack

A stack is a last-in-first-out buffer in which the last stored data is retrieved first. Storing a data word to a stack is known as a *push* operation, and retrieving a data word from a stack is known as a *pop* operation. The I/O signals of a stack are similar to those of a FIFO buffer except that we generally use the push and pop signals in place of the wr and rd signals. Design a stack using a register file and verify its operation.

5.9.9 ROM-based sign-magnitude adder

We can implement any n-input, m-output function with a 2^n-by-m ROM. Consider the sign-magnitude adder discussed in Section 4.6.2 and assume that a and b are 4-bit input signals. Design this circuit as follows:

1. Write a program in a conventional programming language, such as C or Java, to generate a 2^8-by-4 truth table for this circuit.
2. Follow the ROM template in Listing 5.25 to derive the HDL code. Cut and paste the table to the code.
3. Synthesize the circuit and verify its operation.
4. Check the synthesis report and compare the sizes (in terms of the number of logic cells) of the original implementation and the ROM-based implementation.
5. Expand a and b to 8-bit input signals and repeat steps 1 to 4.

5.9.10 ROM-based temperature conversion

Temperature can be measured in Celsius or Fahrenheit scale. Let c and f be a temperature reading in Celsius and Fahrenheit scales. They are related by

$$f = \frac{9}{5} * c + 32$$

The conversion involves multiplication and division operations and direct implementation requires a significant amount of hardware resource. For a simple application, such as a digital thermometer, we can create a lookup table for conversion and store it in a ROM.

Consider a conversion circuit with following specification:

- The range is between 0°C and 100°C (32°F and 212°F).
- The input and output are in 8-bit unsigned format.
- A separate `format` signal indicates whether the input is in Celsius or Fahrenheit scale. The output is to be converted to other scale.

We can create two lookup tables for the two conversions. Note that because of the small size of these tables, it is possible to store the two tables in a single Cyclone II M4K module. Design the circuit and verify its operation.

CHAPTER 6

FSM

An FSM (finite state machine) is a sequential circuit that transits among a finite number of internal states. The transitions depend on the current state and external input and do exhibit a simple, "regular" pattern. In this chapter, we provide an overview of the basic characteristics and representation of FSMs and discuss the derivation of HDL codes.

6.1 INTRODUCTION

An FSM (finite state machine) is used to model a system that transits among a finite number of internal states. The transitions depend on the current state and external input. Unlike a regular sequential circuit, the state transitions of an FSM do not exhibit a simple, repetitive pattern. Its next-state logic is usually constructed from scratch and is sometimes known as "random" logic. This is different from the next-state logic of a regular sequential circuit, which is composed mostly of "structured" components, such as incrementors and shifters.

In practice, the main application of an FSM is to act as the controller of a large digital system, which examines the external commands and status and activates proper control signals to control operation of a *data path*, which is usually composed of regular sequential components. This is known as an FSMD (finite state machine with data path) and is discussed in Chapter 7.

Embedded SOPC Design with Nios II Processor and VHDL Examples. By Pong P. Chu **127**
Copyright © 2011 John Wiley & Sons, Inc.

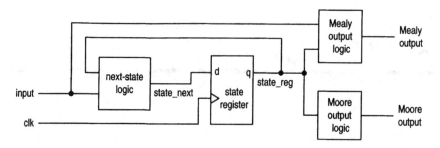

Figure 6.1   Block diagram of a synchronous FSM.

6.1.1   Mealy and Moore outputs

The basic block diagram of an FSM is the same as that of a regular sequential circuit and is repeated in Figure 6.1. It consists of a state register, next-state logic, and output logic. An FSM is known as a *Moore machine* if the output is only a function of state, and is known as a *Mealy machine* if the output is a function of state and external input. Both types of output may exist in a complex FSM, and we simply refer to it as containing a Moore output and a Mealy output. The Moore and Mealy outputs are similar but not identical. Understanding their subtle differences is the key for controller design. The example in Section 6.3.1 illustrates the behaviors and constructions of the two types of outputs.

6.1.2   FSM representation

An FSM is usually specified by an abstract *state diagram* or *ASM chart* (algorithmic state machine chart), both capturing the FSM's input, output, states, and transitions in a graphical representation. The two representations provide the same information. The FSM representation is more compact and better for simple applications. The ASM chart representation is somewhat like a flowchart and is more descriptive for applications with complex transition conditions and actions.

State diagram   A state diagram is composed of *nodes*, which represent states and are drawn as circles, and annotated *transitional arcs*. A single node and its transition arcs are shown in Figure 6.2(a). A logic expression expressed in terms of input signals is associated with each transition arc and represents a specific condition. The arc is taken when the corresponding expression is evaluated **true**.

The Moore output values are placed inside the circle since they depend only on the current state. The Mealy output values are associated with the conditions of transition arcs since they depend on the current state and external input. To reduce clutter in the diagram, only asserted output values are listed. The output signal takes the default (i.e., unasserted) value otherwise.

A representative state diagram is shown in Figure 6.3(a). The FSM has three states, two external input signals (i.e., a and b), one Moore output signal (i.e., y1), and one Mealy output signal (i.e., y0). The y1 signal is asserted when the FSM is in the s0 or s1 state. The y0 signal is asserted when the FSM is in the s0 state and the a and b signals are "11".

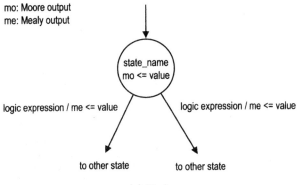

mo: Moore output
me: Mealy output

state_name
mo <= value

logic expression / me <= value   logic expression / me <= value

to other state   to other state

(a) Node

mo: Moore output
me: Mealy output

state entry

state
name

state box

mo <= value

decision box

Boolean
condition

T   F

conditional
output box

me <= value

exit to other ASM   exit to other ASM
block   block

(b) ASM block

Figure 6.2   Symbol of a state.

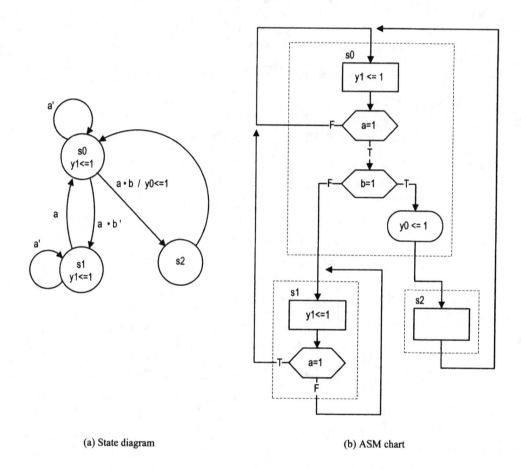

(a) State diagram

(b) ASM chart

Figure 6.3   Example of an FSM.

ASM chart   An ASM chart is composed of a network of ASM blocks. An *ASM block* consists of one *state box* and an optional network of *decision boxes* and *conditional output boxes*. A representative ASM block is shown in Figure 6.2(b).

A state box represents a state in an FSM, and the asserted Moore output values are listed inside the box. Note that it has only one exit path. A decision box tests the input condition and determines which exit path to take. It has two exit paths, labeled T and F, which correspond to the **true** and **false** values of the condition. A conditional output box lists asserted Mealy output values and is usually placed after a decision box. It indicates that the listed output signal can be activated only when the corresponding condition in the decision box is met.

A state diagram can easily be converted to an ASM chart, and vice versa. The corresponding ASM chart of the previous FSM state diagram is shown in Figure 6.3(b).

6.2 FSM CODE DEVELOPMENT

The procedure of developing code for an FSM is similar to that of a regular sequential circuit. We first separate the state register and then derive the code for the combinational next-state logic and output logic. The main difference is the next-state logic. For an FSM, the code for the next-state logic follows the flow of a state diagram or ASM chart.

For clarity and flexibility, we use the VHDL's *enumerated data type* to represent the FSM's states. The enumerated data type can best be explained by an example. Consider the FSM of Section 6.1.2, which has three states: s0, s1, and s2. We can introduce a user-defined enumerated data type as follows:

```
type eg_state_type is (s0, s1, s2);
```

The data type simply lists (i.e., *enumerates*) all symbolic values. Once the data type is defined, it can be used for the signals, as in

```
signal state_reg, state_next: eg_state_type;
```

During synthesis, software automatically maps the values in an enumerated data type to binary representations, a process known as *state assignment*. Although there is a mechanism to perform this manually, it is rarely needed.

The complete code of the FSM is shown in Listing 6.1. It consists of segments for the state register, next-state logic, Moore output logic, and Mealy output logic.

Listing 6.1   FSM example

```
library ieee;
use ieee.std_logic_1164.all;
entity fsm_eg is
4    port(
        clk, reset: in std_logic;
        a, b: in std_logic;
        y0, y1: out std_logic
    );
9 end fsm_eg;

architecture mult_seg_arch of fsm_eg is
    type eg_state_type is (s0, s1, s2);
    signal state_reg, state_next: eg_state_type;
```

```vhdl
14 begin
       -- state register
       process(clk,reset)
       begin
          if (reset='1') then
19           state_reg <= s0;
          elsif (clk'event and clk='1') then
             state_reg <= state_next;
          end if;
       end process;
24     -- next-state logic
       process(state_reg,a,b)
       begin
          case state_reg is
             when s0 =>
29              if a='1' then
                   if b='1' then
                      state_next <= s2;
                   else
                      state_next <= s1;
34                 end if;
                else
                   state_next <= s0;
                end if;
             when s1 =>
39              if (a='1') then
                   state_next <= s0;
                else
                   state_next <= s1;
                end if;
44        when s2 =>
                state_next <= s0;
          end case;
       end process;
       -- Moore output logic
49     process(state_reg)
       begin
          case state_reg is
             when s0|s1 =>
                y1 <= '1';
54           when s2 =>
                y1 <= '0';
          end case;
       end process;
       -- Mealy output logic
59     process(state_reg,a,b)
       begin
          case state_reg is
             when s0 =>
                if (a='1') and (b='1') then
64                 y0 <= '1';
                else
                   y0 <= '0';
                end if;
             when s1 | s2 =>
69              y0 <= '0';
          end case;
       end process;
    end mult_seg_arch;
```

The key part is the next-state logic. It uses a case statement with the **state_reg** signal as the selection expression. The next state (i.e., **state_next** signal) is determined by the current state (i.e., **state_reg**) and external input. The code for each state basically follows the activities inside each ASM block of Figure 6.3(b).

An alternative code is to merge next-state logic and output logic into a single combinational block, as shown in Listing 6.2.

Listing 6.2   FSM with merged combinational logic

```
architecture two_seg_arch of fsm_eg is
   type eg_state_type is (s0, s1, s2);
3   signal state_reg, state_next: eg_state_type;
begin
   -- state register
   process(clk,reset)
   begin
8      if (reset='1') then
         state_reg <= s0;
      elsif (clk'event and clk='1') then
         state_reg <= state_next;
      end if;
13   end process;
   -- next-state/output logic
   process(state_reg,a,b)
   begin
      state_next <= state_reg;  -- default back to same state
18      y0 <= '0';   -- default 0
      y1 <= '0';   -- default 0
      case state_reg is
         when s0 =>
            y1 <= '1';
23            if a='1' then
               if b='1' then
                  state_next <= s2;
                  y0 <= '1';
               else
28                  state_next <= s1;
               end if;
            -- no else branch
            end if;
         when s1 =>
33            y1 <= '1';
            if (a='1') then
               state_next <= s0;
            -- no else branch
            end if;
38         when s2 =>
            state_next <= s0;
      end case;
   end process;
end two_seg_arch;
```

Note that the default output values are listed at the beginning of the code.

The code for the next-state logic and output logic follows the ASM chart closely. Once a detailed state diagram or ASM chart is derived, converting an FSM to HDL code is almost a mechanical procedure. Listings 6.1 and 6.2 can serve as templates for this purpose.

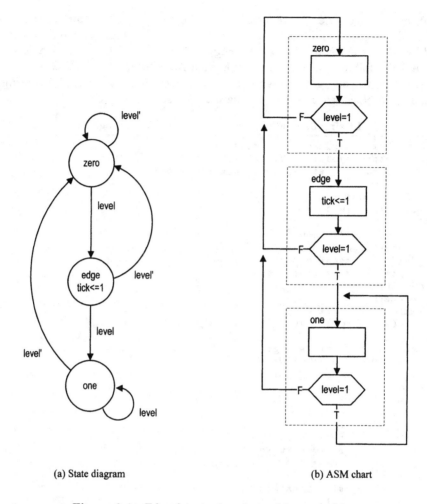

(a) State diagram   (b) ASM chart

Figure 6.4   Edge detector based on a Moore machine.

6.3   DESIGN EXAMPLES

6.3.1   Rising-edge detector

The rising-edge detector is a circuit that generates a short, one-clock-cycle pulse
(we call it a *tick*) when the input signal changes from '0' to '1'. It is usually used to
indicate the onset of a slow time-varying input signal. We design the circuit using
both Moore and Mealy machines, and compare their differences.

Moore-based design   The state diagram and ASM chart of a Moore machine–based
edge detector are shown in Figure 6.4. The `zero` and `one` states indicate that the
input signal has been '0' and '1' for a while. The rising edge occurs when the input
changes to '1' in the `zero` state. The FSM moves to the `edge` state and the output,
`tick`, is asserted in this state. A representative timing diagram is shown at the
middle of Figure 6.5. The code is shown in Listing 6.3.

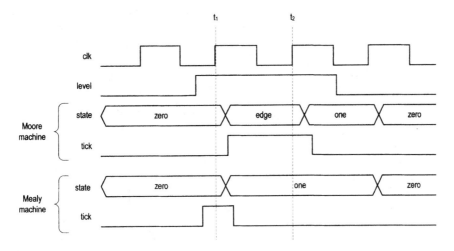

Figure 6.5   Timing diagram of two edge detectors.

Listing 6.3   Moore machine–based edge detector

```
library ieee;
use ieee.std_logic_1164.all;
3 entity edge_detect is
    port(
        clk, reset: in std_logic;
        level: in std_logic;
        tick: out std_logic
8   );
  end edge_detect;

  architecture moore_arch of edge_detect is
    type state_type is (zero, edge, one);
13    signal state_reg, state_next: state_type;
  begin
    — state register
    process(clk,reset)
    begin
18      if (reset='1') then
            state_reg <= zero;
        elsif (clk'event and clk='1') then
            state_reg <= state_next;
        end if;
23  end process;
    — next—state/output logic
    process(state_reg,level)
    begin
        state_next <= state_reg;
28      tick <= '0';
        case state_reg is
            when zero=>
                if level= '1' then
                    state_next <= edge;
33              end if;
            when edge =>
                tick <= '1';
                if level= '1' then
                    state_next <= one;
```

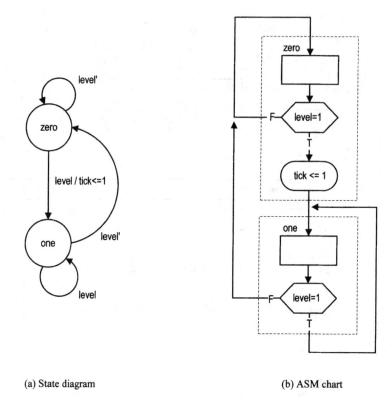

(a) State diagram   (b) ASM chart

Figure 6.6   Edge detector based on a Mealy machine.

```
38              else
                    state_next <= zero;
                end if;
            when one =>
                if level= '0' then
43                  state_next <= zero;
                end if;
        end case;
    end process;
end moore_arch;
```

Mealy-based design   The state diagram and ASM chart of a Mealy machine–based edge detector are shown in Figure 6.6. The **zero** and **one** states have similar meaning. When the FSM is in the **zero** state and the input changes to '1', the output is asserted immediately. The FSM moves to the **one** state at the rising edge of the next clock and the output is deasserted. A representative timing diagram is shown at the bottom of Figure 6.5. Note that due to the propagation delay, the output signal is still asserted at the rising edge of the next clock (i.e., at t_1). The code is shown in Listing 6.4.

Listing 6.4   Mealy machine–based edge detector

```
architecture mealy_arch of edge_detect is
    type state type is (zero, one);
```

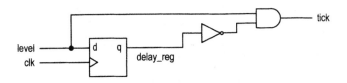

Figure 6.7   Gate-level implementation of an edge detector.

```
3    signal state_reg, state_next: state_type;
   begin
      -- state register
      process(clk,reset)
      begin
8        if (reset='1') then
            state_reg <= zero;
         elsif (clk'event and clk='1') then
            state_reg <= state_next;
         end if;
13   end process;
      -- next-state/output logic
      process(state_reg,level)
      begin
         state_next <= state_reg;
18       tick <= '0';
         case state_reg is
            when zero=>
               if level= '1' then
                  state_next <= one;
23                tick <= '1';
               end if;
            when one =>
               if level= '0' then
                  state_next <= zero;
28             end if;
         end case;
      end process;
   end mealy_arch;
```

Direct implementation   Since the transitions of the edge detector circuit are very simple, it can be implemented without using an FSM. We include this implementation for comparison purposes. The circuit diagram is shown in Figure 6.7. It can be interpreted that the output is asserted only when the current input is '1' and the previous input, which is stored in the register, is '0'. The corresponding code is shown in Listing 6.5.

Listing 6.5   Gate-level implementation of an edge detector

```
architecture gate_level_arch of edge_detect is
   signal delay_reg: std_logic;
begin
4    -- delay register
   process(clk,reset)
   begin
      if (reset='1') then
         delay_reg <= '0';
9       elsif (clk'event and clk='1') then
         delay_reg <= level;
```

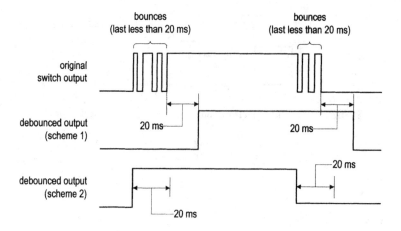

Figure 6.8   Original and debounced waveforms.

```
      end if;
   end process;
   — decoding logic
14    tick <= (not delay_reg) and level;
   end gate_level_arch;
```

Although the descriptions in Listings 6.4 and 6.5 appear to be very different, they describe the same circuit. The circuit diagram can be derived from the FSM if we assign '0' and '1' to the zero and one states.

Comparison   Whereas both Moore machine– and Mealy machine–based designs can generate a short tick at the rising edge of the input signal, there are several subtle differences. The Mealy machine–based design requires fewer states and responds faster, but the width of its output may vary and input glitches may be passed to the output.

The choice between the two designs depends on the subsystem that uses the output signal. Most of the time the subsystem is a synchronous system that shares the same clock signal. Since the FSM's output is sampled only at the rising edge of the clock, the width and glitches do not matter as long as the output signal is stable around the edge. Note that the Mealy output signal is available for sampling at t_1, which is one clock cycle faster than the Moore output, which is available at t_2. Therefore, the Mealy machine–based circuit is preferred for this type of application.

6.3.2 Debouncing circuit

The DE1 board contains four pushbutton switches and ten slide switches. When we move a switch, its contact may bounce back and forth a few times before settling down. The bounces lead to glitches in the signal, as shown at the top of Figure 6.8. The bounces usually settle within 20 ms. The purpose of a *debouncing circuit* is to filter out the glitches associated with switch transitions. The DE1 board provides debouncing circuits for the pushbutton switches but connects the outputs of the slide switches to the FPGA device directly.

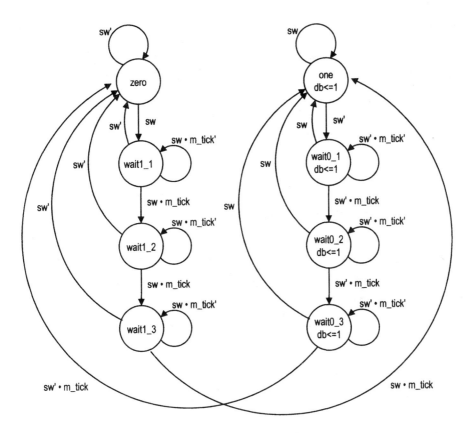

Figure 6.9   State diagram of a debouncing circuit.

We can design a debouncing circuit inside the FPGA device for the slide switches. The debounced output signals from two FSM-based design schemes are shown in the two bottom parts of Figure 6.8. The first design scheme is discussed in this subsection and the second scheme is left as an exercise in Experiment 6.5.2. A better alternative FSMD-based scheme is discussed in Section 7.2.1.

An FSM-based design uses a free-running 10-ms timer and an FSM. The timer generates a one-clock-cycle enable tick (the **m_tick** signal) every 10 ms and the FSM uses this information to keep track of whether the input value is stabilized. In the first design scheme, the FSM ignores the short bounces and changes the value of the debounced output only after the input is stabilized for 20 ms. The output timing diagram is shown at the middle of Figure 6.8. The state diagram of this FSM is shown in Figure 6.9. The **zero** and **one** states indicate that the switch input signal, **sw**, has been stabilized with '0' and '1' values. Assume that the FSM is initially in the **zero** state. It moves to the **wait1_1** state when **sw** changes to '1'. At the **wait1_1** state, the FSM waits for the assertion of **m_tick**. If **sw** becomes '0' in this state, it implies that the width of the '1' value does not last long enough and the FSM returns to the **zero** state. This action repeats two more times for the **wait1_2** and **wait1_3** states. The operation from the **one** state is similar except that the **sw** signal must be '0'.

Since the 10-ms timer is free-running and the m_tick tick can be asserted at any time, the FSM checks the assertion three times to ensure that the sw signal is stabilized for at least 20 ms (it is actually between 20 and 30 ms). The code is shown in Listing 6.6. It includes a 10-ms timer and the FSM.

Listing 6.6   FSM implementation of a debouncing circuit

```
   library ieee;
   use ieee.std_logic_1164.all;
   use ieee.numeric_std.all;
   entity db_fsm is
5    port(
         clk, reset: in std_logic;
         sw: in std_logic;
         db: out std_logic
     );
10 end db_fsm;

   architecture arch of db_fsm is
       constant N: integer:=19;   -- 2^N * 20ns = 10ms
       signal q_reg, q_next: unsigned(N-1 downto 0);
15     signal m_tick: std_logic;
       type eg_state_type is (zero,wait1_1,wait1_2,wait1_3,
                              one,wait0_1,wait0_2,wait0_3);
       signal state_reg, state_next: eg_state_type;
   begin
20 --===========================================
   -- counter to generate 10ms tick
   -- (2^19 * 20ns)
   --===========================================
   process(clk,reset)
25 begin
       if (clk'event and clk='1') then
           q_reg <= q_next;
       end if;
   end process;
30 -- next-state logic
   q_next <= q_reg + 1;
   --output tick
   m_tick <= '1' when q_reg=0 else
             '0';
35 --===========================================
   -- debouncing FSM
   --===========================================
   -- state register
   process(clk,reset)
40 begin
       if (reset='1') then
           state_reg <= zero;
       elsif (clk'event and clk='1') then
           state_reg <= state_next;
45     end if;
   end process;
   -- next-state/output logic
   process(state_reg,sw,m_tick)
   begin
50     state_next <= state_reg; --default: back to same state
       db <= '0';   -- default 0
       case state_reg is
          when zero =>
             if sw='1' then
```

```
                    state_next <= wait1_1;
55                end if;
            when wait1_1 =>
                if sw='0' then
                    state_next <= zero;
60                else
                    if m_tick='1' then
                        state_next <= wait1_2;
                    end if;
                end if;
65            when wait1_2 =>
                if sw='0' then
                    state_next <= zero;
                else
                    if m_tick='1' then
70                        state_next <= wait1_3;
                    end if;
                end if;
            when wait1_3 =>
                if sw='0' then
75                    state_next <= zero;
                else
                    if m_tick='1' then
                        state_next <= one;
                    end if;
80                end if;
            when one =>
                db <='1';
                if sw='0' then
                    state_next <= wait0_1;
85                end if;
            when wait0_1 =>
                db <='1';
                if sw='1' then
                    state_next <= one;
90                else
                    if m_tick='1' then
                        state_next <= wait0_2;
                    end if;
                end if;
95            when wait0_2 =>
                db <='1';
                if sw='1' then
                    state_next <= one;
                else
100                    if m_tick='1' then
                        state_next <= wait0_3;
                    end if;
                end if;
            when wait0_3 =>
105                db <='1';
                if sw='1' then
                    state_next <= one;
                else
                    if m_tick='1' then
110                        state_next <= zero;
                    end if;
                end if;
        end case;
    end process;
115 end arch;
```

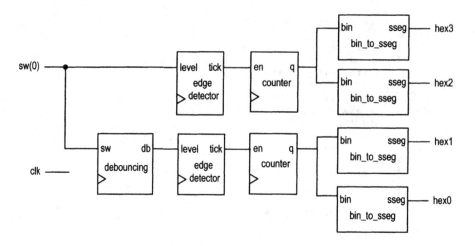

Figure 6.10   Debouncing testing circuit.

6.3.3   Testing circuit

We use a bounce counting circuit to verify operation of the rising-edge detector and the debouncing circuit. The block diagram is shown in Figure 6.10. The input of the verification circuit is from a slide switch. In the lower part, the signal is first fed to the debouncing circuit and then to the rising-edge detector. Therefore, a one-clock-cycle tick is generated each time the switch is moved up and down. The tick in turn controls the enable input of an 8-bit counter, whose content is passed to the seven-segment LED decoders and shown on the left two digits of the prototyping board's seven-segment LED display. In the upper part, the input signal is fed directly to the edge detector without the debouncing circuit, and the number is shown on the right two digits of the prototyping board's seven-segment LED display. The bottom counter thus counts one desired 0-to-1 transition as well as the bounces.

The code is shown in Listing 6.7. It basically uses component instantiation to realize the block diagram. In addition, a pushbutton switch is used to clear the counters to zero.

Listing 6.7   Verification circuit for a debouncing circuit and rising-edge detector

```
library ieee;
use ieee.std_logic_1164.all;
use ieee.numeric_std.all;
entity debounce_test is
   port(
      clk: in std_logic;
      sw: in std_logic_vector(0 downto 0);
      key: in std_logic_vector(0 downto 0);
      hex3, hex2, hex1, hex0: out std_logic_vector(6 downto 0)
   );
end debounce_test;

architecture arch of debounce_test is
   signal q1_reg, q1_next: unsigned(7 downto 0);
   signal q0_reg, q0_next: unsigned(7 downto 0);
   signal b_count, d_count: std_logic_vector(7 downto 0);
```

```
      signal sw_reg, db_reg: std_logic;
      signal db_level, db_tick, btn_tick, clr: std_logic;
   begin
20    -- ========================================================
      -- component instantiation
      -- ========================================================
      -- instantiate debouncing circuit
      db_unit: entity work.db_fsm(arch)
25       port map(
            clk=>clk, reset=>'0',
            sw=>sw(0), db=>db_level);

      -- instantiate four instances of 7-seg LED decoders
30    sseg_unit_0: entity work.bin_to_sseg
         port map(bin=>d_count(3 downto 0), sseg=>hex0);
      sseg_unit_1: entity work.bin_to_sseg
         port map(bin=>d_count(7 downto 4), sseg=>hex1);
      sseg_unit_2: entity work.bin_to_sseg
35       port map(bin=>b_count(3 downto 0), sseg=>hex2);
      sseg_unit_3: entity work.bin_to_sseg
         port map(bin=>b_count(7 downto 4), sseg=>hex3);
      -- ========================================================
      -- edge detection circuits
40    -- ========================================================
      process(clk)
      begin
         if (clk'event and clk='1') then
            sw_reg <= sw(0);
45          db_reg <= db_level;
         end if;
      end process;
      btn_tick <= (not sw_reg) and sw(0);
      db_tick <= (not db_reg) and db_level;
50    -- ========================================================
      -- two counters
      -- ========================================================
      clr <= not key(0);
      process(clk)
55    begin
         if (clk'event and clk='1') then
            q1_reg <= q1_next;
            q0_reg <= q0_next;
         end if;
60    end process;
      -- next-state logic for the counter
      q1_next <= (others=>'0') when clr='1' else
                 q1_reg + 1 when btn_tick='1' else
                 q1_reg;
65    q0_next <= (others=>'0') when clr='1' else
                 q0_reg + 1 when db_tick='1' else
                 q0_reg;
      --output
      b_count <= std_logic_vector(q1_reg);
70    d_count <= std_logic_vector(q0_reg);
   end arch;
```

The seven-segment display shows the accumulated numbers of 0-to-1 edges of bounced and debounced switch input. After moving the slide switch several times, we can determine the average number of bounces for each transition.

6.4   BIBLIOGRAPHIC NOTES

The bibliographic information for this chapter is similar to that for Chapter 4.

6.5   SUGGESTED EXPERIMENTS

6.5.1   Dual-edge detector

A dual-edge detector is similar to a rising-edge detector except that the output is
asserted for one clock cycle when the input changes from 0 to 1 (i.e., rising edge)
and 1 to 0 (i.e., falling edge).

1. Design the circuit based on the Moore machine and draw the state diagram
   and ASM chart.
2. Derive the HDL code based on the state diagram of the ASM chart.
3. Derive a testbench and use simulation to verify operation of the code.
4. Replace the rising detectors in Section 6.3.3 with dual-edge detectors and
   verify their operations.
5. Repeat steps 1 to 4 for a Mealy machine–based design.

6.5.2   Alternative debouncing circuit

One problem with the debouncing design in Section 6.3.2 is the delayed response
of the onset of a switch transition. An alternative is to react to the first edge in
the transition and then wait for a small amount of time (at least 20 ms) to have
the input signal settled. The output timing diagram is shown at the bottom of
Figure 6.8. When the input changes from 0 to 1, the FSM responds immediately.
The FSM then ignores the input for about 20 ms to avoid glitches. After this
amount of time, the FSM starts to check the input for the falling edge. Follow the
design procedure in Section 6.3.2 to design the alternative circuit.

1. Derive the state diagram and ASM chart for the circuit.
2. Derive the HDL code.
3. Derive the HDL code based on the state diagram and ASM chart.
4. Derive a testbench and use simulation to verify operation of the code.
5. Replace the debouncing circuit in Section 6.3.3 with the alternative design
   and verify its operation.

6.5.3   Parking lot occupancy counter

Consider a parking lot with a single entry and exit gate. Two pairs of photo sensors
are used to monitor the activity of cars, as shown in Figure 6.11. When an object
is between the photo transmitter and the photo receiver, the light is blocked and
the corresponding output is asserted to 1. By monitoring the events of two sensors,
we can determine whether a car is entering or exiting or whether a pedestrian is
passing through. For example, the following sequence indicates that a car enters
the lot:

- Initially, both sensors are unblocked (i.e., the a and b signals are 00).
- Sensor a is blocked (i.e., the a and b signals are 10).
- Both sensors are blocked (i.e., the a and b signals are 11).

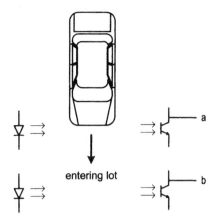

Figure 6.11   Conceptual diagram of gate sensors.

- Sensor a is unblocked (i.e., the a and b signals are 01).
- Both sensors becomes unblocked (i.e., the a and b signals are 00).

Design a parking lot occupancy counter as follows:

1. Design an FSM with two input signals, a and b, and two output signals, car_enter and car_exit. The car_enter and car_exit signals assert one clock cycle when a car enters and one clock cycle when a car exits the lot, respectively.
2. Derive the HDL code for the FSM.
3. Design a counter with two control signals, inc and dec, which increment and decrement the counter when asserted. Derive the HDL code.
4. Combine the counter and the FSM and seven-segment LED decoding circuits. Use two debounced pushbuttons to mimic operation of the two sensor outputs. Verify operation of the occupancy counter.

CHAPTER 7

FSMD

An FSMD (finite state machine with data path) combines an FSM and regular sequential circuits. The FSMD can be used to implement systems described by *RT (register transfer) operation*, which is a methodology to realize a software algorithm in hardware. In this chapter, we provide an overview of the RT operation and extended ASM chart, discuss the derivation of HDL codes, and use several examples to illustrate the development.

7.1 INTRODUCTION

An FSMD (finite state machine with data path) combines an FSM and regular sequential circuits. The FSM, which is sometimes known as a *control path*, examines the external commands and status and generates control signals to specify operation of the regular sequential circuits, which are known collectively as a *data path*. Algorithms described in *RT (register transfer) operation*, in which the operations are specified as data manipulation and transfer among a collection of registers, can be converted to FSMD and realized in hardware.

7.1.1 Single RT operation

An RT operation specifies data manipulation and transfer for a single destination register. It is represented by the notation

$$r_{dest} \leftarrow f(r_{src1}, r_{src2}, \dots, r_{srcn})$$

where r_{dest} is the destination register, r_{src1}, r_{src2}, and r_{srcn} are the source registers, and $f(\cdot)$ specifies the operation to be performed. The notation indicates that the contents of the source registers are fed to the $f(\cdot)$ function, which is realized by a combinational circuit, and the result is passed to the input of the destination register and stored in the destination register at the next rising edge of the clock. Following are several representative RT operations:

- r1 ← 0. A constant 0 is stored in the r1 register.
- r1 ← r1. The content of the r1 register is written back to itself.
- r2 ← r2 >> 3. The r2 register is shifted right three positions and then written back to itself.
- r2 ← r1. The content of the r1 register is transferred to the r2 register.
- i ← i + 1. The content of the i register is incremented by 1 and the result is written back to itself.
- d ← s1 + s2 + s3. The summation of the s1, s2, and s3 registers is written to the d register.
- y ← a*a. The a squared is written to the y register.

A single RT operation can be implemented by constructing a combinational circuit for the $f(\cdot)$ function and connecting the input and output of the registers. For example, consider the a ← a-b+1 operation. The $f(\cdot)$ function involves a subtractor and an incrementor. The block diagram is shown in Figure 7.1(a). For clarity, we use the _reg and _next suffixes to represent the input and output of a register. Note that an RT operation is synchronized by an embedded clock. The result from the $f(\cdot)$ function is not stored to the destination register until the next rising edge of the clock. The timing diagram of the previous RT operation is shown in Figure 7.1(b).

7.1.2 ASMD chart

A circuit based on the RT methodology specifies which RT operations should be executed in each step. Since an RT operation is done on a clock-by-clock basis, its timing is similar to a state transition of an FSM. Thus, an FSM is a natural choice to specify the sequencing of an RT algorithm. We extend the ASM chart to incorporate RT operations and call it an *ASMD* (ASM with data path) chart. The RT operations are treated as another type of activity and can be placed where the output signals are used.

A segment of an ASMD chart is shown in Figure 7.2(a). It contains one destination register, r1, which is initialized with 8, added with content of the r2 register, and then shifted left by two positions. Note that the r1 register must be specified in each state. When r1 is not changed, the r1 ← r1 operation should be used to maintain its current content, as in the s3 state. In the future discussion, we assume that r ← r is the default RT operation for the r register and do not include it in the ASMD chart. Implementing the RT operations of an ASMD chart involves a multiplexing circuit to route the desired next value to the destination register.

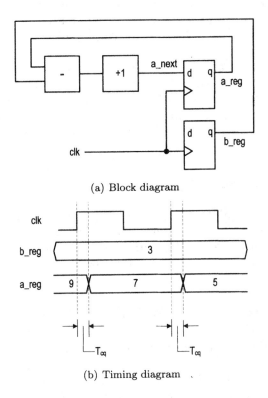

(a) Block diagram

(b) Timing diagram

Figure 7.1   Block and timing diagrams of an RT operation.

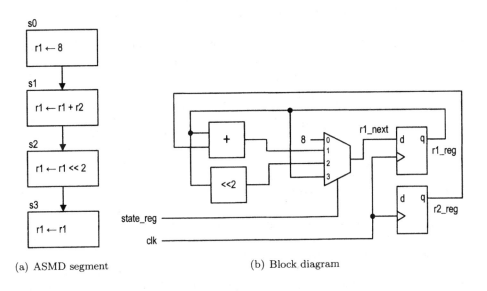

(a) ASMD segment

(b) Block diagram

Figure 7.2   Realization of an ASMD segment.

For example, the previous segment can be implemented by a 4-to-1 multiplexer, as shown in Figure 7.2(b). The current state (i.e., the output of the state register) of the FSM controls the selection signal of the multiplexer and thus chooses the result of the desired RT operation.

An RT operation can also be specified in a conditional output box, as the r2 register shown in Figure 7.3(a). Depending on the a>b condition, the FSMD performs either r2 ← r2+a or r2 ← r2+b. Note that all operations are done in parallel inside an ASMD block. We need to realize the a>b, r2+a, and r2+b operations and use a multiplexer to route the desired value to r2. The block diagram is shown in Figure 7.3(b).

7.1.3 Decision box with a register

The appearance of an ASMD chart is similar to that of a normal flowchart. The main difference is that the RT operation in an ASMD chart is controlled by an embedded clock signal and the destination register is updated *when the FSMD exits the current ASMD block*, but not within the block. The r ← r-1 operation actually means that:

- r_next <= r_reg - 1;
- r_reg <= r_next at the rising edge of the clock (i.e., when the FSMD exits the current block).

This "delayed store" may introduce subtle errors when a register is used in a decision box. Consider the FSMD segment in Figure 7.4(a). The r register is decremented in the state box and used in the decision box. Since the r register is not updated until the FSMD exits the block, the old content of r is used for comparison in the decision box. If the new value of r is desired, we should use the output of the combinational logic (i.e., r_next) in the decision box (i.e., replace the r=0 expression with r_next=0), as shown in Figure 7.4(b). Note that we use the := notation, as in r_next:=r-1, to indicate the immediate assignment of r_next.

Block diagram of an FSMD The conceptual block diagram of an FSMD is divided into a data path and a control path, as shown in Figure 7.5. The data path performs the required RT operations. It consists of:

- *Data registers*: store the intermediate computation results
- *Functional units*: perform the functions specified by the RT operations
- *Routing network*: routes data between the storage registers and the functional units

The data path follows the `control` signal to perform the desired RT operations and generates the `internal status` signal.

The control path is an FSM. As a regular FSM, it contains a state register, next-state logic, and output logic. It uses the external `command` signal and the data path's `status` signal as the input and generates the `control` signal to control the data path operation. The FSM also generates the `external status` signal to indicate the status of the FSMD operation.

Note that although an FSMD consists of two types of sequential circuits, both circuits are controlled by the same clock, and thus the FSMD is still a synchronous system.

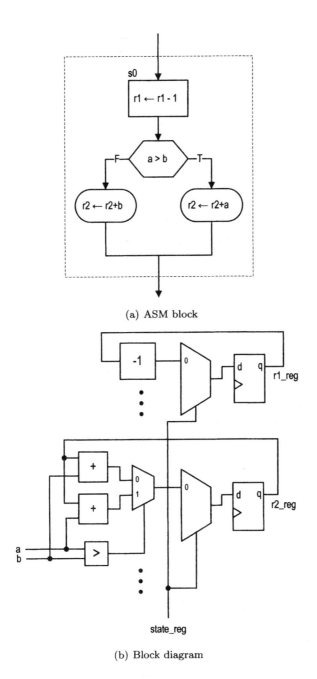

(a) ASM block

(b) Block diagram

Figure 7.3   Realization of an RT operation in a conditional output box.

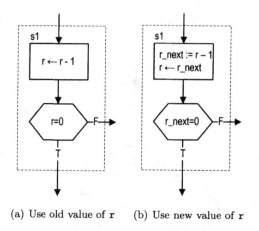

(a) Use old value of **r**   (b) Use new value of **r**

Figure 7.4   ASM block affected by a delayed store.

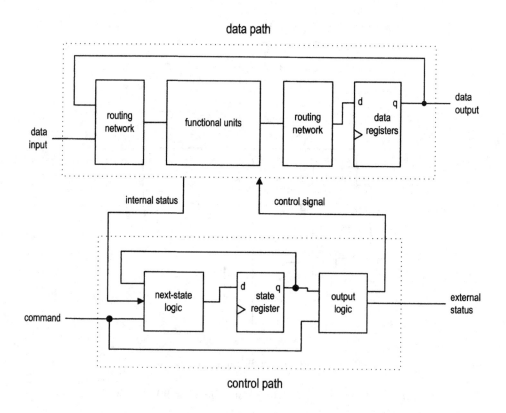

Figure 7.5   Block diagram of an FSMD.

7.2   CODE DEVELOPMENT OF AN FSMD

We use an improved debouncing circuit to demonstrate derivation of the FSMD code. Although the debouncing circuit in Section 6.3.2 uses an FSM and a timer (which is a regular sequential circuit), it is not based on the RT methodology because the two units are running independently and the FSM has no control over the timer. Since the 10-ms enable tick can be asserted at any time, the FSM does not know how much time has elapsed when the first tick is detected in the `wait1_1` or `wait0_1` state. Thus, the waiting period in this design is between 20 and 30 ms but is not an exact interval. This deficiency can be overcome by applying the RT methodology. In this section, we use this improved debouncing circuit to illustrate FSMD code development.

7.2.1   Debouncing circuit based on RT methodology

With the RT methodology, we can use an FSM to control the initiation of the timer to obtain the exact interval. The ASMD chart is shown in Figure 7.6. The circuit is expanded to include two output signals: `db_level`, which is the debounced output, and `db_tick`, which is a one-clock-cycle enable pulse asserted at the zero-to-one transition. The `zero` and `one` states mean that the `sw` input has been stabilized for '0' and '1', respectively. The `wait1` and `wait0` states are used to filter out short glitches. The `sw` signal must be stable for a certain amount of time or the transition will be treated as a glitch. The data path contains one register, q, which is 21 bits wide. Assume that the FSMD is originally in the `zero` state. When the `sw` input signal becomes '1', the FSMD moves to the `wait1` state and initializes q to "1 ⋯ 1". In the `wait1` state, the q decrements in each clock cycle. If `sw` remains as '1', the FSMD returns to this state repeatedly until q reaches "0 ⋯ 0" and then moves to the `one` state.

Recall that the 50-MHz (i.e., 20-ns period) system clock is used on the prototyping board. Since the FSMD stays in the `wait1` state for 2^{21} clock cycles, it is about 40 ms (i.e., $2^{21} * 20$ ns). We can modify the initial value of the q register to obtain the desired wait interval.

There are two ways to derive the HDL code, one with an *explicit description* of the data path components and the other with an *implicit description* of the data path components.

7.2.2   Code with explicit data path components

The first approach to FSMD code development is to separate the control FSM and the key data path components. From an ASMD chart, we first identify the key components in the data path and the associated control signals and then describe these components in individual code segments.

The key data path component of the debouncing circuit ASMD chart is a custom 21-bit decrement counter that can:

- Be initialized with a specific value
- Count downward or pause
- Assert a status signal when the counter reaches 0

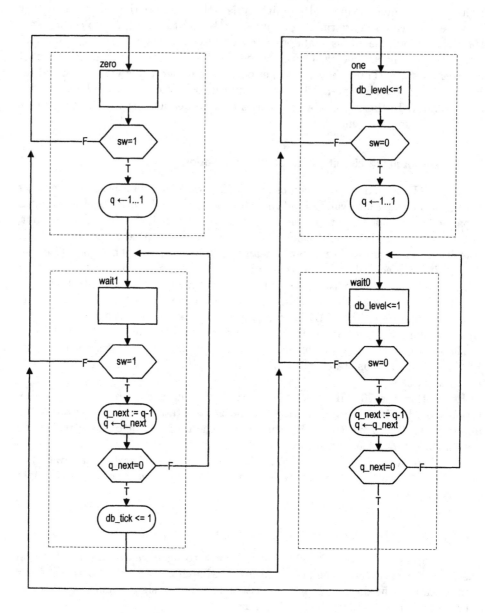

Figure 7.6   ASMD chart of a debouncing circuit.

We can create a binary counter with a `q_load` signal to load the initial value and a `q_dec` signal to enable the counting. The counter also generates a `q_zero` status signal, which is asserted when the counter reaches zero. The complete data path is composed of the q register and the next-state logic of the custom decrement counter. A comparison circuit is included to generate the `q_zero` status signal. The control path consists of an FSM, which takes the `sw` input and the `q_zero` status and asserts the control signals, `q_load` and `q_dec`, according to the desired action in the ASMD chart. The HDL code follows the data path specification and the ASMD chart, and is shown in Listing 7.1.

Listing 7.1 Debouncing circuit with an explicit data path component

```
library ieee;
use ieee.std_logic_1164.all;
use ieee.numeric_std.all;
entity debounce is
    port(
        clk, reset: in std_logic;
        sw: in std_logic;
        db_level, db_tick: out std_logic
    );
end debounce ;

architecture exp_fsmd_arch of debounce is
    constant N: integer:=21;  -- filter of 2^N * 20ns = 40ms
    type state_type is (zero, wait0, one, wait1);
    signal state_reg, state_next: state_type;
    signal q_reg, q_next: unsigned(N-1 downto 0);
    signal q_load, q_dec, q_zero: std_logic;
begin
    -- FSMD state & data registers
    process(clk,reset)
    begin
        if reset='1' then
            state_reg <= zero;
            q_reg <= (others=>'0');
        elsif (clk'event and clk='1') then
            state_reg <= state_next;
            q_reg <= q_next;
        end if;
    end process;

    -- FSMD data path (counter) next-state logic
    q_next <= (others=>'1') when q_load='1' else
              q_reg - 1 when q_dec='1' else
              q_reg;
    q_zero <= '1' when q_next=0 else '0';

    -- FSMD control path next-state logic
    process(state_reg,sw,q_zero)
    begin
        q_load <= '0';
        q_dec <= '0';
        db_tick <= '0';
        state_next <= state_reg;
        case state_reg is
            when zero =>
                db_level <= '0';
                if (sw='1') then
                    state_next <= wait1;
```

```
49                           q_load <= '1';
                        end if;
                  when wait1=>
                     db_level <= '0';
                     if (sw='1') then
54                      q_dec <= '1';
                        if (q_zero='1') then
                           state_next <= one;
                           db_tick <= '1';
                        end if;
59                   else — sw='0'
                        state_next <= zero;
                     end if;
                  when one =>
                     db_level <= '1';
64                   if (sw='0') then
                        state_next <= wait0;
                        q_load <= '1';
                     end if;
                  when wait0=>
69                   db_level <= '1';
                     if (sw='0') then
                        q_dec <= '1';
                        if (q_zero='1') then
                           state_next <= zero;
74                      end if;
                     else — sw='1'
                        state_next <= one;
                     end if;
               end case;
79       end process;
      end exp_fsmd_arch;
```

7.2.3 Code with implicit data path components

An alternative coding style is to embed the RT operations within the FSM control path. Instead of explicitly defining the data path components, we just list RT operations with the corresponding FSM state. The code of the debouncing circuit is shown in Listing 7.2.

Listing 7.2 Debouncing circuit with an implicit data path component

```
   architecture imp_fsmd_arch of debounce is
      constant N: integer:=21;   — filter of 2^N * 20ns = 40ms
      type state_type is (zero, wait0, one, wait1);
      signal state_reg, state_next: state_type;
5     signal q_reg, q_next: unsigned(N-1 downto 0);
   begin
      — FSMD state & data registers
      process(clk,reset)
      begin
10       if reset='1' then
            state_reg <= zero;
            q_reg <= (others=>'0');
         elsif (clk'event and clk='1') then
            state_reg <= state_next;
15          q_reg <= q_next;
         end if;
      end process;
```

```
     — next—state logic & data path functional units/routing
     process(state_reg,q_reg,sw,q_next)
20   begin
        state_next <= state_reg;
        q_next <= q_reg;
        db_tick <= '0';
        case state_reg is
25         when zero =>
              db_level <= '0';
              if (sw='1') then
                 state_next <= wait1;
                 q_next <= (others=>'1');
30            end if;
           when wait1=>
              db_level <= '0';
              if (sw='1') then
                 q_next <= q_reg - 1;
35               if (q_next=0) then
                    state_next <= one;
                    db_tick <= '1';
                 end if;
              else — sw='0'
40               state_next <= zero;
              end if;
           when one =>
              db_level <= '1';
              if (sw='0') then
45               state_next <= wait0;
                 q_next <= (others=>'1');
              end if;
           when wait0=>
              db_level <= '1';
50            if (sw='0') then
                 q_next <= q_reg - 1;
                 if (q_next=0) then
                    state_next <= zero;
                 end if;
55            else — sw='1'
                 state_next <= one;
              end if;
        end case;
     end process;
60 end imp_fsmd_arch;
```

The code consists of a memory segment and a combinational logic segment. The former contains the state register of the FSM and the data register of the data path. The latter basically specifies the next-state logic of the control path FSM. Instead of generating control signals, the next data register values are specified in individual states. The next-state logic of the data path, which consists of functional units and routing network, is created accordingly.

7.2.4 Comparison

Code with implicit data path components essentially follows the ASMD chart. We just convert the chart to an HDL description. Although this approach is simpler and more descriptive, we rely on synthesis software for data path construction and have less control. This can best be explained by an example. Consider the ASMD segment in Figure 7.7. The implicit description becomes

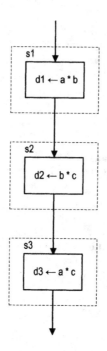

Figure 7.7   ASMD segment with sharing opportunity.

```
case
    when s1
        d1_next <= a * b;
        . . .
    when s2
        d2_next <= b * c;
        . . .
    when s3
        d3_next <= a * c;
        . . .
end case;
```

The synthesis software may infer three multipliers. Since a combinational multiplier is a complex circuit, it is more efficient to share the circuit. We can use an explicit description to isolate the multiplier:

```
case
    when s1
        in1 <= a;
        in2 <= b;
        d1_next <= m_out;
        . . .
    when s2
        in1 <= b;
        in2 <= c;
        d2_next <= m_out;
        . . .
    when s3
```

```
      in1 <= a;
      in2 <= c;
      d3_next <= m_out;
      . . .
end case;
-- explicit  description  of  a  single  multiplier
m_out <= in1 * in2;
```

The code ensures that only one multiplier is inferred during synthesis. The implicit and explicit descriptions can be mixed for a complex FSMD design. We frequently isolate and extract complex data path components for code clarity and efficiency.

7.3   DESIGN EXAMPLES

7.3.1   Fibonacci number circuit

The Fibonacci numbers constitute a sequence defined as

$$
fib(i) = \begin{cases} 0 & \text{if } i = 0 \\ 1 & \text{if } i = 1 \\ fib(i-1) + fib(i-2) & \text{if } i > 1 \end{cases}
$$

One way to calculate $fib(i)$ is to construct the function iteratively, from 0 to the desired i. This approach requires two temporary registers to store the two most recently calculated values [i.e., $fib(i-1)$ and $fib(i-2)$)] and one index register to keep track of the number of iterations. The ASMD chart is shown in Figure 7.8, in which t1 and t0 are temporary storage registers and n is the index register. In addition to the regular data input and output signals, i and f, we include a command signal, **start**, which signals the beginning of operation, and two status signals: **ready**, which indicates that the circuit is idle and ready to take new input, and **done_tick**, which is asserted for one clock cycle when the operation is completed. Since this circuit, like many other FSMD designs, is probably a part of a larger system, these signals are needed to interface with other subsystems.

The ASMD chart has three states. The **idle** state indicates that the circuit is currently idle. When **start** is asserted, the FSMD moves to the **op** state and loads initial values to three registers. The t0 and t1 registers are loaded with 0 and 1, which represent $fib(0)$ and $fib(1)$, respectively. The n register is loaded with i, the desired number of iterations.

The main computation is iterated through the **op** state by three RT operations:

* t1 ← t1 + t0
* t0 ← t1
* n ← n - 1

The first two RT operations obtain a new value and store the two most recently calculated values in t1 and t0. The third RT operation decrements the iteration index. The iteration ended when n reaches 1 or its initial value is 0 [i.e., $fib(0)$]. Unlike a regular flowchart, the operations in an ASMD block can be performed concurrently in the same clock cycle. We put all comparison and RT operations in the **op** state to reduce the computation time. Note that the new values of the t1 and t0 registers are loaded at the same time when the FSMD exits the **op** state

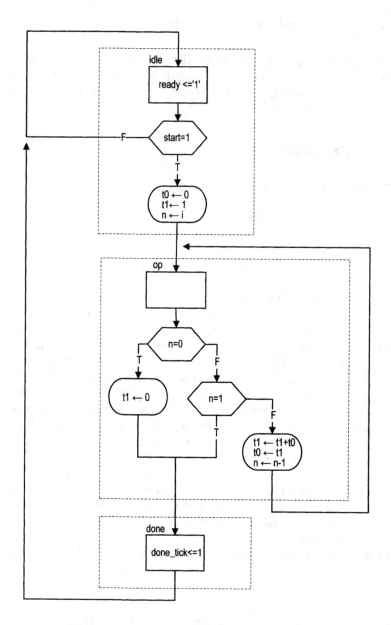

Figure 7.8   ASMD chart of a Fibonacci circuit.

(i.e., at the next rising edge of the clock). Thus, the original value of t1, not t1+t0, is stored to t0. The purpose of the **done** state is to generate the one-clock-cycle done_tick signal to indicate completion of the computation. This state can be omitted if this status signal is not needed.

The code follows the ASMD chart and is shown in Listing 7.3. Note that the Fibonacci function grows rapidly and the output signal should be wide enough to accommodate the desired result.

Listing 7.3   Fibonacci number circuit

```
1 library ieee;
  use ieee.std_logic_1164.all;
  use ieee.numeric_std.all;
  entity fib is
      port(
6         clk, reset: in std_logic;
          start: in std_logic;
          i: in std_logic_vector(4 downto 0);
          ready, done_tick: out std_logic;
          f: out std_logic_vector(19 downto 0)
11     );
  end fib;

  architecture arch of fib is
      type state_type is (idle,op,done);
16     signal state_reg, state_next: state_type;
      signal t0_reg, t0_next: unsigned(19 downto 0);
      signal t1_reg, t1_next: unsigned(19 downto 0);
      signal n_reg, n_next: unsigned(4 downto 0);
  begin
21     -- fsmd state and data registers
      process(clk,reset)
      begin
          if reset='1' then
              state_reg <= idle;
26             t0_reg <= (others=>'0');
              t1_reg <= (others=>'0');
              n_reg <= (others=>'0');
          elsif (clk'event and clk='1') then
              state_reg <= state_next;
31             t0_reg <= t0_next;
              t1_reg <= t1_next;
              n_reg <= n_next;
          end if;
      end process;
36     -- fsmd next-state logic
      process(state_reg,n_reg,t0_reg,t1_reg,start,i,n_next)
      begin
          ready <='0';
          done_tick <= '0';
41         state_next <= state_reg;
          t0_next <= t0_reg;
          t1_next <= t1_reg;
          n_next <= n_reg;
          case state_reg is
46             when idle =>
                  ready <= '1';
                  if start='1' then
                      t0_next <= (others=>'0');
                      t1_next <= (0=>'1', others=>'0');
```

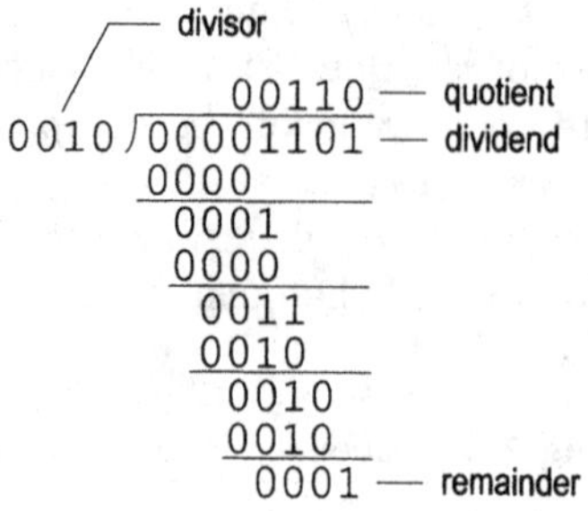

Figure 7.9   Long division of two 4-bit unsigned integers.

```
51                  n_next <= unsigned(i);
                    state_next <= op;
                end if;
            when op =>
                if n_reg=0 then
56                  t1_next <= (others=>'0');
                    state_next <= done;
                elsif n_reg=1 then
                    state_next <= done;
                else
61                  t1_next <= t1_reg + t0_reg;
                    t0_next <= t1_reg;
                    n_next <= n_reg - 1;
                end if;
            when done =>
66              done_tick <= '1';
                state_next <= idle;
        end case;
    end process;
    -- output
71  f <= std_logic_vector(t1_reg);
end arch;
```

7.3.2 Division circuit

Because of complexity, the division operator cannot be synthesized automatically. We use an FSMD to implement the long-division algorithm in this subsection. The algorithm is illustrated by the division of two 4-bit unsigned integers in Figure 7.9. The algorithm can be summarized as follows:

1. Double the dividend width by appending 0's in front and align the divisor to the leftmost bit of the extended dividend.
2. If the corresponding dividend bits are greater than or equal to the divisor, subtract the divisor from the dividend bits and make the corresponding quotient bit 1. Otherwise, keep the original dividend bits and make the quotient bit 0.
3. Append one additional dividend bit to the previous result and shift the divisor to the right one position.
4. Repeat steps 2 and 3 until all dividend bits are used.

The sketch of the data path is shown in Figure 7.10. Initially, the divisor is stored

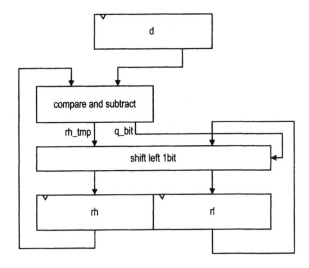

Figure 7.10   Sketch of division circuit's data path.

in the **d** register and the extended dividend is stored in the **rh** and **rl** registers. In each iteration, the **rh** and **rl** registers are shifted to the left by one position. This corresponds to shifting the divisor to the right of the previous algorithm. We can then compare **rh** and **d** and perform subtraction if **rh** is greater than or equal to **d**. When **rh** and **rl** are shifted to the left, the rightmost bit of **rl** becomes available. It can be used to store the current quotient bit. After we iterate through all dividend bits, the result of the last subtraction is stored in **rh** and becomes the remainder of the division, and all quotients are shifted into **rl**.

The ASMD chart of the division circuit is somewhat similar to that of the previous Fibonacci circuit. The FSMD consists of four states, **idle**, **op**, **last**, and **done**. To make the code clear, we extract the *compare and subtract* circuit to separate code segments. The main computation is performed in the **op** state, in which the dividend bits and divisor are compared and subtracted and then shifted left by 1 bit. Note that the remainder should not be shifted in the last iteration. We create a separate state, **last**, to accommodate this special requirement. As in the preceding example, the purpose of the **done** state is to generate a one-clock-cycle **done_tick** signal to indicate completion of the computation. The code is shown in Listing 7.4.

Listing 7.4   Division circuit

```
library ieee;
use ieee.std_logic_1164.all;
use ieee.numeric_std.all;
entity div is
    generic(
        W: integer:=8;
        CBIT: integer:=4    -- CBIT=log2 (W)+1
    );
    port(
        clk, reset: in std_logic;
```

```vhdl
         start: in std_logic;
         dvsr, dvnd: in std_logic_vector(W-1 downto 0);
13       ready, done_tick: out std_logic;
         quo, rmd: out std_logic_vector(W-1 downto 0)
      );
   end div;

18 architecture arch of div is
      type state_type is (idle,op,last,done);
      signal state_reg, state_next: state_type;
      signal rh_reg, rh_next: unsigned(W-1 downto 0);
      signal rl_reg, rl_next: std_logic_vector(W-1 downto 0);
23    signal rh_tmp: unsigned(W-1 downto 0);
      signal d_reg, d_next: unsigned(W-1 downto 0);
      signal n_reg, n_next: unsigned(CBIT-1 downto 0);
      signal q_bit: std_logic;
   begin
28    -- fsmd state and data registers
      process(clk,reset)
      begin
         if reset='1' then
            state_reg <= idle;
33          rh_reg <= (others=>'0');
            rl_reg <= (others=>'0');
            d_reg <= (others=>'0');
            n_reg <= (others=>'0');
         elsif (clk'event and clk='1') then
38          state_reg <= state_next;
            rh_reg <= rh_next;
            rl_reg <= rl_next;
            d_reg <= d_next;
            n_reg <= n_next;
43       end if;
      end process;

      -- fsmd next-state logic and data path logic
      process(state_reg,n_reg,rh_reg,rl_reg,d_reg,
48             start,dvsr,dvnd,q_bit,rh_tmp,n_next)
      begin
         ready <='0';
         done_tick <= '0';
         state_next <= state_reg;
53       rh_next <= rh_reg;
         rl_next <= rl_reg;
         d_next <= d_reg;
         n_next <= n_reg;
         case state_reg is
58          when idle =>
               ready <= '1';
               if start='1' then
                  rh_next <= (others=>'0');
                  rl_next <= dvnd;                  -- dividend
63                d_next <= unsigned(dvsr);         -- divisor
                  n_next <= to_unsigned(W+1, CBIT); -- index
                  state_next <= op;
               end if;
            when op =>
68             -- shift rh and rl left
               rl_next <= rl_reg(W-2 downto 0) & q_bit;
               rh_next <= rh_tmp(W-2 downto 0) & rl_reg(W-1);
               --decrease index
```

```
              n_next <= n_reg - 1;
73            if (n_next=1) then
                  state_next <= last;
              end if;
           when last =>   -- last iteration
              rl_next <= rl_reg(W-2 downto 0) & q_bit;
78            rh_next <= rh_tmp;
              state_next <= done;
           when done =>
              state_next <= idle;
              done_tick <= '1';
83      end case;
     end process;

     -- compare and subtract
     process(rh_reg, d_reg)
88   begin
        if rh_reg >= d_reg then
           rh_tmp <= rh_reg - d_reg;
           q_bit <= '1';
        else
93         rh_tmp <= rh_reg;
           q_bit <= '0';
        end if;
     end process;

98   -- output
     quo <= rl_reg;
     rmd <= std_logic_vector(rh_reg);
  end arch;
```

7.3.3 Binary-to-BCD conversion circuit

We discussed the BCD format in Section 5.6.1. In this format, a decimal number is represented as a sequence of 4-bit BCD digits. A binary-to-BCD conversion circuit converts a binary number to the BCD format. For example, the binary number "0010 0000 0000" becomes "0101 0001 0010" (i.e., 512_{10}) after conversion.

The binary-to-BCD conversion can be processed by a special BCD shift register, which is divided into 4-bit groups internally, each representing a BCD digit. Shifting a BCD sequence to the left requires adjustment if a BCD digit is greater than 9_{10} after shifting. For example, if a BCD sequence is "0001 0111" (i.e., 17_{10}), it should become "0011 0100" (i.e., 34_{10}) rather than "0010 1110". The adjustment requires subtracting 10_{10} (i.e., "1010") from the right BCD digit and adding 1 (which can be considered as a carry-out) to the next BCD digit. Note that subtracting 10_{10} is equivalent to adding 6_{10} for a 4-bit binary number. Thus, the foregoing adjustment can also be achieved by adding 6_{10} to the right BCD digit. The carry-out bit is generated automatically in this process.

In the actual implementation, it is more efficient to first perform the necessary adjustment on a BCD digit and then shift. We can check whether a BCD digit is greater than 4_{10} and, if this is the case, add 3_{10} to the digit. After all the BCD digits are corrected, we can then shift the entire register to the left by one position. A binary-to-BCD conversion circuit can be constructed by shifting the binary input to a BCD shift register bit by bit, from MSB to LSB. Its operation can be summarized as follows:

Table 7.1   Binary-to-BCD conversion example

Operation		Special BCD shift register			Binary input
		BCD digit 2	BCD digit 1	BCD digit 0	
Initial					111 1111
Bit 6	no adjustment shift left 1 bit			1 (1_{10})	11 1111
Bit 5	no adjustment shift left 1 bit			11 (3_{10})	1 1111
Bit 4	no adjustment shift left 1 bit			111 (7_{10})	1111
Bit 3	BCD digit 0 adjustment shift left 1 bit		1 (1_{10})	1010 0101 (5_{10})	111
Bit 2	BCD digit 0 adjustment shift left 1 bit		1 11 (3_{10})	1000 0001 (1_{10})	11
Bit 1	no adjustment shift left 1 bit		110 (6_{10})	0011 (3_{10})	1
Bit 0	BCD digit 1 adjustment shift left 1 bit	1 (1_{10})	1001 0010 (2_{10})	0011 0111 (7_{10})	

1. For each 4-bit BCD digit in a BCD shift register, check whether the digit is greater than 4. If this is the case, add 3_{10} to the digit.
2. Shift the entire BCD register left one position and shift in the MSB of the input binary sequence to the LSB of the BCD register.
3. Repeat steps 1 and 2 until all input bits are used.

The conversion process of a 7-bit binary input, "111 1111" (i.e., 127_{10}), is demonstrated in Table 7.1.

The code of a 13-bit conversion circuit is shown in Listing 7.5. It uses a simple FSMD to control the overall operation. When the **start** signal is asserted, the binary input is stored into the **p2s** register. The FSM then iterates through the 13 bits, similar to the process described in previous examples. Four adjustment circuits are used to correct the four BCD digits. For clarity, they are isolated from the next-state logic and described in a separate code segment.

Listing 7.5   Binary-to-BCD conversion circuit

```
library ieee;
use ieee.std_logic_1164.all;
use ieee.numeric_std.all;
entity bin2bcd is
    port(
```

```
            clk: in std_logic;
            reset: in std_logic;
            start: in std_logic;
 9          bin: in std_logic_vector(12 downto 0);
            ready, done_tick: out std_logic;
            bcd3,bcd2,bcd1,bcd0: out std_logic_vector(3 downto 0)
        );
    end bin2bcd ;
14
    architecture arch of bin2bcd is
        type state_type is (idle, op, done);
        signal state_reg, state_next: state_type;
        signal p2s_reg, p2s_next: std_logic_vector(12 downto 0);
19      signal n_reg, n_next: unsigned(3 downto 0);
        signal bcd3_reg, bcd2_reg, bcd1_reg, bcd0_reg:
              unsigned(3 downto 0);
        signal bcd3_next, bcd2_next, bcd1_next, bcd0_next:
              unsigned(3 downto 0);
24      signal bcd3_tmp, bcd2_tmp, bcd1_tmp, bcd0_tmp:
              unsigned(3 downto 0);
    begin
        -- state and data registers
        process (clk,reset)
29      begin
            if reset='1' then
                state_reg <= idle;
                p2s_reg <= (others=>'0');
                n_reg <= (others=>'0');
34              bcd3_reg <= (others=>'0');
                bcd2_reg <= (others=>'0');
                bcd1_reg <= (others=>'0');
                bcd0_reg <= (others=>'0');
            elsif (clk'event and clk='1') then
39              state_reg <= state_next;
                p2s_reg <= p2s_next;
                n_reg <= n_next;
                bcd3_reg <= bcd3_next;
                bcd2_reg <= bcd2_next;
44              bcd1_reg <= bcd1_next;
                bcd0_reg <= bcd0_next;
            end if;
        end process;

49      -- fsmd next-state logic / data path operations
        process(state_reg,start,p2s_reg,n_reg,n_next,bin,
                bcd0_reg,bcd1_reg,bcd2_reg,bcd3_reg,
                bcd0_tmp,bcd1_tmp,bcd2_tmp,bcd3_tmp)
        begin
54          state_next <= state_reg;
            ready <= '0';
            done_tick <= '0';
            p2s_next <= p2s_reg;
            bcd0_next <= bcd0_reg;
59          bcd1_next <= bcd1_reg;
            bcd2_next <= bcd2_reg;
            bcd3_next <= bcd3_reg;
            n_next <= n_reg;
            case state_reg is
64              when idle =>
                    ready <= '1';
                    if start='1' then
```

```
                        state_next <= op;
                        bcd3_next <= (others=>'0');
69                      bcd2_next <= (others=>'0');
                        bcd1_next <= (others=>'0');
                        bcd0_next <= (others=>'0');
                        n_next <="1101";  -- index
                        p2s_next <= bin;  -- input shift register
74                      state_next <= op;
                  end if;
               when op =>
                  -- shift in binary bit
                  p2s_next <= p2s_reg(11 downto 0) & '0';
79                -- shift 4 BCD digits
                  bcd0_next <= bcd0_tmp(2 downto 0) & p2s_reg(12);
                  bcd1_next <= bcd1_tmp(2 downto 0) & bcd0_tmp(3);
                  bcd2_next <= bcd2_tmp(2 downto 0) & bcd1_tmp(3);
                  bcd3_next <= bcd3_tmp(2 downto 0) & bcd2_tmp(3);
84                n_next <= n_reg - 1;
                  if (n_next=0) then
                      state_next <= done;
                  end if;
               when done =>
89                state_next <= idle;
                  done_tick <= '1';
         end case;
      end process;

94    -- data path function units
      -- four BCD adjustment circuits
      bcd0_tmp <= bcd0_reg + 3 when bcd0_reg > 4 else
                      bcd0_reg;
      bcd1_tmp <= bcd1_reg + 3 when bcd1_reg > 4 else
99                    bcd1_reg;
      bcd2_tmp <= bcd2_reg + 3 when bcd2_reg > 4 else
                      bcd2_reg;
      bcd3_tmp <= bcd3_reg + 3 when bcd3_reg > 4 else
                      bcd3_reg;
104
      -- output
      bcd0 <= std_logic_vector(bcd0_reg);
      bcd1 <= std_logic_vector(bcd1_reg);
      bcd2 <= std_logic_vector(bcd2_reg);
109   bcd3 <= std_logic_vector(bcd3_reg);
   end arch;
```

7.3.4 Period counter

A period counter measures the period of a periodic input waveform. One way to construct the circuit is to count the number of clock cycles between two rising edges of the input signal. Since the frequency of the system clock is known, the period of the input signal can be derived accordingly. For example, if the frequency of the system clock is f and the number of clock cycles between two rising edges is N, the period of the input signal is $N * \frac{1}{f}$.

The design in this subsection measures the period in milliseconds. Its ASMD chart is shown in Figure 7.11. The period counter takes a measurement when the start signal is asserted. We use a rising-edge detection circuit to generate a one-clock-cycle tick, edge, to indicate the rising edge of the input waveform. After

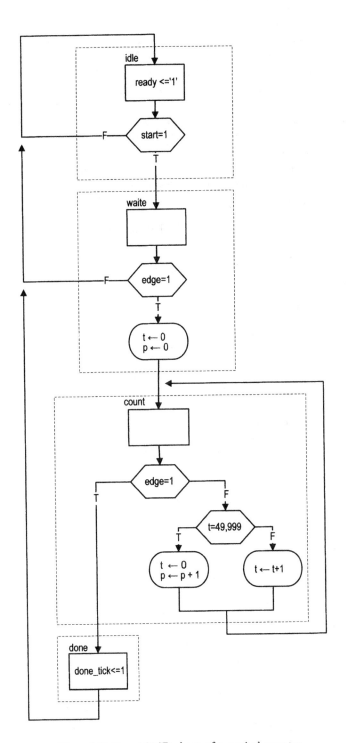

Figure 7.11   ASMD chart of a period counter.

start is asserted, the FSMD moves to the **waite** state to wait for the first rising edge of the input. It then moves to the **count** state when the next rising edge of the input is detected. In the **count** state, we use two registers to keep track of the time. The t register counts for 50,000 clock cycles, from 0 to 49,999, and then wraps around. Since the period of the system clock is 20 ns, the t register takes 1 ms to circulate through 50,000 cycles. The p register counts in terms of milliseconds. It is incremented once when the t register reaches 49,999. When the FSMD exits the **count** state, the period of the input waveform is stored in the p register and its unit is milliseconds. The FSMD asserts the **done_tick** signal in the **done** state, as in previous examples.

The code follows the ASMD chart and is shown in Listing 7.6. We use a constant, CLK_MS_COUNT, for the boundary of the millisecond counter. It can be replaced if a different measurement unit is desired.

Listing 7.6 Period counter

```
library ieee;
use ieee.std_logic_1164.all;
use ieee.numeric_std.all;
entity period_counter is
5    port(
         clk, reset: in std_logic;
         start, si: in std_logic;
         ready, done_tick: out std_logic;
         prd: out std_logic_vector(9 downto 0)
10   );
   end period_counter;

   architecture arch of period_counter is
      constant CLK_MS_COUNT: integer := 50000; -- 1 ms tick
15    type state_type is (idle, waite, count, done);
      signal state_reg, state_next: state_type;
      signal t_reg, t_next: unsigned(15 downto 0);
      signal p_reg, p_next: unsigned(9 downto 0);
      signal delay_reg: std_logic;
20    signal edge: std_logic;
   begin
      -- state and data register
      process(clk,reset)
      begin
25       if reset='1' then
            state_reg <= idle;
            t_reg <= (others=>'0');
            p_reg <= (others=>'0');
            delay_reg <= '0';
30       elsif (clk'event and clk='1') then
            state_reg <= state_next;
            t_reg <= t_next;
            p_reg <= p_next;
            delay_reg <= si;
35       end if;
      end process;

      -- edge detection circuit
      edge <= (not delay_reg) and si;
40
      -- fsmd next-state logic / data path operations
      process(start,edge,state_reg,t_reg,t_next,p_reg)
      begin
```

```
         ready <= '0';
45       done_tick <= '0';
         state_next <= state_reg;
         p_next <= p_reg;
         t_next <= t_reg;
         case state_reg is
50          when idle =>
                ready <= '1';
                if (start='1') then
                    state_next <= waite;
                end if;
55          when waite =>  -- wait for the first edge
                if (edge='1') then
                    state_next <= count;
                    t_next <= (others=>'0');
                    p_next <= (others=>'0');
60              end if;
            when count =>
                if (edge='1') then    -- 2nd edge arrived
                    state_next <= done;
                else  -- otherwise count
65                  if t_reg = CLK_MS_COUNT-1 then  -- 1ms tick
                        t_next <= (others=>'0');
                        p_next <= p_reg + 1;
                    else
                        t_next <= t_reg + 1;
70                  end if;
                end if;
            when done =>
                done_tick <= '1';
                state_next <= idle;
75       end case;
     end process;
     prd <= std_logic_vector(p_reg);
 end arch;
```

7.3.5 Accurate low-frequency counter

A frequency counter measures the frequency of a periodic input waveform. The common way to construct a frequency counter is to count the number of input pulses in a fixed amount of time, say, 1 second. Although this approach is fine for high-frequency input, it cannot measure a low-frequency signal accurately. For example, if the input is around 2 Hz, the measurement cannot tell whether it is 2.123 Hz or 2.567 Hz. Recall that the frequency is the reciprocal of the period (i.e., $frequency = \frac{1}{period}$). An alternative approach is to measure the period of the signal and then take the reciprocal to find the frequency. We use this approach to implement a low-frequency counter in this subsection.

This design example demonstrates how to use the previously designed parts to construct a large system. For simplicity, we assume that the frequency of the input is between 1 and 10 Hz (i.e., the period is between 100 and 1000 ms). The operation of this circuit includes three tasks:

1. Measure the period.
2. Find the frequency by performing a division operation.
3. Convert the binary number to BCD format.

We can use the period counter, division circuit, and binary-to-BCD converter to perform the three tasks and create another FSM as the master control to sequence and coordinate the operation of the three circuits. The block diagram is shown in Figure 7.12(a), and the ASM chart of the master control is shown in Figure 7.12(b). The FSM uses the start and done_tick signals of these circuits to initialize each task and to detect completion of the task. The code is shown in Listing 7.7.

Listing 7.7   Low-frequency counter

```
library ieee;
use ieee.std_logic_1164.all;
use ieee.numeric_std.all;
entity low_freq_counter is
    port(
        clk, reset: in std_logic;
        start: in std_logic;
        si: in std_logic;
        bcd3,bcd2,bcd1,bcd0: out std_logic_vector(3 downto 0)
    );
end low_freq_counter;

architecture arch of low_freq_counter is
    type state_type is (idle, count, frq, b2b);
    signal state_reg, state_next: state_type;
    signal prd: std_logic_vector(9 downto 0);
    signal dvsr, dvnd, quo: std_logic_vector(19 downto 0);
    signal prd_start, div_start, b2b_start: std_logic;
    signal prd_done_tick, div_done_tick, b2b_done_tick:
            std_logic;
begin
    --=====================================
    -- component instantiation
    --=====================================
    -- instantiate period counter
    prd_count_unit: entity work.period_counter
    port map(clk=>clk, reset=>reset, start=>prd_start, si=>si,
            ready=>open, done_tick=>prd_done_tick, prd=>prd);
    -- instantiate division circuit
    div_unit: entity work.div
    generic map(W=>20, CBIT=>5)
    port map(clk=>clk, reset=>reset, start=>div_start,
            dvsr=>dvsr, dvnd=>dvnd, quo=>quo, rmd=>open,
            ready=>open, done_tick=>div_done_tick);
    -- instantiate binary-to-BCD convertor
    bin2bcd_unit: entity work.bin2bcd
    port map
        (clk=>clk, reset=>reset, start=>b2b_start,
         bin=>quo(12 downto 0), ready=>open,
         done_tick=>b2b_done_tick,
         bcd3=>bcd3, bcd2=>bcd2, bcd1=>bcd1, bcd0=>bcd0);
    -- signal width extension
    dvnd <= std_logic_vector(to_unsigned(1000000, 20));
    dvsr <= "0000000000" & prd;

    --=====================================
    -- master FSM
    --=====================================
    process(clk,reset)
    begin
        if reset='1' then
            state_reg <= idle;
```

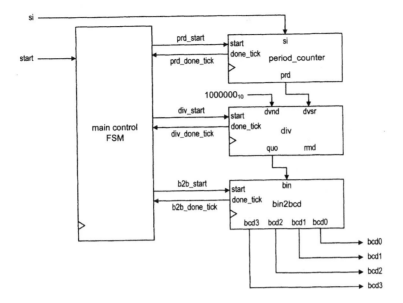

(a) Top-level block diagram

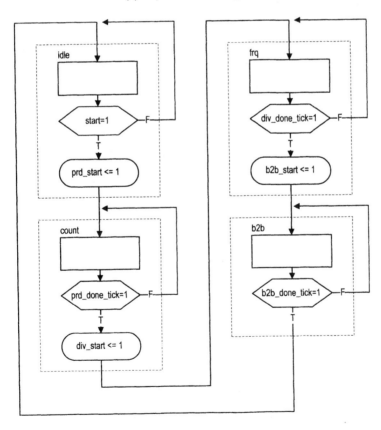

(b) ASM chart of main control

Figure 7.12 Accurate low-frequency counter.

```
              elsif (clk'event and clk='1') then
                  state_reg <= state_next;
              end if;
          end process;
57
          process(state_reg,start,
                  prd_done_tick,div_done_tick,b2b_done_tick)
          begin
              state_next <= state_reg;
62            prd_start <='0';
              div_start <='0';
              b2b_start <='0';
              case state_reg is
                  when idle =>
67                    if start='1' then
                          state_next <= count;
                          prd_start <='1';
                      end if;
                  when count =>
72                    if (prd_done_tick='1') then
                          div_start <='1';
                          state_next <= frq;
                      end if;
                  when frq =>
77                    if (div_done_tick='1') then
                          b2b_start <='1';
                          state_next <= b2b;
                      end if;
                  when b2b =>
82                    if (b2b_done_tick='1') then
                          state_next <= idle;
                      end if;
              end case;
          end process;
87 end arch;
```

7.4 BIBLIOGRAPHIC NOTES

The bibliographic information for this chapter is similar to that for Chapter 4.

7.5 SUGGESTED EXPERIMENTS

7.5.1 Alternative debouncing circuit

Consider the alternative debouncing circuit in Experiment 6.5.2. Redesign the circuit using the RT methodology:

1. Derive the ASMD chart for the circuit.
2. Derive the HDL code based on the ASMD chart.
3. Derive a testing circuit similar to that in Section 6.3.3 with the alternative debouncing circuit and verify its operation.

7.5.2   BCD-to-binary conversion circuit

A BCD-to-binary conversion converts a BCD number to the equivalent binary representation. Assume that the input is an 8-bit signal in BCD format (i.e., two BCD digits) and the output is a 7-bit signal in binary representation. Follow the procedure in Section 7.3.3 to design a BCD-to-binary conversion circuit:

1. Derive the conversion algorithm and ASMD chart.
2. Derive the HDL code based on the ASMD chart.
3. Derive a testbench and use simulation to verify operation of the code.
4. Synthesize the circuit, program the FPGA, and verify its operation.

7.5.3   Fibonacci circuit with BCD I/O: design approach 1

To make the Fibonacci circuit more user friendly, we can modify the circuit to use the BCD format for the input and output. Assume that the input is an 8-bit signal in BCD format (i.e., two BCD digits) and the output is displayed as four BCD digits on the seven-segment LED display. Furthermore, the LED will display "9999" if the resulting Fibonacci number is larger than 9999 (i.e., overflow). The operation can be done in three steps: convert input to the binary format, compute the Fibonacci number, and convert the result back to the BCD format.

The first design approach is to follow the procedure in Section 7.3.5. We first construct three smaller subsystems, which are the BCD-to-binary conversion circuit, Fibonacci circuit, and binary-to-BCD conversion circuit, and then use a master FSM to control the overall operation. Design the circuit as follows:

1. Implement the BCD-to-binary conversion circuit in Experiment 7.5.2.
2. Modify the Fibonacci number circuit in Section 7.3.1 to include an output signal to indicate the overflow condition.
3. Derive the top-level block diagram and the master control FSM state diagram.
4. Derive the HDL code.
5. Derive a testbench and use simulation to verify operation of the code.
6. Synthesize the circuit, program the FPGA, and verify its operation.

7.5.4   Fibonacci circuit with BCD I/O: design approach 2

An alternative to the previous "subsystem approach" in Experiment 7.5.3 is to integrate the three subsystems into a single system and derive a customized FSMD for this particular application. The approach eliminates the overhead of the control FSM and provides opportunities to share registers among the three tasks. Design the circuit as follows:

1. Redesign the circuit of Experiment 7.5.3 using one FSMD. The design should eliminate all unnecessary circuits and states, such as the various done_tick signals and the done states, and exploit the opportunity to share and reuse the registers in different steps.
2. Derive the ASMD chart.
3. Derive the HDL code based on the ASMD chart.
4. Derive a testbench and use simulation to verify operation of the code.
5. Synthesize the circuit, program the FPGA, and verify its operation.
6. Check the synthesis report and compare the number of LEs used in the two approaches.

7. Calculate the number of clock cycles required to complete the operation in the two approaches.

7.5.5 Auto-scaled low-frequency counter

The operation of the low-frequency counter in Section 7.3.5 is very restricted. The frequency range of the input signal is limited between 1 and 10 Hz. It loses accuracy when the frequency is beyond this range. Recall that the accuracy of this frequency counter depends on the accuracy of the period counter of Section 7.3.5, which counts in terms of millisecond ticks. We can modify the t counter to generate a microsecond tick (i.e., counting from 0 to 49) and increase the accuracy 1000-fold. This allows the range of the frequency counter to increase to 9999 Hz and still maintain at least four-digit accuracy.

Using a microsecond tick introduces more than four accuracy digits for low-frequency input, and the number must be shifted and truncated to be displayed on the seven-segment LED. An auto-scaled low-frequency counter performs the adjustment automatically, displays the four most significant digits, and places a decimal point in the proper place. For example, according to their range, the frequency measurements will be shown as "1.234", "12.34", "123.4", or "1234.". Since the decimal point LED of the seven-segment LED display is not used on the DE1 board, use four discrete LEDs to mimic the location of the decimal point of previous measurements.

The auto-scaled low-frequency counter needs an additional BCD adjustment circuit. It first checks whether the most significant BCD digit (i.e., the four MSBs) of a BCD sequence is zero. If this is the case, the circuit shifts the BCD sequence to the left by four positions and increments the decimal point counter. The operation is repeated until the most significant BCD digit is not "0000".

The complete auto-scaled low-frequency counter can be implemented as follows:

1. Modify the period counter to use the microsecond tick.
2. Extend the size of the binary-to-BCD conversion circuit.
3. Derive the ASMD chart for the BCD adjustment circuit and the HDL code.
4. Modify the control FSM to include the BCD adjustment in the last step.
5. Design a simple decoding circuit that uses the decimal point counter's output to activate the desired decimal point of the seven-segment LED display.
6. Derive a testbench and use simulation to verify operation of the code.
7. Synthesize the circuit, program the FPGA, and verify its operation.

7.5.6 Reaction timer

Eye–hand coordination is the ability of the eyes and hands to work together to perform a task. A reaction timer circuit measures how fast a human hand can respond after a person sees a visual stimulus. This circuit operates as follows:

1. The circuit has three input pushbuttons, corresponding to the clear, start, and stop signals. It uses a single discrete LED as the visual stimulus and displays relevant information on the seven-segment LED display.
2. A user pushes the clear button to force the circuit returning to the initial state, in which the seven-segment LED shows a welcome message, "HI", and the stimulus LED is off.

3. When ready, the user pushes the **start** button to initiate the test. The seven-segment LED goes off.

4. After a random interval between 2 and 15 seconds, the stimulus LED goes on and the timer starts to count upward. The timer increases every millisecond and its value is displayed in the format of "0000" millisecond on the seven-segment LED.

5. After the stimulus LED goes on, the user should try to push the **stop** button as soon as possible. The timer pauses counting once the **stop** button is asserted. The seven-segment LED shows the reaction time. It should be around 150 to 300 milliseconds for most people.

6. If the **stop** button is not pushed, the timer stops after 1 second and displays "1000".

7. If the **stop** button is pushed before the stimulus LED goes on, the circuit displays "9999" on the seven-segment LED and stops.

Design the circuit as follows:

1. Derive the ASMD chart.
2. Derive the HDL code based on the ASMD chart.
3. Synthesize the circuit, program the FPGA, and verify its operation.

7.5.7   Babbage difference engine emulation circuit

The Babbage difference engine is a mechanical digital computation device designed to tabulate a polynomial function. It was proposed by Charles Babbage, an English mathematician, in the nineteenth century. The engine is based on Newton's method of differences and avoids the need of multiplication. For example, consider a second-order polynomial $f(n) = 2n^2 + 3n + 5$. We can find the difference between $f(n)$ and $f(n-1)$:

$$f(n) - f(n-1) = 4n + 1$$

Assume that n is an integer and $n \geq 0$. The $f(n)$ can be defined recursively as

$$f(n) = \begin{cases} 5 & \text{if } n = 0 \\ f(n-1) + 4n + 1 & \text{if } n > 0 \end{cases}$$

This process can be repeated for the $4n + 1$ expression. Let $g(n) = 4n + 1$. We can find the difference between $g(n)$ and $g(n-1)$:

$$g(n) - g(n-1) = 4$$

The $g(n)$ can be defined recursively as

$$g(n) = \begin{cases} 5 & \text{if } n = 1 \\ g(n-1) + 4 & \text{if } n > 1 \end{cases}$$

and $f(n)$ can be rewritten as

$$f(n) = \begin{cases} 5 & \text{if } n = 0 \\ f(n-1) + g(n) & \text{if } n > 0 \end{cases}$$

Note that only additions are involved in the recursive definitions of $f(n)$ and $g(n)$.

Based on the definition of the last two recursive equations, we can derive an algorithm to compute $f(n)$. Two temporary registers are needed to keep track of the most recently calculated $f(n)$ and $g(n)$, and two additions are needed to update $f(n)$ and $g(n)$. Assume that n is a 6-bit input and interpreted as an unsigned integer. Design this circuit using the RT methodology:

1. Derive the ASMD chart.
2. Derive the HDL code based on the ASMD chart.
3. Derive a testbench and use simulation to verify operation of the code.
4. Synthesize the circuit, program the FPGA, and verify its operation.
5. Let $h(n) = n^3 + 2n^2 + 2n + 1$. Use the method above to find the recursive representation of $h(n)$ (note that three levels of recursive equations are needed for a three-order polynomial). Repeat steps 1 to 4.

PART II

BASIC NIOS II SOFTWARE DEVELOPMENT

CHAPTER 8

NIOS II PROCESSOR OVERVIEW

Nios II is Altera's proprietary processor targeted for its FPGA devices. It is configurable and can be trimmed to meet specific needs. In this chapter, we examine its basic organization and key components. The emphasis is on the features that may affect future software and I/O peripheral development.

8.1 INTRODUCTION

Nios II is a soft-core processor targeted for Altera's FPGA devices. As opposed to a fixed prefabricated processor, a soft-core processor is described by HDL codes and then mapped onto FPGA's generic logic cells. This approach offers more flexibility. A soft-core processor can be configured and tuned by adding or removing features on a system-by-system basis to meet performance or cost goals.

The Nios II processor follows the basic design principles of a RISC (Reduced Instruction Set Computer) architecture and uses a small, optimized set of instructions. Its main characteristics are:

- Load-store architecture
- Fixed 32-bit instruction format
- 32-bit internal data path
- 32-bit address space
- Memory-mapped I/O space
- 32-level interrupt requests

Embedded SOPC Design with Nios II Processor and VHDL Examples. By Pong P. Chu **181**
Copyright © 2011 John Wiley & Sons, Inc.

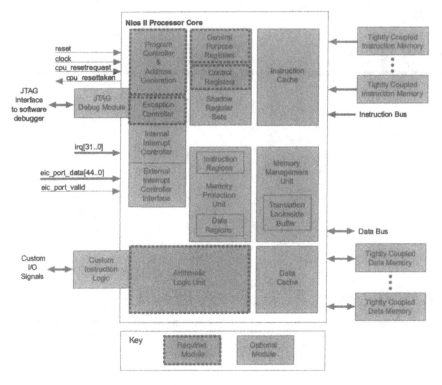

Figure 8.1   Block diagram of a Nios II processor (Courtesy of Altera Corp.).

- 32 general-purpose registers

The conceptual block diagram of a Nios II processor is shown in Figure 8.1. The main blocks are:

- Register file (general purpose registers) and control registers
- ALU (arithmetic and logic unit)
- Exception and interrupt handler
- Optional instruction cache and data cache
- Optional MMU (memory management unit)
- Optional MPU (memory protection unit)
- Optional JTAG debug module

There are three basic versions of Nios II:

- Nios II/f: The fast core is designed for optimal performance. It has a 6-stage pipeline, instruction cache, data cache, and dynamic branch prediction.
- Nios II/s: The standard core is designed for small size while maintaining good performance. It has a 5-stage pipeline, instruction cache, and static branch prediction.
- Nios II/e: The economy core is designed for optimal size. It is not pipelined and contains no cache.

These processors' key characteristics are summarized on the top of Table 8.1 and their sizes and performances (which are based on the Cyclone II family) are listed on the bottom.

Table 8.1   Comparison of Nios II versions

	Nios II/e	Nios II/s	Nios II/f
Processor pipeline	1 stage	5 stages	6 stages
Branch prediction	-	static	dynamic
Multiplication	software	3-cycle multiplier	1-cycle multiplier
Shift	software	3-cycle barrel shifter	1-cycle barrel shifter
Instruction cache	-	0.5 KB to 64 KB	0.5 KB to 64 KB
Data cache	-	-	0.5 KB to 64 KB
MMU/MPU	-	-	optional
Circuit size	540 LEs	1030 LEs	1600 LEs
Max clock rate	195 MHz	110 MHz	140 MHz
Performance	18 MIPS	55 MIPS	145 MIPS

Within each version, the processor can be further configured by including or excluding certain features (such as the JTAG debugging unit) and adjusting the size and performance of certain components (such as cache size). While the performance and size are different, the three versions share the same instruction set. Thus, from the software programmer's point of view, the three versions appear to be identical and the software dose not needed to be modified for a particular core.

Although the Nios II processor is described by HDL codes, the file is encrypted and a user cannot modify its internal organization via the codes. It should be treated as a black box that executes the specified instructions. The main blocks of the processor are examined briefly in following sections. The emphasis is on their impacts on applications rather than their internal implementation.

8.2   REGISTER FILE AND ALU

8.2.1   Register file

A Nios II processor consists of thirty two 32-bit general-purpose registers. The register 0 is hardwired and always returns the value zero and the register 31 is used to hold the return address during a procedure call. The other registers are treated identical by the processor but may be assigned for special meaning by an assembler or compiler. The processor also has several control registers, which report the status and specify certain processor behaviors. Since we use C language for software development in this book, these registers are not directly referenced in codes.

8.2.2   ALU

ALU operates on data stored in general-purpose registers. An ALU operation takes one or two inputs from registers and stores a result back to a register. The relevant instructions are:

- Arithmetic operations: addition, subtraction, multiplication, and division.
- Logical operations: and, or, nor, and xor.
- Shift operations: logic shift right and left, arithmetic shift right and left, and rotate right and left.

Ideally, the ALU should support all these operation. However, the implementation of multiplication, division, and variable-bit shifting operation is quite complex and requires more hardware resources. A Nios II processor can be configured to include or exclude these units. An instruction without hardware support is known as an "unimplemented instruction" in Altera literature. When an unimplemented instruction is issued, the processor generates an exception, which in turn initiates an exception handling routine to emulate the operation in software.

8.3   MEMORY AND I/O ORGANIZATION

8.3.1   Nios II memory interface

A Nios II processor utilizes separate ports for instruction and data access. The instruction master port fetches the instructions and performs only read operation. The data master port reads data from memory or a peripheral in a load instruction and writes data to memory or a peripheral in a store instruction. The two master ports can use two separate memory modules or share one memory module.

8.3.2   Overview of memory hierarchy

In an ideal scenario, a system should have a large, fast, and uniform memory, in which data and instruction can be accessed at the speed of the processor. In reality, this is hardly possible. Fast memory, such as the embedded memory modules within an FPGA device, is usually small and expensive. On the other hand, large memory, such as the external SDRAM (synchronous dynamic RAM) chip, is usually slow. One way to overcome the problem is to organize the storage as a hierarchy and put the small, fast memory component closer to the processor. Because the program execution tends to access a small part of the memory space for a period of time (known as *locality of memory reference*), we can put this portion in a small fast storage. A typical memory hierarchy contains cache, main memory, and a hard disk. A memory management technique known as *virtual memory* is used to make the hard disk appear as part of the memory space.

A Nios II processor supports both cache and virtual memory and can also provide memory protection and tightly coupled memory. The memory and I/O organization of a fully featured configuration is shown in Figure 8.2.

8.3.3   Virtual memory

Virtual memory gives an application program the impression that the computer system has a large contiguous working memory space, while in fact the actual

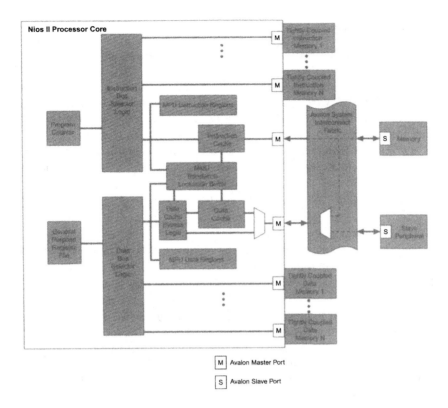

Figure 8.2   Nios II memory and I/O organization (Courtesy of Altera Corp.).

physical memory is only a fraction of the actual size and some data is stored in an external hard disk. Implementing a virtual memory system requires a mechanism to translate a virtual address to a physical address. The task is usually done jointly by an operating system and special hardware. In a Nios II/f processor, an optional MMU (memory management unit) can be included for this purpose.

Utilizing MMU requires an operation system that supports the virtual memory. Hardware alone will not serve any useful purpose.

8.3.4   Memory protection

Modern operation systems include protection mechanisms to restrict user applications to access critical system resource. For example, some operation systems divide the program execution into kernel mode (without restriction) and user mode (with restriction). Implementation of this scheme also requires special hardware support. In a Nios II/f processor, an optional MPU (memory protection unit) can be included for this purpose. As in MMU, a proper operation system is needed to utilize the MPU feature.

In a Nios II configuration, the use of MMU and MPU is mutually exclusive, which means that only one of them can be included.

8.3.5   Cache memory

Cache memory is a small, fast memory between the processor and main memory, as shown in Figure 8.2. In a Nios II processor, the cache is implemented by FPGA's internal embedded memory modules and the main memory is usually composed of external SDRAM devices. As long as most memory accesses are within cached locations, the average access time will be closer to the cache latency than to the main memory latency.

The operation of cache memory can be explained by a simple example. Consider the execution of a loop segment in a large program, which resides on the main memory. The steps are:

- At the beginning, code and data are loaded from the main memory to the cache.
- The loop segment is executed.
- When the execution is completed, the modified data is transferred back from the cache to the main memory.

In this process, the access time at the beginning and end is similar to the main memory latency and the access time for loop execution is similar to the cache latency. Since a typical loop segment iterates through the body many times, the average access time of this segment is closer to the cache latency.

A Nios II processor can be configured to include an instruction cache or both instruction and data cache. The sizes of the caches can be adjusted as well. Unlike MMU and MPU, no special operating system feature is needed to utilize the cache. Cache simply speeds up the average memory access time and is almost transparent to software.

8.3.6   Tightly coupled memory

Tightly coupled memory is somewhat unique to the embedded system. It is a small, fast memory that provides guaranteed low-latency memory access for timing-critical applications. One problem with a cache memory system is that its access time may vary. While its average access time is improved significantly, the worst-case access time can be really large (for example, the data is in SDRAM). Many tasks in an embedded system are time-critical and cannot tolerate this kind of timing uncertainty.

To overcome the problem, a Nios II configuration can add additional master instruction and data ports for tightly coupled memory. While the cache is loaded as needed and its content changes dynamically, tightly coupled memory is allocated for a specific chunk of instruction or data. The assignment is done at the system initialization. One common use of tightly coupled memory is for interrupt service routines. The high-priority interrupts are frequently critical and must be processed within a certain deadline. Putting the routines in a tightly coupled memory removes the timing uncertainty and thus guarantees the response time.

8.3.7   I/O organization

The Nios II processor uses a memory-mapped I/O method to perform input and output between the processor and peripheral devices. An I/O device usually contains a collection of registers for command, status, and data. In the memory-mapped I/O scheme, the processor uses the same address space to access memory and the registers of I/O devices. Thus, the load and store instructions used to access memory can also be used to access I/O devices.

The inclusion of a data cache may cause a problem for this scheme because I/O command and status should not be buffered in an intermediate storage between the processor and I/O devices. A bypass path is needed for this purpose, as implemented by a two-to-one multiplexer in Figure 8.2. The Nios II processor introduces an additional set of load and store instructions for this purpose. When an I/O load or store instruction is issued, the operation bypasses the data cache and data is retrieved from or send to the master port directly.

8.3.8   Interconnect structure

In a traditional system, the main memory module and I/O devices are connected to a common, shared bus structure. Contention on bus sometimes becomes the bottleneck of the system. The Nios II processor utilizes Altera's Avalon interconnect structure. The interconnect is implemented by a collection of decoders, multiplexers, and arbitrators and provides concurrent transfer paths. We discuss this in more detail in Chapter 14.

8.4   EXCEPTION AND INTERRUPT HANDLER

The exception and interrupt handler processes the internal exceptions and external interrupts. The Nios II processor supports up to 32 interrupts and has 32 level-sensitive interrupt request inputs. When an exception or interrupt occurs, the processor transfers the execution to a specific address. An interrupt service routine at this address determines the cause and takes appropriate actions. We discuss the detailed mechanism and software development in Chapter 12.

8.5   JTAG DEBUG MODULE

The debug module connects to the signals inside the processor and can take control over the processor. A host PC can use the FPGA's JTAG port to communicate with the debug module and perform a variety of debugging activities, such as downloading programs to memory, setting break points, examining registers and memory, and collecting execution trace data. The debug module can be included or excluded in the processor and its functionality can be configured. We can include it during the development process and remove it from the final production.

8.6   BIBLIOGRAPHIC NOTES

The architecture and design of the Nios II processor closely resemble the 32-bit MIPS processor. *Computer Organization and Design: The Hardware/Software Interface* by D. A. Patterson and J. L. Hennessy provides a comprehensive coverage of this processor. Chapter 11, titled "Designing Soft-Core Processors for FPGAs" by J. Ball, of *Processor Design: System-on-Chip Computing for ASICs and FPGAs,* edited by J. Nurmi, discusses implementation issues that are unique to FPGA implementation. Altera's *Nios II Processor Reference Handbook* gives the complete description of the processor's architecture and instruction set.

8.7   SUGGESTED PROJECTS

8.7.1   Comparison of Nios II and MIPS

The overall design and architecture of the Nios II processor and the MIPS processor discussed in *Computer Organization and Design* are very similar. Compare the following features of the two processors:

- Instruction set
- Pipeline structure
- Interrupt handling

CHAPTER 9

NIOS II SYSTEM DERIVATION AND LOW-LEVEL ACCESS

A Nios II-based system consists of customized hardware and software. Altera's *SOPC Builder* is used to configure the processor and I/O peripherals and its *Nios II EDS* platform is used to develop software. In this chapter, we use a simple flashing-LED system as a tutorial to demonstrate the process, with emphasis on the hardware and software interface and basic coding techniques to access low-level I/O peripherals.

9.1 DEVELOPMENT FLOW REVISITED

The basic Nios II-based development process is discussed in Section 1.3.1 and the flow is repeated in Figure 9.1.

9.1.1 Hardware development

We examine the hardware development (the left branch of the flow) in Part I and discuss the design of various types of circuits. As the complexity of systems grows, it becomes difficult and time consuming to construct everything from scratch. One way to ease the problem is to use pre-designed modules, either from previous projects or other parties. Since the modules are usually intellectual property of the designing party, they are known as *IP cores* (*intellectual property cores*). An IP core can be delivered in high-level HDL codes, which is known as a *soft core*, or in a detailed transistor-level layout, which is known as a *hard core*.

Embedded SOPC Design with Nios II Processor and VHDL Examples. By Pong P. Chu **189**
Copyright © 2011 John Wiley & Sons, Inc.

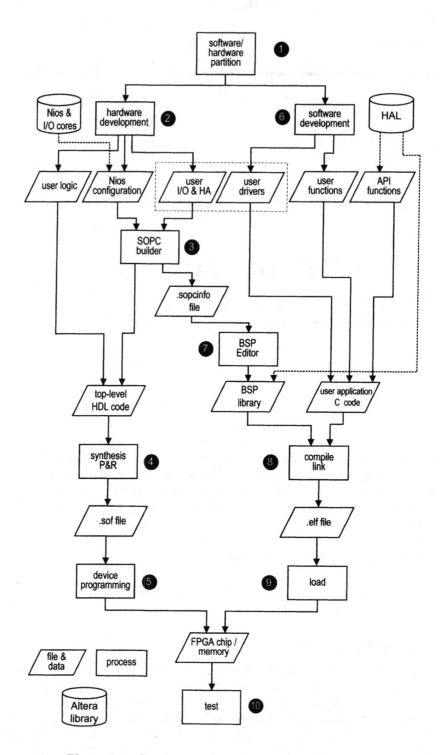

Figure 9.1   Development flow of a system with Nios II.

To facilitate the embedded system design, Altera provides a suite of IP cores, including the Nios II processor and a collection of commonly used I/O peripherals. They are delivered as soft cores. A full-fledged Nios II system may involve a sophisticated bus structure and a variety of I/O peripherals. Altera uses a utility program, known as *SOPC Builder*, to define the system and to generate HDL codes. The codes then can be used as regular HDL components and instantiated in other program. SOPC Builder saves the configuration in a file with an extension of .sopc and also exports the information to a file with an extension of .sopcinfo. The latter is used by the software development.

Note that from software development's point of view, the hardware development provides two key files—an .sof file and an .sopcinfo file. The former is used to program the FPGA device and the latter provides information about the Nios II system configuration.

9.1.2 Software development

The software development is represented by the right branch of Figure 9.1. A platform, known as *Nios II EDS* (*embedded design suite*), is provided by Altera. It is based on a GNU tool chain and customized for the Nios II processor environment. The platform is composed of the following:

- GCC-based compiler with the GNU binary utilities
- Nios II processor-specific port of the newlib C library
- A simple device driver interface know as *HAL* (*hardware abstraction layer*)

These tools can be accessed by the *SBT (software build tools) command-line interface*, by the *SBT GUI*, or by the *IDE GUI*. A simpler utility provided by the Altera University Program, known as *Altera Monitor Program*, can also be used to compile and debug C programs. In this book, we use the SBT GUI for software development.

The SBT GUI is based on the *Eclipse* open development environment and customized for the Nios II software development flow. It basically runs the tools and scripts "behind the scenes" and supports creating, modifying, building, running, and debugging programs targeted for a Nios II system.

A Nios II software projects contains two major parts: *user applications* and *BSP* (*board support package*). The former is the user's programs and the latter is support codes for a specific Nios II configuration. Note that the BSP is based on the information from the .sopcinfo file, as shown in Figure 9.1. The codes from the two parts are compiled and linked into a single software image (i.e., an .elf file) and loaded into Nios II system's main memory.

9.1.3 Flashing-LED system

The overall development process of a Nios II-based system is quite involved. To illustrate the process, we construct a simple flashing-LED system. It contains two discrete LEDs that turn on and off alternatively and uses ten slide switches to control the on-and-off interval. The system contains a "vanilla" Nios II configuration and a minimal amount of application software codes.

The main purpose of the flashing-LED system is to introduce the development procedures and to get familiarized with the software platform. The key steps in the development are:

1. Create a hardware project in Quartus II.
2. Create a Nios II system and generate HDL codes with SOPC Builder.
3. Create a top-level HDL file that instantiates the Nios II system.
4. Compile the top-level HDL code in Quartus II.
5. Program the FPGA device.
6. Create a BSP library.
7. If needed, configure the BSP library with BSP Editor.
8. Develop user application codes.
9. Build and run software.

The steps are demonstrated in two tutorials in the next two sections. The first tutorial covers the hardware development (i.e., steps 1 to 5) with a focus on SOPC Builder and the second tutorial covers the software development (i.e., steps 6 to 9) with a focus on the SBT GUI. The tutorials are based on version 10.0 of the software and some differences may exist in other versions.

9.2 NIOS II HARDWARE GENERATION TUTORIAL

9.2.1 Create a hardware project in Quartus II

This step is similar to that in Section 3.5.1 except that no HDL file is needed at this point.

9.2.2 Create a Nios II system and generate HDL codes

We can use SOPC Builder to configure a Nios II system and then generate HDL codes. There are several tasks in this step:

- Create a new SOPC system.
- Add and configure a Nios II processor.
- Add and configure memory modules.
- Specify the reset and exception vectors.
- Add and configure I/O modules.
- Add a system id module.
- Adjust the memory base address and interrupt.
- Generate HDL and information files.

Create a new SOPC system   A new SOPC system can be created as follows:

1. In the Quartus II GUI, select Tools ≻ SOPC Builder. The SOPC Builder window starts and its Create New System dialog appears.
2. Enter nios_led1 in the System Name field to specify the name of the Nios II system.
3. Click the VHDL button in the Target HDL field to generate VHDL codes later.
4. Click OK button to return to the SOPC Builder window. The initial SOPC Builder window appears and is in the System Contents tab page, as shown in Figure 9.2. The left panel, labeled Component Library, lists the IP categories. A category can be expanded to show the available IP cores. The middle right panel displays the current system configuration.
5. In the Target panel, make sure that the Cyclone II device family is used. This selection matches the FPGA device on the DE1 board.

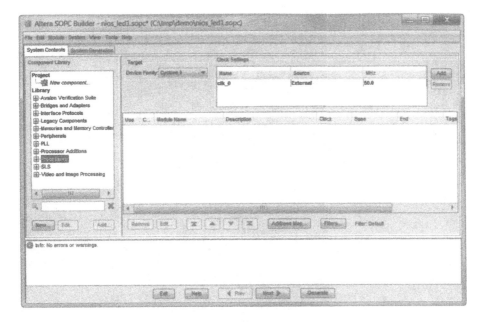

Figure 9.2   Initial SOPC Builder window.

6. In the Check Setting panel, rename clk_0 to clk under the Name column and make sure that clock frequency is 50.0 under the MHz column. This selection matches the oscillator frequency on the DE1 board.

7. Select File ≻ Save to save the system in a file. The default name of the file is nios_led1.sopc. Note that the .sopc extension is used to represent a SOPC system configuration file.

Add and configure a Nios II processor   The procedure to add and configure a Nios II processor module is:

1. In Component Library panel, expand the Processors category and then select Nios II Processor.

2. Click the Add button and the Nios II Processor dialog appears, as shown in Figure 9.3.

3. There are several tabs in the dialog page. In the Core Nios II tab page, click the Nios II/e button to select the "economic core." The configuration process requires to specify the reset and exception addresses in the Reset Vector and Exception Vector fields. However, since no memory module is added at this point, it cannot be done. The default setting can be used for other tab pages. Note that a level-1 JTAG debug module is used in the default setting.

4. Click the Finish button to return to the SOPC Builder window. The processor is added to the middle panel, as shown in Figure 9.4. Note that a warning message is displayed in the bottom message panel regarding the reset and exception vectors.

5. Click the Finish button to return to the SOPC Builder window.

6. Select cpu_0 item, right-click the mouse, and select Rename. Change cpu_0 to cpu.

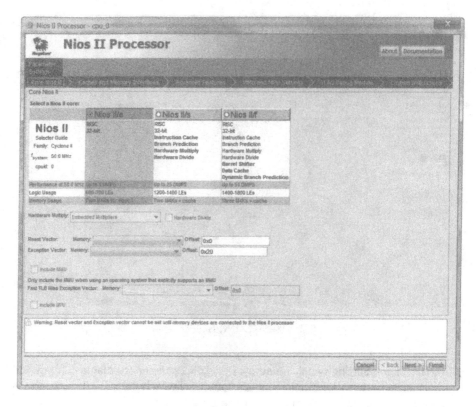

Figure 9.3   Nios II Processor dialog.

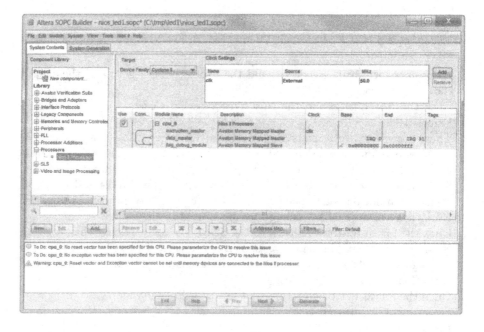

Figure 9.4   Initial SOPC Builder window with Nios II processor added.

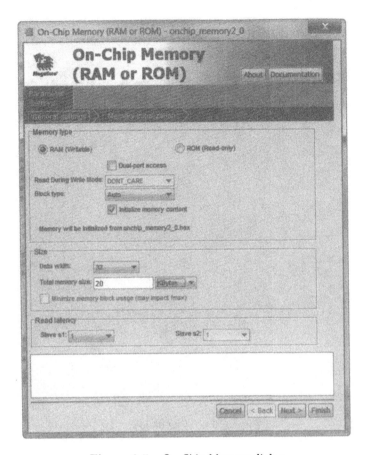

Figure 9.5   On-Chip Memory dialog.

Add and configure memory modules   A variety of memory modules can be incorporated into a Nios II system. Since the flashing-LED system is very simple, it requires only a small amount of memory. The FPGA's internal memory module can be used for this purpose. The procedure to add a 20 KB on-chip memory is:

1. In Component Library panel, expand the Memories and Memory Controllers category, expand the On-Chip category, and then click On-Chip Memory (RAM or ROM).
2. Click the Add button and the On-Chip Memory dialog appears, as shown in Figure 9.5.
3. In the Block Type field, select Auto.
4. In the Total memory size field, enter 20 and select KBytes to specify a memory size of 20 KB.
5. Click the Finish button to return to the SOPC Builder window.
6. Rename the memory module to onchip_mem.

Specify the reset and exception vectors   After adding the memory module, we can specify the reset and exception vectors. The procedure is:

1. In the working area, click the cpu item to bring up the previous Nios II Processor dialog.

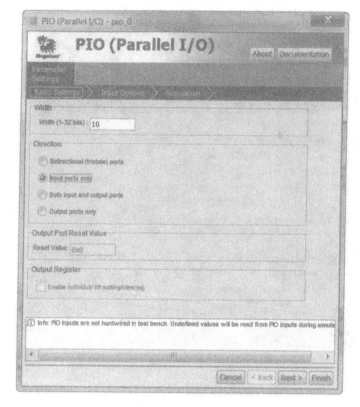

Figure 9.6   PIO dialog.

2. In Reset Vector, select onchip_mem in the Memory field.
3. In Exception Vector, select onchip_mem in the Memory field.
4. Click the Finish button to return to the SOPC Builder window.

Add and configure I/O modules   The flashing-LED system accepts input from ten slide switches and controls two LEDs. Altera provides an IP core, *PIO* (for *parallel I/O*), as a general-purpose I/O port to receive input stimuli and drive output signals. For the flashing-LED system, we can use a 10-bit input port for the switches and a 2-bit output port to drive the LEDs. The procedure to add and configure the two ports is:

1. In Component Library panel, expand the Peripherals category, expand the Microcontroller Peripherals category, and then click PIO (Parallel I/O).
2. Click the Add button and the PIO dialog appears, as shown in Figure 9.6.
3. Enter 10 in the Width field.
4. Click the Input ports only button in the Direction field to specify that this PIO module is an input port.
5. Click the Finish button to return to the SOPC Builder window.
6. Rename the PIO module to switch.
7. Add another PIO instance and set its width to 2 and its direction to Output ports only.
8. Rename the PIO module to led.

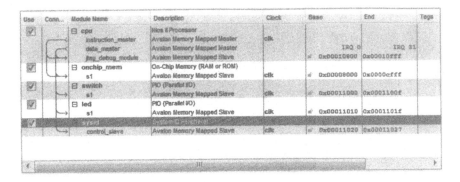

Figure 9.7 Completed Nios II configuration.

Add system id module The *system id IP core* is a special-purpose peripheral used to maintain the consistency between hardware configuration and BSP and is generally transparent in the user's application program. Its purpose and usage is discussed in Section 9.4. The procedure to add the module is:

1. In Component Library panel, expand the Peripherals category, expand the Debug and performance category, and then click System ID Peripheral.
2. Click the Add button and an information box appears.
3. Click the Finish button to return to the SOPC Builder window.
4. Rename the module to sysid.

The complete system is created and the configuration panel is shown in Figure 9.7.

Adjust memory base address and interrupt A Nios II processor uses the memory-mapped-I/O scheme and assigns normal memory space to I/O ports. The memory assignment is done automatically when a memory module or an I/O core is added and the beginning and end of the memory block are shown under the Base and End columns of the SOPC Builder window. For example, the base address of the led module in Figure 9.7 is 0x00011010 and the block ends at 0x0001101f. The initial memory assignment occasionally leads to conflicts. We can adjust the assignment in SOPC Builder by selecting System in the menu and then Auto-Assign Base Addresses. The addresses can also be manually modified if needed.

Some I/O modules may generate interrupt requests and we can adjust the request priority as well. Since no I/O module in this Nios II system generates an interrupt, no IRQ request is shown in Figure 9.7.

Generate HDL and information files After the Nos II system is created, the HDL and information files can be generated. The procedure is:

1. Select the System Generation tab page in the SOPC Builder window, as shown in Figure 9.8.
2. Click the Generation button on the bottom to start the process. The progress is displayed in the middle message panel. After completion, the System generation was successful message appears.
3. Click the Exit button to close the SOPC Builder window.

SOPC Builder generates a collection of files. For our purposes, the following are of interest (recall that the system name is nios_led1):

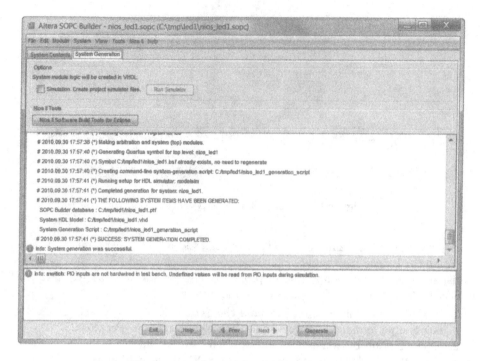

Figure 9.8   System Generation tab page.

- nios_led1.spoc: It is the SOPC Builder design file that contains the system configuration. It can be considered as the "source file" and can be used by SOPC Builder to regenerate other files.
- nios_led1.spocinfo: It contains relevant configuration information and is used by Nios II EDS to generate BSP.
- nios_led1.vhd: It is the top-level VHDL file for the generated Nios II system.
- Other VHDL files: These are VHDL files for the I/O modules and Nios II processor's subsystems. The onchip_memory.vhd, switch.vhd, and led.vhd files are for the memory and I/O modules of the Nios II system and their contents can be examined in a text editor. On the other hand, the key codes for the Nios II processor are encrypted.

9.2.3   Create a top-level HDL file that instantiates the Nios II system

Once the HDL files are generated by SOPC Builder, they can be used and processed like regular HDL codes. We just need to instantiate the top-level Nios II system and include the relevant files in a Quartus II project. The top-level entity declaration can be found in the top-level HDL file, nios_led1.vhd, and the name of the entity is `nios_led1`. The nios_led1.vhd contains multiple design units and is quite large. We can open the file and use the keyword `nios_led1` to search the entity declaration.

In the flashing-LED system, the code for the entity declaration is

```
entity nios_led1 is
   port(
       -- global signals:
       signal clk: in std_logic;
```

```
        signal reset_n: in std_logic;
        -- the_switch
        signal in_port_to_the_switch:
                in std_logic_vector(9 downto 0);
        -- the_led
        signal out_port_from_the_led:
                out std_logic_vector(1 downto 0)
     );
13 end entity nios_led1;
```

It shows that the system, in addition to the clock and reset signals, contains a 10-bit input port and a 2-bit output port. The names of these ports are derived from the module names defined in SOPC Builder.

The Nios II system can be used and instantiated as a regular HDL component and integrated with other parts. Since our demonstration system contains no additional logic, we just need to create a top-level module to wrap the Nios II system. The HDL code is shown in Listing 9.1.

Listing 9.1   Top-level system

```
library ieee;
2 use ieee.std_logic_1164.all;
  entity nios_led1_top is
  port(
     clk: in std_logic;
     sw: in std_logic_vector(9 downto 0);
7    ledg: out std_logic_vector(1 downto 0)
  );
  end nios_led1_top;

  architecture arch of nios_led1_top is
12    component nios_led1
        port(
           clk: in std_logic;
           reset_n: in std_logic;
           in_port_to_the_switch: in std_logic_vector (9 downto 0);
17         out_port_from_the_led: out std_logic_vector (1 downto 0)
        );
     end component;
  begin
     nios_unit: nios_led1
22   port map(
        clk=>clk, reset_n=>'1',
        in_port_to_the_switch=>sw,
        out_port_from_the_led=>ledg
     );
27 end arch;
```

To make the code clear, we include a component declaration within the architecture body, as explained in Section 2.3. The I/O ports of the Nios II system are mapped to ten slide switches and two green LEDs on the DE1 board. Since the system reset feature is not used, the active-low **reset_n** signal is connected to '1'.

9.2.4   Compiling and programming

After the top-level wrapping file is created, the Nios II system can be processed as a normal Quartus II project:

- Set the top-level entity.
- Add all previously generated HDL files to the project.
- Import a pin-assignment constraint file.
- If needed, set the timing constraint for the 50-MHz clock.
- Compile the project.
- Program the FPGA device.

After compiling, a configuration image .sof file is generated. Since the Altera Programmer software is a stand-alone utility, it can be called from Nios II EDS. Thus, after obtaining the .sof and .sopcinfo files, we can develop software without involving HDL files or Quartus II.

9.3 NIOS II SBT GUI TUTORIAL

We use the Nios II SBT GUI for software development, which involves steps 5 to 9 in Section 9.1.3.

9.3.1 Create BSP library

A complete Nios II EDS software project consists of a user application and a BSP supporting library. A BSP library is based on a particular Niso II system configuration. Once a BSP library is created, it can be used for subsequently user applications. The procedure to create a BSP library is:

1. In the Windows **Start** menu, navigate to **Nios II 10.0 Software Build Tools** for **Eclipse** to launch the program. The **Eclipse** initialization screen appears.
2. If the **Workspace Launcher** dialog appears, click **OK** to accept the default workspace location.
3. The **Eclipse SBT GUI** window appears, as shown in Figure 9.9. Verify that the **Nios II** perspective is selected on the top right corner.
4. Select **File ≻ New ≻ Nios II Board Support Package**. The dialog appears, as shown in Figure 9.10.
5. In the **Project name** field, enter led1_bsp to specify the name of the project. The _bsp suffix indicates it is a project to construct a BSP library.
6. In the **SOPC Information File Name** field, navigate to the previous Quartus II project directory and select the nios_led1.spocinfo file.
7. The default directory can be used by checking the **Use default location** box. In the **Type** field, keep Altera HAL. Since there is only one processor in the design, the name of the Nios II module, cpu, appears in the CPU name field automatically.
8. Click the **Finish** button to start the construction process and return to the **Eclipse** window.

When the construction process completes, the BSP directory, led1_bsp, is displayed in the left panel, which is referred to as the **Navigator** or **Project Explorer** subwindow.

9.3.2 Configure the BSP using BSP Editor

Because of the simplicity of the flashing-LED system, the default BSP configuration works fine and this step can be omitted. For a more sophisticated Nios II system, we may need to modify the BSP configuration to adjust certain I/O charateristics

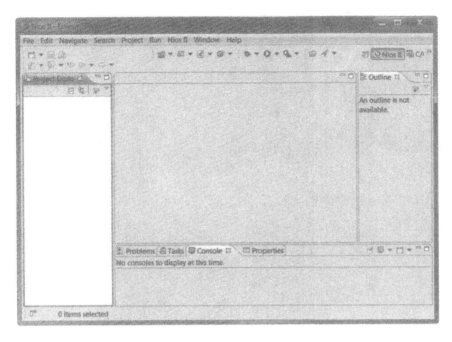

Figure 9.9   Initial Eclipse SBT GUI window.

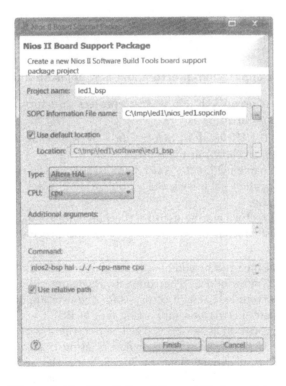

Figure 9.10   Nios II Board Support Package dialog.

Figure 9.11   Nios II Application Dialog.

and fine-tune the software driver setting. The *BSP Editor* utility program can be used to perform this task and the procedure is discussed in Section 11.2.3.

9.3.3 Create user application directory and add application files

A user application is organized in a separate project directory. The procedure to create the application directory is:

1. In Eclipse SBT GUI window, Select File ≻ New ≻ Nios II Application. The dialog appears, as shown in Figure 9.11.
2. In the Project name field, enter led1_test to specify the name of the project.
3. In the BSP Location field, click the ... button and select the previously constructed led1_bsp.
4. The default directory can be used by checking the Use default location box.
5. Click the Finish button to start the construction process and return to the Eclipse window.

When the construction process completes, the application directory is displayed in the left Navigator subwindow. The complete Navigator subwindow is shown in Figure 9.12. The bottom led1_test directory is for the user application and the top led1_bsp directory contains various supporting files of BSP.

Although the application structure is constructed, there is no user program file. We can create new C files from scratch or add existing C files to the application directory. The addition can be done by highlighting the application directory, selecting File ≻ Import..., and then choosing General ≻ File System to navigate to the designated location and files. It also can be done by using Windows Explorer to

Figure 9.12   Navigator window.

drag the desired files to the application directory and then selecting File ≻ Refresh to update the directory. There should be one C file that contains the `main()` routine.

The development of C codes is discussed in Sections 9.5, 9.6, and 9.8 and subsequent chapters. For now, we just use the C code in Section 9.11, which can be downloaded from the companion web site or manually created, and add the file to the application directory.

9.3.4   Build and run software

Building and running the final software consists of the following tasks:

- Compile the C codes in BSP directory to object codes.
- Compile the C codes in user application directory to object codes.
- Link the relevant object files to form the final image, which is an .elf (for *extensible linking format*) file.
- Load the .elf file to the designated memory modules.
- Let the processor start the execution (i.e., run the application program).

In the SBT GUI, a simple way to complete these tasks is to choose to run the application software in the Nios II system. Appropriate utility routines and scripts will be invoked automatically to complete the required tasks. The basic procedure is:

- If needed, download the previous nios_led1_top.sof file to the FPGA device. The Altera programmer program can be invoked in SBT GUI by selecting Nios II ≻ Quartus II Programmer.
- Select the application directory, led1_test, in the Navigator subwindow. Right-click the mouse and select Run As ≻ Nios II Hardware. A Progress Information

Figure 9.13   Screenshot of checking code size.

box appears and displays the progress of the process. The relevant information is also shown on the bottom Console tab page of the Eclipse window.

Since compiling and building the BSP library is quite complicated, it may take some time when a project is first constructed and executed. Subsequent runs will be much faster. After completion, we should be able to observe the two flashing green LEDs and use the slide switches to set the flashing rate.

9.3.5   Check code size

An embedded system frequently has a limited amount of memory and thus knowing the final code size (also known as *footprint*) is important. In Nios II EDS v10, this task can only be done in the command shell. The basic procedure is:

- In the Windows Start menu, navigate to Nios II 10.0 Command Shell to launch the program. A command window appears.
- Type `nios2-elf-size project_directory/project_name.elf`. The screenshot for checking the led1_test.elf file is shown in Figure 9.13.
- The total size is displayed under the dec column, which is 4408 bytes (i.e., 4.4 KB).

Recall that a 20 KB on-chip memory is included in the flashing-LED system. About 15.6 KB is left for run-time stack and heap, which is sufficient for the application.

The 4.4 KB code size of led1_test.elf is quite large for a such simple program. This is due to the fact that certain housekeeping codes are added to the application program to construct the final image. In Section 11.2.3, we discuss some methods to decrease the code size. The resulting footprint of led1_test.elf can be reduced to 0.9 KB.

9.4   SYSTEM ID CORE FOR HARDWARE-SOFTWARE CONSISTENCY

An embedded SOPC design consists of both custom hardware and software. When an Nios II system is created, its configuration is stored in an .sopc file and an .sopcinfo file. For the hardware portion, the .sopc file is used to generate HDL files, which in turn are synthesized into an FPGA configuration file (i.e., .sop file). For the software portion, the .sopcinfo file is used to generate the BSP library, which in turn is linked and integrated into the final software image (i.e., .elf file).

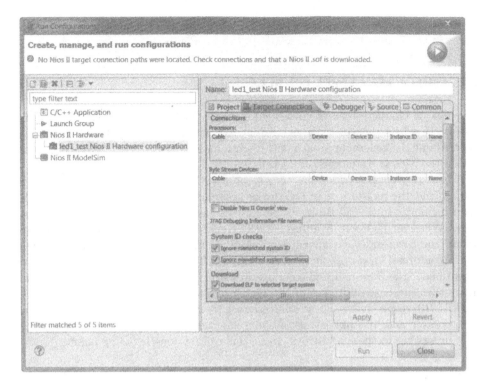

Figure 9.14   Run Configurations error screen.

During the development process, multiple Nios II systems can be generated for exploration and testing purposes. Different versions of .sopcinfo and .sop sometimes get mixed. It is possible to use the FPGA configuration file of one design but use a software image file based on a BSP library with another design. This type of errors is very difficult to debug.

Altera introduces a special *system id core* to maintain the consistency between the hardware configuration file and software image. The id is a unique 32-bit value based on an SOPC design (like a *signature* or check-sum). The system id core is simply a read-only register that stores this signature. When the core is added to a Nios II system, the system has a "hard copy" of the signature. After synthesis, the signature is embedded in the configuration .sof file. The same id value is also stored in the .sopcinfo file. When the .sopcinfo file is processed during software construction, the signature is passed to BSP and eventually integrated into the software image .elf file. Thus, both the .sop and .elf files have the same signature.

While downloading a software image, Nios II EDS checks whether the signature of the system id module in the FPGA device matches the signature embedded within the software image and thus ensures the consistency between the hardware and software. As long as there is no mismatch, this procedure is mostly transparent to users.

If two signatures do not match or no system id IP module is used in the target Nios II system, an error is reported in the **Target Connection** tab page, as shown in Figure 9.14. If the mismatch is not harmful, we can override the default setting by checking the Ignore mismatched system ID and Ignore mismatched system timestamp boxes in the System ID checks row and run the configuration again.

9.5 DIRECT LOW-LEVEL I/O ACCESS

One unique characteristic of embedded system development is that the software program frequently needs to access and interact with low-level I/O devices directly. We examine the general approach in this section and discuss more robust techniques within the Nios II framework in Section 9.6

A Nios II processor uses a *memory-mapped I/O* method to access I/O ports. In this scheme, the registers of an I/O device are mapped into the address space of the main memory. In SOPC Builder, the address assignment is performed automatically. When an I/O core is added, SOPC Builder checks the number of registers within the core and then allocates a chunk of memory space accordingly. The starting address, known as the *base address*, and the end address of the chunk, are shown under the Base and End columns of SOPC Builder. For example, Figure 9.7 shows that the base addresses of the switch and led modules are 0x00011000 and 0x00011010, respectively. These addresses are treated as regular memory addresses by processor and an application program can access the I/O devices by reading from or writing to these addresses. In C, this can be done by using pointer.

9.5.1 Review of C pointer

In C, the pointer data type corresponds to a memory address. The concept of a pointer can be explained by a simple code segment:

```
int x=1, y=5, z=8, *ptr;

ptr = &x;    // ptr gets address of x
y = *ptr;    // content of y gets content pointed by ptr
*ptr = z;    // content pointed by ptr gets content of z
```

In C, a non-pointer variable can be thought as an abstract memory location identified by the name of the variable and a value is stored to the location in an assignment. A pointer variable is designated with *, as in int *ptr, which indicates that ptr is a pointer (i.e., a memory address) and it points to a location with the int data type. A snapshot after the initial declaration and assignment of this segment is shown in Figure 9.15(a). We use an arrow to indicate that ptr is a pointer variable. It is pointed to nowhere (i.e., null) since it is unassigned initially. Two unary operators, & and *, are associated with pointer operations. The & operator returns the address of a variable and is known as the *address-of* operator. For example, in statement ptr = &x, &x returns the address of x, which is then assigned to ptr. The result is shown in Figure 9.15(b). The * operator returns the content pointed by the pointer and is known as the *dereference* operator. For example, in statement y = *ptr, content pointed by ptr is assigned to y and in statement *ptr = z, the value of z is stored to the location pointed by ptr. The graphical representations are shown in Figure 9.15(c) and (d).

The value of a pointer variable is usually manipulated implicitly, as illustrated by the previous segment. The actual value of ptr is system dependent. In a desktop programming environment, we usually do not and need not know the explicit value.

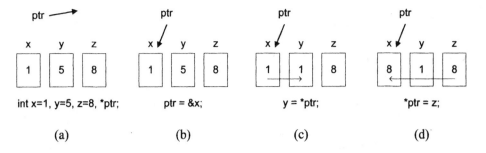

Figure 9.15   Snapshots of pointer operation.

9.5.2   C pointer for I/O register

In the flashing-LED system, an I/O register is assigned with a memory address, which can be thought as a value of a pointer. Unlike a normal desktop program discussed in the previous subsection, we know the explicit value of the address and must use this value to access the register. Recall that the base addresses of the switch and led modules are 0x00011000 and 0x00011010. For example, we can read the value from the switch module and write a pattern to the led module:

```
int sw;
char pattern=0x01;

sw = *(0x00011000);
*(0x00011010) = pattern;
```

The statements are primitive and difficult to comprehend. Several improvements can be made. Let us consider the read statement. First, we can add a type cast, (volatile int *), to describe the nature of this value:

```
sw = * (volatile int *) (0x00011000);
```

The int * portion indicates that the constant value is a pointer that points to an object with the int data type. The keyword volatile gives the compiler the hint that the value of the object may be modified without processor interaction and thus certain optimizations should not be performed. Second, we can define a symbolic constant to replace the hard literal:

```
#define SWITCH_BASE 0x00011000
. . .
sw = * (volatile int *) SWITCH_BASE;
```

To maintain modularity and enhance readability, we can define a macro to encapsulate the type casting and dereference operations. A macro to read an I/O register can be defined as

```
#define SWITCH_BASE      0x00011000
#define demo_iord(addr) (*(volatile int *)(addr))
. . .
sw = demo_iord(SWITCH_BASE);
```

A similar macro can also be used to write an I/O register and the complete segment becomes

```
int sw;
char pattern=0x01;

#define SWITCH_BASE 0x00011000
#define LED_BASE    0x00011010
#define demo_iord(addr)        (*(volatile int *)(addr))
#define demo_iowr(addr, data)  (*(int *)(addr) = (data))

sw = demo_iord(SWITCH_BASE);
demo_iowr(LED_BASE, (int) pattern);
```

Like the names indicate, the two macros are only for demonstration purposes. While this approach works most of the time, it suffers from several subtle problems. Altera provides three simple modules to assist the low-level I/O access and to make the code more robust. This alternative is discussed in the next section.

9.6 ROBUST LOW-LEVEL I/O ACCESS

When a BSP library is constructed, several header files are created. Three important files and their functionalities are:

- `system.h`: provides automatically generated base addresses.
- `alt_types.h`: provides explicitly defined low-level data types.
- `io.h`: provides enhanced I/O register read and write macros.

Using the data type and macros defined in these files make the low-level I/O access more robust.

9.6.1 system.h

We can determine the base addresses of each I/O device by examining the **Base** column of SOPC Builder and define them as constant, as described in the previous section. Since the base addresses are automatically assigned in SOPC Builder, we need to examine and update these addresses for each new Nios II system and any subsequent revision. This is a tedious and error-prone process.

To solve the problem, the Nios II EDS framework automates this process. During the BSP library construction, the BSP Builder examines the .sopcinfo file, extracts information on each module, and creates a file, `system.h`, to record the information. Note that this file is listed under the led1_bsp directory in Figure 9.12.

In `system.h`, each I/O device is identified by the symbolic name given in the **Module Name** column of SOPC Builder and suffixes are used to represent the corresponding properties. For example, the entry for the **switch** module is

```
#define ALT_MODULE_CLASS_switch altera_avalon_pio
#define SWITCH_BASE 0x11000
#define SWITCH_BIT_CLEARING_EDGE_REGISTER 0
#define SWITCH_BIT_MODIFYING_OUTPUT_REGISTER 0
#define SWITCH_CAPTURE 0
#define SWITCH_DATA_WIDTH 10
. . .
```

Note that the constants are always in uppercase in the file. The `system.h` file is regenerated automatically when the .sopcinfo file is updated and thus should not be edited manually. The base address of an instantiated module is specified as _BASE in `system.h`, as in

```
#define SWITCH_BASE 0x11000
```

After including this file, we don't need to define the base address manually.

9.6.2 `alt_types.h`

C has many predefined data types, such as **short**, **int**, **long**, etc. The width (i.e., number of bits) of each data type is left to the compiler and implementation. While interacting with low-level device activities, it is often important to know the exact width and format of registers and data. To facilitate this, Altera provides a simple header file, `alt_types.h`, which explicitly specifies the width and format of each data type. The new data types are:

- `alt_8`: signed 8-bit integer.
- `alt_u8`: unsigned 8-bit integer.
- `alt_16`: signed 16-bit integer.
- `alt_u16`: unsigned 16-bit integer.
- `alt_32`: signed 32-bit integer.
- `alt_u32`: unsigned 32-bit integer.
- `alt_64`: signed 64-bit integer.
- `alt_u64`: unsigned 64-bit integer.

It is good practice to use these data types for low-level coding.

9.6.3 `io.h`

The I/O read and write macros discussed in Section 9.5.2 are common techniques used in embedded system programming. However, these may cause a subtle error for a system with data cache, which can be found in today's high-performance embedded processors, such as the Nios II/f configuration. Since a processor considers an I/O register as regular memory, the relevant data may be temporarily stored in the data cache and can only be written to the I/O register when the corresponding block is deallocated from the cache.

There is no easy way to fix this in C code. However, the Nios II processor contains separate load and store instructions for I/O access. When these instructions are used, read and write operations bypass the cache, as discussed in Section 8.3.7. In the `io.h` file, Altera provides two macros for I/O access:

- `IORD(base, offset)`: read an I/O register with the specified **base** address and **offset**.
- `IOWR(base, offset, data)`: write **data** into an I/O register with the specified **base** address and **offset**.

These macros are implemented by using the proper machine instructions and thus bypass the data cache. The previous I/O access statements

```
sw = demo_iord(SWITCH_BASE);
demo_iowr(LED_BASE, pattern);
```

can be written as

```
sw = IORD(SWITCH_BASE, 0);
IOWR(LED_BASE, 0, pattern);
```

Note that an offset value is also needed for the two new macros.

9.7 SOME C TECHNIQUES FOR LOW-LEVEL I/O OPERATIONS

Because an embedded program interacts with low-level I/O devices, it frequently needs to manipulate a bit or a field of a data object. We briefly examine some relevant techniques in following subsections.

9.7.1 Bit manipulation

C has several bitwise operators, including ~ (not), & (and), | (or), and ^ (xor), which operate on one or two operands at bit levels.

The ~ operator inverts all individual bits. For example, if **d** is 0xb3 (i.e., 1011_0011), ~**d** becomes 0x8c (i.e., 0100_1100). The statement **max** = ~0 inverts all bits from 0's to 1's and **max** becomes the all-one pattern, which corresponds to the largest number in any unsigned data type.

The **&**, **|**, and **^** operators can be used to manipulate a bit or a group of bits in a data object. The operation involves a data operand and a **mask** operand, which specifies the bits to be modified. The operations are shown in the following C segment:

```
alt_u8 mask=0x60;   // 0110_0000; mask; bits 6 and 5 asserted
alt_u8 d=0xb3;      // 1011_0011; data
alt_u8 a0,a1,a2,a3;

a0 = d & mask;      // 0010_0000; isolate bits 6 and 5 from d
a1 = d & ~mask;     // 1001_0011; clear bits 6 and 5 of d to 0
a2 = d | mask;      // 1111_0011; set bits 6 and 5 of d to 1
a3 = d ^ mask;      // 1101_0011; toggle bits 6 and 5 of d
```

In the example, we assume that **d** is an 8-bit data and bits 6 and 5 represent a special 2-bit field. The **mask** variable identifies this field by asserting bits 6 and 5. We can isolate this field from **d** (i.e., clear all other bits to 0) by applying the and operation with the mask, as in **d&mask**. Conversely, we can clear this field and keep the remaining bits intact by using the inverted mask, as in **d&~mask**. Similarly, we can set this field to 11 and keep the remaining bits intact by applying the or operation with the mask, as in **d|mask**.

The toggle operation is based on the observation that for any 1-bit Boolean variable x, $x \oplus 0 = x$ and $x \oplus 1 = x'$. We can toggle the desired field by applying the xor operation with the mask, as in **d^mask**.

9.7.2 Packing and unpacking

To save address space, an I/O register frequently contains multiple fields. These fields are extracted and separated (i.e., *unpacked*) after an application program reads the I/O register. Conversely, these fields needed to be *packed* into one object when they are written to the I/O register. The unpacking and packing processes can be done by using the bitwise manipulation and shift operation.

For example, assume that a 32-bit I/O register contains a 16-bit field (for an integer) and two 8-bit fields (for two characters), as shown in Figure 9.16. The

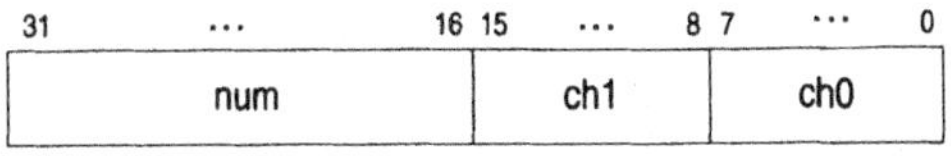

31	...	16	15	...	8	7	...	0

num	ch1	ch0

Figure 9.16 An I/O register with three fields.

code segment to unpack a retrieved I/O word is:

```
alt_u32 iodata;
int num;
char ch1, ch0;

iodata = IORD(...);
num = (int)  ((iodata & 0xffff0000) >> 16);
ch1 = (char) ((iodata & 0x0000ff00) >> 8);
ch0 = (char) ((iodata & 0x000000ff));
```

We first apply an and mask, such as 0xffff0000, to clear the the irrelevant bits, and then shift a proper amount to remove trailing 0's. In this process, the interpretation of a field changes from "a collection of bits" to a specific data type, such as int or char. It is good practice to use type casting to indicate the change of interpretation and data type of the extracted field.

The code segment to pack three fields to an I/O word reverses the previous operation:

```
alt_u32 iodata;
int num;
char ch1, ch0;

iodata = (alt_u32)(num);                 // num in bits 15:0
iodata = (iodata<<8) | (alt_u32) ch1;    // num in bits 23:8
iodata = (iodata<<8) | (alt_u32) ch0;    // num in bits 31:16
IOWR(..., iodata);
```

The first statement puts **num** between bit 15 and bit 0. The second statement first shifts **num** to the left by 8 bits, which makes the 8 LSBs all 0's, and then uses the bitwise or operation to fill the 8 LSBs with the value of **ch1**. The same process is repeated to append the **ch0** field. Again, proper type casting should be used in the process.

9.8 SOFTWARE DEVELOPMENT

9.8.1 Basic embedded program architecture

An embedded application consists of a collection tasks, implemented by hardware accelerators, software routines, or both. Unlike a normal desktop application, an embedded program may run continuously and does not terminate. The top-level program (main program) schedules, coordinates, and manages these tasks. The simplest control architecture is an infinite "super loop," in which the tasks are executed sequentially. The pseudo code for a super-loop architecture is

```
main(){
  sys_init();
  while(1){
    task_1();
    task_2();
    ...
    task_n();
  }
}
```

The system runs the `sys_init()` function once to perform initialization and then enters the infinite loop and invokes the task functions in turn. A task function handles certain I/O activities. Some tasks may have timing constraints and must be processed within the given time limits. This scheme works properly if the overall loop execution time is small and the processor can respond to each task in a timely manner. Additional control architectures and scheduling issues are discussed in Section 12.3.

9.8.2 Main program and task routines

The flashing-LED system turns on and off two LEDs alternatively according to the interval specified by the ten sliding switches. The two main tasks are reading the interval value from the switches and toggling the two LEDs after a specific amount of time. The top-level program of this LED-flashing system is shown in Listing 9.2. It follows the basic program architecture discussed in Section 9.8.1 and consists of two major routines.

Listing 9.2

```
#include "io.h"
#include "alt_types.h"
#include "system.h"

int main(){
  int prd;

  while(1){
    sw_get_command_v0(SWITCH_BASE ,&prd);
    led_flash_v0(LED_BASE , prd);
  }
}
```

The `sw_get_command_v0()` function reads the value of the switch and the code is shown in Listing 9.3. Since the same functionality is repeated in the subsequent chapters with modified codes, the _v0 (for version 0) suffix is added.

Listing 9.3

```
void sw_get_command_v0(alt_u32 sw_base, int *prd)
{
  *prd = IORD(sw_base, 0) & 0x000003ff;  // read flashing period
}
```

Since the switch is 10 bits wide, we use a mask 0x000003ff to clear the unrelated bits to 0's.

The `led_flash_v0()` function waits for the specified interval and toggles two discrete LEDs. The code is shown in Listing 9.4.

Listing 9.4

```
void led_flash_v0(alt_u32 dled_base, int prd)
{
  static alt_u8 led_pattern = 0x01;     // initial pattern
  unsigned long i, itr;

  led_pattern ^= 0x03;                   // toggle 2 LEDs (2 LSBs)
  IOWR(dled_base, 0, led_pattern);       // write LEDs
  itr = prd * 2500;
  for (i=0; i<itr; i ++){}               // dummy loop for delay
}
```

Since 8-bit data is the smallest unit in C, we use an 8-bit variable, led_pattern, to store the LED pattern. It is declared as a static variable so that its value can be kept between function calls. The two LSBs are toggled by xoring with the 0x03 mask when this function is executed. The delay is achieved by a dummy for loop. We assume that each loop iteration takes two instructions, each instruction takes 10 clock cycles in a Nios II/e (economic) configuration, and the system clock is 50 MHz (i.e., 20-ns period). Each iteration will take 400 ns ($2 * 10 * 20$ ns) and it requires 2500 iterations for a 1-ms delay. Delay obtained by this method is just a rough estimation and is not very accurate.

Note that we do not use any global variable or constant in the two task functions and module-dependent constants, such as LED_BASE and SWITCH_BASE, are confined in main program. Thus, only a top-level program needs to be revised if an I/O module is modified.

9.9 BIBLIOGRAPHIC NOTES

SOPC Builder is part of the Quartus II package and its detailed description can be found in *Volume 4: SOPC Builder* of *Quartus II Handbook Version 10.0*. Nios II EDS is documented in Altera's *Nios II Software Developer's Handbook*. Chapter 1 of the handbook, titled *Overview*, provides a general overview of the EDS framework and Chapter 2 of the handbook, titled *Getting Started with the Graphical User Interface*, gives a general introduction to the SBT GUI environment.

9.10 SUGGESTED EXPERIMENTS

9.10.1 Chasing LED circuit

There are 18 discrete LEDs on a DE1 board. A chasing LED circuit turns on one LED at a time sequentially and thus the lit LED appears to move (i.e., chase) along the strip. The detailed specification is:

1. The 16 discrete LEDs are used as output, one lit at a time.
2. The lit LED moves sequentially in either direction. It changes direction when reaching the rightmost or leftmost position.
3. The pushbutton switch 0 (labeled key0) on the DE1 board is used to "initialize" the process. When it is pressed, the lit LED is moved to the rightmost position.

4. The lower five slide switches are used to control the chasing speed of the LED. The highest speed should be slow enough for visual inspection.

Derive and synthesize a Nios II system, develop software, and verify the system operation.

9.10.2 Collision LED circuit

A collision LED circuit is similar to the chasing LED circuit in Section 9.10.1 but turns on two LEDs at a time. The two LEDs move independently and change direction when reaching the rightmost or leftmost position or "colliding" in the middle. The detailed specification is:

1. The 16 discrete LEDs are used as output, two lit at a time.
2. The lit LEDs move sequentially in either direction. They changes direction when reaching the rightmost or leftmost position or "colliding" in the middle.
3. The pushbutton switch 0 (labeled key0) is used to "initialize" the first lit LED. When it is pressed, the first lit LED is moved to the rightmost position.
4. The pushbutton switch 1 (labeled key1) is used to "initialize" the second lit LED. When it is pressed, the second lit LED is moved to the leftmost position.
5. The lower five slide switches are used to control the chasing speed of the first LED and the upper five slide switches are used to control the chasing speed of the second LED. The two chasing speeds are independent.

Derive and synthesize a Nios II system, develop software, and verify the system operation.

9.10.3 Pulse width modulation circuit

A PWM circuit is described in Section 5.9.2. Instead of using custom hardware, we can use an embedded system to perform this task as well. Derive and synthesize a Nios II system, develop software, and verify the system operation.

9.10.4 Rotating square circuit

A rotating square circuit is described in Section 5.9.3. Instead of using custom hardware, we can use an embedded system to perform this task as well. Derive and synthesize a Nios II system, develop software, and verify the system operation.

9.10.5 Heartbeat circuit

A heartbeat circuit is described in Section 5.9.4. Instead of using custom hardware, we can use an embedded system to perform this task as well. Derive and synthesize a Nios II system, develop software, and verify the system operation.

9.11 COMPLETE PROGRAM LISTING

Listing 9.5   chu_main_led1_adhoc.c

```
/***************************************************************************
 *
 * Module:    Simple flashing-LED system
 * File:      chu_main_led1_adhoc.c
 * Purpose:   Flash two LEDs alternatively
 *            and use 10 slide switches to set the flashing period
 *
 ***************************************************************************/
#include "io.h"
#include "alt_types.h"
#include "system.h"


/***************************************************************************
 * function: sw_get_command_v0()
 * purpose:  get flashing period from switches
 * argument:
 *   sw_base: base address of switch PIO
 *   prd: pointer to period
 * return:
 *   updated prd
 * note:
 ***************************************************************************/
void sw_get_command_v0(alt_u32 sw_base, int *prd)
{
  *prd = IORD(sw_base, 0) & 0x000003ff;   // read flashing period
}


/***************************************************************************
 * function: led_flash_v0()
 * purpose:  toggle 2 LEDs according to the given period
 * argument:
 *   led_base: base address of discrete LED PIO
 *   prd: flashing period in ms
 * return:
 * note:
 *   - The delay is done by estimating execution time of a dummy for loop
 *   - Assumption: 400 ns per loop iteration (2500 iterations per ms)
 *       - 2 instructions per loop iteration
 *       - 10 clock cycles per instruction
 *       - 20 ns per clock cycle (50-MHz clock)
 ***************************************************************************/
void led_flash_v0(alt_u32 led_base, int prd)
{
  static alt_u8 led_pattern = 0x01;      // initial pattern
  unsigned long i, itr;

  led_pattern ^= 0x03;                    // toggle 2 LEDs (2 LSBs)
  IOWR(led_base, 0, led_pattern);         // write LEDs
  itr = prd * 2500;
  for (i=0; i<itr; i ++){}                // dummy loop for delay
}
```

```
/*************************************************************************
* function: main()
* purpose:  top-level program
* note:
*************************************************************************/
int main(){
  int prd;

  while(1){
    sw_get_command_v0(SWITCH_BASE ,&prd);
    led_flash_v0(LED_BASE, prd);
  }
}
```

CHAPTER 10

PREDESIGNED NIOS II I/O PERIPHERALS

The I/O peripherals of Nios II are soft cores and can be incorporated into a Nios II system and eventually synthesized into the same FPGA chip. Altera provides a set of commonly used I/O peripherals that can be easily configured and integrated in SOPC Builder. In this chapter, we examine the structure and use of three peripherals for general input and output interface, serial communication, and timing, and use them to construct a more sophisticated Nios II system.

Software complexity grows as I/O devices become more involved. A common approach to alleviate the problem is to confine the low-level I/O transactions in a collection of routines, sometimes know as *device drivers*, and shield the details from application programs. In this chapter, we use an enhanced flashing-LED system to demonstrate the use of new I/O cores and the development of ad hoc drivers. We intentionally avoid Altera's predesigned HAL-compliant drivers and postpone the coverage to Chapter 11.

10.1 OVERVIEWS

Like the Nios II processor, its I/O peripherals are usually described by HDL codes and implemented as soft cores. For a commonly used I/O function, a predesigned core is usually available and we just need to instantiate it when a Nios II system is constructed. For a specialized I/O peripheral, we may need to design it from scratch and then integrate the circuit with the processor. Altera provides a set of I/O cores for commonly used I/O functionalities.

When we select and use a predesigned I/O core, we must pay attention to the following:

- *Function description.* We need to study the functionality carefully and understand the capability and limitation of the core and then decide whether the core matches our need.
- *Configurability.* Because of the programmability of FPGA, many I/O cores can be configured. The configuration is done when a core is instantiated in SOPC Builder. We can include or exclude certain features, such as interrupt capability, or specify the size of certain components, such the size of an FIFO buffer.
- *Register map.* Recall that the Nios II processor utilizes the memory-mapped I/O scheme. From the viewpoint of a processor and application software, an I/O core is represented by a collection of registers. The processor accesses the I/O core by reading or writing the proper bit or field of a register. The register map provides detailed information about these bits and fields.
- *Device driver.* Device driver is a collection of software routines used to access an I/O core. We can construct the routines from scratch or utilize a predesigned software library. The development of a generic driver is discussed in Section 10.7 and the use of Altera's library is covered in Chapter 11.

Altera supplies more than two dozen I/O cores for its SOPC platform. We select three most common peripherals, *PIO*, *JTAG UART*, and *Timer*, and study them in more detail in the subsequent sections. We discuss the use of *SDRAM controller* and *PLL* cores in Chapter 15.

10.2 PIO CORE

A *PIO (parallel input/output) core* provides a memory-mapped interface between a port of the Avalon interconnect and a general-purpose I/O port. We use it as simple input port and output port in Chapter 9. A full-featured PIO core is more sophisticated and its conceptual diagram is shown in Figure 10.1. The I/O port connects either to on-chip user logic or to off-chip external devices via FPGA's I/O pins.

10.2.1 Configuration

The configuration of a PIO core is done in several steps. When a PIO core is instantiated in SOPC Builder, the Basic Settings tab page appears, as shown in Figure 10.2(a). There are four fields in this page:

- Width. This field specifies the number of bits in the I/O port.
- Direction. This field indicates the direction of signal flow and one of the four modes can be selected.
- Output Port Reset Value. If the previous mode consists of an output port, this field specifies the reset value of the output port.
- Output Register. If the previous mode consists of an output port, the Enable individual bit setting/clearing option can be turned on to set or clear individual

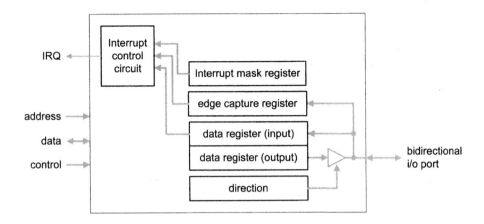

Figure 10.1 Conceptual diagram of a full-featured PIO core.

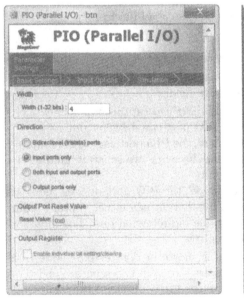

(a) Basic Setting page

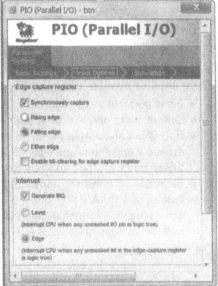

(b) Input Options page

Figure 10.2 Instantiation pages of the PIO core.

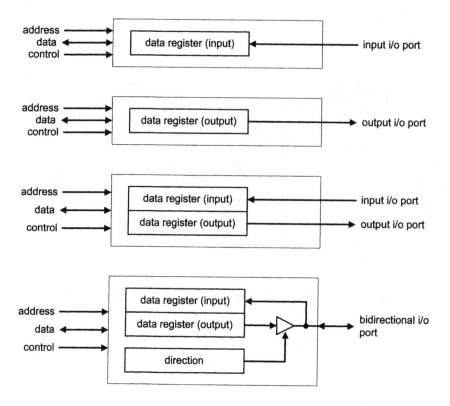

Figure 10.3   Four direction modes of a PIO core.

bits of the output port. Two additional registers, **outset** and **outclear**, are included in the implementation.

There are four possible modes in the **Direction** field:

- **Input ports only.** In this mode, the PIO port can capture input only.
- **Output ports only.** In this mode, the PIO port can drive output only.
- **Both input and output ports.** In this mode, the PIO port can capture input and drive output simultaneously. Note that there are two separate unidirectional buses, one for input and one for output.
- **Bidirectional (tristate) ports.** In this mode, the PIO port utilizes the tristate buffers of FPGA's I/O pins. The individual bit can either capture input or drive output. Note that there is only one bidirectional bus. The direction of each bit is controlled by a separate register.

The conceptual diagram of the four modes is shown in Figure 10.3. Note that in the last two modes the read and write data are stored in two separate registers.

If a PIO port consists of an input port, we need to continue to the **Input Options** tab page, as shown in Figure 10.2(b), for additional settings. There are two main fields:

- **Edge Capture Register** We can turn on the **Synchronously capture** option to introduce an additional register to capture the transition edge of an input signal. The corresponding bit is set to 1 when an edge is detected in the input signal. The processor can later clear the bit to 0 by writing 1 to the

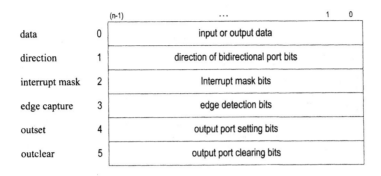

Figure 10.4   Register map of a PIO core.

same register. After turning on the option, we need to specify the following features:

- Select the type of edge to detect, which can be **Rising Edge, Falling Edge,** or **Either Edge.**
- Specify whether to turn on **Enable bit-clearing for edge capture register** to clear individual bits in the edge capture register.

• **Interrupt.** We can turn on the **Generate IRQ** option to include an interrupt circuit. An interrupt request signal is added and it is asserted when a specified event occurs on the input port. We must further specify whether the event is triggered by the input **level** or by the capture of an **edge.** An additional interrupt mask register is included in implementation to indicate whether the request from the corresponding bit is enabled.

If desired, we can continue to the **Simulation** tab page to include additional testbench features and specify the values of the input ports during simulation.

10.2.2   Register map

The processor controls and communicates with a PIO core via a set of registers. The address, name, and fields are summarized in the register map in Figure 10.4.

Address offset 0 is for the input and output data. The PIO core implementation actually consists of two separate registers for this purpose, as shown in the bottom two configurations of Figure 10.3. Although the two registers share the same address offset, they operate independently. Writing to this address stores the value to the output data register and has no effect on the content of the input data register.

The other registers are optional and are inferred according to the configuration. Address offset 1 is for the direction register, which is inferred if the PIO core is configured as a bidirectional mode. If a bit in the direction register is 1, the tristate buffer is enabled and the corresponding I/O pin functions as an output pin. If the bit is 0, the tristate buffer is disabled and the pin functions as an input pin.

Address offset 2 is for the interrupt mask register, which is inferred if the PIO core is configured to include the interrupt circuit. Setting a bit in the register to 1 enables the interrupt request for the corresponding bit of the PIO input port. Address offset 3 is for the edge capture register, which is inferred if this feature is included during configuration. A bit is set to 1 when an edge is detected in the

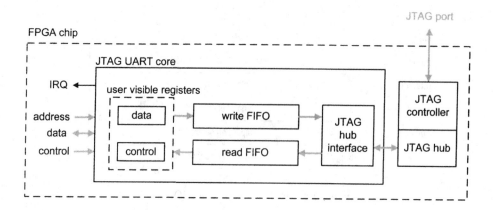

Figure 10.5   Conceptual diagram of the JTAG UART core.

corresponding bit of the PIO input port. Writing 1 to the register clears its content to 0.

Address offsets 4 and 5 are for the output-set and output-clear registers. Setting a bit in these registers sets or clears the corresponding bit in the output port.

10.2.3   Visible register

In an Altera's I/O core, a register of a register map implies a "user-visible register" and uses only one port address. In the actual implementation, it is possible that two physical registers are associated with that port address, as in the data register of the PIO core. This kind of arrangement is quite common in Altera SOPC's I/O cores. In the remaining of the book, we just follow the convention used in Altera documentation and treat the two physical registers as one visible register.

10.3   JTAG UART CORE

UART stands for *universal asynchronous receiver and transmitter*. It contains two serial lines for data communication, one for receiving and one for transmitting. Parallel data (usually 8 bits) are sent bit by bit via the serial lines between two systems. A UART is commonly used in conjunction with the EIA RS-232 interface to form the serial port of a PC or an embedded system.

Altera's *JTAG UART core* is similar to a serial port. Instead of using an RS-232 interface, the data is received and transmitted via FPGA's JTAG controller and JTAG port. This eliminates the need of a separate serial connection between a host PC and the prototyping board. The core handles the internal JTAG interface and control. From the processor and application program's point of view, it can be treated as a regular serial port and used to communicate the serial character stream between the PC and the board. The conceptual diagram is shown in Figure 10.5.

10.3.1   Configuration

To increase the performance and regulate data transmission, a write FIFO buffer and a read FIFO buffer are included in the JTAG UART core. The configura-

Figure 10.6 Instantiation page of a JTAG UART core.

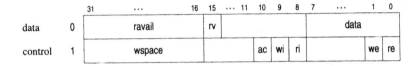

Figure 10.7 Register map of the JTAG UART core.

tion mainly specifies the characteristics of the two buffers. When a JTAG UART core is instantiated in SOPC Builder, the Configuration page appears, as shown in Figure 10.6. There are three fields for each buffer:

- Buffer Depth. This field specifies the number of bytes in the FIFO buffer.
- IRQ Threshold. This field specifies the interrupt condition. The interrupt request signal is asserted when the number of data bytes in the FIFO buffer reaches the specified threshold.
- Construct using registers instead of memory blocks. Turing on this option forces the synthesis software to use the logic elements to implement the buffer.

10.3.2 Register map

The user-visible interface to the JTAG UART core consists of two 32-bit registers, one for data and one for control, as shown in the register map in Figure 10.7. The data register contains the following fields:

- data. This field contains the byte to transfer to or from the JTAG core. During the write operation, it holds a character to be written to the write FIFO buffer. During the read operation, it holds a character read from the read FIFO buffer.

- rv. The bit is 1 if the **data** field is valid.
- ravail. This field contains the number of characters remaining in the read FIFO buffer (after the current read).

The control register contains the following fields:

- re. This bit needs to be set to 1 to enable the read interrupt request.
- we. This bit needs to be set to 1 to enable the write interrupt request.
- ri. This bit indicates whether the read interrupt request is pending.
- wi. This bit indicates whether the write interrupt request is pending.
- ac. This bit indicates whether there has been JTAG activity since the bit was cleared.
- wspace. This field contains the number of spaces available in the write FIFO buffer.

10.4 INTERNAL TIMER CORE

The *internal timer core* supports various timing needs, such as measuring the interval between events and generating periodic pulses. The key part of the core is a counter that counts down from a specific value to 0. The value is known as *timeout period* and stored in the timeout period registers. When the counter reaches 0, a specific bit is set, the optional interrupt request is asserted, and an optional output pulse can be generated as well. After reaching 0, the counter can pause and stay there (in the *count-down-once mode*) or reload with values from the period registers and restart the counting (in the *continuous mode*). Optional control signals can be used to stop, start, or reset the counter operation.

The counter is driven by the system clock and each count corresponds to one clock period. The elapsed time thus is equal to *number of counts * clock period*.

10.4.1 Configuration

The timer core is versatile and can be configured to fit different timing needs. When a timer core is instantiated in SOPC Builder, the **Parameter Settings** page appears, as shown in Figure 10.8. There are several fields:

- **Timeout period.** This field specifies the timeout period and determines the initial value of the timeout period register. It can be specified in terms of a time unit (e.g., **ms**) rather than the number of clocks. In the former, SOPC Builder will use the system clock information to convert it from a time unit to the number of clocks.
- **Timer counter size.** This field specifies the number of bits in the counter, which can be either 32 or 64. In a system with a 50-MHz clock, a 32-bit counter can count up to 85.9 seconds (i.e., $2^{32} * 20$ ns) and a 64-bit counter can count up to more than 10,000 years (i.e., $2^{64} * 20$ ns)
- **Hardware options.** This field specifies which optional features should be included and contains several subfields:
  - **Presets.** This subfield lists several predefined configurations, including **Simple periodic interrupt**, **Full-featured**, **Watchdog**, and **Custom**.
  - **Registers.** This subfield specifies whether certain features should be instantiated, including whether allowing the processor to update (i.e.,

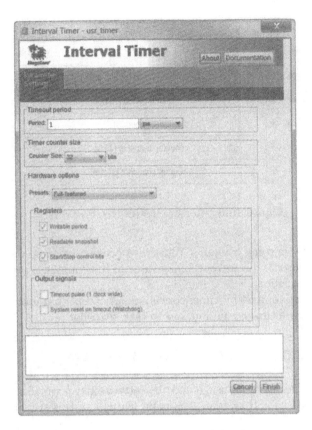

Figure 10.8   Instantiation page of the timer core.

write) timeout period counter, to read the current count from the snapshot registers, and to pause and resume counting.
- Output signals. This field indicates whether to include the optional timeout pulse output or watchdog timer reset output.

10.4.2   Register map

The user-visible register of a 32-bit timer core consists of up to six 16-bit registers,
as shown in the register map in Figure 10.9. The status register contains two fields:
- to. The to (for "time out") bit is set to 1 when the counter reaches zero. It
stays set until a processor writes 0 to this bit to clear it.
- run. This bit reads as 1 when the counter is running.

The control register contains four fields:
- ito. This bit indicates whether interrupt is enabled.
- cont. This bit specifies whether the timer operates in the continuous mode or
count-once mode.
- start. Writing 1 to this bit starts the counter running (counting down).
- stop. Writing 1 to this bit stops the counter.

The periodl and periodh registers store the lower 16 bits and upper 16 bits of the
32-bit timeout period value. The snapl and snaph registers are used to take a

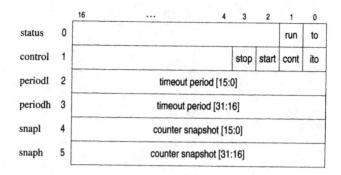

Figure 10.9   Register map of the timer core.

"snapshot" of the current counter. When a processor issues a write instruction (write data is ignored), the current 32-bit value of the counter is copied to the two 16-bit snapshot registers.

The register map for a 64-bit timer core is similar except that four 16-bit registers are used to store the timeout period and to take snapshot.

10.5   ENHANCED FLASHING-LED NIOS II SYSTEM

To demonstrate the I/O core usage and the software development, we construct a more sophisticated Nios II system. The main features of this system are:

- A 512 KB external SRAM device for the main memory.
- A JATG UART to establish a serial link to a console.
- Two full-feature timers, one for the system tasks and one for user application.
- An input port for ten slide switches.
- An input port for four pushbutton switches with edge capture.
- Two output ports for the discreet green and red LEDs.
- An output port for the four seven-segment LED displays.

The conceptual top-level block diagram is shown in Figure 10.10. This Nios II system consists of the general I/O cores and utilizes all simple I/O peripherals (i.e., switches and LEDs) of the DE1 board. It is used in this chapter and Chapters 11 and 12.

We can follow the tutorial in Section 9.2 to construct and synthesize the new Nios II system. The following subsections discuss the creation of the SOPC design and top-level HDL file.

10.5.1   SOPC design

The procedure of generating a simple flashing-LED system is discussed in Section 9.2.2. Several additional steps are needed for the enhanced Nios II system. The detailed tasks are:

- Create a new SOPC system.
- Add and configure a Nios II processor.
- Add the SRAM controller module.
- Specify the reset and exception vectors.

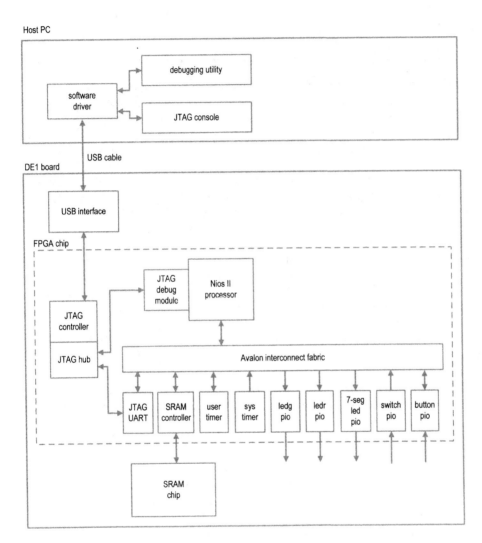

Figure 10.10   Block diagram of an enhanced flashing-LED system.

- Add and configure PIO modules.
- Add and configure two timer modules.
- Add and configure a JTAG UART module.
- Add a system id module.
- Adjust memory base address and interrupt.
- Generate HDL and information files.

Create a new SOPC system   This task is similar to that in Section 9.2.2 except we name this system `nios_led2`.

Add and configure a Nios II processor   This task is similar to that in Section 9.2.2.

Add the external SRAM memory module   The available memory options on the DE1 board are discussed in Section 5.7.1. The EP2C20 device of the DE1 board has 26 KB internal memory. This memory is used to implement the registers and cache of the Nios II processor well as the buffers and lookup tables in I/O cores and user logic. The remaining portion can also be used as Nios II processor's memory module.

In Chapter 9, we use 20 KB internal memory to create an on-chip memory module in the simple flashing-LED system. However, because of its limited capacity, the memory module can only accommodate simple software programs. In the enhance flashing-LED system, we replace it with the external 512 KB SRAM device.

A Nios II processor accesses the external memory device via a *memory controller*. Altera provides several SRAM controller IP cores in SOPC Builder. However, since their configurations do not match the characteristics of DE1 board's SRAM chip, they cannot be used. A custom SRAM controller core is developed for this purpose and its construction is discussed in Chapter 15. For now, we just use it as an existing IP core.

To instantiate the core, we must make it core visible in SOPC Builder's Library panel. The procedure to include our custom cores in SOPC Builder is:

1. Create a directory, say, `chu_ip`, in hard disk.
2. Copy the SRAM controller directory `chu_avalon_sram` and its content, which include the `chu_avalon_sram_hw.tcl` and `chu_avalon_sram.vhd` files, to the `chu_ip` directory.
3. In SOPC Builder window, select Tools ≻ Options.... The Options dialog appears.
4. In dialog, select the IP Search Path page.
5. Click the Add... button, navigate to the `chu_ip` directory, and click the Open button to add the directory to search path. The resulting page is shown Figure 10.11.
6. Click the Finish button. SOPC Builder searches the paths and adds the found IP cores to the left Library panel.
7. In the Block Type field, select Auto.
8. A new category, chu_ip, appears. Expand the category and the `chu_avalon_sram` core should be listed under this category.

Once included, the core can be used like other normal IP cores. We add it to the enhanced flashing-LED system and rename the module `sram`.

Specify the reset and exception vectors   This task is similar to that in Section 9.2.2 except that the `sram` module must be used in the Memory field.

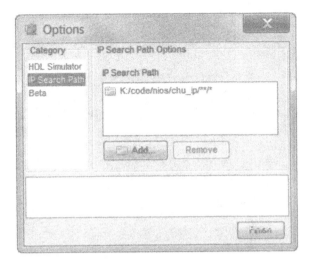

Figure 10.11   IP Search Path page of SOPC Builder **Options** dialog.

Add and configure PIO modules We use PIO cores for the switches and LEDs on the DE1 board. The procedure is:

1. For the eight discrete green LEDs, add a PIO module, configure it as an 8-bit output port, and rename it **ledg**.
2. For the ten discrete red LEDs, add a PIO module, configure it as a 10-bit output port, and rename it **ledr**.
3. For the four seven-segment LED displays, add a PIO module, configure it as a 32-bit output port, and rename it **sseg**. To accommodate the software program development, we allocate 8 bits (i.e., a byte) for each display. Since there are only seven segments on a display, the MSB of the byte is not used.
4. For the ten slide switches, add a PIO module, configure it as a 10-bit input port, and rename it **switch**.
5. For the four pushbutton switches, add a PIO module, configure it as a 4-bit input port in the **Basic Settings** tab page, set up the edge capture and interrupt in the **Input Options** tab page, as shown in Figure 10.2(b), and rename it **btn**. The setting infers an edge capture register in the module to capture the 1-to-0 transitions in the input signal and an interrupt circuit to generate interrupt request.

Add and configure two timer modules We include two timer cores, one for the system-level housekeeping function and one for the user application. The core can be found in the **Peripherals ≻ Microcontroller Peripherals** category and the configuration page is shown in Figure 10.8. The default setting can be used for our purposes. After creating two timers, we rename the two modules sys_timer and usr_timer.

Add and configure a JTAG UART We include a JTAG UART core as a communication link between the host PC and the prototyping board. The core can be found in the **Interface Protocols ≻ Serial** category and the configuration page is shown in Figure 10.6. The default setting can be used for our purposes. We rename the module jtag_uart.

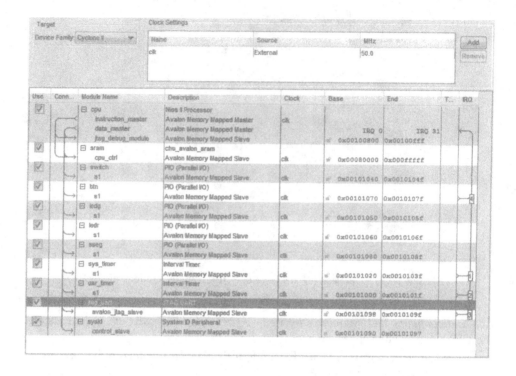

Figure 10.12   Completed SOPC Builder window.

Add system id module   This task is identical to that in Section 9.2.2.

Adjust memory base address and interrupt request   After all the modules are added, we can adjust its memory base addresses, similar to that in Section 9.2.2. In this system, several modules generate interrupt requests and these requests are displayed under the IRQ column, as shown in Figure 10.12. For our purposes, the request priority should be assigned in a specific order: sys_timer, usr_timer, jtag_uart, and btn. We can manually edit the priority number under the IRQ column as needed. Note that the request with the smallest number has the highest priority.

The completed system is shown in Figure 10.12 and the configuration and information are stored in `nios_led2.sopc` and `nios_led2.sopcinfo` files, respectively. We can generate the HDL files accordingly.

10.5.2   Top-level HDL file

To synthesize the Nios II system, we need to create a Quartus II project and include a top-level HDL file to instantiate the Nios II system, as discussed in Section 9.2.3. Since our demonstration system contains no additional logic, we just need to create a top-level module to wrap the Nios II system. The HDL code is shown in Listing 10.1.

Listing 10.1 Top-level system

```vhdl
library ieee;
use ieee.std_logic_1164.all;
entity nios_led2_top is
port(
    clk: in std_logic;
    sw: in std_logic_vector(9 downto 0);
    key: in std_logic_vector(3 downto 0);
    ledg: out std_logic_vector(7 downto 0);
    ledr: out std_logic_vector(9 downto 0);
    hex3, hex2, hex1, hex0: out std_logic_vector(6 downto 0);
    sram_addr: out std_logic_vector (17 downto 0);
    sram_dq: inout std_logic_vector (15 downto 0);
    sram_ce_n, sram_oe_n, sram_we_n: out std_logic;
    sram_lb_n, sram_ub_n: out std_logic
);
end nios_led2_top;

architecture arch of nios_led2_top is
    component nios_led2 is
        port(
            signal clk: in std_logic;
            signal reset_n: in std_logic;
            -- 4-bit pushbutton switch
            signal in_port_to_the_btn: in std_logic_vector (3 downto 0);
            -- 10-bit slide switch
            signal in_port_to_the_switch: in std_logic_vector (9 downto 0);
            -- 4 seven-segment LED displays
            signal out_port_from_the_sseg:
                out std_logic_vector (31 downto 0);
            -- 8-bit green LEDs
            signal out_port_from_the_ledg:
                out std_logic_vector (7 downto 0);
            -- 10-bit red LEDs
            signal out_port_from_the_ledr:
                out std_logic_vector (9 downto 0);
            -- 512K SRAM
            signal sram_addr_from_the_sram:
                out std_logic_vector (17 downto 0);
            signal sram_dq_to_and_from_the_sram:
                inout std_logic_vector (15 downto 0);
            signal sram_ce_n_from_the_sram: out std_logic;
            signal sram_lb_n_from_the_sram: out std_logic;
            signal sram_oe_n_from_the_sram: out std_logic;
            signal sram_ub_n_from_the_sram: out std_logic;
            signal sram_we_n_from_the_sram: out std_logic
        );
    end component nios_led2;
    signal sseg4: std_logic_vector(31 downto 0);

begin
    nios_unit: nios_led2
    port map(
        clk=>clk, reset_n=>'1',
        in_port_to_the_btn=>key,
        in_port_to_the_switch=>sw,
        out_port_from_the_ledg=>ledg,
        out_port_from_the_ledr=>ledr,
        out_port_from_the_sseg=>sseg4,
        -- SRAM
        sram_addr_from_the_sram => sram_addr,
```

```
          sram_dq_to_and_from_the_sram => sram_dq,
          sram_ce_n_from_the_sram => sram_ce_n,
63        sram_lb_n_from_the_sram => sram_lb_n,
          sram_oe_n_from_the_sram => sram_oe_n,
          sram_ub_n_from_the_sram => sram_ub_n,
          sram_we_n_from_the_sram => sram_we_n
       );
68     hex3 <= sseg4(30 downto 24);
       hex2 <= sseg4(22 downto 16);
       hex1 <= sseg4(14 downto 8);
       hex0 <= sseg4(6 downto 0);
   end arch;
```

Note that the 32-bit **sseg** PIO port is mapped to the **sseg4** signal and its interval fields are assigned to four seven-segment LED displays, **hex0**, **hex1**, **hex2**, and **hex3**.

We can create a project, add the top-level module and relevant HDL files, and compile the project to obtain the FPGA configuration file, **nios2_led2_top.sof**.

10.6   SOFTWARE DEVELOPMENT OF ENHANCED FLASHING-LED SYSTEM

The enhanced flashing-LED system includes several commonly used I/O peripherals. We use it to introduce the concept of device driver and illustrate the process of embedded software development.

10.6.1   Introduction to device driver

Unlike a desktop application program, an embedded application program involves more I/O interactions and has direct access to I/O devices. The main task of an embedded program sometimes is just to monitor and coordinate the operation of I/O peripherals. As an I/O device becomes more sophisticated, controlling its operation requires more effort. For example, the data and control registers of the JTAG UART core contain nine fields, each having a unique functionality and read/write access mode. Controlling the UART requires to check status, issue command, retrieve data, and write data from or to the proper fields of the registers. It is a tedious and error-prone process. The software complexity multiplies when multiple I/O devices are used in a system.

A common approach to alleviate the problem is to confine the low-level I/O transactions in a collection of routines, sometimes known as *device drivers*. A higher-level application program does not interact with the I/O devices directly. Instead, it calls a device driver routine to perform the desired operation and thus is shielded from the low-level detail. When an I/O device is modified, only the corresponding driver routines need to be revised and thus the impact is localized. We demonstrate the development and use of a set of ad hoc device drivers via the enhanced flashing-LED system in this section.

Altera provides a collection of device drivers for its SOPC I/O cores and integrates them into the *HAL* (*hardware abstraction layer*) platform. We intentionally avoid this platform in this section and postpone the discussion in Chapter 11.

10.6.2   Program structure of the enhanced flashing-LED system

The enhanced flashing-LED system operates as follows:

- The desired flashing time interval is specified by the ten slide switches. The value is loaded to the system when pushbutton switch 1 (labeled **key1** on the DE1 board) is pressed.
- The flashing can be temporarily stopped. Pressing pushbutton switch 0 pauses and resumes the operation alternatively.
- The four seven-segment LED displays show the current status of the system. The left-most display shows a 'P' pattern if the flashing is paused and the next three displays shown the value of the flashing interval in milliseconds.
- The host console displays a message whenever a new interval value is set. This message is transmitted via the JTAG UART module.
- The timer module keeps track of flashing interval and turns on and off the two discrete LEDs alternatively.

We follow the basic program template discussed in Section 9.8.1 and divide the main program into five tasks. The skeleton of the main program is

```
initialize system;
while(1){
  get commands from slide and pushbutton switches;
  display interval value on console (via JTAG UART);
  display status on seven-segment LED displays;
  toggle two LEDs after the specified interval;
}
```

Since this system is more complex, we use a modular approach and utilize three levels of hierarchy:

- Top level: main program.
- Second level: routines for the five tasks.
- Third level: routines for the device drivers.

10.6.3   Main program

The main program and portion of the header are shown in Listing 10.2.

Listing 10.2   Flash LED with generic device driver

```
#include "system.h"
. . .
typedef struct flash_cmd{
  int pause;
  int prd;
} cmd_type;
. . .
int main(){
  cmd_type sw_cmd={0,100};   // initial value: not pause, 100 ms interval

  flashsys_init_v1(BTN_BASE, USR_TIMER_BASE);
  while(1){
    sw_get_command_v1(BTN_BASE, SWITCH_BASE ,&sw_cmd);
    jtaguart_disp_msg_v1(JTAG_UART_BASE, sw_cmd);
    sseg_disp_msg_v1(SSEG_BASE, sw_cmd);
    led_flash_v1(LEDG_BASE, USR_TIMER_BASE, sw_cmd);
  }
```

```
}
```

We define a structure type, `cmd_type`, that encapsulates the "command," which consists of the pause status and the flashing interval. The program consists of five functions, each corresponding to a task. The implementation of these functions is discussed in Section 10.8.

As discussed in Section 9.6.1, the base address of each instantiated I/O device is defined as a constant with suffix _BASE, such as `USR_TIMER_BASE`, in `system.h` and its name is derived from the name given in SOPC Builder. These names change when the underlying Nios II system configuration is modified. To limit its impact, it is good practice to use the constant names only in the main program.

10.6.4   Function naming convention

In this book, we generally name the I/O routines in the following convention:

```
<device_name>_<action>_<object>
```

The `<action>` term is like a verb and specifies the function to be performed. The `<object>` term is optional and, if it exists, specifies the type of object to be performed on. For example, `timer_init()` means that the timer is initialized and `vga_disp_ch()` means that the VGA monitor displays a character, and `vga_disp_str()` means that the VGA monitor displays a string.

10.7   DEVICE DRIVER ROUTINES

The "device driver" here means the software routines accessing I/O devices. The term is used loosely and a more detailed treatment is covered in Section 11.4. We only include routines that are needed for this particular flashing-LED program and most routines are just for demonstration purposes.

10.7.1   Driver for PIO peripherals

The PIO core provides a simple interface to access an I/O port. The system's slide switches, pushbutton switches, discrete LEDS, and seven-segment LED displays use this core.

Register map   The PIO core may infer up to six registers, as shown in Figure 10.4. A register can be accessed by adding the proper offset value to the base address. The read and write functions in Altera's `io.h` take the base address and the offset as two arguments and perform the address computation:

```
IORD(base, offset)
IOWR(base, offset, data)
```

For example, we can use `IORD(MYPIO_BASE,3)` to read data from the edge capture register of a PIO module named `MYPIO`. Using a hard literal like 3 makes the code difficult to understand and error-prone. One way to alleviate the problem is to show the register map as a set of constant definitions, as in

```
#define PIO_DATA_REG_OFT 0 // data register addr offset
```

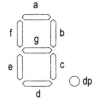

(a) Diagram of a seven-segment LED display

(b) Hexadecimal digit patterns

Figure 10.13   Seven-segment LED display and hexadecimal patterns.

```
#define PIO_DIRT_REG_OFT 1 // direction reg addr offset
#define PIO_INTM_REG_OFT 2 // interrupt mask reg addr offset
#define PIO_EDGE_REG_OFT 3 // edge capture reg addr offset
```

The previous statement becomes IORD(MYPIO_BASE,PIO_EDGE_REG_OFT), which is more expressive.

Basic data access   Although the IORD() and IOWR() functions can be used to access the I/O registers directly, as in Listing 9.11, we can use macros to make the functions more descriptive. Two macros are define to access the data register of a PIO core:

```
#define pio_read(base)         IORD(base,PIO_DATA_REG_OFT)
#define pio_write(base,data)   IOWR(base,PIO_DATA_REG_OFT,data)
```

The two routines can be used to read data from switches and write data to discrete LEDs.

Seven-segment LED display   The DE1 board has four seven-segment displays. The segment naming convention and hexadecimal digit patterns of Figure 4.5 are repeated in Figure 10.13. Recall that the LED segments are configured as active low, which means that a segment is lit if the corresponding control signal is 0.

In a C program, we can use a byte to represent the display pattern and ignore the MSB of the byte. For example, we use 0x40 (i.e., "01000000" pattern), which turns off the g segment and turns on all others, to display the 0 digit. One easy way to generate the hexadecimal value is to store the patterns in a 16-element array and use it as a lookup table. The function is shown in Listing 10.3.

Listing 10.3

```
alt_u8 sseg_conv_hex(int hex)
{
  /* active-low hex digit 7-seg patterns (0-9,a-f); MSB ignored */
  static const alt_u8 SSEG_HEX_TABLE[16] = {
    0x40, 0x79, 0x24, 0x30, 0x19, 0x92, 0x02, 0x78, 0x00, 0x10, // 0-9
    0x88, 0x03, 0x46, 0x21, 0x06, 0x0E};                        // a-f
  alt_u8 ptn;
```

```
  if (hex < 16)
    ptn = SSEG_HEX_TABLE[hex];
  else
    ptn = 0xff;    //blank
  return (ptn);
}
```

To reduce the number of instantiated PIO cores, we group the four displays together and create a 32-bit SSEG PIO module when the Nios II system is constructed. Since a pattern is represented as a byte in a program, we must pack four patterns into a 32-bit word and then output it to the module. A driver routine is constructed for this purpose and is shown in Listing 10.4. The argument `ptn` points to the head of a four-element array. The routine packs four patterns of the array to a 32-bit word and outputs it to the display port.

Listing 10.4

```
void sseg_disp_ptn(alt_u32 base, alt_u8 *ptn){
  alt_u32 sseg_data;
  int i;

  /* form a 32-bit data */
  for (i=0; i<4; i++){
    sseg_data = (sseg_data << 8) | *ptn;
    ptn++;
  }
  pio_write(base, sseg_data);
}
```

For example, we can use the following code segment to show "12EF" on the displays:

```
  alt_u8 msg[4];
  . . .
  msg[0] = sseg_conv_hex(1);    // 1
  msg[1] = sseg_conv_hex(2);    // 2
  msg[2] = sseg_conv_hex(14);   // E
  msg[3] = sseg_conv_hex(15);   // F
  sseg_disp_ptn(SSEG_BASE, msg);
```

Pushbutton switch   In this flashing-LED system, the pushbutton switches are used to load the interval value from the slide switches and to pause and resume the flashing operation. We want to use the transition edge, rather than the level, of the input signal to control the operation and include the edge capture register when configuring the `btn` PIO module.

A bit of the edge capture register is set to 1 when the designated transition edge of the corresponding input signal is detected. It remains as 1 until the processor performs a write operation to clear the register to 0. We define two macros to read and clear the edge register of the `btn` module:

```
  #define btn_read(base)    IORD(base, PIO_EDGE_REG_OFT)
  #define btn_clear(base)   IOWR(base, PIO_EDGE_REG_OFT, 0xf)
```

Note that the `IOWR()` operation does not write 1 to the edge capture register but clears the corresponding bits in the register.

File organization For each I/O core, we put the relevant driver materials together into two files. One is a header file that includes constant declarations, macros, and function definitions and the other file contains the codes for function prototypes. The complete code listing of the PIO core driver is stored in the files chu_avalon_gpio.h and chu_avalon_gpio.c, as shown in Listings 10.12 and 10.13 of Section 10.12. This driver is used in the subsequent chapters as well.

10.7.2 JTAG UART

From the application program's point of view, accessing the JTAG UART module involves reading the receiving FIFO buffer and writing the transmitting FIFO buffer. Since the task routine just needs to display the message on the console, we only discuss the driver routines for writing.

Register map The JTAG UART core contains two registers, as shown in Figure 10.7. For clarity, the offsets are defined as

```
#define JUART_DATA_REG_OFT  0  // data register addr offset
#define JUART_CTRL_REG_OFT  1  // control register addr offset
```

Both data and control registers are composed of multiple fields. A field can be isolated and extracted by masking proper bits and shifting a proper amount, as discussed in Section 9.7.2. For example, assume that **creg** is the data retrieved from the control register. We can extract the WSPACE field, which occupies bits 16 to 31, as follows:

```
wspace = (creg & 0xffff0000) >> 16;
```

As for the symbolic register offset, symbolic constants can be used for the masks and bit offsets. For example, we can define the mask and bit offset as

```
#define JUART_WSPA_MSK      0xffff0000 // mask
#define JUART_WSPA_BIT_OFT  16         // bit offset
```

and the previous statement becomes

```
wspace = (creg & JUART_WSPA_MSK) >> JUART_WSPA_BIT_OFT;
```

This method is preferred if these fields are used in multiple places. Since our drivers are relatively simple, we do not use symbolic constants for masks or bit offsets in this book.

Basic I/O access For this particular flashing-LED application, we need to write a string to the transmitting FIFO buffer of the JTAG UART module. This operation requires to check the available buffer space and write data to the FIFO buffer. We define two macros for these two activities:

```
/* check # slots available in FIFO buffer */
#define jtaguart_rd_wspace(base)  \
    ((IORD(base, JUART_CTRL_REG_OFT) & 0xffff0000) >> 16)
/* write an 8-bit char */
#define jtaguart_wr_ch(base, data)  \
    IOWR(base, JUART_DATA_REG_OFT, data & 0x000000ff)
```

Driver routine   The driver routine transmits a string through the JTAG UART module and is shown in Listing 10.5. The argument **msg** points to the head of a string. Within the loop, we check whether a space is available in FIFO and write a character accordingly. The loop continues until the end of string (i.e., 0) is reached.

Listing 10.5

```
void jtaguart_wr_str(alt_u32 jtag_base, char* msg)
{
  alt_u32 data32;

  while(*msg){
    data32 = (alt_u32) *msg;
    if(jtaguart_rd_wspace(jtag_base)!=0){     // buffer space available
      jtaguart_wr_ch(jtag_base, data32);      // send a char
      msg++;
    } // end if
  } // end while
}
```

This routine uses a simple "busy-waiting" strategy; i.e., it continues looping until a buffer space is available. The routine may halt the program execution if the size of the string exceeds the capacity of the FIFO buffer. Since the message is short and sparse in the flashing-LED system, the strategy does not lead to any serious problem.

10.7.3 Timer

The Altera timer core is very flexible and can be configured to perform a variety of functions. To count the flashing interval in this system, we use the core as follows:

- During initialization, set the timer to count 1 ms continuously.
- During normal operation, keep track of the number of 1-ms ticks and toggle the LEDs as needed.

We implement the driver routines to perform this operation.

Register map   The 32-bit timer core contains up to six 16-bit registers, as shown in Figure 10.9. Only the first four registers are used in the routines and their offsets are defined as

```
#define TIMER_STAT_REG_OFT 0   // status register addr offset
#define TIMER_CTRL_REG_OFT 1   // control reg addr offset
#define TIMER_PRDL_REG_OFT 2   // period low reg addr offset
#define TIMER_PRDH_REG_OFT 3   // period high reg addr offset
```

Basic I/O access   The to (timeout) field of the status register is set to 1 when the counter reaches 0 and it remains 1 until the processor writes 0 to this field. This field can be treated as a "tick" that is asserted periodically. We define two macros to check and clear the to field:

```
#define timer_read_tick(base)   \
  (IORD(base, TIMER_STAT_REG_OFT) & 0x01)
#define timer_clear_tick(base)  \
  IOWR(base, TIMER_STAT_REG_OFT, 0)
```

Driver routine The driver routine sets up the countdown period and activates the counting. The code is shown in Listing 10.6. The 32-bit argument `prd` indicates the timeout period in terms of the number of clock cycles. It is split into two 16-bit data and written to the two period registers. The last statement sets the counter in "continuous" mode, starts the counter operation, and enables interrupt by writing 1's to the cont, start, and ito fields of the control register. The interrupt feature is ignored in this chapter but used in Chapter 11.

Listing 10.6

```
void timer_wr_prd(alt_u32 timer_base, alt_u32 prd){
  alt_u16 high, low;

  /* unpack 32-bit timeout period into two 16-bit half words */
  high = (alt_u16) (prd>>16);
  low = (alt_u16) (prd & 0x0000ffff);
  /* write timeout period */
  IOWR(timer_base, TIMER_PRDH_REG_OFT, high);
  IOWR(timer_base, TIMER_PRDL_REG_OFT, low);
  /* configure timer to start, continuous mode, enabling interrupt */
  IOWR(timer_base, TIMER_CTRL_REG_OFT, 0x0007);
}
```

10.8 TASK ROUTINES

After completing the driver routines, we can drive the five task functions discussed in Listing 10.2. Note that a global variable with the data type of `cmd_type` is used among the routines to pass the current command. The suffix _v1 is used to indicate the version of these functions.

10.8.1 The `flashsys_init_v1()` function

The system initialization function is shown in Listing 10.7. It clears the btn module's edge capture register and sets up the timer for 1 ms. Note that the Nios II system runs at 50 MHz and thus the 1-ms interval requires 50,000 clocks (i.e.,1 ms = 50000 * 20 ns).

Listing 10.7

```
void flashsys_init_v1(alt_u32 btn_base, alt_u32 timer_base)
{
  btn_clear(btn_base);               // clear button edge-capture reg
  timer_wr_prd(timer_base, 50000);   // set 1-ms timeout period
}
```

10.8.2 The `sw_get_command_v1()` function

The `sw_get_command_v1()` function is shown in Listing 10.8. It processes the push-button switch and slide switch inputs and gets the command. We examine the btn module's edge capture register and update the **pause** and **prd** fields of **cmd** if an activity is detected. The register is cleared in the end to capture future activities.

Listing 10.8

```
void sw_get_command_v1(alt_u32 btn_base, alt_u32 sw_base, cmd_type *cmd)
{
  alt_u8 btn;

  btn = (alt_u8) btn_read(btn_base) & 0xf;   // read 4 pushbuttons
  if (btn!=0){                               // a button pressed
    if (btn & 0x01)                          // button 0 pressed
      cmd->pause = cmd->pause ^ 1;           // toggle pause bit
    if (btn & 0x02)                          // button 1 pressed
      cmd->prd = pio_read(sw_base) & 0x03ff; // load new interval
    btn_clear(btn_base);                     // clear edge-capture reg
  }
}
```

10.8.3 The jtaguart_disp_msg_v1() function

The jtaguart_disp_msg_v1() function is shown in Listing 10.9. When a new interval is detected, it sends a message of "Interval: dddd ms" to the console, where dddd is the decimal value of interval.

A static variable, current, is used to save the current interval. It is compared with the prd field of the new command to determine whether the interval is changed.

The message is stored in a string variable, msg. The size of the array is implicitly defined by the initial value, "Interval: 0000 ms\n". The four elements occupied by 0000 are modified according to the value of a new interval. The corresponding digits are obtained by simple division and modulo operations and then converted to the proper ASCII code by adding the ASCII value of character '0'.

Listing 10.9

```
void jtaguart_disp_msg_v1(alt_u32 jtag_base, cmd_type cmd)
{
  static int current=0;                      // current interval
  char msg[] ="Interval: 0000 ms\n";

  if (cmd.prd!=current){                      // new interval detected
    msg[13] = cmd.prd%10 + '0';              // ascii code for 0 digit
    msg[12] = (cmd.prd/10)%10 + '0';         // ascii code for 10 digit
    msg[11] = (cmd.prd/100)%10 + '0';        // ascii code for 100 digit
    msg[10] = cmd.prd/1000 + '0';            // ascii code for 1000 digit
    jtaguart_wr_str(jtag_base, msg);         // send string to console
    current = cmd.prd;                        // update current interval
  }
}
```

The process of constructing the message string is quite tedious. If the stdio library is available, the integer-to-string conversion can be done by calling the fprintf() function. This issue is discussed in Chapter 11.

10.8.4 The sseg_disp_msg_v1() function

The sseg_disp_msg_v1() function shows the current command on the four-digit seven-segment LED display. The most significant digit indicates whether the flashing is paused (a P pattern or blank) and the other three digits show the value of the flashing interval. The three-digit display can accommodate up to 999 and stays at the maximum if the value exceeds 999. The function is shown in Listing 10.10. The LED patterns are stored in a four-element array. Three digits are obtained by division and modulo operations and their corresponding patterns are obtained by calling the **sseg_conv_hex()** function. Since the table only defines patterns for a hexadecimal number, the 'P' and blank patterns of the leftmost seven-segment display are specified explicitly in code.

Listing 10.10

```
void sseg_disp_msg_v1(alt_u32 sseg_base, cmd_type cmd)
{
  int pd;
  alt_u8 hex, msg[4];


  if (cmd.prd > 999)         // 999 is max # to be displayed
    pd = 999;
  else
    pd = cmd.prd;
  hex = pd%10;               // 0 digit
  msg[3] = sseg_conv_hex(hex);
  hex = (pd/10)%10;          // 10 digit
  msg[2] = sseg_conv_hex(hex);
  hex = pd/100;              // 100 digit
  msg[1] = sseg_conv_hex(hex);
  /* specify pattern for the most significant digit */
  if (cmd.pause)
    msg[0] = 0x0c;           // P pattern
  else
    msg[0] = 0xff;           // Blank
  sseg_disp_ptn(sseg_base, msg); // display the whole pattern
}
```

10.8.5 The led_flash_v1() function

The **led_flash_v1()** function toggles the two discrete LEDs according to the specified interval and the code is shown in Listing 10.11. If the **pause** field is not asserted, the routine toggles the LED pattern and enters the while loop. The loop body checks the 1-ms tick continuously and increments the count, **ntick**, when the tick is asserted. The loop exits when **ntick** reaches the designated value.

Listing 10.11

```
void led_flash_v1(alt_u32 led_base, alt_u32 timer_base, cmd_type cmd)
{
  static alt_u8 led_pattern = 0x01;
  int ntick=0;

  if (cmd.pause)                         // no toggle if pause asserted
    return;
  led_pattern ^= 0x03;                   // toggle 2 LSBs of LEDs
  pio_write(led_base, led_pattern);      // write LEDs
  while (ntick < cmd.prd){
    if (timer_read_tick(timer_base)==1){
      timer_clear_tick(timer_base);
      ntick++;
    } // end if
  } // end while
```

}

The new timing loop uses the hardware timer and is much more accurate than the dummy-loop scheme used in Listing 9.8. Since the timer runs continuously, the first tick counted in the loop includes the time spent on the three functions, `sw_get_command_v1()`, `jtaguart_disp_msg_v1()`, and `sseg_disp_msg_v1()`, of the main loop. The execution time of the three functions is likely smaller than the flashing interval and thus has no impact on the timing.

Note that the timing loop of the function uses a "busy-waiting" strategy. The pushbutton switch can only be examined after each flashing interval and some activities may be missed if the interval is extremely long.

10.9   SOFTWARE CONSTRUCTION AND TESTING

We can follow the tutorial in Section 9.3 to build and run the software in the enhanced Nios II system. The main steps are:

- Create a BSP library using the `nios_led2.sopcinfo` file generated by SOPC Builder. We name the library `nios_led2_bsp`.
- Create an application directory `nios_led2_adhoc` (the suffix `_adhoc` is for the ad hoc device drivers) and add the C files to this directory.
- Program the FPGA device on the DE1 board using the `nios_led2_top.sof` file obtained from synthesis.
- Build and run the software image in the Nios II system.

After the successful implementation, the two LEDs system should flash at the default 100-ms rate. We can use switches to change its rate or pause its operation and verify the operation of the physical system.

With the default BSP setting, the code size of the software image file is about 18 KB, which can be easily accommodated by the 512-KB external SRAM chip.

10.10   BIBLIOGRAPHIC NOTES

Developing a robust and versatile driver is quite involved. *Programming Embedded Systems in C and C ++* by M. Barr introduces the basic concept and illustrates the construction of simple device drivers. Altera's *Embedded Peripherals IP User Guide* lists the available IP cores and provides detailed information about the PIO, timer, and JTAG UART cores.

10.11   SUGGESTED EXPERIMENTS

10.11.1   "Uptime" feature in flashing-LED system

"Uptime" time keeps track of how long the flashing-LED system is running. It is reset to 0 when the system is initialized and then counts continuously. This feature operates as follows:

1. The system sends a "`Flashing-LED system has run for MM minutes`" message to the console every minute, where `MM` is uptime in minutes.

2. The uptime is shown on four seven-segment LED displays in "MMSS" format, where "MM" is minutes and "SS" is seconds.

3. Pushbutton switch 2 is used toggle between the uptime and flashing interval on the seven-segment LED displays.

Develop software without using C library or HAL API functions and verify its operation.

10.11.2 Counting with different timer mode

The "continuous" mode of the timer core is used in the flashing-LED system's software development. The same task can also be achieved by using the "count-once" mode. Modify the driver and task functions and verify the operation.

10.11.3 JTAG UART input

For the enhanced flashing circuit, we want to use the host to obtain the flashing interval. This feature operates as follows:

1. When pushbutton switch 2 is pressed, the host console displays a "New interval: " message.

2. The user should enter three digits.

3. The program should check the correctness of the input, convert the three characters to an integer, and set the interval accordingly.

Develop software without using C library or HAL API functions and verify its operation.

10.11.4 Enhanced collision LED circuit

A collision LED circuit is discussed in Section 9.10.2, in which two LEDs move independently and change direction when reaching the rightmost or leftmost position or "colliding" in the middle. An enhanced version can be constructed with the new I/O modules. It operates as follows:

1. The 16 discrete LEDs are used as the output, two lit at a time.

2. The lit LEDs move sequentially in either direction. They changes direction when reaching the rightmost or leftmost position or "colliding" in the middle.

3. Each LED can travel at 99 different speeds.

4. The speed of the first lit LED is read from the slide switches when pushbutton switch 1 is pressed.

5. The speed of the second lit LED is read from the slide switches when pushbutton switch 2 is pressed.

6. The LED movements pause when pushbutton switch 0 is pressed and resume when it is released.

7. The speed of the first lit LED is shown on the right two seven-segment LED displays and the speed of the second lit LED is shown on the left two seven-segment LED displays.

8. The host console displays the new speed when it is first set.

Develop software without using C library or HAL API functions and verify its operation.

10.11.5   Rotating LED banner circuit

A rotating LED banner circuit is described in Section 5.9.5. Instead of using custom hardware, we can use an embedded system to perform this task. Develop software without using C library or HAL API functions and verify its operation.

10.11.6   Enhanced stopwatch

A stopwatch circuit is described in Section 5.9.6. Instead of using custom hardware, we can use an embedded system to perform this task. Develop software without using C library or HAL API functions and verify its operation.

10.11.7   Parking lot occupancy counter

A parking lot occupancy counter circuit is described in Section 6.5.3. Instead of using custom hardware, we can use an embedded system to perform this task. Develop software without using C library or HAL API functions and verify its operation.

10.11.8   Reaction timer with pushbutton switch control

A reaction timer circuit is described in Section 7.5.6. Instead of using custom hardware, we can use an embedded system to perform this task. Develop software without using C library or HAL API functions and verify its operation.

10.11.9   Reaction timer with keyboard control

For the reaction timer in Section 10.11.8, we want to use the host's keyboard (via the JTAG UART module) to replace the pushbutton witches to control the system operation. The key input is used as follows:
- C key: used as the `clear` signal.
- S key: used as the `start` signal.
- P key: used as the `stop` signal.
- All other key activities will be ignored.

Develop software without using C library or HAL API functions and verify its operation.

10.11.10   Communication with serial port

A DE1 board has a serial port and the Altera IP library provides a UART core. We can replace the JTAG UART with a normal UART and use the serial port to communicate with the host computer. Develop this system as follows:
1. Create a new Nios II system that replaces the JTAG UART module with a normal UART module.
2. Develop the top-level HDL code and synthesize the circuit.
3. Study the UART core manual and develop driver functions.
4. Develop application software and verify the system operation.

Note that this experiment requires a computer equipped with a serial port and a terminal program that displays the serial port data stream (such as Window's HyperTerminal) and a regular serial cable.

10.12 COMPLETE PROGRAM LISTING

Listing 10.12   chu_avalon_gpio.h

```
/*****************************************************************
*
* Module:   General-purpose I/O driver header
* File:     chu_avalon_gpio.h
* Purpose: Routines to access switches and LEDs
*
*****************************************************************/
/* file inclusion */
#include "alt_types.h"
#include "io.h"

/*****************************************************************
* constant definitions
*****************************************************************/
#define PIO_DATA_REG_OFT  0  // data register address offset
#define PIO_DIRT_REG_OFT  1  // direction register address offset
#define PIO_INTM_REG_OFT  2  // interrupt mask register address offset
#define PIO_EDGE_REG_OFT  3  // edge capture register address offset

/*****************************************************************
* macro definitions
*****************************************************************/
/* read/write PIO data register */
#define pio_read(base)         IORD(base, PIO_DATA_REG_OFT)
#define pio_write(base, data)  IOWR(base, PIO_DATA_REG_OFT, data)
/* read/clear pushbutton edge capture register */
/* must write 0xf if the write-individual bit option is used in SOPC */
#define btn_read(base)         IORD(base, PIO_EDGE_REG_OFT)
#define btn_clear(base)        IOWR(base, PIO_EDGE_REG_OFT, 0xf)
#define btn_is_pressed(base)   (IORD(base, PIO_EDGE_REG_OFT)!=0)

/*****************************************************************
* function prototypes
*****************************************************************/
alt_u8 sseg_conv_hex(int hex);
void sseg_disp_ptn(alt_u32 base, alt_u8 *ptn);
```

Listing 10.13   chu_avalon_gpio.c

```
/********************************************************************
*
* Module:   General-purpose I/O driver function prototype
* File:     chu_avalon_gpio.c
* Purpose:  Routines to access switches and LEDs
*
********************************************************************/
/* file inclusion */
#include "chu_avalon_gpio.h"


/********************************************************************
* function: sseg_conv_hex( )
* purpose:  convert a hex digit to 7-segment pattern
* argument:
*   hex: hex digit (0 - 15)
* return: 7-segment LED display pattern
* note:
*   - blank pattern returned if hex > 15
********************************************************************/
alt_u8 sseg_conv_hex(int hex)
{
  /* active-low hex digit 7-seg patterns (0-9,a-f); MSB ignored */
  static const alt_u8 SSEG_HEX_TABLE[16] = {
    0x40, 0x79, 0x24, 0x30, 0x19, 0x92, 0x02, 0x78, 0x00, 0x10, // 0-9
    0x88, 0x03, 0x46, 0x21, 0x06, 0x0E};                        // a-f
  alt_u8 ptn;

  if (hex < 16)
    ptn = SSEG_HEX_TABLE[hex];
  else
    ptn = 0xff;
  return (ptn);
}


/********************************************************************
* function: sseg_disp_ptn( )
* purpose:  display pattern in four 7-segment LED display
* argument:
*   base: base address of 7-segment display
*   ptn: pointer to a 4-element pattern
* return:
* note:
********************************************************************/
void sseg_disp_ptn(alt_u32 base, alt_u8 *ptn){
  alt_u32 sseg_data;
  int i;

  /* form a 32-bit data */
  for (i=0; i<4; i++){
    sseg_data = (sseg_data << 8) | *ptn;
    ptn++;
  }
  pio_write(base, sseg_data);
}
```

Listing 10.14   chu_uart_drv.h

```
/****************************************************************************
 *
 * Module: Demo JTAG UART driver header
 * File:   chu_uart_drv.h
 *
 ****************************************************************************/
/* file inclusion */
#include "alt_types.h"
#include "io.h"


/****************************************************************************
 * constant definitions
 ****************************************************************************/
#define JUART_DATA_REG_OFT   0           // data register address offset
#define JUART_CTRL_REG_OFT   1           // control register addr offset

/* check # slots available in FIFO buffer */
#define jtaguart_rd_wspace(base)   \
   ((IORD(base, JUART_CTRL_REG_OFT) & 0xffff0000) >> 16)
/* write an 8-bit char */
#define jtaguart_wr_ch(base, data)   \
   IOWR(base, JUART_DATA_REG_OFT, data & 0x000000ff)

/****************************************************************************
 * function prototypes
 ****************************************************************************/
void jtaguart_wr_str(alt_u32 jtag_base, char* msg);
```

Listing 10.15   chu_uart_drv.c

```
/***********************************************************************
*
* Module:   Demo JTAG UART driver functions
* File:     chu_uart_drv.c
* Purpose:  Function to transmit a string
*
***********************************************************************/
/* file inclusion */
#include "chu_uart_drv.h"


/***********************************************************************
* function: jtaguart_wr_str()
* purpose:  write (transmit) a string to JTAG UART
* argument:
*   jtag_base: base address of JTAG UART
*   msg: pointer to a string message
* return:
* note:
***********************************************************************/
void jtaguart_wr_str(alt_u32 jtag_base, char* msg)
{
  alt_u32 data32;

  while(*msg){
    data32 = (alt_u32) *msg;
    if(jtaguart_rd_wspace(jtag_base)!=0){   // buffer space available
      jtaguart_wr_ch(jtag_base, data32);    // send a char
      msg++;
    } // end if
  } // end while
}
```

Listing 10.16 chu_timer_drv.h

```
/************************************************************************
*
* Module: Demo timer driver header
* File:   chu_timer_drv.h
*
************************************************************************/
/* file inclusion */
#include "alt_types.h"
#include "io.h"

/************************************************************************
* constant definitions
************************************************************************/
#define TIMER_STAT_REG_OFT 0   // status register address offset
#define TIMER_CTRL_REG_OFT 1   // control register address offset
#define TIMER_PRDL_REG_OFT 2   // period reg (lower 16 bits) addr offset
#define TIMER_PRDH_REG_OFT 3   // period reg (upper 16 bits) addr offset

/************************************************************************
* macro definitions
************************************************************************/
/* check "to" field for ms tick */
#define timer_read_tick(base)   (IORD(base, TIMER_STAT_REG_OFT) & 0x01)
/* clear "to" field */
#define timer_clear_tick(base)  IOWR(base, TIMER_STAT_REG_OFT, 0)

/************************************************************************
* function prototypes
************************************************************************/
void timer_wr_prd(alt_u32 timer_base, alt_u32 prd);
```

Listing 10.17   chu_timer_drv.c

```
/***********************************************************************
 *
 * Module:  Demo timer driver functions
 * File:    chu_timer_drv.c
 * Purpose: Functions to set up timer
 *
 ***********************************************************************/
/* file inclusion */
#include "chu_timer_drv.h"


/***********************************************************************
 * function: timer_wr_prd()
 * purpose:  write timer timeout period and configure/start timer
 * argument:
 *    timer_base: base address of time-stamp timer
 *    prd: timeout period value
 * return:
 * note:
 ***********************************************************************/
void timer_wr_prd(alt_u32 timer_base, alt_u32 prd)
{
  alt_u16 high, low;

  /* unpack 32-bit timeout period into two 16-bit half words */
  high = (alt_u16) (prd>>16);
  low = (alt_u16) (prd & 0x0000ffff);
  /* write timeout period */
  IOWR(timer_base, TIMER_PRDH_REG_OFT, high);
  IOWR(timer_base, TIMER_PRDL_REG_OFT, low);
  /* configure timer to start, continuous mode; enable interrupt */
  IOWR(timer_base, TIMER_CTRL_REG_OFT, 0x0007);
}
```

Listing 10.18   chu_main_led2_adhoc.c

```
/************************************************************************
 *
 * Module:   Advanced flashing-LED system using ad hoc driver
 * File:     chu_main_led2_adhoc.c
 * Purpose: Task routines and main program
 *
 ************************************************************************/
/* file inclusion */
#include "system.h"
#include "chu_avalon_gpio.h"
#include "chu_uart_drv.h"
#include "chu_timer_drv.h"

/************************************************************************
 * data type definitions
 ************************************************************************/
typedef struct flash_cmd{
  int pause;
  int prd;
} cmd_type;

/************************************************************************
 * function: flashsys_init_v1()
 * purpose:  system initialization
 * argument:
 *   btn_base: base address of pushbutton PIO
 *   timer_base: base address of user timer
 * return:
 * note:
 ************************************************************************/
void flashsys_init_v1(alt_u32 btn_base, alt_u32 timer_base)
{
  btn_clear(btn_base);                     // clear button edge-capture reg
  timer_wr_prd(timer_base, 50000);         // set 1-ms timeout period
}

/************************************************************************
 * function: sw_get_command_v1()
 * purpose:  get command from pushbuttons and switches
 * argument:
 *   btn_base: base address of pushbutton PIO
 *   sw_base: base address of switch PIO
 *   cmd: pointer to command
 * return:
 *   updated cmd
 * note:
 ************************************************************************/
void sw_get_command_v1(alt_u32 btn_base, alt_u32 sw_base, cmd_type *cmd)
{
  alt_u8 btn;

  btn = (alt_u8) btn_read(btn_base) & 0xf;   // read 4 pushbuttons
  if (btn!=0){                               // a button pressed
    if (btn & 0x01)                          // button 0 pressed
      cmd->pause = cmd->pause ^ 1;           // toggle pause bit
    if (btn & 0x02)                          // button 1 pressed
      cmd->prd = pio_read(sw_base) & 0x03ff; // load new interval
    btn_clear(btn_base);                     // clear edge-capture reg
  }
}
```

```
/****************************************************************
* function: jtag_uart_disp_msg_v1()
* purpose:  display the interval when it is changed
* argument:
*   jtag_base: base address of JTAG UART
*   cmd: command
* return:
* note:
****************************************************************/
void jtaguart_disp_msg_v1(alt_u32 jtag_base, cmd_type cmd)
{
  static int current=0;                   // current interval
  char msg[] ="Interval: 0000 ms\n";

  if (cmd.prd!=current){                   // new interval detected
    msg[13] = cmd.prd%10 + '0';            // ascii code for 0 digit
    msg[12] = (cmd.prd/10)%10 + '0';       // ascii code for 10 digit
    msg[11] = (cmd.prd/100)%10 + '0';      // ascii code for 100 digit
    msg[10] = cmd.prd/1000 + '0';          // ascii code for 1000 digit
    jtaguart_wr_str(jtag_base, msg);       // send string to console
    current = cmd.prd;                     // update current interval
  }
}


/****************************************************************
* function: sseg_disp_msg_v1()
* purpose:  display current pause status and interval on 4-digit 7-seg LED
* argument:
*   sseg_base: base address of seven-segment LED display PIO
*   cmd: command
* return:
* note:
****************************************************************/
void sseg_disp_msg_v1(alt_u32 sseg_base, cmd_type cmd)
{
  int pd;
  alt_u8 hex, msg[4];


  if (cmd.prd > 999)              // 999 is max # to be displayed
    pd = 999;
  else
    pd = cmd.prd;
  hex = pd%10;                    // 0 digit
  msg[3] = sseg_conv_hex(hex);
  hex = (pd/10)%10;               // 10 digit
  msg[2] = sseg_conv_hex(hex);
  hex = pd/100;                   // 100 digit
  msg[1] = sseg_conv_hex(hex);
  /* specify pattern for the most significant digit */
  if (cmd.pause)
    msg[0] = 0x0c;                // P pattern
  else
    msg[0] = 0xff;                // Blank
  sseg_disp_ptn(sseg_base, msg); // display the whole pattern
}
```

```
/*************************************************************************
 * function: led_flash_v1 ()
 * purpose:  toggle 2 LEDs according to the given interval
 * argument:
 *   led_base: base address of discrete LED  PIO
 *   timer_base: base address of user timer
 *   cmd: command
 * return:
 * note:
 *   The delay is done by continuously checking 1 ms tick
 *************************************************************************/
void led_flash_v1(alt_u32 led_base, alt_u32 timer_base, cmd_type cmd)
{
  static alt_u8 led_pattern = 0x01;
  int ntick=0;

  if (cmd.pause)                              // no toggle if pause asserted
    return;
  led_pattern ^= 0x03;                        // toggle 2 LSBs of LEDs
  pio_write(led_base, led_pattern);           // write LEDs
  while (ntick < cmd.prd){
    if (timer_read_tick(timer_base)==1){
      timer_clear_tick(timer_base);
      ntick++;
    } // end if
  } // end while
}


/*************************************************************************
 * function: main()
 * purpose:  advanced flashing-LED system using ad hoc driver
 * note:
 *************************************************************************/
int  main(){
  cmd_type sw_cmd={0,100};   // initial value: not pause, 100 ms interval

  flashsys_init_v1(BTN_BASE, USR_TIMER_BASE);
  while(1){
    sw_get_command_v1(BTN_BASE, SWITCH_BASE ,&sw_cmd);
    jtaguart_disp_msg_v1(JTAG_UART_BASE, sw_cmd);
    sseg_disp_msg_v1(SSEG_BASE, sw_cmd);
    led_flash_v1(LEDG_BASE, USR_TIMER_BASE, sw_cmd);
  }
}
```

CHAPTER 11

PREDESIGNED NIOS II I/O DRIVERS AND HAL API

Along with the set of commonly used I/O cores, Altera also provides predesigned device drivers and software libraries to access the underlying hardware. These drivers and libraries are integrated under the *HAL (hardware abstraction layer)* framework. HAL presents as a coherent interface and shields low-level details from application programs. In this chapter, we examine the basic concepts of HAL, discuss the software development process within this framework, and rewrite the flashing-LED program in Chapter 10 to illustrate its use.

11.1 OVERVIEW OF HAL

In Chapter 10, we demonstrate the derivation of simple ad hoc device drivers that match the specific requirements of the enhanced flashing-LED system. These device drivers need to be modified for each new application and will require a lot of time and effort. For many commonly used I/O peripherals, a better alternative is to develop a set of flexible and robust device drivers that can be shared by multiple applications. Altera provides a framework, known as *HAL*, for this purpose.

The role of HAL is somewhat murky since it does not fit any traditional paradigms. Altera literature vaguely refers to HAL as a *lightweight run-time environment*. To explain its role, we first examine two better understood paradigms and then point out the tasks performed by HAL.

Embedded SOPC Design with Nios II Processor and VHDL Examples. By Pong P. Chu   **255**
Copyright © 2011 John Wiley & Sons, Inc.

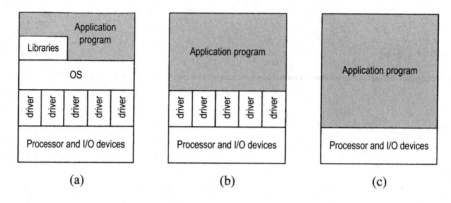

(a)   (b)   (c)

Figure 11.1   Software hierarchy.

11.1.1 Desktop-like and barebone embedded systems

Let us consider a *desktop-like embedded system* and a *barebone embedded system*, two paradigms representing the extremes of the spectrum. We focus on two aspects of the systems: the software hierarchy (when the program is developed) and the initialization process (when the program is executed).

Software hierarchy   The simplified software hierarchy of a desktop-like system is shown in Figure 11.1(a). It contains a full-fledged OS (operating system), such as uClinux, that controls and coordinates the hardware through the device drivers. The OS is the critical part of the software hierarchy. An application program can only access the underlying hardware via the OS and the drivers must comply with the requirements imposed by the OS.

A barebone system here means a simple microcontroller-based system. Because of its simplicity, it does not utilize any layered software model and does not include pre-implemented libraries. The application software controls the hardware directly and is responsible to develop codes to access I/O peripherals. The software hierarchy is shown in Figure 11.1(c). The simple flashing-LED code in Chapter 9 is somewhat like this. To achieve modularity, ad hoc device drivers may be developed as subprograms of the application program, as shown in Figure 11.1(b). Unlike the drivers under the OS, these ad hoc device drivers lack coordination and run independently. The flashing-LED code in Chapter 10 is somewhat like this.

Initialization process   Before running a program, we must set up the hardware, such as flushing the cache and configuring the system stack, and set up software, such as starting device drivers and enabling interrupts. The initialization process is handled differently in a desktop-like system and a barebone system.

In a desktop-like system, the OS is always running. It controls the hardware and schedules and coordinates the application program execution. Starting an application program corresponds to the OS "calling" the corresponding `main()` program. Before calling, the OS prepares the run-time environment, which includes allocating necessary resources and initializing system services. The framework constitutes a *hosted environment* for a C application program. This means that the application program can assume that all libraries and I/O services are ready to use and no

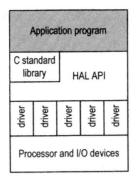

Figure 11.2 HAL-based software hierarchy.

additional work is needed. In this environment, we can simply write a statement like `printf("Hello")` without knowing any low-level detail.

In a barebone system, the loader usually adds a simple start-up code to initialize the processor. An application program is responsible for the other tasks and must prepare the run-time environment by itself. We need to add codes to initialize each I/O device, set up the interrupt service, and coordinate their operations. This is known as a *standing-alone environment* for a C application program. To display the `Hello` message on a console, the program must set up the corresponding I/O device (e.g., an UART) and include an I/O routine to transmit a string.

11.1.2 HAL paradigm

The HAL paradigm lies between a desktop-like system and a barebone system. It provides some features similar to those in a desktop-like system but is without the OS layer.

Software hierarchy The layer model of a HAL-based software hierarchy is shown in Figure 11.2. It consists of the following components:

- *Device drivers*: basic routines to interact with I/O devices.
- *API (application programming interface)*: a set of functions to invoke HAL services, such as timing facilities and interrupt handling, and a collection of Unix-style system functions.
- *C standard library*: ANSI C library functions, such as `printf()`, `fopen()`, etc.

HAL integrates the underlying device drivers to present a coherent interface and provides a shell to shield the complexity of the underlying hardware. For some common I/O functions, an application program can use generic API functions or C library functions without explicitly interacting with the I/O devices.

The appearance of the HAL software hierarchy is somewhat like that in a desktop-like system. However, the HAL interface does not impose rigid complex requirements as an OS does and using the HAL framework is not mandatory. An application program can use functions from HAL API, call the routines from an I/O device driver, or access the I/O registers directly. Similarly, an I/O device can be incorporated into a Nios II system with a HAL-compliant device driver, with an ad hoc driver, or with no driver at all. HAL-compliant device drivers are discussed in more detail in Section 11.1.4.

Initialization process   Unlike an OS, no explicit "HAL process" is presented in memory when the system is running. HAL thus cannot prepare the environment dynamically and call the `main()` program. However, HAL provides a framework to automatically collect hardware information and construct the routines to perform the initialization process. These routines are executed first and set up the run-time environment for the application program. The detailed initialization process is discussed in Section 11.1.6. If all device drivers are compliant with the HAL interface, an application program does not need to perform any explicit initialization task and we can think the development is done in a hosted environment.

11.1.3 Device classes

Although the nature and complexities of I/O devices vary, some devices exhibit certain common characteristics and can be described by a generic model. From the HAL API's point of view, devices can be divided into two types:
- Generic devices.
- Non-generic devices.

A device is called a *generic device* if it can fit into a predefined HAL model and a *non-generic device* otherwise.

Generic device   HAL provides a model for the following classes of devices:
- *Character-mode devices* are peripherals that send and receive a character stream serially, such as the JTAG UART core.
- *Timer devices* are peripherals that count clock ticks and can generate periodic interrupt requests, such as the timer core.
- *Flash memory devices* are nonvolatile memory that uses a special protocol to store data.
- *DMA devices* are peripherals that transfer bulk data between a source and a destination, such as the DMA controller core.
- *Ethernet devices* are peripherals that provide access to an Ethernet connection for a networking stack.
- *File subsystems* provide a mechanism for accessing files stored within physical devices, such as the Altera host-based file system and the Altera zip read-only file system.

For each class, a set of high-level functions is defined in HAL API. An application program can use these functions consistently without worrying the underlying implementation of the device hardware. For example, the `printf()` function is available for a character-mode device. We can use it to send a message to a character-mode device, regardless of whether the device is a JTAG UART, a serial port, or an LCD display.

HAL imposes specific requirements for each class and a device driver must satisfy these requirements to be integrated into the HAL framework.

Non-generic device   A non-generic device has hardware-specific features with usage requirements that do not map well to any class of HAL's generic device models. Its driver usually supplies special routines to perform the desired functionalities. We need to refer to the device's documentation for these features.

11.1.4 HAL-compliant device drivers

The term "device driver" is used loosely in this book for any software routines that access I/O devices. The drivers used in Chapter 10 have a limited scope and just perform certain specific tasks. They run independently and we call them *ad hoc* drivers.

In the HAL environment, a device driver is part of a larger software framework and must comply with requirements and guidelines imposed by this framework. With a compliant device driver, an I/O device becomes accessible to application programs through the C standard library and HAL API functions. In this more rigorous setting, a device driver is defined as a collection of software routines used to interact with a specific I/O peripheral. A HAL-compliant device driver consists of several components:

1. A set of macros to access I/O registers.
2. A set of variables to keep track of the current state of the instantiated device.
3. A routine to initialize the device.
4. A set of routines to control and communicate with the device.
5. A routine to provide interrupt service.
6. A script to direct the BSP integration.

Only the first component is mandatory and the inclusion of the others depends on the nature of an I/O device.

11.1.5 The _regs.h file

The structure and development of a HAL-compliant driver are quite involved and beyond the scope of the book. We only examine the mechanism to access a device's I/O registers. To avoid accidental memory access, HAL defines unique read and write macros for each I/O device and for each register. Their basic format are

```
IORD_dev_reg(base_address)
IOWR_dev_reg(base_address, data)
```

where **dev** and **reg** are the names of an I/O device and a register.

These macros are stored in a header file labeled **dev_regs.h**. For example, the altera_avalon_pio_regs.h file contains the macros for the PIO core. Recall that there are six addressable registers, as discussed in Section 10.2. Part of this file is shown in Listing 11.1.

Listing 11.1

```
#include <io.h>
. . .
#define IORD_ALTERA_AVALON_PIO_DATA(base)            IORD(base, 0)
#define IOWR_ALTERA_AVALON_PIO_DATA(base, data)      IOWR(base,0,data)
. . .
#define IORD_ALTERA_AVALON_PIO_DIRECTION(base)       IORD(base, 1)
#define IOWR_ALTERA_AVALON_PIO_DIRECTION(base, data) IOWR(base,1,data)
. . .
#define IORD_ALTERA_AVALON_PIO_IRQ_MASK(base)        IORD(base, 2)
#define IOWR_ALTERA_AVALON_PIO_IRQ_MASK(base, data)  IOWR(base,2,data)
. . .
#define IORD_ALTERA_AVALON_PIO_EDGE_CAP(base)        IORD(base, 3)
#define IOWR_ALTERA_AVALON_PIO_EDGE_CAP(base, data)  IOWR(base,3,data)
. . .
```

```
#define IORD_ALTERA_AVALON_PIO_SET_BITS(base)         IORD(base, 4)
#define IOWR_ALTERA_AVALON_PIO_SET_BITS(base, data)   IOWR(base, 4, data)
  . . .
#define IORD_ALTERA_AVALON_PIO_CLEAR_BITS(base)       IORD(base, 5)
#define IOWR_ALTERA_AVALON_PIO_CLEAR_BITS(base,data)  IOWR(base, 5, data)
```

For a register with multiple fields, HAL defines constants for bit-field masks and offsets. For example, in altera_avalon_jtag_uart_regs.h, the code segment for data register of the JTAG UART core is

```
#define IOWR_ALTERA_AVALON_JTAG_UART_DATA(base, data) IOWR(base,0,data)
#define IORD_ALTERA_AVALON_JTAG_UART_DATA(base)       IORD(base,0)
  . . .
#define ALTERA_AVALON_JTAG_UART_DATA_DATA_MSK      (0x000000FF)
#define ALTERA_AVALON_JTAG_UART_DATA_DATA_OFST     (0)
#define ALTERA_AVALON_JTAG_UART_DATA_RVALID_MSK    (0x00008000)
#define ALTERA_AVALON_JTAG_UART_DATA_RVALID_OFST   (15)
#define ALTERA_AVALON_JTAG_UART_DATA_RAVAIL_MSK    (0xFFFF0000)
#define ALTERA_AVALON_JTAG_UART_DATA_RAVAIL_OFST   (16)
```

We can extract a field by using these constants to perform mask and shift operations, as discussed in Section 9.7.2. For example, we can extract the rvalid by

```
#include "altera_avalon_jtag_uart_regs.h"
  . . .
d_reg = IORD_ALTERA_AVALON_JTAG_UART_DATA(UART_BASE);
rvalid = (d_reg & ALTERA_AVALON_JTAG_UART_DATA_RVALID_MSK) \
            > ALTERA_AVALON_JTAG_UART_DATA_RVALID_OFST;
```

Note that we must include the proper header file to use the macros and constants.

11.1.6 HAL-based initialization sequence

One of the key functionalities of the HAL framework is to set up the run-time environment for main(). The initialization process first executes the crt0.S file. It contains simple assembly codes and its main task is to set the processor to a known state. The basic steps are:

- Flush the instruction and data cache.
- Configure the stack pointer register.
- Configure the global pointer register.
- Fill the BSS regions with zeros.
- Copy relevant data sections to the designated memory module.
- Call alt_main().

The alt_main() function continues to set up the run-time environment. The basic steps included in the default implementation are:

- Initialize the interrupt controller and enable interrupts.
- Call the alt_sys_init() function, which initializes all device drivers of the instantiated I/O modules.
- Redirect the C standard I/O channels (stdin, stdout, and stderr) to use the appropriate devices.
- Process C++ constructors.
- Call main().
- Call exit().

We discuss the hosted and stand-alone environments in Section 11.1.1. The alt_main() function behaves like a host program that prepares the run-time en-

vironment and calls `main()`. The `main()` function does not need to do general
I/O initialization and thus runs in a hosted environment. From a different point of
view, `alt_main()` can be considered as the main program running in the stand-alone
environment.

11.2 BSP

11.2.1 Overview

An embedded system is built around a specific application and its I/O peripherals
are tailored to support the application. When a program is developed, we must
examine the I/O configuration and then include the proper drivers accordingly. One
way to accomplish this task is to create a "start-up" software library according
to the I/O configuration. It is known as a *BSP (board support package)* since a
traditional embedded system is usually realized by a custom printed circuit board.

In the HAL environment, a BSP provides a run-time software library customized
for an individual Nios II system. During the software construction, the files from the
user application and the supporting files from the BSP library are linked together
to form the final image. When a Nios II system is created, SOPC Builder records
the system configuration in the .sopcinfo file. In the Altera EDS framework, we can
specify the name of this file and invoke the script to generate the BSP library, as
illustrated in Section 9.3.1.

11.2.2 BSP file structure

It is helpful to understand the basic BSP library structure. In Chapter 10, a BSP,
named `nios_led2.bsp`, is generated for the enhanced flashing-LED system. The
top-level structure of the BSP library is shown in Figure 11.3(a). It consists of
three directories and a collection of system-level files. The three directories are:

- drivers: contains the source codes and headers for the instantiated I/O cores.
- HAL: contains the source codes and headers for the HAL API functions.
- obj: contains the compiled object files.

Two of the system-level files are of our interest. The system.h file is a header file
that maintains relevant Nios II system in terms of constant definitions, as explained
in Section 9.6.1. The alt_sys_init.c file contains the customized initialization code,
as discussed in Section 11.1.6.

The expanded structure of the drivers directory and a portion of the HAL directory are shown in Figure 11.3(b). Note that the driver directory contains the header
files and source codes for the instantiated PIO, JTAG UART, timer, and system
id cores, and the HAL directory contains the alt_types.h and io.h files discussed in
Section 9.6.

11.2.3 BSP configuration

The BSP library involves the device drivers and system initialization. Some aspects
of the library can be adjusted and fine-tuned to fit the requirement of an individual
system. The *BSP Editor* utility program can be used to configure the BSP library.

(a) Top-level structure   (b) Expanded structure

Figure 11.3   BSP Library structure.

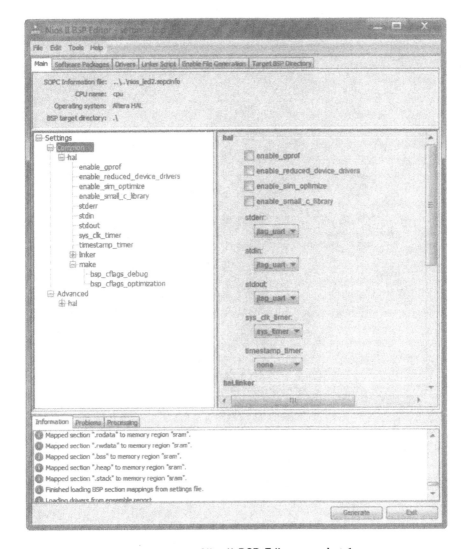

Figure 11.4   Nios II BSP Editor snapshot 1.

For our purposes, we use BSP Editor to set up standard C I/O channels, select full- or reduced-sized device drivers and C library, and configure the initialization code. The procedure to perform these tasks is:

1. In Eclipse SBT GUI window, Select Nios II ≻ BSP Editor. A dialog appears.
2. Select the previously constructed led2_bsp library and click the Ok button. The Nios II BSP Editor window appears.
3. Select the main tab page, and expand Common ≻ hal. The snapshot is shown in Figure 11.4.
4. In the middle of the hal box, there are fields labeled stderr, stdin, and stdout, which indicate the I/O modules that act as standard error, standard input, and standard output channels in C. Select jtag_uart module for all three fields.
5. In the middle of the hal box, there are fields labeled sys_clk_timer and time- stamp_timer, which indicate the timer modules used for the general system

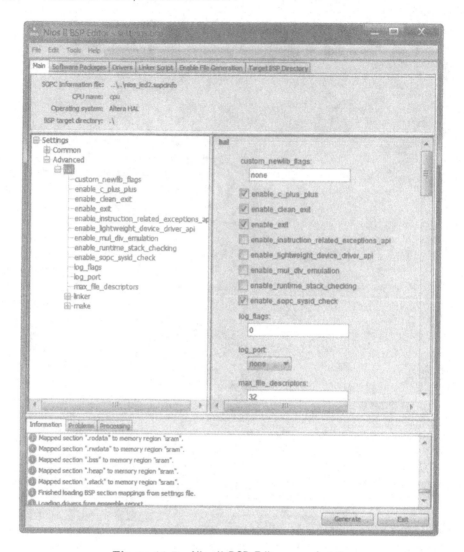

Figure 11.5 Nios II BSP Editor snapshot 2.

timing and time stamp API routines. Select the sys_timer module for the sys_clk_timer field and leave the timestamp_timer field blank (since we are not use the time stamp function).

6. In the top of the hal box, there are several optional items. If needed, we can check enable_reduced_device_drivers to select smaller but slower drivers and can check enable_small_c_library to select the smaller "watered-down" C library functions.

7. In the main tab page, expand Advanced ≻ hal. The snapshot is shown in Figure 11.5.

8. In the top of the hal box, there are several optional items, which are enable_c_plus_plus, enable_clean_exit, enable_exit, enable_lightweight_device_driver_api. Check these items as needed.

9. In the bottom of the hal box, there is a field labeled max_file_descriptors, which specifies that the maximal number of files can be used. Enter the desired number. It must be at least 4 for proper HAL API operation.

10. Click the bottom Generate button to generate the BSP library.

11. Click the Exit button to close the program.

For our purposes, the default BSP setting works fine as long as a Nios II system has adequate main memory. We just need to verify the stderr, stdin, stdout, sys_clk_timer, and timestamp_timer fields to ensure that the proper I/O modules are selected.

Sometimes we may need to reduce the code footprint to accommodate a small memory module. To do this, we can enable the reduced device drivers, use the small C library, disable C++ support if not used, remove the exit feature if the application program never exits, and use a minimal number of file descriptors. For example, the code size of the flashing-LED program of Chapter 9 is about 18 KB. It can be reduced to about 8 KB if these options are used.

11.3 HAL-BASED FLASHING-LED PROGRAM

With the availability of HAL, we can rewrite the codes in Chapter 10 by using functions from the C standard library and HAL API. These functions can be divided according to the *generic* or *non-generic* types of I/O devices defined in Section 11.1.3.

A generic device is usually set up when the BSP library is generated and generic high-level functions, such as those in C library, are constructed based on these devices. An application program can call these functions without worrying underlying low-level details. A non-generic device is more specialized and thus no generic function is available. We need to use the routines or register-access macros defined in the I/O device driver.

In this enhanced flashing-LED system, the JTAG UART and timer modules are generic devices and thus we can use the generic API and library functions for the two devices. The PIO modules are non-generic devices and we must use the proper driver routines to access the I/O registers. The ad hoc driver routines and task functions of Chapter 10 can be modified to take advantages of HAL API. The revisions of these routines are discussed in the following subsections. If a new version is needed, we add a suffix _v2 to distinguish them from the earlier versions.

11.3.1 Functions using generic I/O devices

Function using the JTAG UART core   The JTAG UART core is a character-mode device that can send or receive a serial character stream. In the HAL framework, the device driver of this type of device is integrated with the C standard library and thus the functions in the stdio library, including **printf()** and **scanf()**, can be used. Recall that the instantiated jtag_uart module is selected as the C standard I/O channels when the nios_led2_bsp BSP library is generated, as discussed in Section 11.2.3. In this system, we can use the **printf()** function in C library to display a message. No low-level information, such as the name and base address of the instantiated I/O device, the I/O registers, or the device driver routines, is needed. Similarly, only the generic high-level header files, such as `stdio.h`, are included in the code.

In the flashing-LED codes, the `jtag_uart` module is used to send a message of "Interval: dddd ms" to the host console when a new interval is detected, as discussed in Section 10.8.3. After directing C's `stdout` channel to `jtag_uart`, we can use the print functions in C's `stdio` library to write to the console directly. The modified function, `jtaguart_disp_msg_v2()`, is shown in Listing 11.2.

Listing 11.2

```
#include <stdio.h>
...
void jtag_uart_disp_msg_v2(cmd_type cmd)
{
  static int old=0;

  if (cmd.prd!=old){
    printf("Interval: %03u ms \n", cmd.prd);
    old = cmd.prd;
  }
}
```

Note that the integer-to-string conversion is automatically performed within the `printf()` function.

Function using the timer core   The HAL framework provides a collection of high-level timing related functions in its API. Three useful functions are:

- `usleep (unsigned int t)`: it forces the system execution to halt for `t` microseconds.
- `alt_nticks()`: it returns the number of elapsed system clock ticks since reset.
- `alt_ticks_per_second()`: it returns the number of system clock ticks per second.

Implementing this function requires a timer core. Recall that the instantiated sys_timer module is selected for this purpose when the nios_led2_bsp library is generated, as discussed in Section 11.2.3. After specifying the proper system timer module in BSP, we can use the API timing functions without worrying about low-level details.

In Chapter 10, we use the usr_timer module and write low-level codes to wait a specific amount of time before toggling the two discrete LEDs, as discussed in Section 10.8.5. With HAL API, the same task can be done with the `usleep()` function. The modified function, `led_flash_v2()`, is shown in Listing 11.3.

Listing 11.3

```
#include <unistd.h>                      // header file for usleep()
...
void led_flash_v2(alt_u32 led_base, cmd_type cmd)
{
  static alt_u8 led_pattern = 0x01;

  if (cmd.pause)                         // no toggle if pause asserted
    return;
  led_pattern ^= 0x03;                   // toggle 2 LSBs of LEDs
  pio_write(led_base, led_pattern);      // write LEDs
  usleep(1000*cmd.prd);                  // delay for cmd.prd ms
}
```

The `led_flash_v2()` function follows the design principle of `led_flash_v1()` in Chapter 10 and waits for a specific amount of time (i.e., flashing interval). During the waiting period, the program execution is suspended and the processor is largely idle. Instead of the "busy-waiting" scheme, a better alternative is to use a function to check the elapsed time and toggle the LEDs as needed. This can be done by using HAL API's `alt_nticks()` and `alt_ticks_per_second()` functions. The modified flashing function, `led_flash_v3()`, is shown in Listing 11.4.

Listing 11.4

```
#include <sys/alt_alarm.h>      // header file for alt_nticks() and
                                //             alt_ticks_per_second()
...
void led_flash_v3(alt_u32 led_base, cmd_type cmd)
{
  static alt_u8 led_pattern = 0x01;
  static int last=0;
  int now;

  if (cmd.pause)                         // no toggle if pause asserted
    return;
  now = (int) (alt_nticks() * alt_ticks_per_second() / 1000);
  if ((now - last)<cmd.prd)              // interval not reached
    return;
  last = now;
  led_pattern ^= 0x03;                   // toggle 2 LSBs of LEDs
  pio_write(led_base, led_pattern);      // write LEDs
}
```

The function uses a static variable, `last`, to record the last time when the LEDs toggle and calculates the current time, `now`, with `alt_ticks_per_second()` and `alt_nticks()`. When the elapsed time reaches the designated interval, the LEDs toggle and `last` is updated. Both `last` and `now` are represented in term of millisecond.

11.3.2 Functions using non-generic I/O devices

The LED and switch modules of the enhanced flashing-LED system are PIO cores. Since the PIO core is a non-generic devices, we need to use core-specific driver routines to control and communicate with these cores.

The PIO core is fairly simple and the key function of the driver is to read and write the core's I/O registers. In Chapter 10, we define several macros to read data from switches and write data to LEDs. The corresponding segment in chu_avalon_gpio.h is shown in Listing 11.5.

Listing 11.5

```
#include "io.h"

#define PIO_DATA_REG_OFT  0  // data register addr offset
#define PIO_EDGE_REG_OFT  3  // edge capture register addr offset
/* read/write PIO data register */
#define pio_read(base)          IORD(base, PIO_DATA_REG_OFT)
#define pio_write(base, data)   IOWR(base, PIO_DATA_REG_OFT, data)
/* read/clear pushbutton edge capture register */
#define btn_read(base)          IORD(base, PIO_EDGE_REG_OFT)
#define btn_clear(base)         IOWR(base, PIO_EDGE_REG_OFT, 0xf)
```

The code uses symbolic constants and the basic IORD() and IOWR() functions.

In the HAL framework, direct I/O register access is performed by low-level macros in the header file, as discussed in Section 11.1.5. We can use the macros defined in altera_avalon_pio_regs.h, which is shown in Listing 11.1, to access the designated PIO register. The revision of the previous segment is shown in Listing 11.6.

Listing 11.6

```
#include "altera_avalon_pio_regs.h"

/* read/write PIO data register */
#define pio_read(base)        IORD_ALTERA_AVALON_PIO_DATA(base)
#define pio_write(base, data) IOWR_ALTERA_AVALON_PIO_DATA(base, data)
/* read/clear pushbutton edge capture register */
#define btn_read(base)        IORD_ALTERA_AVALON_PIO_EDGE_CAP(base)
#define btn_clear(base)       IOWR_ALTERA_AVALON_PIO_EDGE_CAP(base, 0xf)
```

Note that these IORD_ALTERA_AVALON_PIO... macros are based on the IORD() and IOWR() functions as well. This approach is somewhat "wordy" and we do not use this type of macro definitions in this book.

The sw_get_command_v1() and sseg_disp_msg_v1() functions of Chapter 10 only use the macros defined in chu_avalon_gpio.h to access I/O ports and thus can be kept without modification.

11.3.3 Initialization routine and main program

Initialization routine An embedded application usually includes a system initialization routine. It is executed in the beginning to set up I/O peripherals to the desired condition. For example, the flashsys_init_v1() function in Section 10.8 clears the btn module's edge capture register and sets up the usr_timer module for 1-ms interval.

The HAL framework initializes the Nios II system and prepares the run-time environment for the application's the main() program, as discussed in Section 11.1.6. One step in the initialization process is to call the alt_sys_init() function to set up the I/O modules. This function basically assembles initialization routines supplied by the instantiated device drivers when the BSP library is constructed.

The alt_sys_init.c file can be found in the BSP directory. For the enhanced enhanced flashing-LED Nios II system, the function is

```
void alt_sys_init(void)
{
  ALTERA_AVALON_TIMER_INIT(SYS_TIMER, sys_timer);
  ALTERA_AVALON_TIMER_INIT(USR_TIMER, usr_timer);
  ALTERA_AVALON_JTAG_UART_INIT(JTAG_UART, jtag_uart);
  ALTERA_AVALON_SYSID_INIT(SYSID, sysid);
}
```

An application program generally does not need to repeat the initialization process for peripherals with HAL-compliant device drivers. However, if a device is not used in a "normal" way or the driver is not HAL-compliant, the application program may need to add additional initialization code to configure the device.

In the enhanced flashing-LED program, the JTAG UART and timer cores supply proper device drivers. Their initialization is handled by the HAL framework and is transparent to the application program. On the other hand, the device driver of the PIO core contains only the register-access macros and has no initialization routine. To clear the pushbutton's edge capture register in the beginning, we must do it manually within the `main()` program. The modified initialization routine is shown in Listing 11.7.

Listing 11.7

```
void flashsys_init_v2(alt_u32 btn_base)
{
  btn_clear(btn_base);                // clear button edge-capture reg
}
```

Main program The revised top-level `main()` program is shown in Listing 11.8. Either the `led_flash_v2()` or `led_flash_v3()` function can be used to perform the flashing-LED task.

Listing 11.8

```
int main(){
  cmd_type sw_cmd={0,100};   // not pause; 100 ms interval

  flashsys_init_v2(BTN_BASE);
  while(1){
    sw_get_command_v1(BTN_BASE, SWITCH_BASE ,&sw_cmd);
    jtag_uart_disp_msg_v2(sw_cmd);
    sseg_disp_msg_v1(SSEG_BASE, sw_cmd);
    led_flash_v2(LEDG_BASE,sw_cmd);
  }
}
```

Its basic organization is similar to that in Section 10.6.3 but the task functions are modified to take advantage of the HAL framework. The suffix _v2 indicates that a new version of the function is used.

Note that no `SYS_TIMER_BASE` or `JTAG_UART_BASE` is shown in the code. The operation of these modules is embedded in the HAL API functions and thus is transparent to the user application program.

11.3.4 Software construction and testing

We can follow the tutorial in Section 9.3 to build and run the software in the enhanced Nios II system. The main steps are:

- If continued from Chapter 10, the BSP library, `nios_led2_bsp`, should already be constructed. If not, create a BSP library using the `nios_led2.sopcinfo` file generated in Chapter 10.
- Invoke the BSP Editor to verify that the jtag_uart module is assigned to stdout and the sys_timer module is assigned to system_timer.
- Create a user directory `nios_led2_hal` (the suffix _hal for HAL API) and add the C files to this directory.
- Program the FPGA device on the DE1 board using the `nios_led2_top.sof` file obtained from Chapter 10.
- Build and run the software image in the Nios II system.

After the successful implementation, the two LEDs system should flash at the default 100-ms rate. We can use switches to change its rate or pause its operation and verify the operation of the physical system.

With the default BSP setting, the code size of the image file is about 57 KB, which can be easily accommodated by the 512-KB external SRAM chip. In comparison, the image file of Chapter 10 is about 18 KB. The difference is due to the complexity of the HAL driver routines and API functions. The code size can be decreased by selecting proper options in BSP editor, as discussed in 11.2.3. The reduced sizes are 11 KB and 8 KB, respectively.

11.4 DEVICE DRIVER CONSIDERATION

We compare the software hierarchy of the HAL framework and a desktop-like system in Section 11.1.2. One main difference is that the HAL framework is not as rigid as an OS. Using a HAL-compliant device driver is not mandatory and an application program can access the I/O device without going through the API. We examine and compare various methods in this section.

11.4.1 I/O access methods

For our discussion purposes, we can divide the I/O access into several methods:
- Method 1: via standard library functions.
- Method 2: via generic API functions.
- Method 3: via device-specific HAL-compliant driver functions.
- Method 4: via device-specific ad hoc driver functions.
- Method 5: via direct I/O register read and write.

Method 1 is the most general and most abstract. The HAL framework initializes the device and "hooks up" the driver routines with the functions in C's standard stdio and Unix-style unistd library. This method can be used for generic character mode devices and file systems. The use of the `print()` function in Listing 11.2 is a representative example. Following are several observations about this method:
- Standard C functions, such as `print()`, are called in the application program.
- Generic C header files, such as stdio.h, are included in the application program.
- The system-dependent device base addresses, such as `JTAG_UART_BASE`, are never referenced in the application program.
- No manual device initialization is needed in the application program.

This method thus completely shields the low-level I/O details. From the application program's point of view, it is running under a generic computer platform. The codes can be developed, ported, and implemented in any system.

Method 2 is used for other generic I/O devices. A collection of functions is defined in HAL API. For example, the `alt_dma_txchan_send()` function posts a transmit request to a DMA transmit channel. Following are the properties of this method:
- API functions are called in the application program.
- Relevant Altera API header files, such as `<sys/alt_dma.h>`, are included in the application program.
- The system-dependent device base addresses are usually not referenced in the application program.
- No manual device initialization is needed in the application program.

These API functions are associated with the Altera Nios II HAL framework and thus the corresponding codes cannot be ported to the other platform.

Method 3 is used for I/O devices that do not map well into HAL's generic models. The HAL framework initializes the device and provides proper driver routines to perform the specific functions. For example, Altera's SPI core implements the standard SPI serial interface protocols. It provides a function, `alt_avalon_spi_command()`, to issue a control sequence on the SPI bus. Following are the properties of this method:

- Device-specific functions are called in the application program.
- Device-specific header files, such as `altera_avalon_spi.h`, are included in the application program.
- The system-dependent base addresses are usually required in the functions.
- No manual device initialization is needed in the application program.

The first three methods use HAL-compliant device drivers and they are integrated within the HAL framework. Note that the compliance is not mandatory. Method 4 uses routines from ad hoc non-HAL-compliant device drivers. One main difference between methods 3 and 4 is that in the latter the application program must set up the device and perform initialization, as in the standalone environment. The drivers discussed in Chapter 10 fall into this category.

Method 5 accesses and manipulates the I/O device directly. The codes discussed in Chapter 9 use this method. Except for simplistic I/O devices, a device driver should be developed and this method should be avoided.

11.4.2 Comparisons

The benefits and drawbacks of these methods can be examined from different perspectives.

Application programmer's perspective An application programmer is concerned with the overall functionality of a system. A more abstract method handles more low-level details and presents a better and more consistent interface. This will shorten the development time and make the codes less error-prone, easier to maintain, and more portable. From this perspective, an application programmer usually likes to have the most abstract method implemented whenever possible.

One drawback of using the abstract method is the potential code size and timing issues. An abstract high-level function must be general, flexible, and robust and needs to accommodate various scenarios and error conditions. This leads to a larger code size. For example, the `printf()` function consists of sophisticated codes to convert various data types, such as floating point, to a formatted string. The compiled code size is about 50 KB, which exceeds the capacity of Cyclone 2C20 device's internal memory. If we just use this function to send a one-line message, as in the flashing-LED example in Section 11.3, most functionalities is left unused. The simple `jtaguart_disp_msg2()` routine in Section 10.8 can perform the same task but requires only few hundred bytes.

Using a high-level function may also have an impact on timing. In many embedded applications, a system must respond in a specific amount of time and we need to know the timing characteristics of individual routines. These types of information may not be available from a predesigned high-level function. Furthermore, these functions may contain additional polling loops or interrupt service routines that may complicate the overall timing.

In general, the benefits of an abstract method outweigh its drawbacks. Unless there is a compelling size or timing constraint, the most abstract method is preferred. We can also adjust certain BSP parameters to reduce the code footprint and to fine tune character-mode device drivers and the standard library. These can be tried before we switch to a less abstract method.

Device driver developer's perspective   A device driver developer derives the software routines used in methods 2, 3, and 4 and is concerned with the complexity of various methods. Clearly, a more abstract method demands more time and efforts. Developing a flexible and robust device driver is by no means a simple task. We must have comprehensive knowledge of the HAL specification and the low-level characteristics of the I/O device and consider various operation scenarios and possible error conditions. On the other hand, it is relatively simpler to develop a set of ad hoc non-HAL compliant driver functions used in method 4.

In general, if the I/O device is to be used repeatedly and widely, it is worthwhile to invest more resources to develop a comprehensive driver. The subsequent savings will offset the high initial cost.

Embedded SOPC system designer's perspective   In a desktop system environment, the application program and device driver are usually developed by separate groups. The former is done by users and the latter is provided by OS or device manufactures. This is also the case for a traditional embedded system. Since the I/O peripherals of a processor is fixed, the manufacture can afford to develop a comprehensive set of device drivers and libraries, similar to those used in methods 1, 2, and 3, and users only need to develop application programs.

An embedded SOPC system presents a different scenario. This platform allows us to integrate custom I/O peripherals and hardware accelerators into a system. A designer is now responsible for the development of hardware as well as the software device driver and application program and can distribute the functionalities between the driver and application program. We need to weigh the software complexity and the potential of future reuse and determine the type and completeness of the driver. For example, a simple ad hoc device driver should be satisfactory for a special-purpose I/O peripheral that is used only in one specific system. On the other hand, a more comprehensive driver should be developed for a peripheral used in multiple projects.

Note that an application program can use different access methods at the same time, including both HAL-compliant and non-HAL compliant drivers. For example, we can use Method 1 to send a message to a console but use Method 4 to control a special hardware accelerator. The general conceptual diagram is shown in Figure 11.6.

11.4.3   Device drivers in this book

Deriving HAL-compliant device drivers requires a significant amount of time and effort. Since the focus of this book is on the development and integration of custom hardware, we use the ad hoc drivers (method 4) for the I/O peripherals and hardware accelerators designed later in this book. Information for HAL-compliant drivers can be found in the bibliographic section.

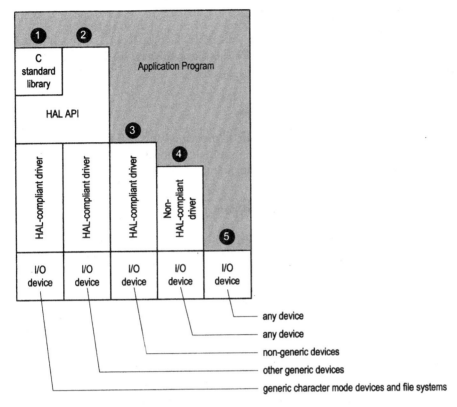

Figure 11.6   All-inclusive software hierarchy.

11.5   BIBLIOGRAPHIC NOTES

Chapters 5, 6, and 7 of Altera's *Nios II Software Developer's Handbook*, titled "Overview of the Hardware Abstraction Layer," "Developing Programs Using the Hardware Abstraction Layer," and "Developing Device Drivers for the Hardware Abstraction Layer," provide a comprehensive review of the HAL framework. Chapter 14, titled "HAL API Reference," documents the available functions of HAL API. The application note AN459, titled "Guidelines for Developing a Nios II HAL Device Driver," uses the UART as an example to explain the process of developing and debugging a HAL-compliant device driver. The source codes of HAL API functions and drivers are available in the HAL and drivers directories of the BSP library.

11.6   SUGGESTED EXPERIMENTS

11.6.1   "Uptime" feature in flashing-LED system

The "uptime" feature is described in Section 10.11.1. Instead of using low-level routines, we can use HAL API functions to perform this task. Develop software and verify its operation.

11.6.2 Enhanced collision LED circuit

The enhanced collision LED circuit is discussed in Section 10.11.4. Instead of using low-level routines, we can use HAL API functions to perform this task. Develop software and verify its operation.

11.6.3 Parking lot occupancy counter

A parking lot occupancy counter circuit is described in Section 6.5.3. Instead of using custom hardware, we can use an embedded system to perform this task. Develop software with HAL API functions and verify its operation.

11.6.4 Reaction timer with keyboard control

A reaction timer with host keyboard control is described in Section 10.11.9. Develop software with C library and HAL API functions and verify its operation.

11.6.5 Digital alarm clock

We want to design a digital alarm clock. The time is displayed on the seven-segment and discrete LEDs and its control is done by the host's keyboard (via JTAG UART). It operates as follows:

1. The minute and second of the clock are shown on four seven-segment LED displays in "MMSS" format, where "MM" is minutes and "SS" is seconds.
2. The hour of the clock are shown by four discrete red LEDs in hexadecimal format.
3. When the designated alarm time is reached, the discrete green LEDs flash.
4. The S key (for setup) is used to set up the current time. When it is pressed, the user is asked to enter the current time via host's console.
5. The A key (for alarm) is used to set up the alarm time. When it is pressed, the user is asked to enter the alarm time via host's console.
6. The C key (for clear) is used to clear the alarm. If the alarm is on (i.e., the green LEDs flash), it will be turned off (i.e., the green LEDs stop flashing) when the C key is pressed.
7. All other key activities will be ignored.

Develop software and verify its operation.

11.7 COMPLETE PROGRAM LISTING

Listing 11.9   chu_main_led2_hal.c

```
/************************************************************************
 *
 *  Module:   Advanced flashing-LED system using Altera HAL/C stdlib
 *  File:     chu_main_led2_hal.c
 *  Purpose:  Task routines and main program
 *
 ************************************************************************/
/* include section */
/* General C library */
#include <stdio.h>
#include <unistd.h>
/* Altera-specific library */
#include "alt_types.h"
#include "system.h"
/* Module-specific library */
#include "chu_avalon_gpio.h"

/************************************************************************
 * data type definitions
 ************************************************************************/
typedef struct flash_cmd{
  int pause;
  int prd;
} cmd_type;

/************************************************************************
 * function: flashsys_init_v2()
 * purpose:  system initialization
 * argument:
 *   btn_base: base address of pushbutton PIO
 * return:
 * note:
 ************************************************************************/
void flashsys_init_v2(alt_u32 btn_base)
{
  btn_clear(btn_base);                    // clear button edge-capture reg
}

/************************************************************************
 * function: jtag_uart_disp_msg_v2()
 * purpose:  display the interval when it is changed
 * argument:
 *   cmd: command
 * return:
 * note:
 ************************************************************************/
void jtag_uart_disp_msg_v2(cmd_type cmd)
{
  static int old=0;

  if (cmd.prd!=old){
    printf("Interval: %03u ms \n", cmd.prd);
    old = cmd.prd;
  }
}
```

```
/*****************************************************************
* function: led_flash_v2()
* purpose:
* argument:
*    cmd: command
* return:
* note:
*    The delay is done by continuously checking 1-ms tick
*****************************************************************/
void led_flash_v2(alt_u32 led_base, cmd_type cmd)
{
  static alt_u8 led_pattern = 0x01;

  if (cmd.pause)                          // no toggle if pause asserted
    return;
  led_pattern ^= 0x03;                    // toggle 2 LSBs of LEDs
  pio_write(led_base, led_pattern);       // write LEDs
  usleep(1000*cmd.prd);                   // delay for cmd.prd ms
}



/*****************************************************************
* function: sw_get_command_v1()
*    same as the one in Listing 10.18
*****************************************************************/



/*****************************************************************
* function: sseg_disp_msg_v1()
*    same as the one in Listing 10.18
*****************************************************************/



/*****************************************************************
* function: main()
* purpose:  advanced flashing-LED system using HAL
* note:
*****************************************************************/
int  main(){
  cmd_type sw_cmd={0,100};   // not pause; 100 ms interval

  flashsys_init_v2(BTN_BASE);
  while(1){
    sw_get_command_v1(BTN_BASE, SWITCH_BASE ,&sw_cmd);
    jtag_uart_disp_msg_v2(sw_cmd);
    sseg_disp_msg_v1(SSEG_BASE, sw_cmd);
    led_flash_v2(LEDG_BASE,sw_cmd);
  }
}
```

CHAPTER 12

INTERRUPT AND ISR

An *interrupt* is an important external I/O event. When an interrupt occurs, the processor suspends normal program execution and temporarily transfers control to the designated *ISR (interrupt service routine)*. The HAL framework utilizes a single top-level exception handling routine to oversee and coordinate all interrupt activities. In this chapter, we examine the basic concepts of interrupt-driven design and modify the flashing-LED program of Chapter 11 to illustrate the construction of an ISR.

12.1 INTERRUPT PROCESSING IN THE HAL FRAMEWORK

An *exception* is a special condition that requires a processor's immediate attention. An exception can be raised by an abnormal internal event, such as division by zero or an unimplemented instruction, or by an important external event that has priority over normal program execution. The latter is referred to as a *hardware interrupt*. Since the book focuses on the I/O interface, our discussion is primarily on hardware interrupt.

Processing an interrupt involves three basic tasks:

1. Suspend current program execution and save the current system state.
2. Transfer control to a special routine to handle the exception.
3. Restore the system state and resume the normal program execution.

Embedded SOPC Design with Nios II Processor and VHDL Examples. By Pong P. Chu   **277**
Copyright © 2011 John Wiley & Sons, Inc.

In the HAL framework, these tasks are divided between hardware and software and are accomplished by a coordinated effort.

12.1.1 Overview

The interrupt processing in the HAL framework consists of three elements:

- Nios II processor.
- A collection of ISRs, each processing an individual hardware interrupt.
- A *top-level exception handler* routine, which saves and restores the system state and dispatches the proper ISR.

The detailed procedure is:

1. The processor completes the current instruction.
2. The processor disables further interrupts, saves the content of the status register, and saves the content of the program counter, which is the next address of normal program execution.
3. The processor transfers execution to the top-level exception handler by loading the program counter with the predetermined exception address.
4. The exception handler saves the contents of the processor's registers.
5. The exception handler determines the cause of the interrupt.
6. The exception handler calls the proper ISR according to the cause.
7. The ISR clears the associated interrupt condition.
8. The ISR performs the designated device-specific function and then returns the control to the exception handler.
9. The exception handler restores the contents of the processor's registers and enables future interrupts.
10. The exception handler exits by issuing the **eret** (exception return) instruction.
11. The processor executes the **eret** instruction, which restores the status register and loads the previous saved return address to the program counter.
12. The processor resumes the normal program execution from the interrupted point.

Note that the software portion (i.e., exception handler and ISRs) is imposed by the HAL framework. It can be replaced if the system is constructed in a different software developing platform.

12.1.2 Interrupt controller of the Nios II processor

The Nios II processor provides a simple, non-vectored interrupt controller. It follows the RISC design principle, which keeps the hardware small and fast, and delegates most processing to software. During the interrupt process, the processor only saves and restores the contents of the status register and program counter.

The Nios II interrupt controller supports up to 32 external hardware interrupts. Each interrupt has a unique level-sensitive *interrupt request* (*IRQ*) input signal. The processor can enable and disable an individual interrupt or all interrupts by writing proper control registers. The Nios II processor's exceptions and interrupts are not vectored, which means that, when an interrupt occurs, the execution is transferred to the same exception address for all types of interrupts. The addresses are specified in SOPC Builder at system generation time. The Nios II processor does not impose any inherent priority over the interrupt requests. The priority is established by the software exception handler.

12.1.3   Top-level exception handler

The HAL framework uses a single routine to oversee all interrupt activities and perform the needed housekeeping tasks. It is known as *top-level exception handler* or simply *exception handler*. The routine's starting address corresponds to the exception address and the execution is transferred to this routine when an interrupt occurs. The exception handler has three major functions:

- Save the content of the registers at the beginning of the interrupt processing.
- Restore the original content of the registers at the end of the interrupt processing.
- Determine the cause of the interrupt, resolve multiple requests according to the established priorities, and dispatch the proper ISR.

The exception handler is automatically included in the HAL's initialization routine.

Priority and dispatch table   In the HAL framework, an interrupt request is assigned a number between 0 and 31 and the priority corresponds inversely to the IRQ number, which means that the request 0 represents the highest priority interrupt and the request 31 is the lowest. When a system is constructed, the interrupt request number can be assigned within SOPC Builder. For example, the enhanced flashing-LED system constructed in Section 10.5 consists of four interrupt requests, which are generated from the btn module, the jtag_uart module, and two timer modules. These requests are shown under the IRQ column in Figure 10.12. The desired priorities for I/O modules can be assigned by editing the corresponding interrupt request numbers, as discussed in Section 10.5.1. The assigned number will be recorded in the .sopcinfo file and later transferred to the system.h file. HAL's exception handler uses this information to sort the requests' priorities.

The exception handler uses a lookup table to record the requests in a particular system configuration and the starting addresses of the corresponding ISRs. To be included in the table, we must explicitly *register* the IRQ of the I/O module. The registration process is done by calling an API function:

```
int alt_irq_register (alt_u32 id,
                      void* context,
                      void (*isr)(void*, alt_u32));
```

The id argument is the assigned interrupt request number of the I/O module. The isr argument is the name of the corresponding function to be called in response to this interrupt request number. Note that it is actually a pointer pointing to the starting address of the function. The context argument is a pointer used to pass context-specific information to the ISR. It is opaque to the exception handler and provided entirely for the benefit of the user-defined ISR.

HAL API functions   The exception handler is invoked when an interrupt occurs and thus cannot be explicitly called from a program. However, there several relevant functions defined in HAL API to manage interrupt requests:

```
alt_irq_register ()
alt_irq_disable ()
```

```
alt_irq_enable()
alt_irq_disable_all()
alt_irq_enable_all()
alt_irq_interruptible()
alt_irq_non_interruptible()
alt_irq_enabled()
```

Except for the `alt_irq_register()` function, these functions are used to enable or disable a specific or all interrupt requests.

12.1.4 Interrupt service routines

An *ISR* (*interrupt service routine*) is a software routine that handles an individual hardware interrupt. In the HAL framework, the top-level exception handler performs general housekeeping, and an ISR is responsible only for clearing the associated devices interrupt condition and running the device-specific function. The procedure for clearing an interrupt condition usually involves resetting a specific flag or status register or clearing the corresponding counter. The device's corresponding IRQ signal will be deasserted accordingly.

ISR prototype To be properly registered by the `alt_irq_register()` function, an ISR must match the prototype defined in HAL framework. The prototype function is in the form of

```
void isr(void* context, alt_u32 id)
```

The `context` and `id` arguments are the same as for the `alt_irq_register()` function. Two examples are illustrated in the following section.

Restriction on ISR Since the HAL framework is intended for embedded systems, its exception handling process is not as robust and complete as that in a desktop operation system. An ISR runs in a "restricted environment," which prevents the use of many HAL API functions. By default, the top-level exception handler disables further interrupts while processing the current exception. If an ISR calls a function that involves other ISRs and waits for another interrupt, the execution becomes deadlocked.

The Nios II architecture actually supports nested interrupt (i.e., an interrupt occurring an ISR) and technically we can enable further interrupts by calling proper HAL API interrupt enable functions. This approach complicates the software development and is not recommended. The ISRs usually involve critical system operation. It is good practice to keep the code *simple and fast*. This practice can simplify the software development and the timing analysis, as discussed in Section 12.3.

12.2 INTERRUPT-BASED FLASHING-LED PROGRAM

Because the top-level exception handler processes most housekeeping tasks, developing an ISR in HAL framework is not very involved. It consists of the following steps:

1. In SOPC Builder, instantiate the I/O module with the proper interrupt feature.

2. In SOPC Builder, assign the desired IRQ priorities in the REQ column. Recall that IRQ 0 has the highest priority.
3. Derive the ISR for the device, including codes to clear the interrupt condition.
4. Register the ISR during the system initialization.

To demonstrate the use of interrupt and the development of an ISR, we rewrite the enhanced flashing-LED program with a custom interrupt-driven timer routine. Two versions are discussed in the following subsections.

12.2.1 Interrupt of timer core

The timer core is discussed in Section 10.4. Its interrupt operation involves two bits:

- The ito bit of the control register. This bit indicates whether interrupt is enabled.
- The to (for "timeout") bit of the status register. This bit is set to 1 when the counter reaches zero and stays set until a processor writes 0 to this bit to clear it. It is used as timer's interrupt request signal.

We instantiate two timer modules when constructing the enhanced LED-flashing Nios II system in Section 10.5. The sys_timer module is configured in BSP as a system timer, as discussed in Section 11.3. We use the usr_timer module for the demonstration. Note that in SOPC Builder the usr_timer module is configured to include interrupt capability and an IRQ number is assigned in the IRQ column, as shown in Figure 10.12.

12.2.2 Driver of timer core

An embedded system frequently includes various timing tasks and requires a real-time system-wide clock tick. One method is to use a timer to generate a periodic interrupt tick. For the Altera timer core, this can be achieved by configuring it in continuous mode, in which the counter automatically reloads the predetermined value when reaching zero. Since the to bit is set to 1 as counter reaches zero, the interrupt request is asserted periodically.

Two routines are required to facilitate the timer ISR construction:

- An initialization routine to set up the timer to generate the interrupt request at the designated interval.
- A routine to clear the current interrupt request.

We develop a simple driver in Section 10.7.3 and the `timer_wr_prd()` function and the `timer_clear_tick()` macro can be used for this purpose.

The `timer_wr_prd()` function is shown in Listing 10.6. In addition to write the desired interval to two period registers, note that the last statement

```
IOWR(timer_base, TIMER_CTRL_REG_OFT, 0x0007);
```

sets the counter in "continuous" mode, starts the counter operation, and enables interrupt by writing 1's to the cont, start, and ito fields of the control register. Although the interrupt request of the usr_timer module is enabled, we do not register it in the code of previous chapters and thus the interrupt has no effect.

The `timer_clear_tick()` macro is defined as

```
IOWR(base, TIMER_STAT_REG_OFT, 0)
```

It resets the to field of the status register to 0. Since the to field is also used as an interrupt request, resetting the to field clears the current request and enables future interrupts.

12.2.3   ISR version 1

We discuss the software developments of the enhanced flashing-LED system in HAL framework in Chapters 10 and 11 and the top-level main program is

```
flash_led_init();
while(1){
  sw_get_command();
  jtag_uart_disp_msg();
  sseg_disp_msg();
  led_flash();
}
```

In this scheme, the tasks are executed in a round-robin manner. Since the main purpose of the system is to flash LEDs, the **led_flash()** function should be given higher priority. This can be accomplished with the availability of timer interrupt.

In the first version of the ISR, we make the entire **led_flash()** function "interrupt driven" and implement all the functionalities in an ISR. The main program becomes

```
flash_led_init();
alt_irq_register(..., flash_led_isr_v1());
while(1){
  sw_get_command();
  jtag_uart_disp_msg();
  sseg_disp_msg();
}
```

Note that the ISR is invoked when the IRQ signal is asserted but not called by the main program.

A HAL ISR can communicate with other functions using the **context** argument or global variables. We use the former in this version of the ISR. Because the ISR replaces the entire **led_flash()** function, it needs the command variable as well as the base addresses of the LED PIO module and timer module. The information constitutes the "context" for this ISR. A structure can be defined for this purpose:

```
typedef struct ctxt1{
  cmd_type *cmd_ptr;
  alt_u32 timer_base;
  alt_u32 led_base;
} ctxt1_type;
```

Note that a pointer type is used for the command because the content of the command can be updated by the **sw_get_command()** function. A pointer to this structure can be passed to the ISR function.

In this design, we use the 1-ms tick to assert the interrupt request. When the ISR is invoked, it updates the elapsed time and toggles the LEDs as needed. The code is shown in Listing 12.1.

Listing 12.1

```
static void flash_led_isr(void* context, alt_u32 id)
{
  ctxt1_type *ctxt;
  cmd_type *cmd;
  static int ntick = 0;
  static unsigned char led_pattern = 0x01;

  /* type casting */
  ctxt = (ctxt1_type *) context;
  cmd = ctxt->cmd_ptr;
  /* clear "to" flag; also enable future interrupt */
  timer_clear_tick(ctxt->timer_base);
  if (cmd->pause)
    return;
  if (ntick < cmd->prd)
    ntick++;
  else {
    ntick = 0;
    led_pattern ^= 0x03;        // invert 2 LSBs
    pio_write(ctxt->led_base, led_pattern);
  }
}
```

The ISR gets the base address and command via the `context` argument, which is recast to the `ctxt1_type` type. Various parameters are extracted accordingly. The `ntick` variable is used to keep track of elapsed time. Invoking the ISR corresponds to the progress of 1 ms. If the system is not paused, `ntick` is incremented each time. When `ntick` reaches the designated flashing period, it is then reset to 0 and the LED pattern is toggled as well.

The main program is shown in Listing 12.2.

Listing 12.2

```
...
typedef struct flash_cmd{
  int pause;
  int prd;
} cmd_type;

typedef struct ctxt1{
  cmd_type *cmd_ptr;
  alt_u32 timer_base;
  alt_u32 led_base;
...
int main(){
  cmd_type sw_cmd={0,100};   // not pause; 100 ms interval
  ctxt1_type ctxt1;

  /* initialization */
  flash_led_init_v1(BTN_BASE, USR_TIMER_BASE);
  /* construct the "context" structure */
  ctxt1.led_base = LEDG_BASE;
  ctxt1.timer_base = USR_TIMER_BASE;
  ctxt1.cmd_ptr = &sw_cmd;
  /* register ISR */
  alt_irq_register(USR_TIMER_IRQ, (void *) &ctxt1, flash_led_isr);
  /* main loop */
  while(1){
    sw_get_command_v1(BTN_BASE, SWITCH_BASE ,&sw_cmd);
```

```
      jtag_uart_disp_msg_v2(sw_cmd);
      sseg_disp_msg_v1(SSEG_BASE, sw_cmd);
   }
}
```

The program first calls the `flash_led_init_v1()` function to initialize the usr_timer module and clear the edge capture register of the btn module. Since the required setup is identical to that in Chapter 10, the same function in Listing 10.7 is used again. It then constructs the context structure, `ctxt1`, and registers the ISR using the HAL's `alt_irq_register()` function. Note that `USR_TIMER_IRQ` corresponds to the interrupt request number assigned to the usr_timer module in SOPC Builder. It is a constant defined in the `system.h` file when the BSP library is generated. The routines in the main loop are identical to those in Section 11.3.

12.2.4 ISR version 2

A general principle of designing an ISR is to keep it simple and fast and to include only the most essential operation within the ISR. The second version of the timer ISR follows this principle. We construct the ISR to establish a real-time clock with a millisecond tick and then utilize this information in the main program to handle the flashing interval.

In this version, we use global variables to exchange information between the ISR and other functions. Two variables are defined in the file:

```
   alt_u32 isr_timer_base;    // base address of the timer module
   alt_u32 sys_ms_tick;       // elapsed ms ticks
```

The new ISR is shown in Listing 12.3.

Listing 12.3

```
static void ms_clock_isr(void* context, alt_u32 id)
{
   /* clear "to" flag; also enable future interrupt */
   timer_clear_tick(isr_timer_base);
   /* increment ms tick */
   sys_ms_tick++;
}
```

The ISR simply clears the timer's interrupt condition and increments `sys_ms_tick`. Since the ISR is invoked every millisecond, `sys_ms_tick` keeps tracks of elapsed time in terms of milliseconds. Since `sys_ms_tick` is defined as a 32-bit unsigned data type, this real-time clock can count up to 2^{32} ms (about 50 days). Because `sys_ms_tick` is a global variable, its value can be accessed by any function in the file.

The new LED flashing routine, `led_flash_v4()`, is shown in Listing 12.4.

Listing 12.4

```
void led_flash_v4(alt_u32 led_base, cmd_type cmd)
{
   static alt_u8 led_pattern = 0x01;
   static alt_u32 last=0;

   if (cmd.pause)                          // no toggle if pause asserted
      return;
```

```
  if ((sys_ms_tick - last)<cmd.prd)      // interval not reached
    return;
  last = sys_ms_tick;
  led_pattern ^= 0x03;                    // toggle 2 LSBs of LEDs
  pio_write(led_base, led_pattern);       // write LEDs
}
```

The `led_flash_v4()` function is almost identical to `led_flash_v3()` of Listing 11.4 except that the `now` variable (obtained by the `alt_nticks()` function) is replaced by the `sys_ms_tick` millisecond tick generated by a user-defined ISR.

The main program is shown in Listing 12.5.

Listing 12.5

```
int main(){
  cmd_type sw_cmd={0,100};   // not pause; 100 ms interval

  flash_led_init_v1(BTN_BASE, USR_TIMER_BASE);
  /* assign initial values to global variables */
  isr_timer_base = USR_TIMER_BASE;
  sys_ms_tick = 0;
  /* register ISR */
  alt_irq_register(USR_TIMER_IRQ, NULL, ms_clock_isr);
  /* main loop */
  while(1){
    sw_get_command_v1(BTN_BASE, SWITCH_BASE ,&sw_cmd);
    jtag_uart_disp_msg_v2(sw_cmd);
    sseg_disp_msg_v1(SSEG_BASE, sw_cmd);
    led_flash_v4(LEDG_BASE, sw_cmd);
  }
}
```

The other routines remain unchanged. Note that since the context is not used, the NULL pointer is used in `alt_irq_register()` function.

12.3 INTERRUPT AND SCHEDULING

12.3.1 Scheduling

An embedded system continuously executes a collection of individual tasks. *Scheduling* is the method by which tasks are given access to the processor. The simplest scheduling scheme is the super-loop architecture discussed in Section 9.8.1:

```
main(){
  sys_init();
  while(1){
    task_1();
    task_2();
    ...
    task_n();
  }
}
```

In this scheme, each task is polled and, if needed, executed in a round-robin manner. To ensure proper operation, the overall loop execution time must be relatively small so that the processor can respond to each task in a timely manner.

An embedded system may contain some special conditions that need immediate attention and the corresponding tasks should be invoked within a small interval. Since there is no preferential treatment in the super loop, these tasks must wait to be polled. Unless the overall loop time is very short, the super-loop architecture may not be able to meet the imposed timing constraints and some critical tasks may miss the deadlines.

One way to alleviate the problem is to use the interrupt mechanism to bypass the normal polling order. For a critical task, we can include an interrupt circuit in the I/O module, assign a high-priority interrupt request to the module, and move its functionalities to the corresponding ISR. When its interrupt is asserted (i.e., the task needs to be executed), the processor suspends the normal execution of the polling loop and transfers control to the exception handler, which in turn invokes the ISR. This interrupt mechanism essentially works as a scheduler that gives preferential treatment to the critical task and postpones the execution of other tasks in the polling loop. The example in Section 12.2.3 uses this scheme. The original led_flash() function is moved to an ISR and thus has higher priority than other tasks. It is executed immediately when the 1-ms tick interrupt request is asserted. This idea can be extended to multiple interrupts. Recall that the HAL framework supports up to 32 prioritized interrupts. It is possible to include interrupt requests for several I/O devices, assign priorities according to the importance of the corresponding tasks, and implement the desired functionalities in their ISRs. The interrupt mechanism and the exception handler now implicitly establish priorities among the tasks and perform scheduling accordingly.

Despite its simplistic appearance, scheduling with interrupt can be very involved and may complicate the software development. First, ISRs are the most error-prone portion in embedded software. Since an ISR can be invoked any time during program execution, it is difficult to develop, debug, test, and maintain. Second, ISRs complicate the timing analysis of the main polling loop. We must consider the frequency of occurrence and duration of the ISRs to ensure that normal tasks within the polling loop can be executed in a timely manner. When multiple interrupts are used, an ISR may effect and interfere with other ISRs and the complexity multiplies. To ease the problem, it is good practice to *keep ISRs simple and fast*. We should use an ISR to perform the most essential functionalities, such as setting event flags, updating counters, sending a message to a queue, etc., and leave the non-critical computation to a task within the main polling loop. This simplifies the ISR development and timing analysis. The example in Section 12.2.4 follows this principle. The ISR only increments the sys_ms_tick variable and the led_flash_v4() function in the main polling loop examines the variable and performs the remaining computation.

A super-loop plus simple-and-fast ISRs can satisfy the scheduling requirement of many simple embedded systems. However, as an application becomes more complex, the number of critical tasks and timing constraints grows. The interrupt structure becomes involved and difficult to handle. An alternative is to use a *real-time OS* to coordinate the operation. A real-time OS contains a scheduler that monitors the overall system status, suspends and activates tasks according the predefined prioritizing schemes, and allocates the system resource and processor to the selected tasks. A real-time OS, *MicroC/OS-II*, is ported to the Nios II processor by Altera and is integrated into the HAL framework. The detailed discussion is

Table 12.1   Exception handler timing data

Core	Latency	Response time	Recovery time
Nios II/f	10	105	62
Nios II/s	10	128	130
Nios II/e	15	485	222

beyond the scope of this book and the reference materials can be found in the bibliographic section.

12.3.2 Performance

Since interrupt has a significant impact on scheduling and timing, its performance plays a critical role on the design of an embedded system. The performance is effected by the system's memory organization, the top-level exception handler, and the ISRs.

Memory residence   The top-level exception handler and each ISR executes certain number of instructions. The execution time of an instruction partially depends on where the instructions and the corresponding data reside. An Nios II system can incorporate various types of memory, including on-chip memory, external SRAM and external DRAM, and their memory access times are different. The inclusion cache organization further complicates the picture. Since the access can occur either in a fast internal cache or in slow external memory, the exact execution time becomes indeterministic.

To alleviate the problem, the Nios II configuration allows to include an optional tightly coupled memory, as discussed in Section 8.3.6. It is a small in-chip memory providing guaranteed low-latency memory access. We can place the exception handler and ISRs in this region to obtain the best performance.

Timing of top-level exception handler   The top-level exception handler timing includes three parts:

- *Hardware interrupt latency*: from the time when an interrupt is asserted to the time when the processor executes the instruction at the exception address.
- *Response time*: from the time when an interrupt is asserted to the time when the processor executes the first instruction in the ISR. It includes the time for the exception handler to determine the cause of the interrupt and save the content of register file.
- *Recovery time*: the time taken from the last instruction in the ISR to return to normal processing.

The number of clock cycles for each part is shown in Table 12.1 (assume that code and data are stored in on-chip memory). Note that only the interrupt latency is intrinsic to the Nios II processor. The response time and recover time are associated with the exception handler, which is part of the HAL framework. It is possible to replace the exception handler for better performance. For example, if a system only has a simple timer interrupt, we can customize the exception handler not to check the cause of the interrupt and to save only the affected registers. In fact, we can just discard the exception handler and add code to the ISR to save and restore the affected registers.

ISR performance enhancement   Keeping ISRs simple and fast is one of the most important design principles in embedded software development. Traditionally, this is done by moving the non-critical computation out of the interrupt context, as discussed in Section 12.3.1. With the availability of programmable logic, the SOPC platform provides an alternative to enhance ISR performance. We can invest additional hardware resource to reduce the software processing time. Some techniques are:

- Adjust the buffer size or implement advanced features, such as double buffering.
- Utilize DMA (direct memory access) controllers to transfer data between memory modules.
- Implement custom hardware accelerator to perform computation intensive processing.

The SOPC platform introduces a new dimension of flexibility. We can examine the performance criteria and design complexity and find the best trade-off between software and hardware resources.

12.4   BIBLIOGRAPHIC NOTES

Chapter 8 of Altera's *Nios II Software Developer's Handbook*, titled "Exception Handling," explains the construction and use of HAL's top-level exception handler and ISRs. *Programming Embedded Systems in C and C++* by M. Barr introduces the basic concept of real-time OS. *MicroC/OS II: The Real Time Kernel* by J. J. Labrosse describes the implementation of the MicroC/OS, illustrates its usage, and provides detailed documentation on the data structure and function calls.

12.5   SUGGESTED EXPERIMENTS

12.5.1   Flashing-LED system with pushbutton switch ISR

The btn PIO module for the pushbutton switches is configured to include the edge capture register. The register asserts an interrupt request when the designated edge is detected. We can replace the sw_get_command_v1() function with an ISR. Derive the ISR, modify other task functions and main program as needed, and verify its operation.

12.5.2   ISR-driven flashing-LED system

In the enhanced flashing-LED system, it is possible to use ISRs to replace all task functions in the main loop. Derive the ISRs, modify the main program, and verify the operation. Note that the printf() function should be avoided in an ISR.

12.5.3 "Uptime" feature in flashing-LED system

The "uptime" feature is described in Section 10.11.1. Redevelop the software using the timer ISR of this chapter and verify its operation.

12.5.4 Reaction timer with keyboard control

A reaction timer with host keyboard control is described in Section 10.11.9. Develop ISRs for the timer and JTAG UART. Redevelop the software with these ISRs (and without C library or HAL API function) and verify its operation.

12.5.5 Digital alarm clock

A digital alarm clock is described in Section 11.6.5. Develop ISRs for the timer and JTAG UART. Redevelop the software with these ISRs (and without C library or HAL API function) and verify its operation.

12.6 COMPLETE PROGRAM LISTING

Listing 12.6   chu_main_led2_isr_v1.c

```
/******************************************************************
*
* Module:   Advanced flashing-LED system using ISR with "context" argument
* File:     chu_main_led2_isr_v1.c
* Purpose: Task routines and main program
*
******************************************************************/
/* include section */
/* General C library */
#include <stdio.h>
/* Altera-specific library */
#include "alt_types.h"
#include <sys/alt_irq.h>
#include "system.h"
/* Module-specific library */
#include "chu_avalon_gpio.h"
#include "chu_timer_drv.h"

/******************************************************************
* data type definitions
******************************************************************/
typedef struct flash_cmd{
  int pause;
  int prd;
} cmd_type;

typedef struct ctxt1{
  cmd_type *cmd_ptr;
  alt_u32 timer_base;
  alt_u32 led_base;
} ctxt1_type;


/******************************************************************
* function: flash_led_init_v1()
*    same as the one in Listing 10.18
******************************************************************/

/******************************************************************
* function: sw_get_command_v1()
*    same as the one in Listing 10.18
******************************************************************/


/******************************************************************
* function: jtag_uart_disp_msg_v2()
*    same as the one in Listing 11.9
******************************************************************/


/******************************************************************
* function: sseg_disp_msg_v1()
*    same as the one in Listing 10.18
******************************************************************/
```

```
/***********************************************************************
 * function: flash_led_isr()
 * purpose:  isr for flash led
 * argument:
 *   context: pointer to context
 *   id:
 * return:
 * note:
 *   - cmd passed within context
 ***********************************************************************/
static void flash_led_isr(void* context, alt_u32 id)
{
  ctxt1_type *ctxt;
  cmd_type *cmd;
  static int ntick = 0;
  static unsigned char led_pattern = 0x01;

  /* type casting */
  ctxt = (ctxt1_type *) context;
  cmd = ctxt->cmd_ptr;
  /* clear "to" flag; also enable future interrupt */
  timer_clear_tick(ctxt->timer_base);
  if (cmd->pause)
    return;
  if (ntick < cmd->prd)
    ntick++;
  else {
    ntick = 0;
    led_pattern ^= 0x03;        // invert 2 LSBs
    pio_write(ctxt->led_base, led_pattern);
  }
}

/***********************************************************************
 * function: main()
 * purpose:  advanced flashing-LED system using ISR with "context" argument
 * note:
 ***********************************************************************/
int main(){
  cmd_type sw_cmd={0,100};  // not pause; 100 ms interval
  ctxt1_type ctxt1;

  flash_led_init_v1(BTN_BASE, USR_TIMER_BASE);
  ctxt1.led_base = LEDG_BASE;
  ctxt1.timer_base = USR_TIMER_BASE;
  ctxt1.cmd_ptr = &sw_cmd;
  alt_irq_register(USR_TIMER_IRQ, (void *) &ctxt1, flash_led_isr);
  while(1){
    sw_get_command_v1(BTN_BASE, SWITCH_BASE ,&sw_cmd);
    jtag_uart_disp_msg_v2(sw_cmd);
    sseg_disp_msg_v1(SSEG_BASE, sw_cmd);
  }
}
```

Listing 12.7 chu_main_led2_isr_v2.c

```
/*******************************************************************
 *
 * Module:   Advanced flashing-LED system using ISR with global variables
 * File:     chu_main_led2_isr_v2.c
 * Purpose:  Task routines and main program
 *
 *******************************************************************/
/* include section */
/* General C library */
#include <stdio.h>
/* Alt era-specific library */
#include "alt_types.h"
#include <sys/alt_irq.h>
#include "system.h"
/* Module-specific library */
#include "chu_avalon_gpio.h"
#include "chu_timer_drv.h"

/*******************************************************************
 * data type definitions
 *******************************************************************/
typedef struct flash_cmd{
  int pause;
  int prd;
} cmd_type;

/*******************************************************************
 * global variables
 *******************************************************************/
alt_u32 isr_timer_base;    // base address of the timer module
alt_u32 sys_ms_tick;       // elapsed ms ticks


/*******************************************************************
 * function: flash_led_init_v1()
 *    same as the one in Listing 10.18
 *******************************************************************/

/*******************************************************************
 * function: sw_get_command_v1()
 *    same as the one in Listing 10.18
 *******************************************************************/


/*******************************************************************
 * function: jtag_uart_disp_msg_v2()
 *    same as the one in Listing 11.9
 *******************************************************************/


/*******************************************************************
 * function: sseg_disp_msg_v1()
 *    same as the one in Listing 10.18
 *******************************************************************/
```

```
/* ******************************************************************
 * function: ms_clock_isr()
 * purpose:  isr for ms clock tick
 * argument:
 *   context:
 *   id:
 * return:
 ****************************************************************** */
static void ms_clock_isr(void* context, alt_u32 id)
{
  /* clear "to" flag; also enable future interrupt */
  timer_clear_tick(isr_timer_base);
  /* increment ms tick */
  sys_ms_tick++;
}


/* ******************************************************************
 * function: led_flash_v4()
 * purpose:  toggle 2 LEDs according to the given interval
 * argument:
 *   led_base: base address of discrete LED PIO
 *   timer_base: base address of user timer
 *   cmd: command
 * return:
 ****************************************************************** */
void led_flash_v4(alt_u32 led_base, cmd_type cmd)
{
  static alt_u8 led_pattern = 0x01;
  static alt_u32 last=0;

  if (cmd.pause)                          // no toggle if pause asserted
    return;
  if ((sys_ms_tick - last)<cmd.prd)       // interval not reached
    return;
  last = sys_ms_tick;
  led_pattern ^= 0x03;                    // toggle 2 LSBs of LEDs
  pio_write(led_base, led_pattern);       // write LEDs
}

/* ******************************************************************
 * function: main()
 * purpose:  advanced flashing-LED system using ISR with global variables
 * note:
 ****************************************************************** */
int main(){
  cmd_type sw_cmd={0,100};  // not pause; 100 ms interval

  flash_led_init_v1(BTN_BASE, USR_TIMER_BASE);
  /* assign initial values to global variables */
  isr_timer_base = USR_TIMER_BASE;
  sys_ms_tick = 0;
  /* register ISR */
  alt_irq_register(USR_TIMER_IRQ, NULL, ms_clock_isr);
  /* main loop */
  while(1){
    sw_get_command_v1(BTN_BASE, SWITCH_BASE ,&sw_cmd);
    jtag_uart_disp_msg_v2(sw_cmd);
    sseg_disp_msg_v1(SSEG_BASE, sw_cmd);
    led_flash_v4(LEDG_BASE, sw_cmd);
  }
}
```

PART III

CUSTOM I/O PERIPHERAL DEVELOPMENT

CHAPTER 13

CUSTOM I/O PERIPHERAL WITH PIO CORES

The Altera SOPC platform includes a sophisticated structure to create custom I/O interfaces to integrate I/O peripherals into Nios II systems. The process is somewhat tedious. For an I/O peripheral with just a few ports and simple timing requirement, it is possible to utilize multiple PIO cores to function as I/O buffers rather than creating a new core. In this chapter, we demonstrate this scheme by interfacing the previous division circuit with a Nios II processor.

13.1 INTRODUCTION

A Nios II processor utilizes a memory-mapped I/O scheme to access I/O peripherals. The simplest interface between a processor and an I/O peripheral is a collection of registers. The processor treats these registers as memory locations and reads and writes data accordingly. For example, the JTAG UART core in Section 10.3 contains two interface registers and the timer core in Section 10.4 contains six interface registers. Altera SOPC platform includes a sophisticated structure, known as *Avalon interconnect*, to create custom I/O interfaces to accommodate the buffering and timing requirements of I/O peripherals and hardware accelerators. The Avalon interconnect is discussed and used in the remaining chapters of Part III and the chapters of Part IV.

Creating a new IP core in the SOPC platform is a somewhat tedious procedure. An alternative is to utilize the existing PIO cores as interface registers and instantiate a PIO module for each I/O port. Since each individual port requires a PIO module and timing cannot be adjusted, this approach is only feasible for an I/O peripheral with just a few ports and simple timing requirement. In the following sections, we use the previous division circuit to demonstrate this scheme.

13.2 INTEGRATION OF DIVISION CIRCUIT TO A NIOS II SYSTEM

A division circuit is discussed in Section 7.3.2. We can add a 32-bit division circuit to a Nios II system and use it as an accelerator to speed up the division operation. This is just for demonstration purposes since a dedicated division hardware unit can be included in a Nios II processor when configured.

The division circuit has two input data ports (dvnd and dvsr), two output data ports (quo and rmd), one control signal (start), and two status signals (ready and done_tick). The ready signal is 1 when the circuit is ready to take new input data (i.e., is not in use). The external master circuit should place the dividend and divisor data in dvnd and dvsr ports and asserts the start signal for one clock cycle to initiate the operation. When the calculation is completed, the quotient and remainder are sent to the quo and rmd ports and the done_tick signal is asserted for one clock cycle.

13.2.1   PIO modules

We can instantiate an individual PIO module for each I/O port of the division circuit and give it a similar name. The following modules are needed:
- dvnd module: configured as a 32-bit output-only port.
- dvsr module: configured as a 32-bit output-only port.
- start module: configured as a 1-bit output-only port.
- quo module: configured as a 32-bit input-only port.
- rmd module: configured as a 32-bit input-only port.
- ready module: configured as a 1-bit input-only port.
- done_tick module: configured as a 1-bit input-only port with an edge capture register.

The first six modules are configured as simple one-direction registers that drive output or capture input.

The nature of the done_tick signal is somewhat different from the division circuit's other output signals. It is asserted for one clock cycle when the division operation is completed. We add an additional edge capture register for the corresponding module. The register will be set to 1 when the 0-to-1 transition of the done_tick signal is detected. The register can be used as a "flag" between the processor and the division circuit. The division circuit sets this flag to inform the processor that the computation is complete. After retrieving the results, the processor then can clear the flag to 0 by writing dummy data to this module. Furthermore, the interrupt request signal associated with the edge capture register can be treated as the interrupt request from the division circuit.

For clarity, we use separate PIO modules for an individual I/O port. If desired, we can reduce the number of PIO modules by packing some signals (e.g., ready and done_tick signals) into one module and configure the PIO core to include both input and output registers.

13.2.2 Integration

To integrate a division circuit to a Nios II system, we need to instantiate seven PIO modules specified in the previous subsection. It is a somewhat tedious process. It will become more involved and error-prone when multiple instances are needed.

13.3 TESTING

To illustrate the use of the division circuit, we construct a Nios II system that includes the PIO modules and develop software to verify its operation. The procedure is:

1. Create a Nios II system that contains the seven PIO modules and supporting peripherals.
2. Create a top-level HDL file that instantiates the Nios II system and the division circuit.
3. Develop testing software.
4. Build and run software.

Nios II system   The testing Nios II system contains the following parts:
- A Nios II/e processor.
- Seven PIO modules associated with the division circuit.
- An SRAM controller core to utilize the external SRAM device.
- A JTAG UART core to obtain input operands and display the division results.
- A PIO core to interface the four seven-segment LED displays.
- A system id core.

It can be constructed following the procedure in Section 10.5.1 and the completed SOPC configuration is shown in Figure 13.1.

Top-level HDL file   After the HDL files are generated by SOPC Builder, we can create a top-level module. The module incorporates the Nios II system and the division circuit and its code is shown in Listing 13.1.

Listing 13.1   Top-level system

```
library ieee;
use ieee.std_logic_1164.all;
entity nios_div1_top is
port(
    clk: in std_logic;
    ledg: out std_logic_vector(7 downto 0);
    hex3, hex2, hex1, hex0: out std_logic_vector(6 downto 0);
    sram_addr: out std_logic_vector (17 downto 0);
    sram_dq: inout std_logic_vector (15 downto 0);
    sram_ce_n, sram_oe_n, sram_we_n: out std_logic;
    sram_lb_n, sram_ub_n: out std_logic
);
end nios_div1_top;

architecture arch of nios_div1_top is
    component nios_div1 is
        port (
            signal clk: in std_logic;
            signal reset_n: in std_logic;
```

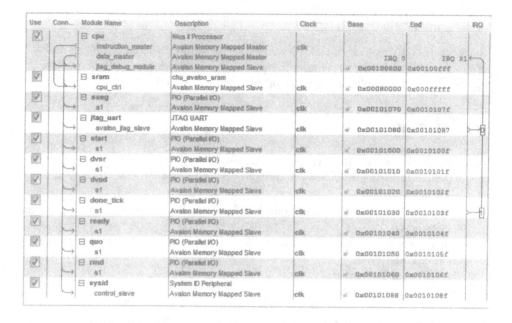

Use	Conn...	Module Name	Description	Clock	Base	End	IRQ
☑		⊟ cpu	Nios II Processor				
		instruction_master	Avalon Memory Mapped Master	clk			
		data_master	Avalon Memory Mapped Master		IRQ 0	IRQ 31	
		jtag_debug_module	Avalon Memory Mapped Slave		0x00100800	0x00100fff	
☑		⊟ sram	chu_avalon_sram				
		cpu_ctrl	Avalon Memory Mapped Slave	clk	0x00080000	0x000fffff	
☑		⊟ sseg	PIO (Parallel I/O)				
		s1	Avalon Memory Mapped Slave	clk	0x00101070	0x0010107f	
☑		⊟ jtag_uart	JTAG UART				
		avalon_jtag_slave	Avalon Memory Mapped Slave	clk	0x00101080	0x00101087	
☑		⊟ start	PIO (Parallel I/O)				
		s1	Avalon Memory Mapped Slave	clk	0x00101000	0x0010100f	
☑		⊟ dvsr	PIO (Parallel I/O)				
		s1	Avalon Memory Mapped Slave	clk	0x00101010	0x0010101f	
☑		⊟ dvnd	PIO (Parallel I/O)				
		s1	Avalon Memory Mapped Slave	clk	0x00101020	0x0010102f	
☑		⊟ done_tick	PIO (Parallel I/O)				
		s1	Avalon Memory Mapped Slave	clk	0x00101030	0x0010103f	
☑		⊟ ready	PIO (Parallel I/O)				
		s1	Avalon Memory Mapped Slave	clk	0x00101040	0x0010104f	
☑		⊟ quo	PIO (Parallel I/O)				
		s1	Avalon Memory Mapped Slave	clk	0x00101050	0x0010105f	
☑		⊟ rmd	PIO (Parallel I/O)				
		s1	Avalon Memory Mapped Slave	clk	0x00101060	0x0010106f	
☑		⊟ sysid	System ID Peripheral				
		control_slave	Avalon Memory Mapped Slave	clk	0x00101088	0x0010108f	

Figure 13.1 The completed SOPC window.

```
        signal out_port_from_the_sseg: out
           std_logic_vector (31 downto 0);
        signal out_port_from_the_start: out std_logic;
23      signal out_port_from_the_dvnd: out
           std_logic_vector (31 downto 0);
        signal out_port_from_the_dvsr: out
           std_logic_vector (31 downto 0);
        signal in_port_to_the_done_tick: in std_logic;
28      signal in_port_to_the_ready: in std_logic;
        signal in_port_to_the_quo: in
           std_logic_vector (31 downto 0);
        signal in_port_to_the_rmd: in
           std_logic_vector (31 downto 0);
33      signal sram_addr_from_the_sram: out
           std_logic_vector (17 downto 0);
        signal sram_ce_n_from_the_sram: out std_logic;
        signal sram_dq_to_and_from_the_sram : inout
           std_logic_vector (15 downto 0);
38      signal sram_lb_n_from_the_sram: out std_logic;
        signal sram_oe_n_from_the_sram: out std_logic;
        signal sram_ub_n_from_the_sram: out std_logic;
        signal sram_we_n_from_the_sram: out std_logic
     );
43  end component nios_div1;
    signal dvnd, dvsr, quo, rmd: std_logic_vector(31 downto 0);
    signal sseg4: std_logic_vector(31 downto 0);
    signal start, ready, done_tick: std_logic;

48 begin
    -- instantiate processor
    nios_unit: nios_div1
    port map(
```

```
        clk=>clk, reset_n=>'1',
53      out_port_from_the_sseg=>sseg4,
        -- division circuit
        out_port_from_the_dvnd=>dvnd,
        out_port_from_the_dvsr=>dvsr,
        out_port_from_the_start=>start,
58      in_port_to_the_quo=>quo,
        in_port_to_the_rmd=>rmd,
        in_port_to_the_ready=>ready,
        in_port_to_the_done_tick=>done_tick,
        -- SRAM
63      sram_addr_from_the_sram=>sram_addr,
        sram_dq_to_and_from_the_sram=>sram_dq,
        sram_ce_n_from_the_sram=>sram_ce_n,
        sram_lb_n_from_the_sram=>sram_lb_n,
        sram_oe_n_from_the_sram=>sram_oe_n,
68      sram_ub_n_from_the_sram=>sram_ub_n,
        sram_we_n_from_the_sram=>sram_we_n
        );
        -- instantiate division circuit
        div_unit: entity work.div
73      generic map(W=>32, CBIT=>5)
        port map(clk=>clk, reset=>'0', start=>start,
                 dvsr=>dvsr, dvnd=>dvnd, quo=>quo, rmd=>rmd,
                 ready=>ready, done_tick=>done_tick);
        -- LEDs
78      hex3 <= sseg4(30 downto 24);
        hex2 <= sseg4(22 downto 16);
        hex1 <= sseg4(14 downto 8);
        hex0 <= sseg4(6 downto 0);
        ledg <= rmd(7 downto 0);
83 end arch;
```

Note that the seven I/O ports of the division circuit are shown as the I/O ports
of the Nios II processor and connected to the division unit via the internal signals.
We can compile this system to obtain the configuration (i.e., .sof) file.

Application program   After the creation of a new I/O peripheral, the driver routines
should be developed to access and communicate with the peripheral. However, since
this division circuit is only used for demonstration purposes, we just put all low-level
access codes in the main program. The complete program is shown Listing 13.2.

Listing 13.2

```
#include <stdio.h>
#include "system.h"
#include "chu_avalon_gpio.h"

int main()
{
  alt_u32 a, b, q, r, ready, done;
  alt_u8 di1_msg[4]={sseg_conv_hex(13),0xfb,0xff,sseg_conv_hex(1)};

  sseg_disp_ptn(SSEG_BASE, di1_msg);                    // display "di 1"
  printf("Division accelerator test #1: \n\n");
  while (1){
    printf("Perform division a / b = q remainder r\n");
    printf("Enter a: ");
    scanf("%d", &a);
    printf("Enter b: ");
```

```
  scanf("%d", &b);
  /* send data to division accelerator */
  pio_write(DVND_BASE, a);
  pio_write(DVSR_BASE, b);
  /* wait until the division accelerator is ready */
  while (1) {
    ready = pio_read(READY_BASE)& 0x00000001;
    if (ready==1)
      break;
  }
  /* generate a start pulse */
  printf("Start ...\n");
  pio_write(START_BASE, 1);
  pio_write(START_BASE, 0);
  /* wait for completion */
  while (1) {
    done = IORD(DONE_TICK_BASE, PIO_EDGE_REG_OFT) & 0x00000001;
    if (done==1)
      break;
  }
  /* clear done_tick register */
  pio_write(DONE_TICK_BASE, 1);
  /* retrieve results from division accelerator */
  q = pio_read(QUO_BASE);
  r = pio_read(RMD_BASE);
  printf("Hardware: %u / %u = %u remainder %u\n", a, b, q, r);
  /* compare results with built-in C operators */
  printf("Software: %u / %u = %u remainder %u\n\n\n", a, b, a/b, a%b);
  } // end while
}
```

The main loop of the program prompts a user to enter the dividend and divisor via the console, sends data to the division core, retrieves the quotient and remainder, and displays them on the console. To verify the correctness of the circuit, the same operation is also performed with C's built-in operators (i.e., a/b, a%b) in the last statement.

The code includes two busy-waiting loops to check whether the division circuit is available and to check whether the calculation is completed. Other mechanisms, such as an interrupt or timer, can be used to coordinate the hardware accelerator's operation as well. In this particular example, the division core takes about 35 clock cycles to complete the execution, which is much faster than the execution time of the scanf() and printf() functions. Thus, these waiting loops are not actually needed and are included just for demonstration purposes.

13.4 SUGGESTED EXPERIMENTS

13.4.1 Division core ISR

Utilize the interrupt capability of the done_tick module of the interface to develop an ISR routine for the division circuit and verify its operation.

13.4.2 Division core with eight-bit data

Assume that only 8-bit data is needed for the division circuit. Since the data width of a Nios II processor is 32 bits, we can pack the data and status signals into a single word. Redesign the interface circuit with a minimal number of PIO modules, derive a Nios II system, develop testing software, and verify its operation.

13.4.3   Division core with 64-bit data

We want to increase the data width of the division circuit to 64 bits. Since the data width of Nios II processor is 32 bits, two words are needed to access an input and output data. Redesign the interface circuit with adequate PIO modules, derive a Nios II system, develop testing software, and verify its operation.

13.4.4   Fibonacci number circuit

The Fibonacci number circuit is discussed in Section 7.3.1. We can modify the circuit to accommodate a 64-bit result (i.e., the output f port is increased to 64 bits wide). Based on this circuit, design the interface circuit with adequate PIO modules, derive a Nios II system, develop testing software, and verify its operation.

13.4.5   Period counter

An accurate low-frequency counter is discussed in Section 7.3.5. In addition to the division circuit, it consists of the binary-to-BCD conversion circuit in Section 7.3.3 and the period counter in Section 7.3.4. Create two new interface circuits with adequate PIO modules for the conversion circuit and the period counter, reconstruct the accurate low-frequency counter in Section 7.3.5 with these PIO modules, derive a Nios II system, develop testing software, and verify its operation.

CHAPTER 14

AVALON INTERCONNECT AND SOPC COMPONENT

The Altera SOPC platform utilizes the *Avalon interconnect* structure to connect the processor and various IP core modules. For a custom I/O peripheral or hardware accelerator, we can add a wrapping circuit that complies with the Avalon specification and convert it into an SOPC component using Altera's *SOPC Editor* utility program. The component then can be used as a normal IP core and integrated into a Nios II system. In this chapter, we provide an overview of Avalon interconnect, discuss the design of the wrapping circuit, and use the previous division circuit to demonstrate the creation and use of an SOPC component.

14.1 INTRODUCTION

An embedded system consists of a processor, memory modules, general I/O peripherals, and hardware accelerators. In a traditional embedded system, the data transfer is designed around the processor and the system usually uses a *shared bus*, which is a collection of wires conveying address, data, and control signals, to connect the processor with other components. Since the bus is a centralized resource, it becomes a bottleneck as the amount of data transfer increases. The degradation is more serious for a memory-mapped I/O architecture since the memory modules and I/O peripherals share the same bus.

The Altera SOPC platform uses a different approach. It defines a set of standardized interfaces, known as *Avalon interfaces*, to accommodate various communication needs and to connect components within an FPGA chip. After we map

Embedded SOPC Design with Nios II Processor and VHDL Examples. By Pong P. Chu   **305**
Copyright © 2011   John Wiley & Sons, Inc.

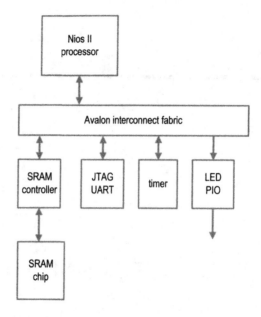

Figure 14.1 Conceptual block diagram of a Nios II system.

an I/O peripheral's ports to the interface signals and set the timing properties, it can be integrated into an Avalon-based system. These interfaces are then realized by *interconnect fabric*, which consists of decoding, multiplexing, arbitration, and timing logic. Although the fabric is usually depicted as a shared entity in the diagram, as in Figure 14.1, the logic and routing structures takes advantage of the FPGA's programmability and are constructed in a distributive fashion. The interconnect fabric is automatically generated by the SOPC Builder software tool and customized to match the individual system configuration. This approach eliminates the contention for a centralized resource and improves the performance and scalability of the system. The conceptual implementation of the previous interconnect fabric is shown in Figure 14.2.

The Avalon standard consists of following types of interfaces:

- *Avalon memory mapped interface (Avalon MM)*: This interface defines an address-based master-slave connection. An *Avalon MM master* uses an address to identify an *Avalon MM slave* and can read data from or write data to the slave.

- *Avalon streaming interface (Avalon-ST)*: This interface defines a dedicated unidirectional link between two components. An *Avalon-ST source* transmits data to an *Avalon-ST sink* continuously.

- *Avalon memory mapped tristate slave interface*: This interface can be considered as a special Avalon MM slave and is used to drive off-chip tristate buses and peripherals.

- *Avalon clock*: This interface defines the clock and reset signals used by a component. An *Avalon clock output interface* generates the clock signal and an *Avalon clock input interface* receives the clock signal.

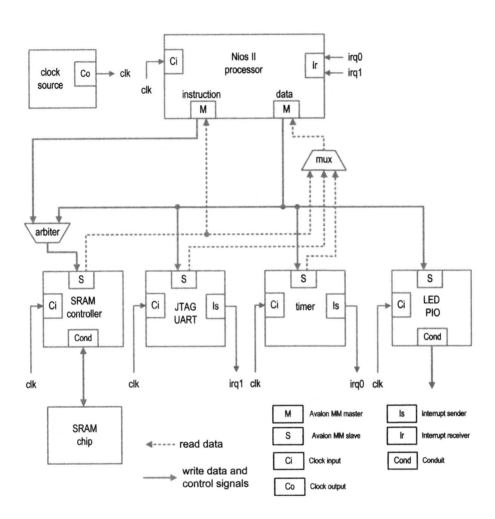

Figure 14.2   Conceptual implementation of Avalon interconnect fabric.

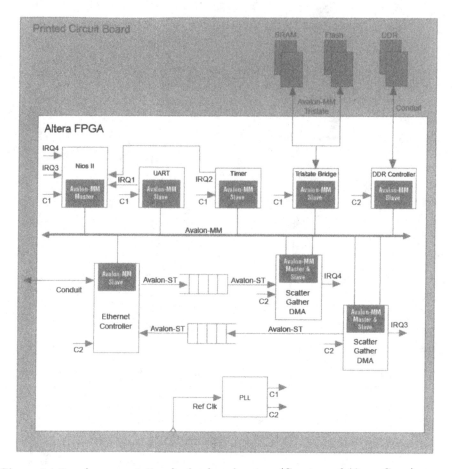

Figure 14.3   A representative Avalon-based system (Courtesy of Altera Corp.).

- *Avalon interrupt*: This interface allows slave components to signal events to master components. An *Avalon interrupt sender interface* generates the interrupt request and an *Avalon interrupt receiver interface* accepts and processes the requests.
- *Avalon conduit*: This interface groups and exports signals to the outside of an SOPC Builder system.

A single component can include any number of these interfaces and can also include multiple instances of the same interface type.

A representative system is shown in Figure 14.3. The Avalon MM interface is the main structure. The Nios II processor uses it to access the control and status registers of on-chip components. The two "scatter gather DMAs" also function as masters and establish direct data transfer between the Ethernet data stream and external memory devices. The Avalon-ST interfaces provide dedicated links between the DMAs and Ethernet controller and thus offload the traffic from the main Avalon MM interface. The interrupt interfaces are included in several components. Four peripheral components generate the requests, which are connected to and serviced by the Nios II processor. The clock interfaces are included in all components.

The two clock signals are generated by a PLL from an external reference clock and then distributed to various components. Finally, an Avalon MM tristate interface provides off-chip access to SRAM devices and flash memory devices, which share the same I/O lines, and a conduit interface for off-chip DDR memory access.

The Avalon MM interface is the key interconnect for a Nios II system and its characteristics and use are discussed in the subsequent sections.

14.2 AVALON MM INTERFACE

The Avalon MM interface basically defines a collection of *interface signals* and a set of *properties*. The latter mainly specifies the timing characteristics of ports. An *Avalon MM master* can initiate a transaction to read or write data and an *Avalon MM slave* responds to the request from the master. In an embedded system, a Nios II processor functions as a master and most I/O peripheral devices and hardware accelerators are slaves. The interface standard is intended to support a wide variety of components and includes many advanced features and options. It can accommodate the slow slave device by stretching the enable signals and utilizing flow-control protocols and can facilitate a high data rate with bursting and pipelined transfers. These features and options are not mandatory and an interface only needs to include the desired features. Our coverage focuses on a subset that is relevant to our designs in Parts III and IV.

14.2.1 Avalon MM slave interface signals

The Avalon MM slave interface defines more than a dozen signals. The fundamental signals are:
- `read` (`read_n`): it is a 1-bit signal asserted in a read transfer (i.e., a master retrieves data from a slave). The `_n` version is used for the active-low signal; i.e., `read_n` is asserted when it is 0.
- `write` (`write_n`): it is a 1-bit signal asserted in a write transfer (i.e., a master writes data to slave). The `_n` version is used for the active-low signal.
- `address`: it is used to specify an offset in the slave address space. Each value identifies a memory location in the slave address space. Its width can be defined by the designer, which ranges from 1 to 32 bits.
- `readdata`: it is the data provided by a slave in a read operation. The width can be 8, 16, 32, 64, 128, 256, 512, or 1024.
- `writedata`: it is the data written to a slave in a write operation. The width selection is similar to that in `readdata`.
- `byteenable` (`byteenable_n`): it enables specific byte lane(s) during transfers. For examples, a 4-bit `byteenable` signal can be used to select byte lanes of a 32-bit `writedata` signal. The value 1111 writes full 32 bits, the values 1100 and 0011 write the upper and lower two bytes, respectively, and the values 1000, 0100, 0010, and 0001 write the byte 3, 2, 1, and 0, respectively.
- `chipselect`: it is a 1-bit signal asserted when the slave device is selected.

The `chipselect` signal is specified in an Avalon memory mapped tristate interface but is frequently included in the Avalon MM slave design. The other unlisted signals are mainly for flow control and pipelined and burst data transfers.

14.2.2   Avalon MM slave interface properties

The Avalon MM slave interface defines about a dozen of properties. Most of them relate to timing of the interface's `read` and `write` signals. The key properties are:

- `timingUnits`: it specifies whether to use *nanoseconds* or *clock cycles* in timing specifications.
- `readWaitTime`: it controls the length of the `read` signal. It allows us to prolong the `read` signal to accommodate a slow I/O device in a read operation.
- `writeWaitTime`: it controls the length of the `write` signal. It allows us to prolong the `write` signal to accommodate a slow I/O device in a write operation.
- `setupTime`: it specifies the time interval between the assertion of the `address` and `data` signals and the assertion of the `read` or `write` signal.
- `holdTime`: it specifies the time interval between the deassertion of the `address` and `data` signals and the deassertion of the `write` signal.
- `readLatency`: it specifies the time interval (i.e., latency) between the assertion of the `read` signal and the availability of data.

The other unlisted properties are mainly for pipelined and burst data transfers.

The Avalon MM interface is a *synchronous* interface and all transactions are performed on a clock-by-clock basis. When a parameter is specified in a nanosecond unit, it will be converted to clock cycles during processing. For example, in a system with a 20-ns clock, a parameter of 50 ns will be converted to 3 clock cycles.

14.2.3   Avalon MM slave timing

The transactions of an Avalon MM interface are controlled by a common clock signal. The various properties can be modified to accommodate the timing characteristics of an I/O device.

In the simplest timing scenario, the transaction is completed in one clock cycle. In this scenario, the `readWaitTime` and `writeWaitTime` properties are set to zeros, which means that there is no extra wait time for the `read` and `write` signals and thus their length is only one clock period. The `setupTime`, `holdTime`, and `readLatency` properties are also set to zeros, which means that the `read`, `write`, `address`, and `data` signals are asserted and deasserted at the same time.

The timing diagram of this scenario is shown in Figure 14.4. A read operation is shown on the left. At the rising edge of the `clk` signal, t_1, a master initiates a read operation by placing the desired address value, `addr0`, on `address` and asserting the `read` signal. After a small delay, the signals are settled at t_2. After the designated slave device detects the commands, it places the data, `data0`, on `readdata` at t_3 and the master latches the data at the rising edge of the next clock, t_4. The master can issue the next read operation at t_4 if desired. This scenario imposes a tight timing constraint on the slave device since the `address` and `read` signals are settled after the first clock edge and the `readdata` signal must be available before the next clock edge. The slave device must return data within the same clock period without using the clock edge to sample `address` and `read`.

A write operation is shown on the right. At the rising edge, t_5, a master initiates a write operation by placing the desired address value, `addr1`, and data value, `data1`, on `address` and `writedata` and asserting the `write` signal. After a small delay, the signals are settled at t_6. It deactivates the signals at the next rising edge,

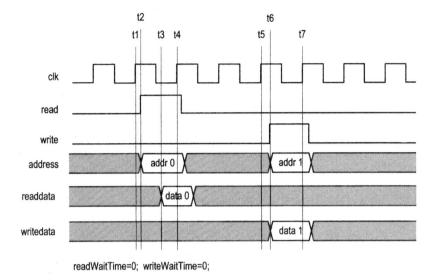

readWaitTime=0; writeWaitTime=0;

Figure 14.4   Timing diagram with no wait state.

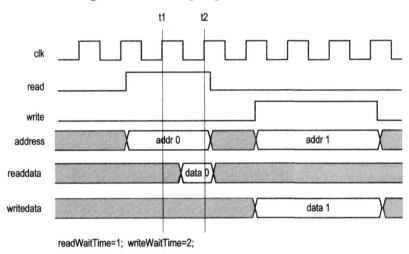

readWaitTime=1; writeWaitTime=2;

Figure 14.5   Timing diagram with one read wait states and two write wait states.

t_7. Due to the propagation delays, these signals remain stable for a small interval. The slave device usually uses the edge at t_7 to check **write** and latch the address and data.

By adjusting the values of **readWaitTime** and **writeWaitTime** properties, we can prolong the length of the **read** and **write** signals. For example, we can set **readWaitTime** to 1 and set **writeWaitTime** to 2 to expand the **read** signal to two clock cycles and expand the **write** signal to three clock cycles. The timing diagram is shown in Figure 14.5. In the read operation, the **address** and **read** signals remain activated for two clock cycles. The master waits one extra clock cycle and retrieves the data at the clock edge labeled t_2. The additional time allows the slave

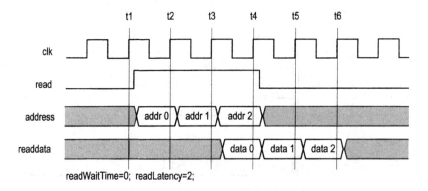

Figure 14.6   Timing diagram of pipelined read operation with a latency of 2.

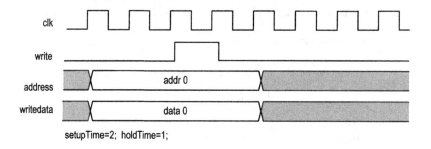

Figure 14.7   Timing diagram of write operation with non-zero setup time and hold time.

to detect the **read** signal latch address at t_1 and thus facilitates the synchronous slave operation. In the write operation, the **write**, **address**, and **writedata** signals last for three clock periods. In addition to specifying a fixed number of wait states, we can also include an additional slave-generated **waitrequest** signal in the Avalon MM interface to let the slave control the length of the wait time.

In some devices, the read operation can be done in a "pipelined" fashion, in which the data are provided at a high rate but there is a latency between the activation of the **read** signal and the availability of the data. We can adjust the **readLatency** to accommodate this scenario. For example, we can set **readLatency** to 2 and **readWaitTime** to 0 and a representative timing diagram is shown in Figure 14.6. Three consecutive read operations are issued at t_1, t_2, and t_3. The data become available after two clock cycles, at t_4, t_5, and t_6, respectively. Note the difference between this and the prolonged **read** signal in Figure 14.5. Despite the initial delay, a master can issue a read operation every clock cycle in the pipelined operation. On the other hand, a master can only issue a read operation every two clock cycles in the prolonged read operation.

In some memory devices, the data and address must be stable before and after the enabling **read** or **write** signal. The intervals are defined as the *setup time* and *hold time* in Avalon specification (note that these are not the same setup time and hold time defined for an FF in Section 5.1.1). We can include the intervals by setting the **setupTime** and **holdTime** properties. For example, we can set **setupTime** to 2 and **holdTime** to 1 and a representative timing diagram is shown in Figure 14.7.

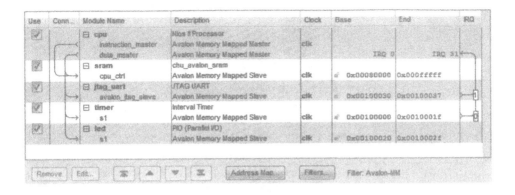

Figure 14.8   Screen capture with Avalon MM connection.

14.3   SYSTEM INTERCONNECT FABRIC FOR AVALON INTERFACE

When constructing an embedded system in the Avalon framework, we simply select the desired IP cores and connect the interface ports. The software tool, SOPC
Builder, will automatically generate the *interconnect fabric* according to the configuration. In a typical Avalon MM-based system, the fabric usually consists of
logic to support address decoding, dynamic bus sizing, data multiplexing, multi-
master arbitration, wait state generation, burst and pipelined transfer control, and
interrupt processing.

The structure of the Avalon interconnect fabric can be explained by an example.
Consider a Nios II system with an external SRAM device, a JTAG UART, a timer,
and a PIO for LEDs. The conceptual top-level diagram is shown in Figure 14.1. It
is similar to the enhanced flashing-LED system discussed in Figure 10.10 but with
fewer I/O peripherals. We can derive the system following the procedure outlined
in Section 10.5.1. The completed screenshot is shown in Figure 14.8.

The system consists of several IP cores: a Nios II processor (without the JTAG
debug module), an SRAM memory controller module, a JTAG UART module, a
timer module, a PIO module, and an implicit clock source. These cores follow
the Avalon specification and their I/O ports are defined through proper interfaces.
Following are the interfaces used:

- Clock source: clock output.
- Nios II processor: two Avalon MM masters (one for instruction and one for
  data), interrupt receiver, and clock input.
- SRAM controller module: Avalon MM slave, conduit, and clock input.
- JTAG UART module: Avalon MM slave, interrupt sender, and clock input.
- Timer module: Avalon MM slave, interrupt sender, and clock input.
- LED PIO module: Avalon MM slave, conduit, and clock input.

The detailed connections among the interfaces are shown in Figure 14.2. Note that
a multiplexing circuit, labeled mux, is used to select and route the designated read
data to the data port of Nios II, and an arbitration circuit, labeled arbiter, is used
to coordinate the SRAM access from two Avalon MM masters.

In the SOPC Builder environment, the system connection is represented in table
format, as in Figure 14.8. The Description column shows the names and interfaces

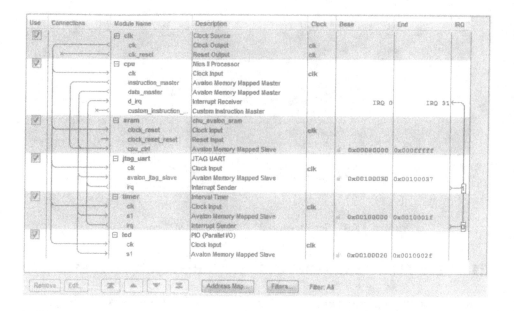

Figure 14.9   Screen capture with detailed connection.

of the IP cores. The connection between the interfaces is depicted under the **Connections** (or **Conn...**) column. By default, it just displays connections of the Avalon MM interface. We can include the connections of other interface types by clicking on the bottom **Filter...** button and selecting the desired types. The column in Figure 14.9 shows all connections, including the clock and interrupt interfaces. It is similar to the diagram in Figure 14.2. To reduce the clutter in the **Connections** column, SOPC Builder includes separate columns for the clock interface, labeled **clock** in the table, and for the interrupt interface, labeled **IRQ**. The **IRQ** column shows the interrupt receiver and sender. A number is associated with each interrupt sender to indicate the priority of the corresponding interrupt request.

The remaining two columns, **Base** and **End**, in Figure 14.8 show the address assignment. Recall that the Nios II processor uses the memory-mapped I/O scheme, which means the same 32-bit address space is assigned to both memory modules and I/O peripherals. The two columns show the address space allocated to each module. For example, the JTAG UART core has two 32-bit addressable registers (totaling 8 bytes) and thus requires 8 address locations. The table shows that the memory address space between 0x00100030 and 0x00100037 is assigned to this module.

When we construct this Nios II system, the default interface connections, memory mapping, and interrupt priority are automatically generated when a component is instantiated. The default setting is generally correct for a simple system. Modifying the setting is quite simple. To revise a connection, we can move the mouse pointer over the **Connections** column to toggle any crossing point in the connection matrix. Similarly, we can move the mouse pointer to the designated fields to select a different clock signal, modify the address map, or assign a different interrupt priority.

14.4   SOPC I/O COMPONENT WRAPPING CIRCUIT

A custom hardware accelerator or I/O peripheral is usually implemented using the FSMD scheme. The circuit typically contains multiple input and output data ports as well as various control and status signals. For example, the division circuit discussed in Section 7.3.2 has two input data ports (`dvnd` and `dvsr`), two output data ports (`quo` and `rmd`), one control signal (`start`), and two status signals (`ready` and `done_tick`).

On the other hand, a Nios II processor utilizes the Avalon interconnect for data transfer and control. To connect a custom circuit to a Nios II system, we need to covert the circuit to an IP core (i.e., an SOPC component) with adequate Avalon interfaces. This usually involves adding a wrapping circuit on top of the FSMD circuit to make its I/O ports compatible with the Avalon specification. The wrapping circuit usually consists of interface buffers, output decoding circuits, and input multiplexing and decoding circuits.

14.4.1   Interface I/O buffer

In the memory-mapped I/O scheme, an I/O port is treated as a memory location and the processor reads and writes the port directly. To achieve this, we can include a register within the I/O port and make it function like one regular memory word. The original FSMD circuit may or may not use registers with its I/O signals and the wrapping circuit must provide buffers as needed.

There is a difference between a regular memory location and a register of an I/O port. For a memory location, the processor performs both read and write operations and thus always knows the data "validness" of this location. For a register of an I/O port, on the other hand, the processor is only responsible for "half" of the access. In a write port, the processor writes (i.e., produces) the data and the external I/O circuit reads (i.e., consumes) the data. In a read port, the processor reads (i.e., consumes) the data and the external I/O circuit writes (i.e., produces) the data.

By the nature of the I/O data, an I/O port can be classified as a *continuous-access* or *one-time-access port*. For a continuous-access port, the data is produced and consumed continuously, such as the switch input and the LED output of Section 9.2. On the other hand, the availability of data of a one-time-access port is triggered by a single discrete event, such as receiving a character in an UART buffer. After the data is consumed, it must be removed from the buffer to prevent the same data from being processed again. Since the production rate and consumption rate are different, the buffer needs a mechanism to signal the availability of *new* data and to prevent the old data from being retrieved multiple times.

Two commonly used schemes can be used to coordinate the data production and consumption:

- A flag FF and a one-word buffer
- A FIFO buffer

To facilitate these schemes, the producing subsystem will assert a status signal, `wr_tick` (similar to the `done_tick` of the division circuit), for one clock cycle when generating a new data and the consuming subsystem will assert a control signal, `clr_tick`, for one clock cycle after retrieving the data. For clarity, we ignore the

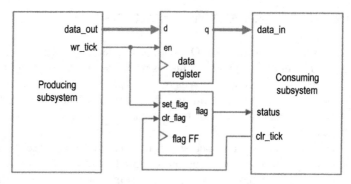

(a) One-word buffer with a flag FF

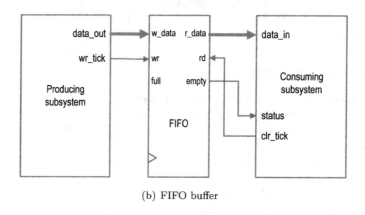

(b) FIFO buffer

Figure 14.10   Interface buffering circuit.

decoding and multiplexing logic for now and assume that the buffer is a dedicated link between two subsystems.

Flag FF scheme   This scheme uses a one-word buffer and a *flag* FF and its top-level block diagram is shown in Figure 14.10(a). The flag FF keeps track of whether a new data word is available. The FF has two input signals. One is set_flag, which sets the flag FF to 1, and the other is clr_flag, which clears the flag FF to 0. When the producing subsystem generates a new data word, it asserts the en signal of the buffer to load the data and sets the flag FF to 1 to indicate that a new data word is available. The consuming subsystem checks the output of the flag FF. When the flag is 1, it retrieves the data and asserts the clr_flag signal to clear the flag FF to 0 to indicate that the data has been processed. The code for this scheme is shown in Listing 14.1.

Listing 14.1   Interface with a flag FF and buffer

```
library ieee;
use ieee.std_logic_1164.all;
entity flag_buf is
   generic(W: integer:=8);
   port(
      clk, reset: in std_logic;
      clr_flag, set_flag: in std_logic;
```

```
          din: in std_logic_vector(W-1 downto 0);
 9        dout: out std_logic_vector(W-1 downto 0);
          flag: out std_logic
       );
    end flag_buf;

14  architecture arch of flag_buf is
       signal buf_reg, buf_next: std_logic_vector(W-1 downto 0);
       signal flag_reg, flag_next: std_logic;
    begin
       -- FF & register
19     process(clk,reset)
       begin
          if reset='1' then
             buf_reg <= (others=>'0');
             flag_reg <= '0';
24        elsif (clk'event and clk='1') then
             buf_reg <= buf_next;
             flag_reg <= flag_next;
          end if;
       end process;
29     -- next-state logic
       process(buf_reg,flag_reg,set_flag,clr_flag,din)
       begin
          buf_next <= buf_reg;
          flag_next <= flag_reg;
34        if (set_flag='1') then
             buf_next <= din;
             flag_next <= '1';
          elsif (clr_flag='1') then
             flag_next <= '0';
39        end if;
       end process;
       -- output logic
       dout <= buf_reg;
       flag <= flag_reg;
44  end arch;
```

The buffer register can be omitted if the producing subsystem already contains an output register.

FIFO scheme   In the flag FF scheme, only one-word buffer space is provided. If the producing subsystem generates a new data word before the consuming subsystem processes the old data word (i.e., the flag FF is still asserted), the old word will be overwritten, an error known as *data overrun*.

To provide some cushion, we can use a FIFO buffer discussed in Section 5.6.2. The block diagram is shown in Figure 14.10(b). The **wr_tick** signal of the producing subsystem is connected to the **wr** signal of the FIFO. When a new data word is generated, the **wr** signal is asserted one clock cycle and the corresponding data is written to the FIFO. The consuming subsystem obtains the data from FIFO's read port. The **clr_tick** signal is connected to the **rd** signal of the FIFO. After retrieving a word, it asserts the **rd** signal of the FIFO to remove the corresponding item. The **empty** signal of the FIFO can be used to indicate whether any received data word is available. The producing subsystem can continue the operation without destroying the previous data. Data overrun will not occur as long as the consuming subsystem retrieves the data word before the FIFO is full. We can adjust the size of the FIFO to accommodate the processing condition of the two subsystems.

Note that the basic characteristics of the flag FF and FIFO schemes are quite similar. The flag FF and one-word register function like a special one-word FIFO buffer with a "full" status signal.

14.4.2 Memory alignment

Nios II is a 32-bit processor, which means that data is processed in a 32-bit unit (commonly known as a *word*) within the processor. It also contains a 32-bit address bus. However, Nios II's address space is represented in terms of *bytes* and the 32-bit address bus implies an addressable space of 2^{32} bytes. To accommodate 32-bit data, four bytes of memory are grouped together to form a word. Within the main memory, the words are aligned (i.e., the address's two LSBs of the starting byte in a word is always is 00) for easy access and a word can be accessed by using 30 MSBs of the address. When needed, we use the term "word-address" for the 30-bit word addressable space and the term "byte-address" for the 32-bit byte addressable space.

Recall that the widths of the read and write data buses in Avalon MM interfaces are not fixed. SOPC Builder adds additional circuits to accommodate the needed alignment and conversion. For clarity, we assume that the I/O data widths are 32 bits and use the 30 bit "word-address" bus in the subsequent discussion.

14.4.3 Output decoding from an Avalon MM master

During a basic write operation, an Avalon MM master places the data and address on the Avalon interface's writedata and address lines and activates the write signal. A basic wrapping circuit consists of a collection of buffers and a decoding circuit. Consider a custom I/O peripheral circuit with four 32-bit input ports. For simplicity, we assume that these are continuous-access ports. The conceptual diagram is shown in Figure 14.11. We also assume that the Avalon MM master has a 32-bit data bus and a 30-bit word-address bus, similar to those used in a Nios II processor. The key input signals to the wrapping circuits are:

- io_writedata: This is the 32-bit data to be written into a register.
- io_address: This is the 2-bit address used to identify which register is to be written.
- io_chipselect: This is a 1-bit control signal used to enable (i.e., "select") the I/O circuit. The operation of the I/O circuit is disabled if it is not asserted.
- io_write: This is a 1-bit write enable signal. When it is asserted, the data will be written into the designated register.

We choose names similar to those in the Avalon interface definition but add an io_ prefix in front. This eases the mapping later when an SOPC component is created.

The key part is the decoding circuit, whose function table is shown in Table 14.1. It is basically a 2-to-2^2 decoder.

Note that the word-address width of a Nios II processor is 30 bits but the width of the wrapping circuit is only 2 bits. Recall that when a Nios II processor is constructed, SOPC Builder assigns a base address to each I/O module. If the assigned base is 12001000_{16} (in 32-bit byte-address format), the addresses range for this I/O peripheral becomes 12001000_{16} to $1200100f_{16}$ (i.e., 16 bytes or 4 words). During the construction, the Avalon interconnect fabric automatically includes a decoding circuit for this address space, as shown in Figure 14.11. The decoding

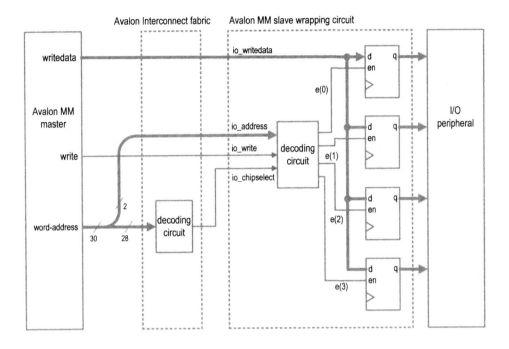

Figure 14.11   Output decoding in wrapping circuit.

Table 14.1   Functional table of a decoding circuit

io_chipselect	input io_write	io_address	output e
0	–	–	0000
1	0	–	0000
1	1	00	0001
1	1	01	0010
1	1	10	0100
1	1	11	1000

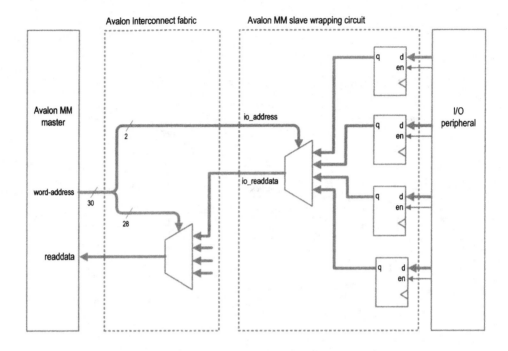

Figure 14.12   Input multiplexing in wrapping circuit (with continuous-access ports).

circuit outputs 1 when the 28 MSBs of the 32-bit byte-address bus match the 28 MSBs of 12001000_{16}, which in turn activates the io_chipselect signal and selects the I/O peripheral.

14.4.4   Input multiplexing to an Avalon MM master

During a basic read operation, an Avalon MM master places the address on the Avalon interface's **address** line, activates the **read** signal, and retrieves data from the **readdata** line. A basic wrapping circuit consists of a collection of buffers and a multiplexing circuit. Consider a custom I/O peripheral circuit with four 32-bit continuous-access output ports. The conceptual diagram is shown in Figure 14.12. The key input signals of the wrapping circuits are:

- **io_readdata**: This is the 32-bit data to be read from a register.
- **io_address**: This is the 2-bit address used to identify which register is to be read.

The multiplexing circuit is a standard 2^2-to-1 multiplexer. The **io_address** signal is used as the selection signal to route one of the inputs to output. Since the word-address width of a Nios II processor is 30 bits, the four I/O registers only occupies a small memory space. When a Nios II system is constructed, the Avalon interconnect fabric automatically includes an additional multiplexing circuit, as shown in Figure 14.12. The 28 MSBs from the Nios II processor byte-address bus are used to construct the selection signal and determine whether to route this peripheral's output to the processor.

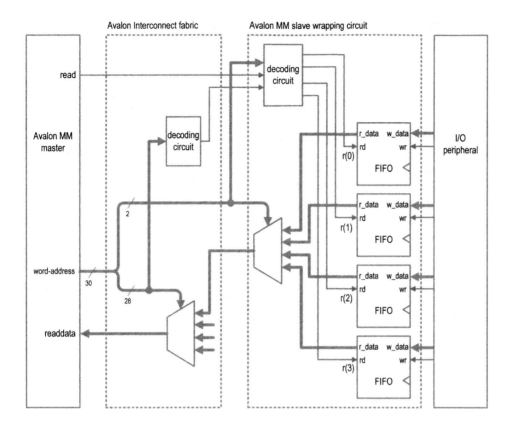

Figure 14.13   Input multiplexing in wrapping circuit (with one-time-access ports).

For a one-time access port, the processor must clear the FF flag or remove the item from the FIFO buffer after reading. This can be achieved by using the Avalon interface's **read** signal and a decoding circuit. The conceptual diagram of a custom I/O peripheral circuit with four FIFO output ports is shown in Figure 14.13. The input multiplexing circuit routes the selected input to the Avalon MM master, as shown in Figure 14.12. The decoding circuit is identical to that in Figure 14.11 except that the **write** signal is replaced by the **read** signal. The decoded output can be interpreted as a "removal" signal, which removes the previously retrieved data when asserted.

Alternatively, we can include an additional input port in the wrapping circuit. The port can be connected to a decoding circuit whose output controls the **rd** signals of the FIFOs. A separate write instruction can be issued to assert the designated **rd** signal to remove a data item from a FIFO.

14.4.5   Practical consideration

A real I/O peripheral contains input and output signals of different widths, access characteristics, and timing constraints. We can add necessary buffers, decoding circuits, and multiplexing circuits in the wrapping circuit to match the Avalon MM slave interface specification.

Note that the registers for the write operation and read operation are usually separated in the I/O wrapping circuit. The read port and the write port may have the same address even if the two ports are unrelated.

14.5   SOPC COMPONENT CONSTRUCTION TUTORIAL

SOPC Builder basically automates the busing and interconnecting process. The system interconnect fabric can connect any combination of components as long as these components conform to the Avalon interface specification. The process of developing an SOPC component consists of the following steps:

1. Design and develop the digital system.
2. Add wrapping logic and I/O signals to accommodate the Avalon interface requirement.
3. Use SOPC Builder's *Component Editor* to create the component, which involves the creation of interfaces and the specification of timing properties.

We use the division circuit discussed in Section 7.3.2 as an example to demonstrate the process of constructing a new SOPC component. The circuit can be considered as a hardware accelerator that speeds up the integer division. We can add the necessary interface logic and convert the circuit to an SOPC component. Note that since a Nios II processor can be configured to include a division unit, this circuit is only used for the demonstration purposes.

The division circuit is designed in Section 7.3.2 and the code is shown in Listing 7.4. The following subsections illustrate the next two steps.

14.5.1   Avalon interfaces

The division circuit has two input data ports (**dvnd** and **dvsr**), two output data ports (**quo** and **rmd**), one control signal (**start**), and two status signals (**ready** and **done_tick**). The **ready** signal is 1 when the circuit is ready to take new input data (i.e., is not in use). The external master circuit should place the dividend and divisor data in **dvnd** and **dvsr** ports and asserts the **start** signal for one clock cycle to initiate the operation. When the calculation is completed, the quotient and remainder are sent to the **quo** and **rmd** ports and the **done_tick** signal is asserted for one clock cycle.

In addition to the original outputs, we want to connect the eight LSBs the remainder to the eight green LEDs on the DE1 board. This is an artificial requirement used to demonstrate the conduit interface.

To develop a wrapping circuit, we examine the characteristics of the I/O peripherals and determine the required Avalon interfaces. The following are needed for the wrapping circuit for the division circuit:

- One clock input interface for the system clock.
- One Avalon MM slave interface for general data access.
- One interrupt sender interface to issue an interrupt request when the operation is done.
- One conduit interface for eight discreet LEDs.

Since the data width of Nios II is 32 bits, it is reasonable to make the division circuit 32 bits wide as well. However, to demonstrate the use of parameters in an

SOPC component, we keep the W and CBIT generics of the division circuit in the top-level wrapping circuit.

14.5.2 Register map

The main part of the design is the Avalon MM slave interface. We first determine the *register map*, which defines the addresses for the relevant I/O ports. From the perspective of a Nios II processor (i.e., an Avalon MM master), these addresses are the *offsets* relative to the base address of the I/O module and thus we call it offset in general. After determining the register map, we can add the necessary buffers, decoding logic, and multiplexing logic accordingly.

There is no specific requirement to define the offsets. For clarity, we generally use separate register offsets for individual signals. One possible assignment is shown below. The registers, their address offsets, and fields are:

- Write addresses (data from cpu)
    - offset 0 (dividend register)
        * bits W-1 to 0: dividend data
    - offset 1 (divisor register)
        * bits W-1 to 0: divisor data
    - offset 2 (start register)
        * Dummy data used to generate an enable pulse
    - offset 6 (done_tick register)
        * Dummy data used to clear the done_tick flag
- Read addresses (data to cpu)
    - offset 3 (quotient register)
        * bits W-1 to 0: quotient data
    - offset 4 (remainder register)
        * bits W-1 to 0: remainder data
    - offset 5 (ready register)
        * bit 0: ready status
    - offset 6 (done_tick register)
        * bit 0: done_tick flag

The dividend and divisor registers store the input data and two actual registers are required in the wrapping circuit. The start register is a "dummy" register, which means that there is no physical register associated with this offset and the write data is irrelevant. It is included to obtain a decoded write pulse, which is connected to the start signal of the division circuit. When an Avalon MM master writes this address, the decoded pulse is asserted and initiates the division operation. The quotient, remainder, and ready registers store the output data and status signal. Since the division circuit contains registers for the data and the ready signal does not change until the next operation, no register is needed in the wrapping circuit. The done_tick register is implemented as a flag register. It is set by the done_tick signal of the division circuit and cleared by the decoded write pulse. The Avalon MM master can read this register to check whether the calculation is done and write the register to clear it to 0 after retrieving the data. To match the data width of Nios II processor, we usually treat these registers as 32-bit registers. The unused bits will be removed automatically during synthesis and will not introduce additional hardware.

An I/O register is usually used as an input register or an output register but not both. Thus, it is possible for a read register to have the same offset of an irrelevant write register. For example, the offset of quotient register can be assigned to 0. For clarity, we don't use the same offset for the read and write registers unless the operations are related (such as the done_tick register).

The output of the done_tick register is also used as an interrupt request signal. If this feature is used, an ISR routine should clear the interrupt condition by writing a dummy data to this address.

14.5.3 Wrapped division circuit

The HDL code of the wrapped division circuit, which includes an instantiated division circuit and wrapping logic for buffering, decoding, and multiplexing, is shown in Listing 14.2. The I/O ports use names similar to those in the Avalon interfaces but include a div_ prefix.

Listing 14.2 Wrapped division circuit

```
1 library ieee;
  use ieee.std_logic_1164.all;
  entity chu_avalon_div_demo is
     generic(
        W: integer:=32;
6       CBIT: integer:=6    -- CBIT=log2(W)+1
     );
     port (
        -- to be connected to Avalon clock input interface
        clk, reset: in   std_logic;
11      -- to be connected to Avalon MM slave interface
        div_address: in std_logic_vector(2 downto 0);
        div_chipselect: in std_logic;
        div_write: in std_logic;
        div_writedata: in std_logic_vector(W-1 downto 0);
16      div_readdata: out std_logic_vector(W-1 downto 0);
        -- to be connected to Avalon interrupt sender interface
        div_irq: out std_logic;
        -- to be connected to Avalon conduit interface
        div_led: out std_logic_vector(7 downto 0)
21   );
  end chu_avalon_div_demo;

  architecture arch of chu_avalon_div_demo is
     signal div_start, div_ready: std_logic;
26   signal set_done_tick, clr_done_tick: std_logic;
     signal dvnd_reg, dvsr_reg: std_logic_vector(W-1 downto 0);
     signal done_tick_reg: std_logic;
     signal quo, rmd: std_logic_vector(W-1 downto 0);
     signal wr_en, wr_dvnd, wr_dvsr: std_logic;
31 begin
     -- ================================
     -- instantiation
     -- ================================
     -- instantiate division circuit
36   div_unit: entity work.div
        generic map(W=>W, CBIT=>CBIT)
        port map(clk=>clk, reset=>'0', start=>div_start,
```

```
                dvsr=>dvsr_reg, dvnd=>dvnd_reg,
                quo=>quo, rmd=>rmd,
41              ready=>div_ready, done_tick=>set_done_tick);

    -- registers

    process (clk, reset)
46  begin
        if reset='1' then
            dvnd_reg <= (others=>'0');
            dvsr_reg <= (others=>'0');
            done_tick_reg <= '0';
51      elsif (clk'event and clk='1') then
            if wr_dvnd='1' then
                dvnd_reg <= div_writedata;
            end if;
            if wr_dvsr='1' then
56              dvsr_reg <= div_writedata;
            end if;
            if (set_done_tick='1') then
                done_tick_reg <= '1';
            elsif (clr_done_tick='1') then
61              done_tick_reg <= '0';
            end if;
        end if;
    end process;

66  -- write decoding logic

    wr_en <= '1' when div_write='1' and div_chipselect='1' else '0';
    wr_dvnd <= '1' when div_address="000" and wr_en='1' else '0';
    wr_dvsr <= '1' when div_address="001" and wr_en='1' else '0';
71  div_start <= '1' when div_address="010" and wr_en='1' else '0';
    clr_done_tick <= '1' when div_address="110" and wr_en='1' else '0';

    -- read multiplexing logic (assume W=32)

76  div_readdata <=
        quo when div_address="011" else
        rmd when div_address="100" else
        x"0000000"&"000"&div_ready when div_address="101" else
        x"0000000"&"000"&done_tick_reg;
81
    -- conduit signal

    div_led <= rmd(7 downto 0);   -- assume that W> 7

86  -- interrupt request signal

    div_irq <= done_tick_reg;
end arch;
```

The code includes six main segments. The first segment is an instance of the division circuit. The second segment is the registers for the input dividend and divisor data and the done_tick flag.

The third segment is the write decoding logic. The logic consists of a common wr_en signal and four individual write enable signals. The wr_en is asserted when both div_write and div_chipselect are asserted and the other enable signal is asserted when wr_en is 1 and div_address matches the designated offset value. The wr_dvnd and wr_dvsr signals enable the dividend and divisor registers, respectively. The decoded `div_start` signal is connected to the `start` signal of the division circuit and thus initiates the division operation when asserted. The decoded `clr_done_tick` signal is used with the done_tick register and clears it to 0 when asserted.

The fourth segment is the read multiplexing logic. It uses `div_address` as the selection signal and routes the designated data to the read data bus, `div_readdata`. Note that the padding zeros are added as needed.

The fifth segment connects the eight LSBs of `rem` to the external `div_led` port and the sixth segment connects `done_tick_reg` to an interrupt request.

14.5.4   SOPC component creation

An SOPC component can be created and configured by a software utility program known as *Component Editor*, which can be invoked from SOPC Builder. The procedure to create an SOPC component is:

1. Start Component Editor for a new SOPC component.
2. Specify the HDL files.
3. Create interfaces and map signals.
4. Configure interfaces.
5. Define HDL parameters.
6. Edit Library information.
7. Save the component file.

Start Component Editor   Component Editor can be invoked as follows:

1. In Quartus II GUI, select Tool ≻ SOPC Builder. The SOPC Builder window appears.
2. In SOPC Builder window, select File ≻ New Component.... The Component Editor window appears. There are several tab pages in the window and the default Introduction page is invoked first.
3. Click the Next button or select the HDL Files tab to go to the HDL Files page.

Specify the HDL files   The initial HDL Files page is empty and we need to add the relevant HDL files. Recall that the division circuit consists of two files, `div.vhd` and `chu_avalon_div_demo.vhd`. The procedure to add files is:

1. Click the Add... button and navigate to the directory.
2. Select the two files.
3. Component Editor automatically analyzes the files and determines the top-level file. The completed page is shown in Figure 14.14. Note that the `chu_avalon_div_demo.vhd` is correctly identified as the top-level file.
4. Click the Next button or select the Signals tab to go to the Signals page.

Create and map signals   When analyzing the HDL files, Component Editor guesses the purposes of the I/O signals in the top-level module, creates the needed interfaces, and maps the module's I/O signals to the interfaces' signals. The guess may not always be correct and we must adjust the mapping and create additional interfaces as needed. The mapping is listed in the Signals page.

The initial Signals page of the division circuit is shown in Figure 14.15. The Name column lists the I/O signals of the `chu_avalon_div_demo` module. The Interface and Signal Type columns show the interfaces assigned to the I/O signals and

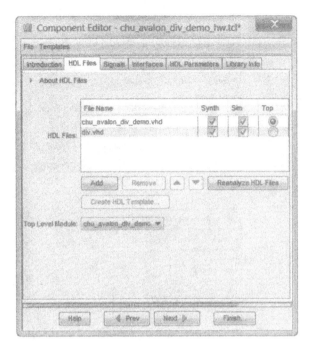

Figure 14.14   The HDL Files page.

Figure 14.15   The initial Signals page.

Figure 14.16   The Signals page with pull-down menu.

the mappings between the I/O signals and interface's signals. For example, the div_write signal is assigned to the div interface (which is an Avalon MM slave interface) and mapped to the div interface's write signal.

Recall that we intend to include four Avalon interfaces in the top-level wrapping circuit. Component Editor correctly identifies two of them, which are the clock output interface, labeled clock_reset in the page, and the Avalon MM slave interface, labeled div in the page. Since the HDL construction follows closely to the Avalon specification and uses a similar naming convention, the signal mapping within the two interfaces is correct.

On the other hand, Component Editor mistakenly maps the div_irq and div_led signals to another Avalon MM slave interface, labeled avalon_slave_0. To correct the problems, we need to add an interrupt sender interface and a conduit interface and reassign the two I/O signals. The procedure to add interfaces and adjust mapping is:

1. Select the Interface field of the div_irq signal and click the right arrow to bring the pull-down menu, which is shown in Figure 14.16. Note that the first three items are the existing interfaces and the remaining new ... items are used to add a new interface.

2. Select the new Interrupt Sender... item to add the new interface and assign the div_irq signal to this interface.

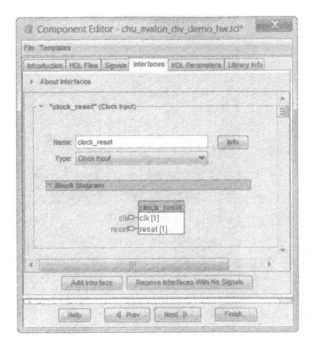

Figure 14.17   The initial Interfaces page.

3. Select the **Signal Type** field of the div_irq signal and click the right arrow to bring the pull-down menu, which lists the available signals of the interrupt sender interface.
4. Select the irq signal.
5. Repeat the two previous steps for the div_led signal to add a conduit interface and map it to the **export** signal of the interface.
6. Click the **Next** button or select the **Interfaces** tab to go to the **Interface** page.

Configure interfaces   The default setting is used when an interface is added. For some interfaces, we may need to modify the setting to meet an individual subsystem's requirement. In the **Interfaces** tab page, we can examine an interface's property and adjust its parameters as needed. For our purposes, we must pay close attention to the timing properties of the Avalon MM slave interface to ensure the correct data access between the processor and an I/O subsystem.

In the **Interfaces** page, each interface is encompassed in an individual entry. Various properties are listed under the entry. The screenshot of the initial **Interfaces** page of the division circuit is shown in Figure 14.17. It displays the clock input interface entry.

There are four interfaces in the division components. The procedure to configure these interfaces is:

1. Component Editor incorrectly includes the **avalon_slave_0** interface in the previous step. It is empty (i.e., no signals connected) after the remapping of the div_irq and div_led signals and is no longer needed. Click the **Remove Interfaces With No Signals** button on the bottom to remove this interface.
2. The clock input interface **clock_reset**, as shown in Figure 14.17, is adequate and can be kept unchanged.

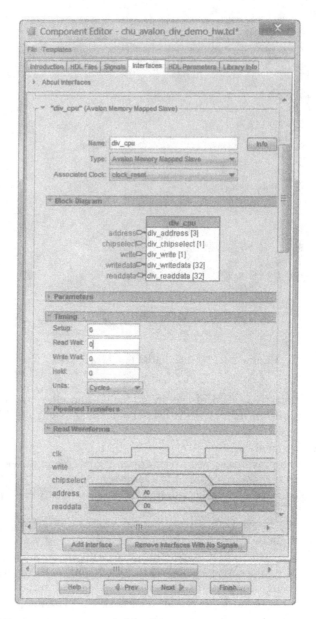

Figure 14.18 The Avalon MM slave entry in Interfaces page.

3. Scroll the right scroll bar to reach the Avalon MM slave interface entry. Rename it to div_cpu (representing the connection between the division circuit and the processor).

4. Enter 0 in the **Read Wait** box of the **Timing** field (the default is 1) since the read data is readily available in the division circuit's registers and the read operation requires no additional wait cycle. The completed entry is shown in Figure 14.18.

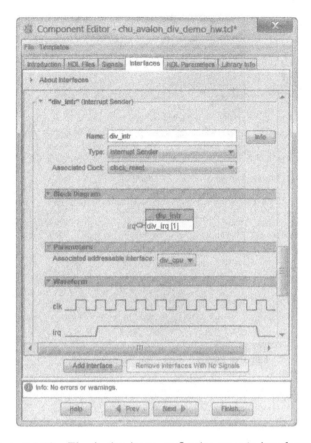

Figure 14.19   The Avalon Interrupt Sender entry in Interfaces page.

5. Scroll the right scroll bar to reach the interrupt sender interface entry and rename it div_intr.
6. Select div_cpu from the pull-down menu of the **Associated addressable interface** field. The completed entry is shown in Figure 14.19.
7. Scroll the right scroll bar to reach the conduit interface entry and rename it div_ledg.
8. Click the **Next** button or select the **HDL Parameters** tab to go to the HDL Parameters page.

If desired, we can return to the Signals page to verify the final mapping, as shown in Figure 14.20.

Define HDL parameters   The top-level HDL module may contain generics (i.e., parameters) and they can be configured in the **HDL Parameters** page. Recall that the generics W and CBITS are included in the chu_avalon_div_demo module, as shown in Listing 14.2. Component Editor extracts the generics and their default values and the corresponding page is shown in Figure 14.21. Note that the **Edit** boxes are checked. This allows the user to edit these parameters when the component is instantiated in SOPC Builder.

Figure 14.20 The final **Signals** page.

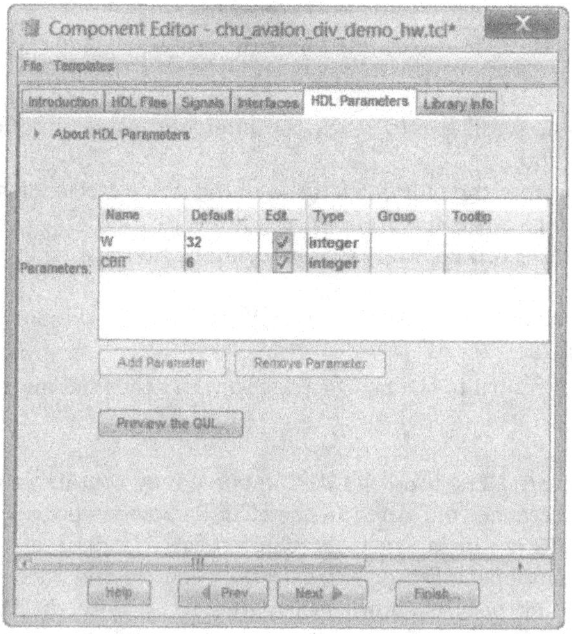

Figure 14.21 The HDL Parameters page.

Figure 14.22   The Library Info page.

Edit Library information   We can enter relevant component information in the Library Info tab page, as shown in Figure 14.22. Recall that the SOPC cores are organized in different categories. This can be done by entering a new group name or selecting an existing group from the pull-down menu of the Groups: field. All the SOPC components developed in this book are organized under the chu_ip group.

Save the component file   The description of an SOPC component is saved in a Tcl script file. The name of the file is similar to that of the top-level HDL file but with an ending of _hw.tcl and the file is stored in the directory where the top-level HDL file is located. For example, the top-level HDL file of the division circuit is chu_avalon_div_demo.vhd and a Tcl file named chu_avalon_div_demo_hw.tcl will be created for the SOPC component.

We can select File ≻ Save to save the file and then select File ≻ Exit to close Component Editor. The chu_ip group and the newly created chu_avalon_div_demo component appears on the left Component Library panel of the SOPC Builder, as shown in Figure 14.23 (note that other components may or may not exist in the chu_ip category).

It is good practice to keep the HDL files and the component Tcl file in the same directory. The entire directory can be moved to a new location or a different computer. We can follow the procedure discussed in Section 10.5.1 to add the component to the SOPC Builder library.

14.5.5   SOPC component instantiation

Once the chu_avalon_div_demo component is created and included in SOPC Builder's library, it can be used and processed as a normal IP core. To include the core in

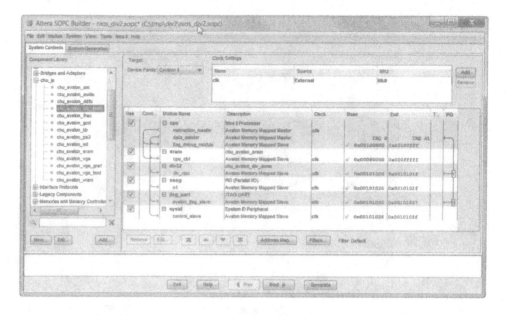

Figure 14.23   The completed SOPC window.

a Nios II system in SOPC Builder, we just select the core and then click the **Add** button. The **chu_avalon_div_demo** dialog appears, as shown in Figure 14.24. The **W** and **CBIT** fields are shown in the **Parameters** box and can be configured as needed.

An instantiated module, labeled div32, is shown in the working area of in Figure 14.23. Note that SOPC Builder automatically assigns its base address according its address width and includes its interrupt request.

14.6   TESTING

To illustrate the use of the division component, we construct a Nios II system that includes the core and develop software to verify its operation. The procedure is:

1. Create a Nios II system that contains the **chu_avalon_div_demo** core and supporting peripherals.
2. Create a top-level HDL file that instantiates the Nios II system and compile the design.
3. Develop testing software.
4. Build and run software.

Nios II system   The Nios II system contains the division core and I/O peripherals to facilitate the testing. Its main parts are:

- A Nios II/e processor.
- The **chu_avalon_div_demo** core.
- An SRAM controller core to utilize the external SRAM device.
- A JTAG UART core to obtain input operands and display the division results.
- A PIO core to interface the four seven-segment LED displays.

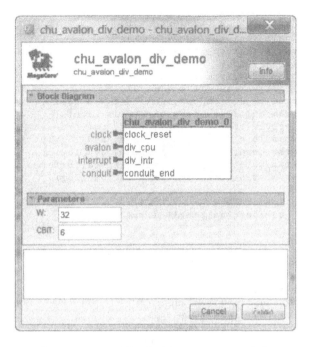

Figure 14.24   The division component dialog.

- A PIO core for the eight discrete green LEDs to display the eight LSBs of the remainder.
- A system id core.

It can be constructed following the procedure in Section 10.5.1 and the completed SOPC configuration is shown in Figure 14.23.

Top-level HDL file   After the HDL files are generated, we can create a top-level module that incorporates the Nios II system. The HDL code is shown in Listing 14.3.

Listing 14.3   Top-level system

```
library ieee;
use ieee.std_logic_1164.all;
entity nios_div2_top is
port(
    clk: in std_logic;
    ledg: out std_logic_vector(7 downto 0);
    hex3, hex2, hex1, hex0: out std_logic_vector(6 downto 0);
    sram_addr: out std_logic_vector (17 downto 0);
    sram_dq: inout std_logic_vector (15 downto 0);
    sram_ce_n, sram_oe_n, sram_we_n: out std_logic;
    sram_lb_n, sram_ub_n: out std_logic
);
end nios_div2_top;

architecture arch of nios_div2_top is
    component nios_div2 is
        port(
            signal clk: in std_logic;
            signal reset_n: in std_logic;
```

```
                 signal div_led_from_the_div32: out
21                   std_logic_vector (7 downto 0);
                 signal out_port_from_the_sseg: out
                     std_logic_vector (31 downto 0);
                 signal sram_addr_from_the_sram: out
                     std_logic_vector (17 downto 0);
26               signal sram_ce_n_from_the_sram: out std_logic;
                 signal sram_dq_to_and_from_the_sram: inout
                     std_logic_vector (15 downto 0);
                 signal sram_lb_n_from_the_sram: out std_logic;
                 signal sram_oe_n_from_the_sram: out std_logic;
31               signal sram_ub_n_from_the_sram: out std_logic;
                 signal sram_we_n_from_the_sram: out std_logic
             );
         end component nios_div2;
         signal sseg4: std_logic_vector(31 downto 0);
36
  begin
         -- instantiate processor
         nios_unit: nios_div2
         port map(
41           clk=>clk, reset_n=>'1',
             div_led_from_the_div32 => ledg,
             out_port_from_the_sseg=>sseg4,
             sram_addr_from_the_sram=>sram_addr,
             sram_dq_to_and_from_the_sram=>sram_dq,
46           sram_ce_n_from_the_sram=>sram_ce_n,
             sram_lb_n_from_the_sram=>sram_lb_n,
             sram_oe_n_from_the_sram=>sram_oe_n,
             sram_ub_n_from_the_sram=>sram_ub_n,
             sram_we_n_from_the_sram=>sram_we_n
51       );
         hex3 <= sseg4(30 downto 24);
         hex2 <= sseg4(22 downto 16);
         hex1 <= sseg4(14 downto 8);
         hex0 <= sseg4(6 downto 0);
56 end arch;
```

Note that the I/O ports of the division circuit are connected to the Avalon interconnect and thus are no longer visible. Only the signal from the conduit interface of the chu_avalon_div_demo module, div_led, is shown as an output port, div_led_from_the_div32, in the nios_div2 module. We can compile this system to obtain the configuration (i.e., .sof) file.

Application program After the creation of a new I/O core, the driver routines should be developed to access and communicate with the core. However, since the chu_avalon_div_demo module is only used for demonstration purposes, we just put all low-level access codes in the main program. The complete program is shown in Listing 14.4.

Listing 14.4

```
#include <stdio.h>
#include "system.h"
#include "chu_avalon_gpio.h"

/* register offset definitions */
#define DVND_REG_OFT  0  // dividend register address offset
#define DVSR_REG_OFT  1  // divisor register address offset
#define STRT_REG_OFT  2  // start register address offset
```

```
#define QUOT_REG_OFT   3   // quotient register address offset
#define REMN_REG_OFT   4   // remainder register address offset
#define REDY_REG_OFT   5   // ready signal register address offset
#define DONE_REG_OFT   6   // done-tick register address offset

/* main program */
int main()
{
  alt_u32 a, b, q, r, ready, done;
  alt_u8 di1_msg[4]={sseg_conv_hex(13),0xfb,0xff,sseg_conv_hex(2)};

  sseg_disp_ptn(SSEG_BASE, di1_msg);                   // display "di 2"
  printf("Division accelerator test #2: \n\n");
  while (1){
    printf("Perform division a / b = q remainder r\n");
    printf("Enter a: ");
    scanf("%d", &a);
    printf("Enter b: ");
    scanf("%d", &b);
    /* send data to division accelerator */
    IOWR(DIV32_BASE, DVND_REG_OFT, a);
    IOWR(DIV32_BASE, DVSR_REG_OFT, b);
    /* wait until the division accelerator is ready */
    while (1) {
      ready = IORD(DIV32_BASE, REDY_REG_OFT) & 0x00000001;
      if (ready==1)
        break;
    }
    /* generate a 1-pulse */
    printf("Start ...\n");
    IOWR(DIV32_BASE, STRT_REG_OFT, 1);
    /* wait for completion  */
    while (1) {
      done = IORD(DIV32_BASE, DONE_REG_OFT) & 0x00000001;
      if (done==1)
        break;
    }
    /* clear done_tick register */
    IOWR(DIV32_BASE, DONE_REG_OFT, 1);
    /* retrieve results from division accelerator */
    q = IORD(DIV32_BASE, QUOT_REG_OFT);
    r = IORD(DIV32_BASE, REMN_REG_OFT);
    printf("Hardware: %u / %u = %u remainder %u\n", a, b, q, r);
    /* compare results with built-in C operators */
    printf("Software: %u / %u = %u remainder %u\n\n\n", a, b, a/b, a%b);
  } // end while
}
```

The basic program structure is similar to that in Listing 13.2. Note that the previous seven symbolic base addresses associated with the division circuit are replaced by one symbolic base address (DIV32_BASE) plus offsets (_OFT). Also, instead of writing 1 then 0 to generate a **start** pulse, as in

```
        pio_write(START_BASE, 1);
        pio_write(START_BASE, 0);
```

we use the enable pulse associated with a write operation

```
        IOWR(DIV32_BASE, STRT_REG_OFT, 1);
```

to activate the **start** signal for one clock cycle.

14.7   BIBLIOGRAPHIC NOTES

The detailed specification and definition of Avalon interfaces are documented in Altera's *Avalon Interface Specifications*. The implementation and realization of the Avalon MM interface are explained in Chapter 2 (titled *System Interconnect Fabric for Memory-Mapped Interfaces*), Volume 4 (titled *SOPC Builder*) of *Quartus II Handbook*. A detailed description of SOPC Component Editor and the Tcl file format can be found in Chapters 4, 6, and 7 of *SOPC Builder*.

14.8   SUGGESTED EXPERIMENTS

14.8.1   Division core ISR

Develop an ISR routine for the division core in Section 14.5 to verify the operation of the interrupt interface.

14.8.2   Alternative buffering scheme for the division core

The division wrapping circuit in Section 14.5 includes a divisor register to store the divisor data and a dummy start register to generate a pulse to initiate the division operation. Closer examination shows that the division circuit already contains an internal register, d_reg, to store the divisor data. Thus, the divisor register of the wrapping circuit is not actually needed. We can revise the wrapping circuit to load the divisor data directly to the division circuit and initiate the division operation at the same time. Derive the modified codes for the wrapping circuit, create a new component, derive a Nios II system, develop testing software, and verify the core's operation.

14.8.3   Division core with eight-bit data

The data width of the division core in Section 14.5 can be configured. We want to redesign the core with a fixed data width of eight bits. Since the data width of the Nios II processor is 32 bits, we can pack the data and status signals into a single word. Redesign the wrapping circuit with a minimal address space and buffering circuits, create a new component, derive a Nios II system, develop testing software, and verify the core's operation.

14.8.4   Division core with 64-bit data

The maximal data width of the division core in Section 14.5 is 32 bits. We want to increase the data width to 64 bits. Since the data width of the Nios II processor is 32 bits, two words are needed to access input and output data. Redesign the wrapping circuit, create a new component, derive a Nios II system, develop testing software, and verify the core's operation.

14.8.5   Fibonacci number circuit

The Fibonacci number circuit is discussed in Section 7.3.1. We can modify the circuit to accommodate a 64-bit result (i.e., the output f port is increased to 64 bits wide). Based on this circuit, design a wrapping circuit, create a new component, derive a Nios II system, develop testing software, and verify the core's operation.

14.8.6   Period counter

An accurate low-frequency counter is discussed in Section 7.3.5. Based on the binary-to-BCD conversion circuit in Section 7.3.3 and the period counter in Section 7.3.4, two new components can be created. We can reconstruct the accurate low-frequency counter in Section 7.3.5 by using the two new cores plus the division core. Derive a Nios II system that includes these cores, develop testing software, and verify the cores' operation.

CHAPTER 15

SRAM AND SDRAM CONTROLLERS

While an FPGA device contains dedicated memory modules within the chip, its capacity is relatively small. Separate external memory devices are needed to support the larger storage requirement. On the DE1 board, there are an SRAM (static random access memory) device and an SDRAM (synchronous dynamic random access memory) device. In this chapter, we provide an overview of various types of memory and discuss the design and use of a *memory controller*, which is the circuit to read and write the external memory devices. Since the off-chip access introduces new types of delays, we also provide a brief overview of timing issues and clock management.

15.1 MEMORY RESOURCES OF DE1 BOARD

The Altera EP2C20 FPGA device and DE1 board provide several options for storage elements:

- *EP2C20's D FFs* (for registers): about 20K bits embedded in logic cells (LEs).
- *EP2C20's embedded RAM*: about 200K bits, configured as 52 4K-bit modules.
- *off-chip SRAM device*: about 4,000K bits, arranged as a 256K-by-16 cell array.
- *off-chip SDRAM device*: about 64,000K bits, arranged as a 4M-by-16 cell array.

Embedded SOPC Design with Nios II Processor and VHDL Examples. By Pong P. Chu   **341**
Copyright © 2011 John Wiley & Sons, Inc.

These memory options exhibit a trade-off between cost and performance. A D FF is the fastest and most versatile option but requires the most silicon area and thus has the highest per-bit cost. It is only feasible for small, fast buffers. On the other hand, an SDRAM cell occupies the smallest silicon area and has the lowest per-bit cost but has the slowest access speed. Thus, it is feasible for a system that requires massive storage but can tolerate relatively slower performance.

It is a good idea to keep in mind the capacities of these options and to select the proper type that is most suitable for an application at hand.

15.2 BRIEF OVERVIEW OF TIMING AND CLOCK MANAGEMENT

As discussed in Section 5.1.2, the single most fundamental design principle is the synchronous methodology, in which all registers are driven by a single global clock. This methodology implicitly assumes that the rising edge of the clock signal can arrive in all registers at the same time. In reality, this assumption is only true for an intermediate-sized circuit within the FPGA device. Non-ideal clocking must be taken into consideration in many designs, especially for a system with high-speed off-chip access. In this section, we provide a brief overview of relevant timing issues and clock management schemes.

15.2.1 Clock distribution network

In a digital gate, the output stage "drives" the input ports of connected components. The number of input ports that can be driven is known as *fan-out*. A typical gate can drive around half a dozen ports (i.e., a fan-out of 6). Since all registers are connected to the same clock signal in a synchronous system, the fan-out of the clock signal is the number of FFs in the system, which can reach thousands or even tens of thousands in a large design.

To facilitate the requirement, an FPGA device contains special *clock distribution networks* to route the clock signal. A network is composed of multiple levels of buffers to increase the driving capability and is carefully placed and routed to balance and minimize the propagation delays. A conceptual three-level clock distribution network is shown in Figure 15.1, in which the fan-out of an individual buffer is four. To provide design flexibility, FPGA devices usually provide multiple clock distribution networks. There are 16 distribution networks in the EP2C20 device. A distribution network reaches all resources within the device and can be used for a global clock as well as control signals, such as a clear or enable signal.

In a real system, the clock's sampling edge may reach FFs at different times and the difference between the arrival times is known as *clock skew*. Because of the propagation delay of buffers, the clock skew between the clock source and a leaf FF can be quite large. However, the skews between the leaf FFs are small since the FFs experience similar delays. Thus, for a synchronous system implemented completely within an FPGA chip (i.e., not considering off-chip signals), we can assume that it is driven by the ideal clock source.

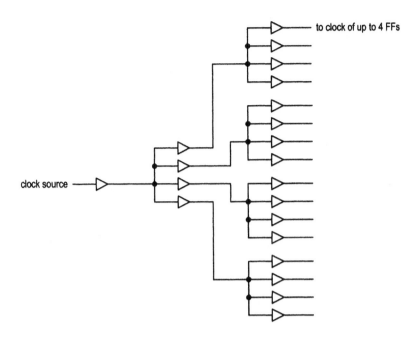

to clock of up to 4 FFs

clock source

Figure 15.1 Conceptual clock distribution network.

15.2.2 Timing consideration of off-chip access

The timing analysis for off-chip signals is more complicated because it involves an I/O buffer delay, an I/O pad delay, and additional routing delays and can be effected by the external load and PCB (printed circuit board) routing.

One important timing parameter of a synchronous system is t_{CO}, which defines the clock-to-output delay (i.e., the time required to obtain a stable output signal after the clock's sampling edge), and we use this to illustrate off-chip timing issues. The simplified timing path to determine the device-level clock-to-output delay is shown in Figure 15.2.

The system within the FPGA chip can be considered as an ideal synchronous system. Its clock-to-output delay is labeled t_{CO} in Figure 15.2 and its value is equal to t_{CQ} plus t_{OUTPUT}, as discussed in Section 5.5. On the other hand, the clock-to-output delay in the device level is the delay from the clock pin to the output pin. It is labeled t_{CO1} in Figure 15.2. t_{CO1} involves additional propagation delays:

- *I/O input delay* of the clock signal: the delays of pad, package pin routing, and I/O buffer.
- *clock routing delay* of the clock signal: the delay of the clock distribution network.
- *logic array to IO buffer delay* of the output signal: the routing delay from an logic element to the I/O buffer.
- *I/O output delay* of the output signal: the delays of pad, package pin routing, and I/O buffer.

The I/O output delay is affected by the load of the pin. During the timing analysis, Quartus Timing Analyzer uses a default value to estimate the value. For more

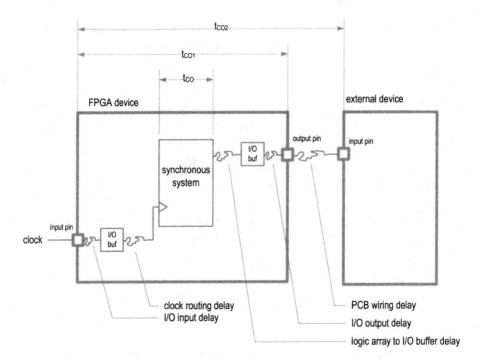

Figure 15.2 Conceptual diagram of off-chip delay.

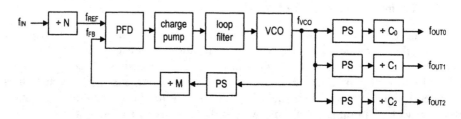

Figure 15.3 Conceptual diagram of Cyclone II PLL.

accurate computation, we need to consider the actual PCB wiring and even the effect of the transmission line. It is labeled t_{CO2} in Figure 15.2.

15.2.3 PLL

To further facilitate clock and timing management, Cyclone II devices also contain PLL (phase-locked loop) circuits. The simplified block diagram of a Cyclone II PLL circuit is shown in Figure 15.3. It consists of a PFD (phase-frequency detector), a charge pump, a loop filter, a VCO (voltage controlled oscillator), and several frequency dividers and PS (phase selection) circuits. The key part of a PLL is the closed feedback loop. The PFD compares the phases of the reference input clock and feedback clock and outputs their difference. The charge pump and loop filter convert the difference to a voltage level. Based on the voltage level, the VCO oscillates at a higher or lower frequency, which affects the phase and frequency of the feedback clock. The negative feedback mechanism eventually forces the feedback clock and the reference input clock to have the same frequency and phase, which is said to be *phase locked*.

There are several frequency dividers in PLL and we can perform *frequency synthesis* by adjusting the values of these dividers. Because of the PLL loop, $f_{REF} = f_{FB}$. Since $f_{REF} = \frac{f_{IN}}{N}$ and $f_{FB} = \frac{f_{VCO}}{M}$, we have

$$f_{VCO} = \frac{M}{N} f_{IN}$$

In a Cyclone II PLL, the VCO output is fed to three separate frequency dividers and phase selection circuits to obtain three output clocks. For example, the frequency of the output clock 0 is

$$f_{OUT0} = \frac{f_{VCO}}{C_0} = \frac{M}{N * C_0} f_{IN}$$

We can also adjust the PS circuit to adjust the phase for the output clocks (i.e., to make the sampling edge of the output clock ahead or behind the sampling edge of the input clock).

The output of a Cyclone II PLL can be connected to a clock distribution network or an output pin. The PLL can be used to change the system clock rate with a fixed external oscillator and drive different subsystems with different clock rates. It can also be used to reduce clock skew and adjust the arrival time of a clock's sampling edge to meet special timing requirements. There are four PLLs in an EP2C20 device.

15.3   OVERVIEW OF SRAM

SRAM (*static random access memory*) is a type of semiconductor memory. The term "static" indicates that the data is retained as long as power is being supplied and thus not changed "dynamically."

15.3.1   SRAM cell

The basic organization of an SRAM cell is shown on the top right corner of Figure 15.4. The two invertors form a latch that stores one bit of information and the two pass transistors function like switches that can be either "closed" (i.e., short circuit) or "open" (i.e. open circuit). The two pass transistors are controlled by a signal commonly referred to as word line. When the word line signal is high, the cell is enabled for access and the latch is connected to the data lines, labeled bit line and $\overline{\text{bit line}}$. As the label indicated, $\overline{\text{bit line}}$ always carries the complemented value of bit line. Although the $\overline{\text{bit line}}$ line is not strictly needed, it is used to improve the noise margins and thus increase the reliability. The bit line and $\overline{\text{bit line}}$ are shown as a single line in the cell array in Figure 15.4. In a read operation, the stored data is connected to the two bit lines and passed to a *sense amplifier* to generate the final value. In a write operation, the desired value and its complement are placed on bit line and $\overline{\text{bit line}}$ to set the latch to the desired value.

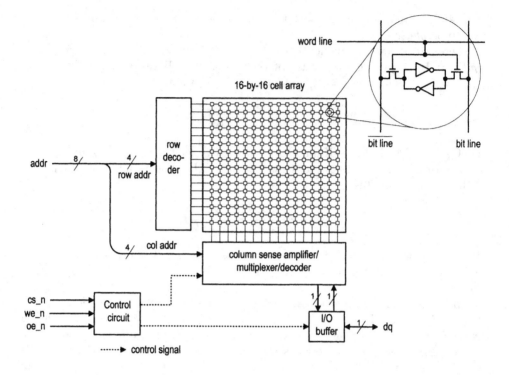

Figure 15.4 Conceptual diagram of a 256-by-1 SRAM.

15.3.2 Basic organization

The memory cells on a memory device are usually arranged as a rectangular matrix and use *two-dimensional decoding and multiplexing* to access the designated cell. The conceptual diagram of a 256-by-1 SRAM (i.e., 256 (2^8) address locations with one bit in each location) is shown in Figure 15.4. The SRAM cells are arranged as a 16-by-16-array. The address bus, addr, is split into two parts. The 4 MSBs are connected to the row decoder and the 4 LSBs are connected to the column multiplexer and decoder. The row decoder is a 4-to-2^4 binary decoder and enables a single row of the cell array. In a read operation, the row's 16 bits of data are retrieved and passed to the 16-by-1 column multiplexer. The desired bit is selected according to the 4 LSBs of addr and routed to the I/O buffer. The column decoder is also a 4-to-2^4 binary decoder. In a write operation, it enables a single cell from the selected row and the data is stored into that cell. The cell array and column multiplexer and decoder can be duplicated multiple times to support a wider data width.

To save I/O pins from a chip, the read data and write data usually share the same physical pins. An I/O buffer is associated with each pin. It contains tristate buffers and control logic to coordinate the bidirectional operation.

In addition to the address and data buses, a typical SRAM chip contains at least three control signals. These signal are:

- ce_n (chip enable or chip select): disables or enables the chip

Table 15.1   Functional table of SRAM control signals

Operation	ce_n	we_n	oe_n	dq
chip disabled	1	-	-	Z
output disabled	0	1	1	Z
read	0	1	0	data out
write	0	0	-	data in

- we_n (write enable): disables or enables the write operation
- oe_n (output enable): disables or enables the output

All these signals are active low and the _n suffix is used to emphasize this property. The functional table is shown in Table 15.1. The ce_n signal can be used to accommodate memory expansion, and the we_n and oe_n signals are used for write and read operations.

15.3.3   Timing

Our discussion in Part I focuses on the *synchronous system*, in which the registers are driven by the same clock signals and the input signals are sampled at the rising edge of the clock. A system will function correctly and reliably as long as the data is stable around the sampling edge. An SRAM device, on the other hand, is *asynchronous*. It does not contain a clock signal and its operation is based on the duration and level of the address, data, and control signals. An SRAM device thus is sometimes referred to as *asynchronous SRAM*.

The timing characteristics of an asynchronous SRAM are quite complex and involve more than two dozen parameters. We concentrate on only a few key parameters that are relevant to our design.

The simplified timing diagrams for two types of read operations are shown in Figure 15.5(a) and (b). The relevant timing parameters are:

- t_{RC}: read cycle time, the minimal elapsed time between two read operations. It is about the same as t_{AA} for SRAM.
- t_{AA}: address access time, the time required to obtain stable output data after an address change.
- t_{OHA}: output hold time, the time that the output data remains valid after the address changes. This should not be confused with the hold time of an edge-triggered FF, which is a constraint for the **d** input.
- t_{DOE}: output enable access time, the time required to obtain valid data after oe_n is activated.
- t_{HZOE}: output enable to high-Z time, the time for the tristate buffer to enter the high-impedance state after oe_n is deactivated.
- t_{LZOE}: output enable to low-Z time, the time for the tristate buffer to leave the high-impedance state after oe_n is activated. Note that even when the output is no longer in the high-impedance state, the data is still invalid.

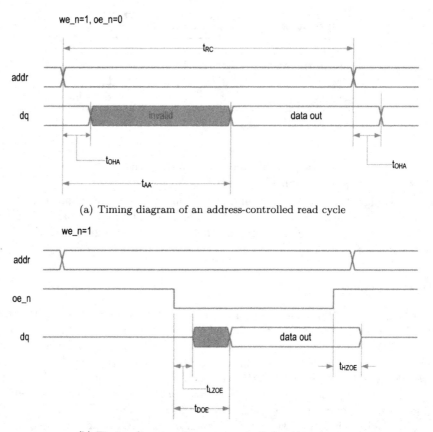

(a) Timing diagram of an address-controlled read cycle

(b) Timing diagram of an oe_n-controlled read cycle

parameter		min	max
t_{RC}	read cycle time	10	–
t_{AA}	address access time	–	10
t_{OHA}	output hold time	2	–
t_{DOE}	output enable access time	–	4
t_{HZOE}	output enable to high-Z time	–	4
t_{LZOE}	output enable to low-Z time	0	–

(c) ISSI IS61LV25616AL timing parameters (in ns)

Figure 15.5   Timing diagrams and parameters of a read operation.

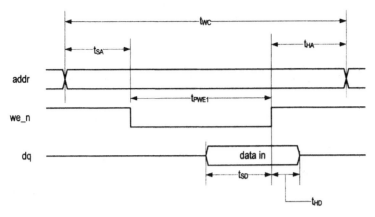

(a) Timing diagram of a write cycle

parameter		min	max
t_{WC}	write cycle time	10	–
t_{SA}	address setup time	0	–
t_{HA}	address hold time	0	–
t_{PWE1}	we_n pulse width	8	–
t_{SD}	data setup time	6	–
t_{HD}	data hold time	0	–

(b) ISSI IS61LV25616AL timing parameters (in ns)

Figure 15.6   Timing diagram and parameters of a write operation.

The simplified timing diagram for a we_n-controlled write operation is shown in Figure 15.6(a). The relevant timing parameters are:

- t_{WC}: write cycle time, the minimal elapsed time between two write operations.
- t_{SA}: address setup time, the minimal time that the address must be stable before we_n is activated.
- t_{HA}: address hold time, the minimal time that the address must be stable after we_n is deactivated.
- t_{PWE1}: we_n pulse width, the minimal time that we_n must be asserted.
- t_{SD}: data setup time, the minimal time that data must be stable before the latching edge (the edge in which we_n moves from 0 to 1).
- t_{HD}: data hold time, the minimal time that data must be stable after the latching edge.

15.3.4   IS61LV25616AL SRAM device

The DE1 board has an IS61LV25616AL device, which is a 256K-by-16 SRAM module manufactured by Integrated Silicon Solution, Inc. (ISSI). This device has an

Table 15.2   Functional table of IS61LV25616AL control signals

Operation	ce_n	we_n	oe_n	lb_n	ub_n	dq (lower)	dq (upper)
disabled	1	-	-	-	-	Z	Z
	0	1	1	-	-	Z	Z
	0	-	-	1	1	Z	Z
read	0	1	0	0	1	data out	Z
	0	1	0	1	0	Z	data out
	0	1	0	0	0	data out	data out
write	0	0	-	0	1	data in	Z
	0	0	-	1	0	Z	data in
	0	0	-	0	0	data in	data in

18-bit address bus, addr, a 16-bit bidirectional data bus, dq, and five control signals. The data bus is divided into upper and lower bytes, which can be accessed individually. In addition to ce_n, we_n, and oe_n, it includes two signals to facilitate the byte-oriented configuration:

- lb_n (lower byte enable): disables or enables the lower byte of the data bus
- ub_n (upper byte enable): disables or enables the upper byte of the data bus

The extended functional table is shown in Table 15.2.

The values of relevant parameters for the read and write operations are shown in Figures 15.5(c) and 15.6(b).

15.4   SRAM CONTROLLER IP CORE

An SRAM controller is a circuit used to access the external SRAM chip. It generates proper control signals, issues the address, and places and retrieves data according to the device's timing specification.

15.4.1   Avalon interfaces

In the Avalon framework, an SRAM controller can be configured as an SOPC component with an Avalon MM slave interface or Avalon MM tristate slave interface. We use the former to illustrate finer timing issues. In addition to the Avalon MM slave interface, the controller contains an clock input interface for the system clock and a conduit interface to connect the SRAM device's I/O pins.

Avalon MM slave signal mapping   By examining the specification of the SRAM functional table, we can relate the SRAM's signals with the Avalon MM slave interface signals:

- SRAM's oe_n: read_n
- SRAM's we_n: write_n

- SRAM's ce_n: chipselect_n
- SRAM's ub_n and lb_n: byteenable_n (two bits)
- SRAM's addr: address (18 bits)
- SRAM's dq: readdata (16 bits) and writedata (16 bits)

Note that SRAM's dq signal (data bus) is bidirectional and thus carries both read and write data. We can use a tristate buffer in the SRAM controller to resolve the situation. The corresponding HDL segment looks like:

```
port(
    dq: inout std_logic_vector(...);
    . . .
)
. . .
dq <= writedata when we_n='0' else (others=>'Z');   -- write
readdata <= dq;                                      -- read
. . .
```

Basic timing analysis   A main task of designing an SRAM controller is to generate a properly timed control signals. In the Avalon framework, the first step is to adjust the timing properties of Avalon MM slave interface's read_n and write_n signals according to the device's specification. We assume that the 50 MHz (i.e., 20 ns) clock signal is used for the Nios II system.

Let us first consider the read operation. In a read cycle, an Avalon MM master issues the address and asserts the read_n signal at the same time. The SRAM device returns the valid data after 10-ns t_{AA}. Since the clock period is 20 ns, the Avalon MM master can sample and retrieve the data from readdata at the rising edge of the next clock. No extra wait state is needed.

The timing diagram of SRAM write operation in Figure 15.6(a) suggests that the write_n signal should include a setup time to accommodate the SRAM's t_{SA} requirement and a hold time to accommodate SRAM's t_{HA} and t_{HD} requirements. However, since IS61LV25616AL is a newer chip and all three parameters are zero. It is feasible to issue the address and data and assert write_n at the same clock cycle. The 20-ns clock period is larger than 10-ns t_{PWE1} and thus no extra wait state is needed.

Additional timing consideration   The previous timing analysis is based on a somewhat ideal scenario. There are several subtle issues in practice. First, the Avalon MM interface is a synchronous protocol and only assures the signals are stable around the clock edge. It does not specify or guarantee the clock-to-output delay or interval of these signals. Thus, the generated signals can be short or may contain glitches. On the other hand, the SRAM timing specification requires the signals to be stable for a specific amount of time (e.g., the 10-ns t_{PWE1} interval). Second, the propagation delays of signals may vary and the variance can be significant because accessing an external device involves various types of I/O propagation delays discussed in Section 15.2.2. The variance can complicate the timing of read since it involves a "round-trip" operation, (i.e., first transmitting address to SRAM and then retrieving data from SRAM). Third, the delay variance may also jeopardize the write operation since it is possible that the data is removed before the deassertion of the we_n signal.

One way to mitigate the problem is to use registers to buffer the incoming and outgoing signals and place these registers within the I/O buffers. The registered output signals are glitch-free and stable for one clock period and thus resolve the first issue. Utilizing the registers in I/O buffers eliminates the logic-array-to-I/O-buffer delay and reduces the main variance in calculating the propagation delays. Due to the relatively large 10-ns slack in read operation (i.e., 20-ns period minus 10-ns t_{AA}), this approach should resolve the timing difficulty in read operation.

Simultaneous data removal and we_n deassertion can still pose a problem because of the potential variance of external wiring delays. To be really safe, we can prolong the data and address for one clock cycle by setting the Avalon MM slave interface's holdtime property to 1. This degrades the performance of the write operation from one clock cycle to two clock cycles. An alternative is to use PLL to generate a slightly leading clock signal to drive the we_n register so that it deasserts slightly before the removal of the data and address signals. However, our test program in Section 15.8 shows that the controller works fine with the original design. It implies that the printed circuit board is designed properly and these features are not needed.

While the registers resolve some timing issues, they introduce *latency* for data access. In a read operation, it takes one clock cycle to register the outgoing address and oe_n and another clock cycle to register the incoming data. Thus, an Avalon MM master experiences a latency of two clock cycles. Note that the Avalon MM master can still issue read in every clock cycle and the data is returned in a pipelined fashion. In a write operation, an Avalon MM master issues address, data, and write_n in the original way. However, because of the buffering registers, the actual SRAM write is delayed by one clock cycle.

15.4.2 Controller circuit

The SRAM controller just consists of registers for the outgoing and incoming signals and a tristate buffer for the bidirectional data bus. Its code is shown in Listing 15.1.

Listing 15.1   SRAM controller

```
library ieee;
2 use ieee.std_logic_1164.all;
  use ieee.numeric_std.all;
  entity chu_avalon_sram is
     port (
        clk, reset: in  std_logic;
7       -- Avalon MM slave interface
        address: in std_logic_vector(17 downto 0);
        chipselect_n: in std_logic;
        byteenable_n: in std_logic_vector(1 downto 0);
        read_n: in std_logic;
12      write_n: in std_logic;
        writedata: in std_logic_vector(15 downto 0);
        readdata: out std_logic_vector(15 downto 0);
        -- Conduit to/from SRAM
        sram_addr: out std_logic_vector (17 downto 0);
17      sram_dq: inout std_logic_vector (15 downto 0);
        sram_ce_n: out std_logic;
        sram_lb_n: out std_logic;
        sram_ub_n: out std_logic;
        sram_oe_n: out std_logic;
22      sram_we_n: out std_logic
     );
  end chu_avalon_sram;
```

```
   architecture arch of chu_avalon_sram is
27     signal addr_reg: std_logic_vector(17 downto 0);
       signal rdata_reg, wdata_reg: std_logic_vector(15 downto 0);
       signal ce_n_reg, lb_n_reg, ub_n_reg: std_logic;
       signal oe_n_reg, we_n_reg: std_logic;
   begin
32     -- registers
       process (clk, reset)
       begin
          if reset='1' then
             addr_reg <= (others=>'0');
37           rdata_reg <= (others=>'0');
             wdata_reg <= (others=>'0');
             ce_n_reg <= '1';
             lb_n_reg <= '1';
             ub_n_reg <= '1';
42           oe_n_reg <= '1';
             we_n_reg <= '1';
          elsif (clk'event and clk='1') then
             addr_reg <= address;
             rdata_reg <= sram_dq;
47           wdata_reg <= writedata;
             ce_n_reg <= chipselect_n;
             lb_n_reg <= byteenable_n(0);
             ub_n_reg <= byteenable_n(1);
             oe_n_reg <= read_n;
52           we_n_reg <= write_n;
          end if;
       end process;
       -- to Avalon
       readdata <= rdata_reg;
57     -- to SRAM
       sram_addr <= addr_reg;
       sram_ce_n <= ce_n_reg;
       sram_lb_n <= lb_n_reg;
       sram_ub_n <= ub_n_reg;
62     sram_oe_n <= oe_n_reg;
       sram_we_n <= we_n_reg;
       -- SRAM tristate data bus
       sram_dq <= wdata_reg when we_n_reg='0' else (others=>'Z');
   end arch;
```

During synthesis, we can set the **Fast Input Register** and **Fast Output Register** options in Quartus to ensure that the registers are placed in I/O buffers. However, software usually can detect the off-chip signals and perform this task automatically.

15.4.3 SOPC component creation

With the HDL file, we can follow the procedure outlined in Section 14.5.4 and create a component in SOPC Builder. The component consists an Avalon MM slave interface (named cpu_ctrl), a clock interface (named clock_reset), and a conduit interface for the SRAM signals (named ctrl_sram). Note that the active-low version of the control signals (i.e., with the _n suffix) should be used. The complete signal mapping is shown in Figure 15.7.

The Avalon MM slave interface needs to be configured to match the timing characteristics of the SRAM controller. Both **Read Wait** and **Write Wait** fields should be 0 and the **Read latency** field should be 2 to accommodate the two-clock delay

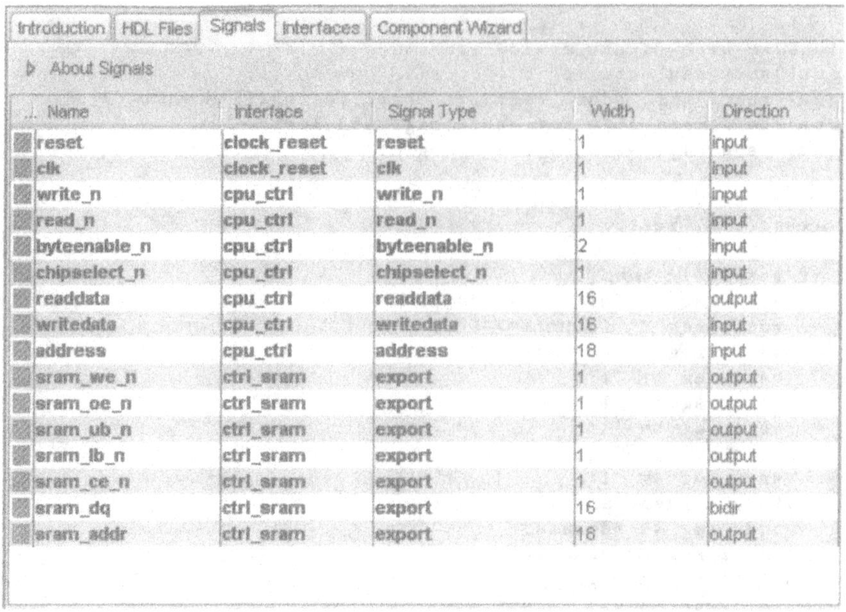

Introduction | HDL Files | Signals | Interfaces | Component Wizard

About Signals

Name	Interface	Signal Type	Width	Direction
reset	clock_reset	reset	1	input
clk	clock_reset	clk	1	input
write_n	cpu_ctrl	write_n	1	input
read_n	cpu_ctrl	read_n	1	input
byteenable_n	cpu_ctrl	byteenable_n	2	input
chipselect_n	cpu_ctrl	chipselect_n	1	input
readdata	cpu_ctrl	readdata	16	output
writedata	cpu_ctrl	writedata	16	input
address	cpu_ctrl	address	18	input
sram_we_n	ctrl_sram	export	1	output
sram_oe_n	ctrl_sram	export	1	output
sram_ub_n	ctrl_sram	export	1	output
sram_lb_n	ctrl_sram	export	1	output
sram_ce_n	ctrl_sram	export	1	output
sram_dq	ctrl_sram	export	16	bidir
sram_addr	ctrl_sram	export	18	output

Figure 15.7 Signal mapping of SRAM controller.

introduced by the input and output registers. Furthermore, we need to expand the Deprecated segment and check the Memory device box to indicate that the interface is a general-purpose memory module. The finished screen shot is shown in Figure 15.8.

Since the SRAM controller core is used earlier in Section 10.5, it may already be included in the SOPC Builder's library.

15.5 OVERVIEW OF DRAM

DRAM (dynamic random access memory) stores a bit of data in a capacitor within an integrated circuit. Because of the charge leakage, the memory cell needs to be refreshed periodically. A DRAM cell is very simple and thus a DRAM chip can reach very high density. It is the "main memory" used in today's computer system.

15.5.1 DRAM cell

The basic organization of a DRAM cell is shown on the top right corner of Figure 15.9. The data bit is stored in the capacitor. Its voltage level, which can be close to 0 V (ground) or V_{DD} (supplied voltage), indicates whether the bit is 0 or 1. The pass transistor functions like a switch and can be turned on or off by the word line signal.

The procedure to write a value to a DRAM cell is:

1. Activate word line to turn on the pass transistor.
2. Set bit line 0 V or V_{DD} to store 0 or 1.
3. Deactivate word line to turn off the pass transistor.

The procedure to read a bit from a DRAM cell is:

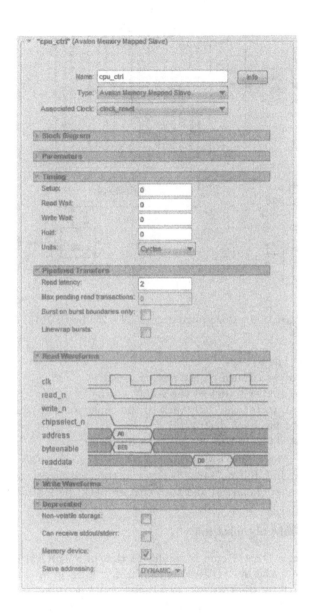

Figure 15.8 Avalon MM slave interface of SRAM controller.

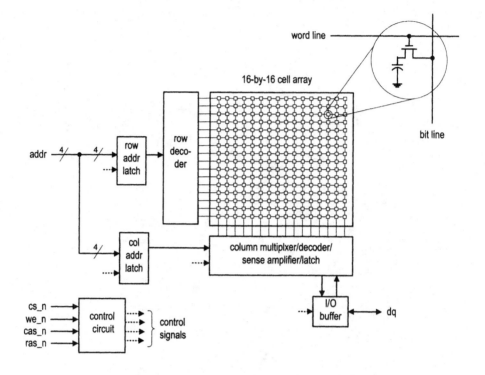

Figure 15.9   Conceptual diagram of a 256-by-1 DRAM.

1. *Precharge* bit line to $\frac{V_{DD}}{2}$
2. Activate word line to turn on the pass transistor.
3. Use a sense amplifier to detect the voltage swing.
4. *Restore* (i.e., rewrite) the data back to the capacitor cell (since the original charge content has been destroyed).
5. Deactivate word line to turn off the pass transistor.

Note that the precharge and restoration steps are needed to accommodate the capacitive storage. In addition, since charges in the capacitor leak gradually, its data must be *refreshed* (i.e., read and then written back) periodically.

15.5.2   Basic DRAM organization

The basic layout of a DRAM chip is similar to that of an SRAM chip. The conceptual diagram of a 256-by-1 DRAM is shown in Figure 15.9. The DRAM cells are arranged as a 16-by-16-array and two-dimensional decoding is used to access the designated cell. However, the DRAM addressing scheme is different. The row address is issued first and stored into a latch (i.e., row address latch) and then the column address issued afterward. This scheme reduces the address I/O pins by half. Note that the width of the **addr** signal is only four bits, half of the size of the SRAM chip. Two additional control signals, **ras_n** (row address strobe) and **cas_n** (column address strobe), are used to indicate the type and validity of the **addr** bus.

The basic procedure to read a bit is listed below. We assume that the bit lines have been already precharged.

1. The external controller places the row value on **addr**.
2. The DRAM stores the address in a row address latch and the row decoder enables a row. The entire row of data is retrieved and stored in a data latch.
3. The external controller places the column value on **addr**.
4. The DRAM stores the address in a column address latch and the column multiplexer routes the selected data bit to data bus, **dq**. The retrieved data is also restored back to the original row.
5. The DRAM precharges the bit lines for the next operation.

The write procedure is identical except for the step 4. During a write operation, the column decoder enables the selected bit in data latch and the input value from dq is written to that bit. The data latch with the updated bit is then written back to the original row, effectively writing that bit to the cell array.

15.5.3   DRAM timing

The DRAM does not contain a clock signal and its operation is based on the duration and level of the address, data, and control signals. Because of address time multiplexing and the need of precharging and restoration, DRAM timing is very involved. The simplified timing diagrams of read, write, and refreshing are shown in Figure 15.10. A typical read operation is shown in Figure 15.10(a). The external controller first places the row address on **addr** and then activates **ras_n** (i.e., makes it 0). The DRAM latches the address at the falling edge and reads the designated row. After an interval of t_{RCD}, the row data is stored to the data latch, the controller places column address on **addr** and then activates **cas_n**. The DRAM latches the address at the falling edge and the column multiplexer routes the selected bit to **dq**. The controller then can deactivate **cas_n** and remove the data from **dq** and deactivate **ras_n** after the row data is restored. After the restoration, DRAM precharges the bit lines for next access. After precharging completes, the controller can start the next access.

The key timing parameters are:

- t_{RC}: read cycle time, the minimal elapsed time between two read operations.
- t_{RAS}: **ras_n** pulse width, the time interval that **ras_n** must be asserted.
- t_{CAS}: **cas_n** pulse width, the time interval that **cas_n** must be asserted.
- t_{RCD}: **ras_n** to **cas_n** delay time, the minimal delay between the assertion of **ras_n** and the assertion of **cas_n**. It represents the time to retrieve data from a row.
- t_{CAC}: **cas_n** access time, time required to obtain stable output data after **cas_n** assertion.
- t_{RP}: row precharge time, time required to precharge a row for another access.

A typical write operation is shown in Figure 15.10(b). The **ras_n** and **cas_n** assertions and precharge interval are similar but **we_n** is activated to write data. In addition to the read and write cycles, a DRAM chip also includes a *refresh cycle*, as shown in Figure 15.10(c). In this cycle, a row of data is read and then stored back. The external controller usually uses a counter and a timer to keep track of the row number and launch refresh cycles periodically.

Accessing a data item in DRAM requires to read and restore an entire row and to precharge all bit lines. These operations are time consuming and the speed of DRAM is thus relatively slow. One way to improve the performance is to allow

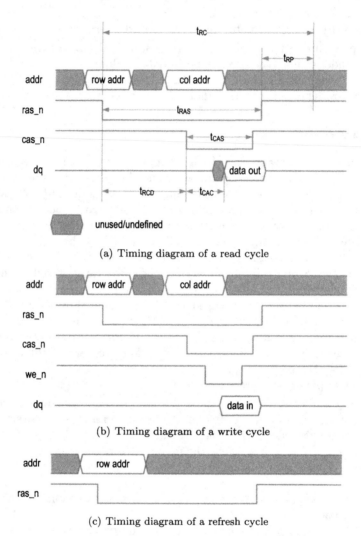

(a) Timing diagram of a read cycle

(b) Timing diagram of a write cycle

(c) Timing diagram of a refresh cycle

Figure 15.10   Timing diagrams of DRAM operations.

multiple column accesses each time a row is retrieved, as in *page-mode* or *burst-mode* operation.

15.6 OVERVIEW OF SDRAM

The DRAM timing is very involved and its operation is *asynchronous*. A signal must be stable for a specific interval of time, not just around a clock edge. To simplify the controller design, modern devices usually include a synchronous control circuit wrapped around the DRAM. The circuit samples the address, data, and control signals at the clock edges and then generates the needed DRAM signals. This type of device is known as *SDRAM* (*synchronous DRAM*).

15.6.1 Basic SDRAM organization

The conceptual SDRAM diagram is shown in Figure 15.11. The major "synchronous wrapping circuit" includes a clock generator, registers (blocks with clock input triangle) for the data, address, and commands, and a control circuit that generates DRAM control signals. The data, address, and command signals are sampled at the rising edge of the clk signal and the memory operation can be perform in a pipelined fashion. The device also includes the dqm_n (for "dq mask") signal to mask or enable the data access.

Current SDRAM devices utilize *multiple banks*, each with its own row decoder and column circuitry, to have several accesses operated in parallel and thus increase overall throughput. The ba signal is used to select a bank.

SDRAM is more versatile than the previous DRAM or SRAM. Various parameters, such as the burst length and burst type (sequential or interleaved), can be adjusted as needed and programmed into the *mode register*. It can perform a concurrent auto precharge and includes a *refresh counter* that automatically tracks and increments the refreshing row address.

The SDRAM device keeps the original names of cs_n, ras_n, cas_n, and we_n but groups them together as a 4-bit *command*. The commands are listed in Table 15.3.

15.6.2 SDRAM timing

Because of the modes and options, SDRAM timing covers many different scenarios. We illustrate the basic concepts by a simple read operation and a simple write operation, in which the burst length is set to 1, the CAS latency is set to 2, and the auto precharge is disabled. The timing diagram of a read operation is shown in Figure 15.12(a). Note that the command and address only need to be stable around the rising edge of the clock. The basic sequence is:

1. At t_0, the SDRAM controller issues an **activate** command and places the row address on addr to initiate the operation. The SDRAM activates the row, retrieves the entire row of data, and stores it to the data latch.

2. At t_2, the SDRAM controller issues a **read** command and places the column address on addr. The SDRAM enables the column multiplexer to route the selected data to output.

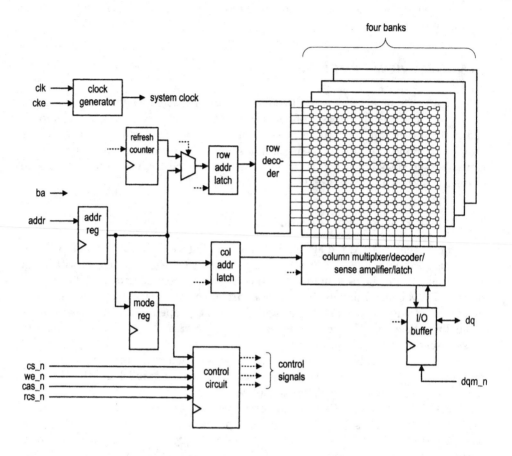

Figure 15.11   SDRAM conceptual diagram.

Table 15.3   SDRAM commands

Command	cs_n	ras_n	cas_n	we_n	Description
inhibit	1	-	-	-	device deselected
no op	0	1	1	1	no operation
activate	0	0	1	1	activate row
read	0	1	0	1	load column and start read
write	0	1	0	0	load column and start write
precharge	0	0	1	0	deactivate row and precharge bit lines
refresh	0	0	0	1	enter auto refresh mode
set mode	0	0	0	0	load data into mode register
burst stop	0	1	1	0	terminate burst operation

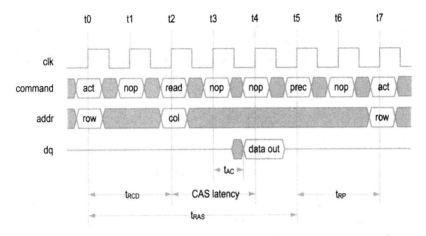

CAS latency=2; burst length=1; with manual precharge

(a) Timing diagram of a read cycle

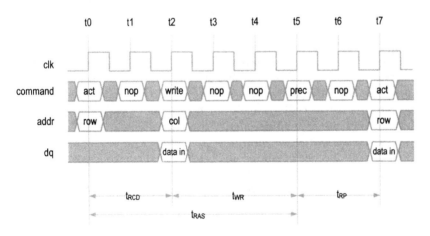

burst length=1; with manual precharge

(b) Timing diagram of a write cycle

Figure 15.12   Timing diagrams of SDRAM operations.

3. At t_4, the data becomes available in dq and the SDRAM controller retrieves data.

4. At t_5, the SDRAM controller issues a precharge command.

5. At t_7, the precharge operation is completed and the SDRAM controller can initiate a new operation at this point.

The key timing parameters are:

- t_{AC}: access time from the clock edge, time required to obtain stable output data after the rising edge of clock.
- *CAS latency*: cas_n to output data delay, the delay, in terms of *number of clock cycles*, between the assertion of a read command and the availability of output data.
- t_{RAS}: ras_n pulse width, time interval that ras_n must be asserted.
- t_{RCD}: row to column command delay time, the minimal delay between the assertion of an activate command (i.e., activate a row) and the assertion of a read or write command (i.e., activate a column).
- t_{RP}: precharge command period, time required to complete precharge operation after a precharge command is issued.

Note the *CAS latency* is specified in terms of clock cycles. It is usually 2 or 3 and is programmed into the mode register in advance. All other parameters are represented in terms of nanoseconds. While designing the controller, we need to translates a time interval to a proper multiple of clock periods. For example, if t_{RCD} is 50 ns and the clock period is 20 ns, the controller must wait at least 3 clock cycles (i.e., 60 ns) before issuing a read command.

The timing diagram of a write operation is shown in Figure 15.12(b). Its basic sequence is:

1. At t_0, the SDRAM controller issues an activate command and places the row address on addr to initiate the operation. The SDRAM activates the row, retrieves the entire row of data, and stores it to the data latch.

2. At t_2, the SDRAM controller issues a write command, places the column address on addr, and places data on the dq bus. The SDRAM retrieves the data, enables the column decoder, writes the selected locations in the data latch, and then writes the entire row of data back to the designated row.

3. At t_5, the row has been updated and the SDRAM controller issues a precharge command.

4. At t_7, the precharge operation is completed and the SDRAM controller can initiate a new operation at this point.

There is one new timing parameter:

- t_{WR}: write recovery time, the time interval required between the end of the write data burst and the start of a precharge command.

15.6.3   ICSI IS42S16400 SDRAM device

The DE1 board has an IS42S16400 device, which is a 4M-by-16 SDRAM module manufactured by Integrated Circuit Solution, Inc. The device is organized as four banks, each containing a 2^{20}-by-16 memory array. It uses a 2-bit ba signal to identify a bank and has a 12-bit address bus and a 16-bit bidirectional data bus.

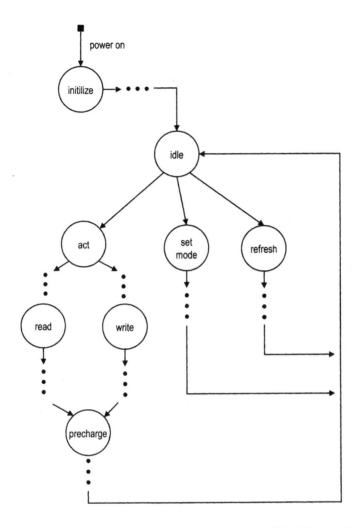

Figure 15.13   Simplified FSM state diagram of an SDRAM controller.

15.7 SDRAM CONTROLLER AND PLL

15.7.1 Basic SDRAM controller

An SDRAM controller accepts request from the main system (such as an Avalon-MM master or a processor) and generates proper SDRAM control signals. It consists of buffers, a timer that issues refreshing ticks, and an FSM that follows the SDRAM timing specification and generates properly sequenced SDRAM commands. The sketch of the FSM state diagram is shown in Figure 15.13. When the power is first applied, the FSM goes through the SDRAM startup process, which consists of a mandatory wait period and several refresh cycles, and then moves to the `idle` state. Four possible actions can be taken. The main system can issue a request to set the mode register, to read data from SDRAM, or to write data to SDRAM and the timer can trigger the controller to initiate a refresh cycle. The

FSM examines the request and branches to the designated path and generates the SDRAM command sequence accordingly.

Because of many possible modes and options of SDRAM, designing a comprehensive and robust SDRAM controller is an involved and tedious task. We use Altera's SDRAM controller IP core for our purposes.

15.7.2   SDRAM controller IP core

Most of today's SDARM devices conform to the standard set by the JEDEC Solid-State Technology Association. Although the bus width, capacity, and speed differ, the devices utilize similar interfaces, commands, and protocols. It is possible to construct a basic SDRAM controller "skeleton" and then adjust it to match to the system clock speed and specification of a specific SDRAM device. This is the approach used by Altera's SDRAM controller IP core.

An SDRAM controller IP core can be instantiated in SOPC Builder. The process basically requires us to consult the data sheet of a given device and determine the values of the relevant parameters. The timing parameters of IS42S16400 can be found in the DE1 board's accompanying data sheets. The steps of instantiating an SDRAM controller core are:

1. In SOPC Builder, select the System Contents tab.
2. In Component Library panel, expand the Memories and Memory Controllers category and then the SDRAM category and then select SDRAM Controller. The SDRAM Controller subwindow appears. It is in Memory Profile tab page as a default, as shown on the left of Figure 15.14.
3. In the Presets field, select Custom since there is no pre-configured setting for the IS42S16400 device.
4. Enter the following for IS42S16400:
   - Bits: 16
   - Chip select: 1
   - Banks: 4
   - Row: 12
   - Column: 8

   The completed page is shown on the left of Figure 15.14.
5. Select the Timing tab page to enter relevant timing parameters. The page is shown on the right of Figure 15.14.
6. Enter the following for the DE1 board's SDRAM:
   - CAS latency cycles: 3
   - Initialization refresh cycles: 8
   - Issue one refresh command every: 15.625 us
   - Delay after power up, before initialization: 200 us
   - Duration of refresh command (t_rfc): 70 ns
   - Duration of precharge command (t_rp): 20 ns
   - ACTIVE to READ or WRITE delay (t_rcd): 20 ns
   - access time (t_ac): 6 ns
   - Write recovery time (t_wr, no auto precharge): 14 ns

   The completed page is shown on the right of Figure 15.14.
7. Click finish to complete the process.

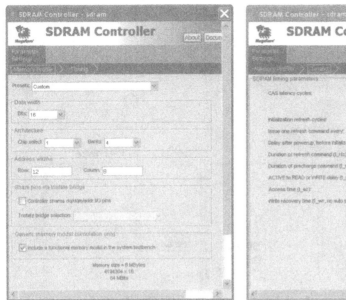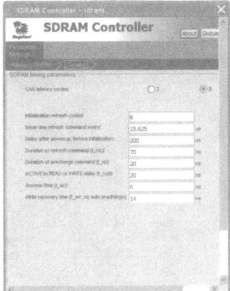

Figure 15.14   Screenshot of SOPC SDRAM controller.

15.7.3   SOPC PLL IP core

In a synchronous transaction, the relevant signals must be valid for a small window of time, during which the clock signal must toggle to capture the correct values. This condition translates to a time constraint on the maximal operating clock rate, as discussed in Section 5.5. The analysis can be applied to a system *within the FPGA chip* since the internal clock skew is small. However, accessing SDRAM involves additional issues. First, since the SDRAM controller and SDRAM reside on two separate devices, the timing parameters, such as clock-to-q delay, setup time, and hold time, are not identical. Second, as discussed in Section 15.2.2, off-chip access introduces additional delays, and has a significant impact on the controller's clock to output time. Finally, clock skew may also exist because the rising edge may not be able to arrive at the SDRAM controller (within the FPGA chip) and the SDRAM device at the same time.

On the DE1 board, the clock signal of SDRAM is connected to an output pin of the FPGA device. We can connect this pin directly to the external 50-MHz clock, as shown in Figure 15.15(a). This clocking scheme does not work reliably because of the timing issues. One way to mitigate the problem is to adjust the phase between the controller's clock signal and the SDRAM's clock signal. The basic idea is to determine a window of time, in which the data, address, and control signals are valid at the SDRAM pins, and align the clock's sampling edge in the middle of the window. The exact boundaries of the window depend on SDRAM device's input hold time, setup time, and output hold time, FPGA register's hold time and setup time, the controller's clock-to-output delay, and clock skew. A detailed analysis procedure can be found in the reference of the bibliographic section. DE1

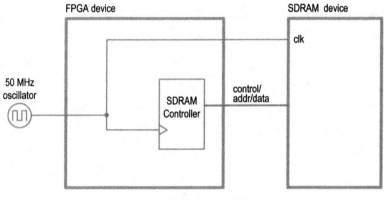

(a) Clocking without PLL

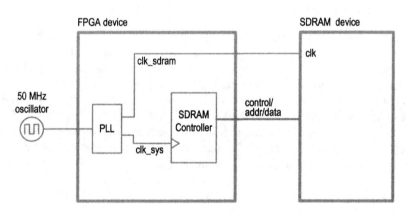

(b) Clocking with PLL

Figure 15.15   SDRAM clocking schemes.

documentation suggests to use a −3 ns shift for the SDRAM clock (i.e., the rising edge of the SDRAM clock is ahead of the rising edge of the controller by 3 ns).

The required SDRAM clock adjustment can be done by using FPGA's internal PLL circuit. The improved clocking scheme is shown in Figure 15.15(b). The external 50-MHz clock is fed to the PLL, which generates a clk_sys clock to drive the FPGA's internal system and a clk_sdram clock, which leads by 3 ns, to drive the external SDRAM device. Using PLL output (i.e., clk_sys) to drive the FPGA's internal logic can reduce the clock skew within the FPGA. Furthermore, we can adjust the system's clock frequency with the same 50-MHz external oscillator.

The PLL circuit can be treated as an SOPC Builder component and integrated into a Nios II system or instantiated as a Quartus megafunction module. We use the former for our system. The steps of instantiating a PLL core in SOPC Builder are:

1. In SOPC Builder, select the System Contents tab.
2. In Component Library panel, expand PLL category and then select Avalon ALTPLL. The MegaWizard Plug-In Manage window appears.
3. The window should display the General/Modes sub-page of the Parameter Setting tab page. Enter 50.0 MHz in the What is the frequency of the inclock0 input? since the external 50-MHz oscillator is used for the PLL input.
4. Select the Output Clocks tab page. There are three sub-pages, labeled clk c0, clk c1, and clk c2, which represent three possible output clock signals discussed in Figure 15.3. The tab page should display the clk c0 sub-page. Its screen capture is shown in Figure 15.16.
5. Check the Use this clock box to activate this clock. Keep the default values for the field. Since both Clock multiplication factor and Clock division factor fields are 1, the clock frequency remains 50 MHz. This clock output is used for the FPGA internal system clock, clk_sym.
6. Select the clk c1 sub-page to configure the second clock. Check the Use this clock box to activate this clock. Enter -3 ns in the Clock phase shift field. This clock output is 50 MHz but with a −3 ns phase shift. It is used to drive the SDRAM device's clock, clk_sdram.
7. Click the Finish button to save the configuration and close the MegaWizard Plug-In Manage window.

15.8   TESTING SYSTEM

We create a simple Nios II system to verify the operation of memory modules. The top-level diagram is shown in Figure 15.17. The system consists of a PLL core to generate the clock signal, a JTAG UART core to display messages on the console, and three memory cores, including an embedded on-chip memory, an SRAM controller, and an SDRAM controller. The testing program is stored in the on-chip memory. It first writes data to the SRAM and SDRAM modules and then reads it back to check errors.

15.8.1   Testing hardware configuration

Nios II system   We can create the testing Nios II system in SOPC Builder. The procedure is:

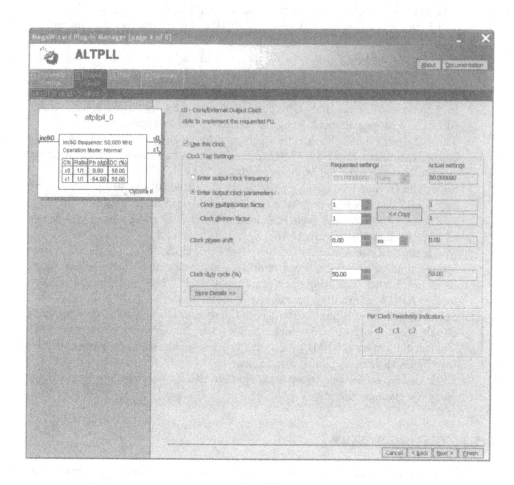

Figure 15.16   Screen capture of MegaWizard ALTPLL megafunction.

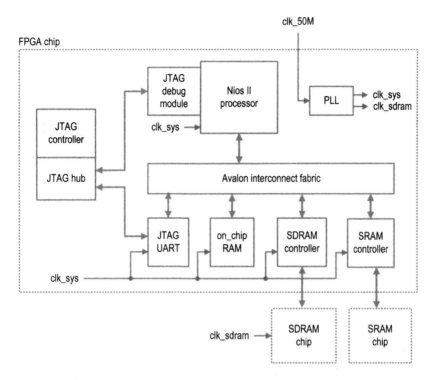

clk_50M

FPGA chip

JTAG debug module

Nios II processor

PLL

clk_sys
clk_sdram

JTAG controller

clk_sys

JTAG hub

Avalon interconnect fabric

JTAG UART

on_chip RAM

SDRAM controller

SRAM controller

clk_sys

clk_sdram

SDRAM chip

SRAM chip

Figure 15.17   Top-level diagram of memory testing system.

1. Add and configure a Nios II/e processor and rename it **cpu**.
2. Add an on-chip memory module of 16 KB and rename it **onchip_mem**.
3. Add the SRAM controller module and rename it **sram**.
4. Follow the procedure in Section 15.7 to add an SDRAM controller module and rename it **sdram**.
5. Select the on-chip memory module for the reset and exception vectors.
6. Add and configure a JTAG UART module and rename it **jtag_uart**.
7. Add a system id module and rename it **sysid**.
8. Follow the procedure in Section 15.7 to add an ALTPLL module and rename it **pll**. The Clock Setting panel of the SOPC Builder is updated to include two additional clocks, labeled **pll_c0** and **pll_c1**, which are the two outputs of the **pll** module.
9. Rename the the external clock **clk_50M** and rename two PLL clocks **clk_sys** and **clk_sdram**, as shown in the top of Figure 15.18.
10. In Clock column, verify that the external **clk_50M** clock is the clock source for the **pll** module and select **clk_sys** as the clock source for all other SOPC modules.
11. Generate HDL and information files.

Although the basic steps are similar to those in Section 10.5.1, this procedure has several unique features. First, there are three memory modules in the system. Since we want to store testing program in the on-chip memory module later, we must select it for the reset and exception vectors in step 5.

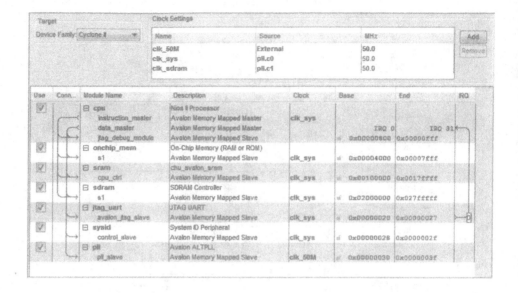

Figure 15.18 Screen capture of SOPC Builder.

Second, because of the inclusion of a PLL module, the system contains three clock sources (one external clock and two PLL-generated clocks). If we click the Clock column of a module, a pull-down menu with three clock sources appears. Because we want to use the PLL-generated clk_sys to drive the Nios II system, we must change the clock source for all modules, as in step 10. The screen capture of the completed system is shown in Figure 15.18.

Finally, the economy core (i.e., Nios II/e) must be selected for the testing system because the data cache of fast core may interfere with the memory testing operation.

Top-level HDL file After the HDL files are generated by SOPC Builder, we can create a top-level module that incorporates the Nios II system. The HDL code is shown in Listing 15.2.

Listing 15.2 Top-level memory testing circuit

```
library ieee;
use ieee.std_logic_1164.all;
entity mem_top is
   port(
      clk: in std_logic;
      -- LEDs
      ledg: out std_logic_vector(7 downto 0);
      ledr: out std_logic_vector(9 downto 0);
      hex3, hex2, hex1, hex0: out std_logic_vector(6 downto 0);
      -- to/from SRAM
      sram_addr: out std_logic_vector (17 downto 0);
      sram_dq: inout std_logic_vector (15 downto 0);
      sram_ce_n: out std_logic;
      sram_lb_n: out std_logic;
      sram_oe_n: out std_logic;
      sram_ub_n: out std_logic;
      sram_we_n: out std_logic;
      -- to/from SDRAM side
```

```
19      dram_clk: out std_logic;
        dram_cs_n, dram_cke: out std_logic;
        dram_ldqm, dram_udqm: out std_logic;
        dram_cas_n, dram_ras_n, dram_we_n: out std_logic;
        dram_addr: out std_logic_vector(11 downto 0);
24      dram_ba_0, dram_ba_1: out std_logic;
        dram_dq: inout std_logic_vector(15 downto 0)
      );
  end mem_top;

29 architecture structure of mem_top is
    component nios_ram is
    port(
        -- clock and reset
        signal clk_50M: in std_logic;
34      signal clk_sdram: out std_logic;
        signal clk_sys: out std_logic;
        signal reset_n: in std_logic;
        -- SDRAM
        signal zs_addr_from_the_sdram: out std_logic_vector (11 downto 0);
39      signal zs_ba_from_the_sdram: out std_logic_vector (1 downto 0);
        signal zs_dq_to_and_from_the_sdram:
              inout std_logic_vector (15 downto 0);
        signal zs_cke_from_the_sdram: out std_logic;
        signal zs_cs_n_from_the_sdram: out std_logic;
44      signal zs_ras_n_from_the_sdram: out std_logic;
        signal zs_cas_n_from_the_sdram: out std_logic;
        signal zs_we_n_from_the_sdram: out std_logic;
        signal zs_dqm_from_the_sdram: out std_logic_vector (1 downto 0);
        -- SRAM
49      signal sram_addr_from_the_sram: out std_logic_vector (17 downto 0);
        signal sram_dq_to_and_from_the_sram:
              inout std_logic_vector (15 downto 0);
        signal sram_ce_n_from_the_sram: out std_logic;
        signal sram_we_n_from_the_sram: out std_logic;
54      signal sram_oe_n_from_the_sram: out std_logic;
        signal sram_ub_n_from_the_sram: out std_logic;
        signal sram_lb_n_from_the_sram: out std_logic
      );
    end component nios_ram;
59 begin
    nios: nios_ram
    port map(
        clk_50M=>clk,
        clk_sys=>open,
64      clk_sdram => dram_clk,
        reset_n=>'1',
        -- SDRAM
        zs_addr_from_the_sdram=>dram_addr,
        zs_ba_from_the_sdram(1)=>dram_ba_1,
69      zs_ba_from_the_sdram(0)=>dram_ba_0,
        zs_cas_n_from_the_sdram=>dram_cas_n,
        zs_cke_from_the_sdram=>dram_cke,
        zs_cs_n_from_the_sdram=>dram_cs_n,
        zs_dq_to_and_from_the_sdram=>dram_dq,
74      zs_dqm_from_the_sdram(1)=> dram_udqm,
        zs_dqm_from_the_sdram(0)=> dram_ldqm,
        zs_ras_n_from_the_sdram=>dram_ras_n,
        zs_we_n_from_the_sdram=>dram_we_n,
        -- SRAM
79      sram_addr_from_the_sram => sram_addr,
```

```
        sram_ce_n_from_the_sram  => sram_ce_n,
        sram_dq_to_and_from_the_sram => sram_dq,
        sram_lb_n_from_the_sram  => sram_lb_n,
        sram_oe_n_from_the_sram  => sram_oe_n,
84      sram_ub_n_from_the_sram  => sram_ub_n,
        sram_we_n_from_the_sram  => sram_we_n
    );
    -- turn off all LEDs
    ledr <= (others=>'0');
89  ledg <= (others=>'0');
    hex3 <= (others=>'1');
    hex2 <= (others=>'1');
    hex1 <= (others=>'1');
    hex0 <= (others=>'1');
94 end structure;
```

The code performs mapping between the ports of the instantiated Nios II system and the I/O signals on the DE1 board. Note that one input clock and two output clocks of the PLL module are also shown as explicit ports in the instantiated Nios II system. The PLL's clk_50M clock is mapped to the clk port of the entity, which is connected to DE1 board's 50-MHz oscillator, and the PLL's clk_sdram clock is mapped to the dram_clk port of the entity, which is connected to the external SDRAM device's clock input. The PLL's clk_sys clock is not used and thus mapped to **open**. If the system inside the FPGA device contains other synchronous subsystems, their clock signals should be connected to clk_sys.

15.8.2   Testing software

The testing program basically writes data the SRAM and SDRAM modules and then reads it back to check errors. Since the content of the SRAM and SDRAM devices will be destroyed during the process, the testing program must be placed in the on-chip memory module. The BSP configuration and the program code are discussed in the following subsections.

BSP setting   We must adjust the BSP setting to accommodate two special requirements of this application:

- The testing code must be located in the on-chip memory module.
- The code size must be small enough to be stored in the on-chip memory module.

When the final software image (i.e., the .elf file) is generated, it is divided into several sections (for code text, stack, heap, etc.). We can specify the locations of these sections in BSP. The procedure is

1. Invoke Nios II BSP Editor.
2. Select the Linker Script tab page. The Linker Section Mappings panel is displayed on the top, as shown in Figure 15.19. Eight sections are included.
3. Select the .bss section (the first row). Click the Linker Region Name column and a pull-down menu of three memory modules (i.e., onchip_mem, sram, and sdram) appears. Select the onchip_mem module.
4. Repeat the process for the other sections.
5. Click the Generate button on the bottom.

This procedure puts the software image in the on-chip memory and frees the SRAM and SDRAM modules for testing.

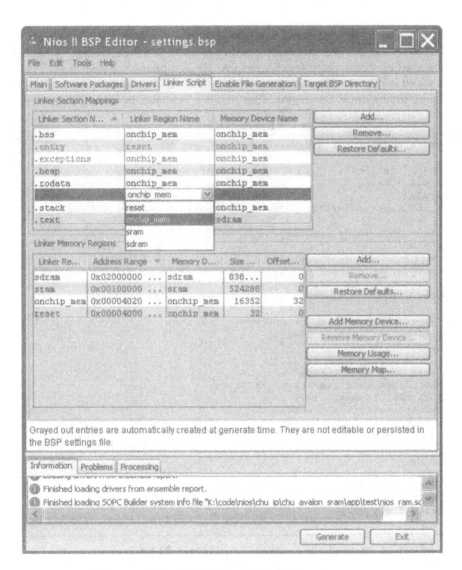

Figure 15.19 BSP Editor **Linker Script** tab page.

The 16 KB size of on-chip memory is fairly small. To accommodate its limited capacity, we must reduce the footprint of the software image. This can be done by enabling certain options in BSP Editor to use the smaller drivers and library functions. The procedure is discussed in the end of Section 11.2.3. We can follow the procedure in Section 9.3.5 to check the size of the software image and it shows that the footprint of testing program in Listing 15.4 is about 8 KB.

Testing program   We can verify the operation of a memory module by first writing data to the module and then reading it back to check errors. A function, check_mem() is constructed for this purpose. Its arguments are mem_base, min, and max, which specify the base address of the memory module and the lower and upper offsets of the intended test region. The function is shown in Listing 15.3.

Listing 15.3

```
int check_mem(alt_u32 mem_base, int min, int max)
{
  int err, real_err;
  alt_u32 *pbase;                    // pointer to the base address
  alt_u32 i;                         // index used to generated data
  alt_u32 t_pattn=0xfa30fa30;        // toggling pattern for data write

  pbase = (alt_u32 *)mem_base;
  err = 0;
  /* write entire test range */
  printf("write started ... \n");
  for (i=min; i<(max-3); i++){
    pbase[i] = i ^ t_pattn;          // invert certain bits
  }
  /* inject 4 errors in the end */
  for (i=max-3; i<=max; i++){
    pbase[i] = i;
  }
  /* read back entire range */
  printf("read back started ...\n");
  for (i=min; i<=max; i++){
    if (pbase[i]!=(i^t_pattn)) {
      err++;
      // printf("  error at address %x: 0x%08x (0x%08x expected) \n",
      //        (int)i, (int)pbase[i], (int)i^t_pattn);
    }
  }
  real_err= err-4;
  printf("completed with %d actual errors.\n", real_err);
    return(real_err);
}
```

In the function, we treat the designated region as a 32-bit word array and cast the pointer to the beginning of the array, pbase, to mem_base. The ith element of the array is filled with $i \wedge$ t_pattn, where t_pattn is a 32-bit toggling pattern. If a bit of t_pattn is 1, the corresponding bit of i is inverted after the xor operation. For verification purposes, we also deliberately inject four errors in the end and these errors should be detected while the array is read back.

The main program is shown in Listing 15.4. It calls the **check_mem()** function to check the entire memory range.

Listing 15.4

```
int main()
{
  printf("DE1 external SRAM/SDRAM test \n\n");
  printf("SRAM test: \n");
  check_mem(SRAM_BASE, 0, 0x0001ffff);    // 128K word address space
  printf("\n\nSDRAM test: \n");
  check_mem(SDRAM_BASE, 0, 0x001fffff);    // 2M word address space
}
```

Since the memory contents are overwritten during the memory test, the testing program must reside on a separate memory module. We include an embedded on-chip module in the Nios II system for this purpose.

15.9   BIBLIOGRAPHIC NOTES

The detailed specification of the DE1 board's SRAM and SDRAM can be found in the *ISSI IS61LV25616 data sheet* and *ICSI IS42S16400 data sheet*, which are included in the DE1 board distribution CD. The SDRAM data sheets from Micron Technology provide a better explanation for the SDRAM operation and timing and can be found in its web site. A text, titled *Memory Systems: Cache, DRAM, Disk* by B. Jacob et al., has a detailed discussion of SRAM and SDRAM operation and organization. The use of the SOPC's SDRAM controller core is covered in Chapter 1, titled *SDRAM Controller Core*, of *Embedded Peripherals IP User Guide*. An example is provided at the end of the chapter to demonstrate the calculation of SDRAM clock's phase shift.

The general I/O timing consideration of Altera FPGA is discussed in Application note 336, titled *Understanding I/O Output Timing for Altera Devices*. The use of the SOPC's PLL core is covered in Chapter 36, titled *PLL Cores*, of *Embedded Peripherals IP User Guide*. The configuration of PLL is documented separately in *Phase-Locked Loop (ALTPLL) Megafunction User Guide* and the specification of Cyclone II PLL is covered in *Cyclone II Device Handbook*.

15.10   SUGGESTED EXPERIMENTS

15.10.1   SRAM controller without I/O register

The SRAM controller designed in Section 15.4 includes registers to buffer all off-chip signals. Redesign the controller by removing all registers and modify the Avalon MM slave interface as needed. Resynthesize the circuit, run the testing program, and examine the error rate.

15.10.2   SRAM controller speed test

We can change the system clock rate by adjusting the parameters of the PLL module. Gradually increase the system clock rate from 50 MHz to the maximal allowable clock rate, which can be found in the report of Quartus's Classic Timing Analysis tool. Resynthesize the circuit, run the testing program, and examine the error rate. Note that the output clock frequency of Cyclone II PLL cannot be modified dynamically and thus the system must be resynthesized repeatedly. If a different board is used and the board contains a newer device, a Magafunction known as *ALTPLL_RECONFIG* can be included to facilitate real-time PLL reconfiguration.

15.10.3 SRAM controller with Avalon MM tristate interface

Redesign the SRAM controller using the Avalon MM tristate interface. Resynthesize the circuit, run the testing program, and verify its operation.

15.10.4 SDRAM controller clock skew test

For the SDRAM controller, we can change the phase of the SDRAM clock (i.e., the clk_sdram clock) by adjusting the parameters of the PLL module. Change the phase from -10 ns to $+10$ ns at 1-ns increments. For each phase, resynthesize the circuit, run the testing program, and examine the error rate.

15.10.5 Memory performance comparison

The HAL platform provides a time stamp utility. It can be used to keep track of the execution time of a program segment. Its use is discussed in Section 20.4. Reconstruct the testing system to include a timer module for time stamping. To compare the performance of three memory modules, develop a testing program that reads 10,000 words from the embedded on-chip memory, from SRAM, and from SDRAM separately and records the execution times. Repeat the procedure for the write operation.

15.10.6 Effect of cache memory

The HAL platform provides a time stamp utility. It can be used to keep track of the execution time of a program segment. Its use is discussed in Section 20.4. Reconstruct the testing system with a "performance" core (i.e., Nios II/f), which contains a data cache, and with a timer module for time stamping. The data cache is used in normal data accesses, such as those in the C program, but can be bypassed by using IORD() and IOWR() functions. Develop a testing program to perform 10,000 data accesses using the cache and 10,000 data accesses without using cache. Observe the effect on data cache. Note that the access pattern can have a significant impact on the hit ratio of the cache and thus effect the average access time.

15.10.7 SDRAM controller from scratch

Instead of using the SOPC Builder's SDRAM controller IP core, design the SDRAM controller from scratch and verify its operation.

15.11 COMPLETE PROGRAM LISTING

Listing 15.5   chu_main_ram_test.c

```
/****************************************************************
 *
 * Module:  Off-chip SRAM and SDRAM test
 * File:    chu_main_ram_test.c
 * purpose: Test external memory modules.
 * IP core base addresses:
 *   - SRAM_BASE: SRAM
 *   - SDRAM_BASE: SDRAM
 *
 ****************************************************************/
/* file inclusion */
/* General C library */
#include <stdio.h>
/* Altera-specific library */
#include <alt_types.h>
#include "system.h"

/****************************************************************
 * function: check_mem()
 * purpose:  check memory within specified region
 * argument:
 *   mem_base: base address of memory
 *   min: lower boundary (address offset) of the test region
 *   max: upper boundary (address offset) of the test region
 * return:
 *   # of errors
 * note:
 *   - the routine is designed for 32-bit data
 *   - 4 artificial errors is ejected in the writing process
 ****************************************************************/
int check_mem(alt_u32 mem_base, int min, int max)
{
  int err, real_err;
  alt_u32 *pbase;                 // pointer to the base address
  alt_u32 i;                      // index used to generated data
  alt_u32 t_pattn=0xfa30fa30;     // toggling pattern for data write

  pbase = (alt_u32 *)mem_base;
  err = 0;
  /* write entire test range */
  printf("write started ... \n");
  for (i=min; i<(max-3); i++){
    pbase[i] = i ^ t_pattn;       // invert certain bits
  }
  /* inject 4 errors in the end */
  for (i=max-3; i<=max; i++){
    pbase[i] = i;
  }
  /* read back entire range */
  printf("read back started ...\n");
  for (i=min; i<=max; i++){
    if (pbase[i]!=(i^t_pattn)) {
      err++;
      // printf(" error at address %x: 0x%08x (0x%08x expected) \n",
      //        (int)i, (int)pbase[i], (int)i^t_pattn);
    }
  }
```

```
    real_err= err-4;
    printf("completed with %d actual errors.\n", real_err);
      return(real_err);
}


/****************************************************************************
* function: main()
* purpose:  test entire ranges of off-chip and SRAM and SDRAM
* note:
*   - BSP configuration
*       - put code in on-chip memory in linker-script page
*       - enable small-C lib / reduce device driver options
*****************************************************************************/
int main()
{
  printf("DE1 external SRAM/SDRAM test \n\n");
  printf("SRAM test: \n");
  check_mem(SRAM_BASE, 0, 0x0001ffff);    // 128K word address space
  printf("\n\nSDRAM test: \n");
  check_mem(SDRAM_BASE, 0, 0x001fffff);    // 2M word address space
}
```

CHAPTER 16

PS2 KEYBOARD AND MOUSE

PS2 port is an interface used by a mouse or a keyboard. The device's activities are embedded in a stream of packets and transmitted to the host via two serial lines. Integrating a PS2 device to a Nios II system requires a custom controller and a proper software driver. In this chapter, we discuss the development of the hardware and software. The hardware includes a PS2 controller to transmit and receive data packets and an Avalon interface wrapping circuit. The software consists of driver routines to process the packets and to decode the keyboard or mouse activities.

16.1 INTRODUCTION

PS2 port was introduced in IBM's Personal System/2 personnel computers. It is a widely supported interface for a keyboard or mouse to communicate with the host. The PS2 port contains two wires for communication purposes. One wire is for data, which is transmitted in a serial stream. The other wire is for the clock information, which specifies when the data is valid and can be retrieved. Although a host receives data from a device most of the time, it occasionally sends a command to the keyboard or mouse to set certain parameters. Thus, the communication of the PS2 port is bidirectional.

The information in a PS2 interface is transmitted as an 11-bit "packet" that contains a start bit, 8 data bits, an odd parity bit, and a stop bit. Whereas the basic format of the packet is identical for a keyboard and a mouse, the interpretation for the data bits is different. A keyboard data stream contains the scan codes of keys

Embedded SOPC Design with Nios II Processor and VHDL Examples. By Pong P. Chu **379**

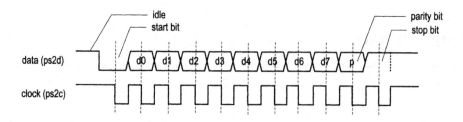

Figure 16.1   PS2-device-to-host timing diagram of a PS2 port.

and a mouse data stream contains the movement information and button status. Thus, separate software drivers are needed.

The DE1 board has a PS2 port. Two FPGA I/O pins are connected to the data and clock lines of the PS2 port. The Nios II system acts as a host. It needs to transmit and receive data in PS2 packet format, control the device, and extract the needed information. We use a custom PS2 controller to perform the first task and a custom software driver to perform the other two tasks. The development involves the following steps:

- Construct a PS2 receiving subsystem.
- Construct a PS2 transmitting subsystem.
- Integrate the two subsystems and buffer to construct a complete PS2 controller.
- Construct the Avalon interface wrapping circuit for the controller and create an SOPC component.
- Develop a software driver to interface with the PS2 controller.
- Develop a software driver to process the keyboard data stream.
- Develop a software driver to initiate a mouse and process its data stream.

These steps are discussed in the subsequent sections.

16.2   PS2 RECEIVING SUBSYSTEM

16.2.1   PS2-device-to-host communication protocol

A PS2 device and its host communicate via packets. The basic timing diagram of transmitting a packet from a PS2 device to a host is shown in Figure 16.1, in which the data and clock signals are labeled **ps2d** and **ps2c**, respectively.

The data is transmitted in a serial stream. Transmission begins with a start bit, followed by 8 data bits and an odd parity bit, and ends with a stop bit. The clock information is carried in a separate clock signal, **ps2c**. The falling edge of the **ps2c** signal indicates that the corresponding bit in the **ps2d** line is valid and can be retrieved. The clock period of the **ps2c** signal is between 60 and 100 μs (i.e., 10 kHz to 16.7 kHz), and the **ps2d** signal is stable at least 5 μs before and after the falling edge of the **ps2c** signal.

16.2.2   Design and code

The basic design of the PS2 port receiving subsystem consists of a falling edge
detection circuit, which generates a one-clock-cycle tick at the falling edge of the
ps2c signal, and a shift circuit, which shifts in and assembles the serial bits. An
FSMD is used to coordinate the overall operation.

The edge detection circuit detects the falling edge and generates an enable tick.
Because of the potential noise and slow transition, a simple filtering circuit is added
to eliminate glitches. Its code is

```
-- register
process (clk, reset)
   filter_reg <= filter_next;
      . . .
end process;

-- 1-bit shifter
filter_next <= ps2c & filter_reg(7 downto 1);
-- "filter"
f_ps2c_next <= '1' when filter_reg="11111111" else
               '0' when filter_reg="00000000" else
               f_ps2c_reg;
```

The circuit is composed of an 8-bit shift register and returns a 1 or 0 when eight
consecutive 1's or 0's are received. Any glitch shorter than eight clock cycles will
be ignored (i.e., filtered out). The filtered output signal is then fed to the regular
falling-edge detection circuit.

The ASMD chart of the receiver is shown in Figure 16.2. The receiver is initially
in the idle state. It includes an additional control signal, rx_en, which is used to
enable or disable the receiving operation. The purpose of the signal is to coordinate
the receiving subsystem operation. After the first falling-edge tick and the rx_en
signal are asserted, the FSMD shifts in the start bit and moves to the dps state. In
the dps state, ten bits, which include eight data bits, one parity bit, and one stop
bit, are sampled at the falling edge of ps2c and the first nine bits are shifted into
the b register. The FSMD then moves to the load state, in which one extra clock
cycle is provided to complete the shifting of the stop bit.

There are two output signals. The rx_idle signal indicates whether the receiving
subsystem is idle. The rx_done_tick signal is asserted in the load state for one
clock cycle to indicate the completion of receiving a packet. The HDL code consists
of the filtering circuit and an FSMD, which follows the ASMD chart. It is shown
in Listing 16.1.

Listing 16.1   PS2 port receiver

```
library ieee;
use ieee.std_logic_1164.all;
use ieee.numeric_std.all;
entity ps2_rx is
   port (
      clk, reset: in std_logic;
      ps2d, ps2c: in std_logic;   -- key data, key clock
      rx_en: in std_logic;
      rx_done_tick: out  std_logic;
      rx_idle: out std_logic;
      dout: out std_logic_vector(7 downto 0)
```

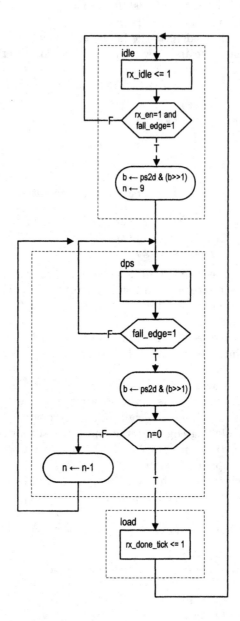

Figure 16.2   ASMD chart of the PS2 port receiver.

```
    );
13 end ps2_rx;

   architecture arch of ps2_rx is
      type statetype is (idle, dps, load);
      signal state_reg, state_next: statetype;
18    signal filter_reg, filter_next: std_logic_vector(7 downto 0);
      signal f_ps2c_reg,f_ps2c_next: std_logic;
      signal b_reg, b_next: std_logic_vector(10 downto 0);
      signal n_reg,n_next: unsigned(3 downto 0);
      signal fall_edge: std_logic;
23 begin

      —— filter and falling edge tick generation for ps2c

      process (clk, reset)
28    begin
         if reset='1' then
            filter_reg <= (others=>'0');
            f_ps2c_reg <= '0';
         elsif (clk'event and clk='1') then
33          filter_reg <= filter_next;
            f_ps2c_reg <= f_ps2c_next;
         end if;
      end process;

38    filter_next <= ps2c & filter_reg(7 downto 1);
      f_ps2c_next <= '1' when filter_reg="11111111" else
                     '0' when filter_reg="00000000" else
                     f_ps2c_reg;
      fall_edge <= f_ps2c_reg and (not f_ps2c_next);

43
      —— fsmd to extract the 8—bit data                •

      —— registers
48    process (clk, reset)
      begin
         if reset='1' then
            state_reg <= idle;
            n_reg   <= (others=>'0');
53          b_reg <= (others=>'0');
         elsif (clk'event and clk='1') then
            state_reg <= state_next;
            n_reg <= n_next;
            b_reg <= b_next;
58       end if;
      end process;
      —— next—state logic
      process(state_reg,n_reg,b_reg,fall_edge,rx_en,ps2d)
      begin
63       rx_idle <='0';
         rx_done_tick <='0';
         state_next <= state_reg;
         n_next <= n_reg;
         b_next <= b_reg;
68       case state_reg is
            when idle =>
               rx_idle <='1';
               if fall_edge='1' and rx_en='1' then
                  —— shift in start bit
```

```
73              b_next <= ps2d & b_reg(10 downto 1);
                n_next <= "1001";
                state_next <= dps;
             end if;
          when dps =>   -- 8 data + 1 pairty + 1 stop
78           if fall_edge='1' then
             b_next <= ps2d & b_reg(10 downto 1);
                if n_reg = 0 then
                    state_next <=load;
                else
83                  n_next <= n_reg - 1;
                end if;
             end if;
          when load =>
             -- 1 extra clock to complete the last shift
88           state_next <= idle;
             rx_done_tick <='1';
       end case;
    end process;
    -- output
93  dout <= b_reg(8 downto 1); -- data bits
 end arch;
```

There is no error detection circuit in the description. A more robust design should check the correctness of the start, parity, and stop bits and include a watchdog timer to prevent the keyboard or mouse from being locked in an incorrect state.

16.3 PS2 TRANSMITTING SUBSYSTEM

16.3.1 Host-to-PS2-device communication protocol

The host-to-PS2-device communication protocol involves bidirectional data exchange. The mouse's data and clock lines actually are *open-collector* circuits. For our design purposes, we treat them as tristate lines. The basic timing diagram of transmitting a packet from a host to a PS2 device is shown in Figure 16.3, in which the data and clock signals are labeled **ps2d** and **ps2c**. For clarity, the diagram is split into two parts to show which activities are generated by the host (i.e., the FPGA-based controller) and which activities are generated by the device (e.g., a mouse). The basic operation sequence is as follows:

1. The host forces the **ps2c** line to be 0 for at least 100 μs to inhibit any mouse activity. It can be considered that the host requests to send a packet.
2. The host forces the **ps2d** line to be 0 and disables the **ps2c** line (i.e., makes it high impedance). This step can be interpreted as the host sending a start bit.
3. The PS2 device now takes over the **ps2c** line and is responsible for future PS2 clock signal generation. After sensing the starting bit, the PS2 device generates a 1-to-0 transition.
4. Once detecting the transition, the host shifts out the least significant data bit over the **ps2d** line. It holds this value until the PS2 device generates a 1-to-0 transition in the **ps2c** line, which essentially acknowledges retrieval of the data bit.
5. Repeat step 4 for the remaining 7 data bits and 1 parity bit.

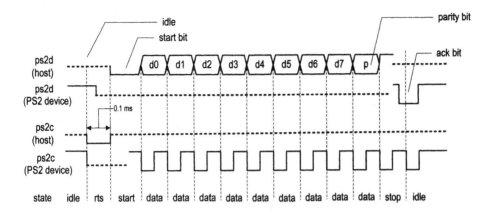

Figure 16.3   Host-to-PS2-device timing diagram of a PS2 port.

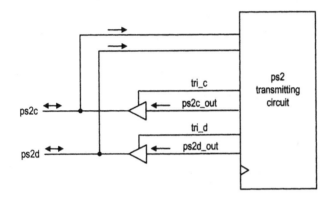

Figure 16.4   Tristate buffers of the PS2 transmission subsystem.

6. After sending the parity bit, the host disables the **ps2d** line (i.e., makes it high impedance). The PS2 device now takes over the **ps2d** line and acknowledges completion of the transmission by asserting the **ps2d** line to 0. If desired, the host can check this value at the last 1-to-0 transition in the **ps2c** line to verify that the packet is transmitted successfully.

16.3.2   Design and code

Unlike the receiving subsystem, the **ps2c** and **ps2d** signals communicate in both directions. A tristate buffer is needed for each signal. The tristate interface is shown in Figure 16.4. The **tri_c** and **tri_d** signals are enable signals that control the tristate buffers. When they are asserted, the corresponding **ps2c_out** and **ps2d_out** signals will be routed to the output ports.

To design the transmitting subsystem, we can follow the sequence of the preceding protocol to create an ASMD chart, as shown in Figure 16.5. The FSMD is initially in the **idle** state. To start the transmission, the main system (e.g., a Nios II processor) asserts the **wr_ps2** signal and places the data on the **din** bus. The

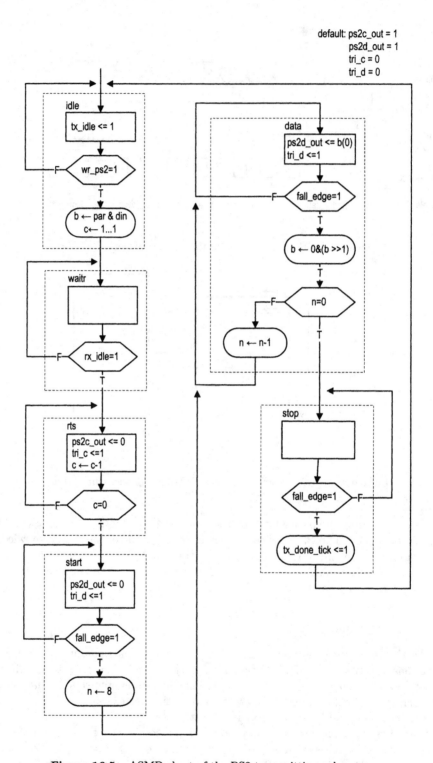

Figure 16.5 ASMD chart of the PS2 transmitting subsystem.

FSMD loads din, along with the parity bit, **par** to the shift_reg register, loads $1 \cdots 1$ to c_reg, and moves to the **waitr** (for "wait receiving") state. In this state, it examines the rx_idle signal to determine whether any receiving operation is in progress and waits there until the operation is completed. The FSMD then moves to the **rts** (for "request to send") state. In this state, the ps2c_out is set to 0 and the corresponding tri_c is asserted to enable the corresponding tristate buffer. The c_reg is used as a 13-bit counter to generate a 164-μs delay. The FSMD then moves to the **start** state, in which the PS2 clock line is disabled and the data line is set to 1. The PS2 device now takes over and generates a clock signal over the ps2c line. After detecting the falling edge of the ps2c signal through the fall_edge signal, the FSMD goes to the **data** state and shifts 8 data bits and 1 parity bit. The n register is used to keep track of the number of bits shifted. The FSMD then moves to the **stop** state, in which the data line is disabled. It returns to the **idle** state after sensing the last falling edge.

Similar to those of the receiving subsystem, the tx_idle signal indicates whether the transmission subsystem is idle and the tx_done_tick signal is asserted for one clock cycle when transmission operation is completed. The code follows the ASMD chart and is shown in Listing 16.2. A filtering circuit is also used to generate the fall_edge signal.

Listing 16.2   PS2 port transmitter

```
1 library ieee;
  use ieee.std_logic_1164.all;
  use ieee.numeric_std.all;
  entity ps2_tx is
      port (
6         clk, reset: in  std_logic;
          din: in std_logic_vector(7 downto 0);
          wr_ps2: in std_logic;
          rx_idle: in std_logic;
          ps2d, ps2c: inout std_logic;
11        tx_idle: out std_logic;
          tx_done_tick: out std_logic
      );
  end ps2_tx;

16 architecture arch of ps2_tx is
      type statetype is (idle, waitr, rts, start, data, stop);
      signal state_reg, state_next: statetype;
      signal filter_reg, filter_next: std_logic_vector(7 downto 0);
      signal f_ps2c_reg,f_ps2c_next: std_logic;
21    signal fall_edge: std_logic;
      signal b_reg, b_next: std_logic_vector(8 downto 0);
      signal c_reg,c_next: unsigned(12 downto 0);
      signal n_reg,n_next: unsigned(3 downto 0);
      signal par: std_logic;
26    signal ps2c_out, ps2d_out: std_logic;
      signal tri_c, tri_d: std_logic;
  begin
      -- filter and falling edge tick generation for ps2c
31
      process (clk, reset)
      begin
          if reset='1' then
              filter_reg <= (others=>'0');
```

```
36              f_ps2c_reg <= '0';
           elsif (clk'event and clk='1') then
              filter_reg <= filter_next;
              f_ps2c_reg <= f_ps2c_next;
           end if;
41      end process;

        filter_next <= ps2c & filter_reg(7 downto 1);
        f_ps2c_next <= '1' when filter_reg="11111111" else
                       '0' when filter_reg="00000000" else
46                     f_ps2c_reg;
        fall_edge <= f_ps2c_reg and (not f_ps2c_next);


        --- fsmd
51      --- registers
        process (clk, reset)
        begin
           if reset='1' then
56             state_reg <= idle;
              c_reg <= (others=>'0');
              n_reg  <= (others=>'0');
              b_reg <= (others=>'0');
           elsif (clk'event and clk='1') then
61             state_reg <= state_next;
              c_reg <= c_next;
              n_reg <= n_next;
              b_reg <= b_next;
           end if;
66      end process;
        -- odd parity bit
        par <= not (din(7) xor din(6) xor din(5) xor din(4) xor
                    din(3) xor din(2) xor din(1) xor din(0));
        -- next-state logic
71      process(state_reg,n_reg,b_reg,c_reg,wr_ps2,din,par,fall_edge,rx_idle)
        begin
           state_next <= state_reg;
           c_next <= c_reg;
           n_next <= n_reg;
76         b_next <= b_reg;
           tx_done_tick <='0';
           ps2c_out <= '1';
           ps2d_out <= '1';
           tri_c <= '0';
81         tri_d <= '0';
           tx_idle <='0';
           case state_reg is
              when idle =>
                 tx_idle <= '1';
86               if wr_ps2='1' then
                    b_next <= par & din;
                    c_next <= (others=>'1'); -- 2^13-1
                    state_next <= waitr;
                 end if;
91            when waitr =>
                 -- wait if receiving in progress
                 if rx_idle='1' then
                    state_next <= rts;
                 end if;
96            when rts =>  -- request to send
```

COMPLETE PS2 SYSTEM

```
                  ps2c_out <= '0';
                  tri_c <= '1';
                  c_next <= c_reg - 1;
                  if (c_reg=0) then
101                   state_next <= start;
                  end if;
               when start => -- assert start bit
                  ps2d_out <= '0';
                  tri_d <= '1';
106               if fall_edge='1' then
                     n_next <= "1000";
                     state_next <= data;
                  end if;
               when data => -- 8 data + 1 parity
111               ps2d_out <= b_reg(0);
                  tri_d <= '1';
                  if fall_edge='1' then
                     b_next <= '0' & b_reg(8 downto 1);
                     if n_reg = 0 then
116                      state_next <= stop;
                     else
                        n_next <= n_reg - 1;
                     end if;
                  end if;
121            when stop => -- assume floating high for ps2d
                  if fall_edge='1' then
                     state_next <= idle;
                     tx_done_tick <='1';
                  end if;
126         end case;
         end process;
         -- tristate buffers
         ps2c <= ps2c_out when tri_c ='1' else 'Z';
         ps2d <= ps2d_out when tri_d ='1' else 'Z';
131 end arch;
```

As in the receiving subsystem, there is no error detection circuit in this code.

16.4 COMPLETE PS2 SYSTEM

The top-level diagram of a complete PS2 system is shown in Figure 16.6. It consists of the receiving subsystem, the transmitting subsystem, and a FIFO buffer. The tx_idle, rx_idle, and rx_en signals are used to coordinate the transmitting and receiving operations so that only one type of operation can be performed at a time. The FIFO buffer is inserted after the receiving subsystem to provide some cushion space since a PS2 device may send packets continuously as we move a mouse or type on a keyboard. On the other hand, since the main system (e.g., a Nios II processor) only issues commands occasionally and it can control the rate, the transmitting subsystem does not need a buffer. The HDL code follows the block diagram and is shown in Listing 16.3.

Listing 16.3 Complete PS2 system

```
library ieee;
use ieee.std_logic_1164.all;
use ieee.numeric_std.all;
4 entity ps2_tx_rx_buf is
     generic(W_SIZE: integer:=2); -- 2^W_SIZE bytes in FIFO
```

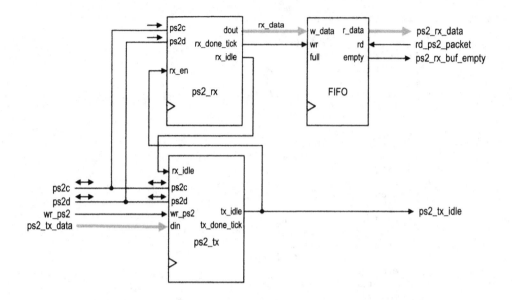

Figure 16.6   Top-level block diagram of a complete PS2 system.

```
    port(
        clk, reset: in  std_logic;
        wr_ps2: in std_logic;
9       rd_ps2_packet: in std_logic;
        ps2_tx_data: in std_logic_vector(7 downto 0);
        ps2_rx_data: out std_logic_vector(7 downto 0);
        ps2_tx_idle: out std_logic;
        ps2_rx_buf_empty: out std_logic;
14      ps2d, ps2c: inout std_logic
    );
  end ps2_tx_rx_buf;

  architecture arch of ps2_tx_rx_buf is
19    signal rx_data: std_logic_vector(7 downto 0);
      signal rx_done_tick: std_logic;
      signal rx_idle, tx_idle: std_logic;

  begin
24    ps2_tx_unit: entity work.ps2_tx(arch)
        port map(clk=>clk, reset=>reset, wr_ps2=>wr_ps2,
                 rx_idle=>rx_idle, din=>ps2_tx_data,
                 ps2d=>ps2d, ps2c=>ps2c,
                 tx_idle=>tx_idle, tx_done_tick=>open);
29    ps2_rx_unit: entity work.ps2_rx(arch)
        port map(clk=>clk, reset=>reset, rx_en=>tx_idle,
                 ps2d=>ps2d, ps2c=>ps2c, rx_idle=>rx_idle,
                 rx_done_tick=>rx_done_tick,
                 dout=>rx_data);
34    rx_fifo_unit: entity work.fifo(arch)
        generic map(B=>8, W=>W_SIZE)
        port map(clk=>clk, reset=>reset, rd=>rd_ps2_packet,
                 wr=>rx_done_tick, w_data=>rx_data,
                 empty=>ps2_rx_buf_empty, full=>open,
```

```
39              r_data=>ps2_rx_data);
    ps2_tx_idle <= tx_idle;
end arch;
```

16.5 PS2 CONTROLLER IP CORE DEVELOPMENT

16.5.1 Avalon interfaces

We can follow the procedure in Section 14.5 to add a wrapping circuit for the PS2 system and create an SOPC component. In this design, we include an Avalon MM slave interface to interact with the Nios II processor, a clock input interface for the system clock, and a conduit interface for PS2 port's clock and data lines.

16.5.2 Register map

A Nios II processor is the Avalon MM master and interacts with the PS2 system as follows:

- receive (i.e., read) an 8-bit data packet from the PS2 controller's receiving FIFO buffer.
- generate (i.e., write) a pulse to remove a packet from the receiving FIFO buffer.
- issue (i.e., write) an 8-bit command to a PS2 device.
- check (i.e., read) the ps2_rx_buf_empty signal to determine whether a packet is in the receiving FIFO buffer.
- check (i.e., read) the ps2_tx_idle signal to determine whether the transmitting subsystem is available.

Based on these interactions, we can define the register map of the Avalon MM slave interface. It is possible to pack all the information to a 32-bit word and to treat the word as a single register address. For clarity, we separate it into three registers, one for reading a packet, one for reading status, and one for writing a packet. The registers, their address offsets, and fields are:

- Read addresses (data to cpu)
  - offset 0 (read data register)
    * bits 7 to 0: 8-bit PS2 packet
  - offset 1 (status register)
    * bit 1: asserted (i.e., 1) when the PS2 transmitting subsystem is idle
    * bit 0: asserted (i.e., 1) when the PS2 receiving FIFO is empty
- Write addresses (data from cpu)
  - offset 0 (read data register)
    * Dummy data used to generate a pulse to remove a packet from the receiving FIFO buffer
  - offset 2 (write data register)
    * bits 7 to 0: 8-bit PS2 command data

To match the data width of a Nios II processor, we treat these registers as 32-bit registers. The unused bits will be removed automatically during synthesis and will not introduce additional hardware.

16.5.3   Wrapped PS2 system

The HDL code of the wrapped PS2 system is shown in Listing 16.4. The I/O ports use names similar to those in the Avalon MM interface but include a ps2_ prefix.

Listing 16.4   Wrapped PS2 system

```vhdl
library ieee;
use ieee.std_logic_1164.all;
use ieee.numeric_std.all;
entity chu_avalon_ps2 is
    generic(W_SIZE: integer:=4);   -- 2^W_SIZE words in FIFO
    port (
        clk, reset: in  std_logic;
        ps2d, ps2c: inout  std_logic;
        -- avalon interface
        ps2_address: in std_logic_vector(1 downto 0);
        ps2_chipselect: in  std_logic;
        ps2_write: in std_logic;
        ps2_writedata: in std_logic_vector(31 downto 0);
        ps2_readdata: out std_logic_vector(31 downto 0)
    );
end chu_avalon_ps2;

architecture arch of chu_avalon_ps2 is
    signal ctrl_reg: std_logic_vector(0 downto 0);
    signal ps2_rx_data: std_logic_vector(7 downto 0);
    signal rd_fifo, ps2_rx_buf_empty: std_logic;
    signal wr_ps2, ps2_tx_idle: std_logic;
begin
    -- instantiation
    ps2_unit: entity work.ps2_tx_rx_buf
        generic map (W_SIZE=>W_SIZE)
        port map(clk=>clk, reset=>reset,
                 ps2d=>ps2d, ps2c=>ps2c, wr_ps2=>wr_ps2,
                 rd_ps2_packet=>rd_fifo,
                 ps2_tx_data=>ps2_writedata(7 downto 0),
                 ps2_rx_data=>ps2_rx_data,
                 ps2_tx_idle=>ps2_tx_idle,
                 ps2_rx_buf_empty=>ps2_rx_buf_empty);
    -- read data multiplexing
    ps2_readdata <= x"000000" & ps2_rx_data when ps2_address="00" else
                    x"0000000" & "00" & ps2_tx_idle & ps2_rx_buf_empty;
    -- remove an item from FIFO
    rd_fifo <= '1' when ps2_chipselect='1' and ps2_address="00" and
                        ps2_write='1' else
               '0';
    -- write data to PS2 transmitting subsystem
    wr_ps2 <= '1' when ps2_chipselect='1' and ps2_address="10" and
                       ps2_write='1' else
              '0';
end arch;
```

The codes for read multiplexing and write decoding logic are similar to those in Listing 14.2. However, the PS2 system contains a receiving FIFO buffer and its characteristic is somewhat different. When reading a data item from the FIFO buffer, we can either remove the item or keep it intact. The latter approach provides more flexibility for software development but requires a separate instruction to perform the removal operation. We use this approach in the design. The segment

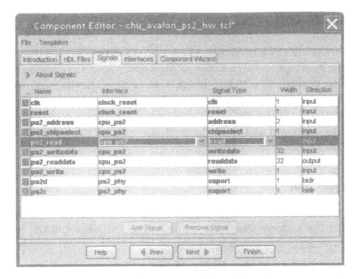

Figure 16.7   Component Editor **Signals** page.

```
rd_fifo <=
   '1' when ps2_chipselect='1' and ps2_address="00"
          and ps2_write='1' else
   '0';
```

generates a one-clock pulse, which in turn removes an item from the FIFO buffer. The rd_fifo signal is asserted when the address is "00" and the ps2_write signal is 1. This can be done by writing a dummy data to the address "00" since the Avalon MM master asserts the corresponding write signal during a write operation. If we replace the ps2_write='1' expression with ps2_read='1', the data item will be automatically removed during a read operation and no separate write instruction is needed.

16.5.4   SOPC component creation

After developing the wrapping circuit, we can convert the wrapped PS2 controller to an SOPC component following the procedure discussed in Section 14.5.4. We name the clock input, Avalon MM slave, and conduit interfaces clock_reset, cpu_ps2, and ps2_phy, respectively. The signal mapping is shown in Figure 16.7.

One key task in the procedure is to adjust the timing property of the Avalon MM slave interface to match the timing characteristic of the FIFO buffer. When the FIFO is not empty, the data is available and thus a Nios II processor can retrieve the data in a one-clock read cycle. It requires no wait states or setup or hold time and thus the corresponding fields in the Avalon MM slave interface should be cleared to 0's, just like these in Figure 14.18.

After the SOPC component is created, it is listed in the chu_ip category of SOPC Builder, as shown in the left panel of Figure 16.9. It can be instantiated as a normal IP core and integrated into a Nios II system.

16.6 PS2 DRIVER

The PS2 driver consists of a collection of routines that perform low-level I/O operations, including routines to process the "write" operations (i.e., issuing commands) and routines to process the "read" operations (i.e., retrieving and managing receiving FIFO). In addition, the ps2_reset_device() function resets a PS2 device to its initial state. It is mainly used with mouse operation and discussed in Section 16.8.3. The keyboard and mouse drivers are built on top of the PS2 driver and are discussed in Sections 16.7 and 16.8

16.6.1 Register map

The register map of the PS2 driver is specified in Section 16.5.2. To make the code clear, we define the address offsets as symbolic register names in the header file:

```
#define CHU_PS2_DATA_REG        0
#define CHU_PS2_CONTROL_REG     1
#define CHU_PS2_WR_DATA_REG     2
```

16.6.2 Write routines

There are two functions related to issuing a command packet to a PS2 device:

- ps2_tx_is_idle(): check whether the PS2 transmitting subsystem is idle.
- ps2_wr_cmd(): issue (i.e., write) a command packet.

The two routines are shown in Listing 16.5.

Listing 16.5

```
int ps2_tx_is_idle(alt_u32 ps2_base)
{
  alt_u32 ctrl_reg;
  int idle_bit;

  ctrl_reg = IORD(ps2_base, CHU_PS2_CONTROL_REG);
  idle_bit = (ctrl_reg & 0x00000002) >> 1;
  return idle_bit;
}

void ps2_wr_cmd(alt_u32 ps2_base, alt_u8 cmd)
{
  IOWR(ps2_base, CHU_PS2_WR_DATA_REG, (alt_u32) cmd);
}
```

Although the two routines are very simple and can be defined as macros, we choose the function format to make the code easier to understand. Since the functions involves low-level interaction, the predefined data types in alt_types.h are used to specify the data widths explicitly.

16.6.3 Read routines

There are several functions to manipulate the FIFO and retrieve the received packets:

- ps2_is_empty(): check whether the receiving FIFO is empty.

- ps2_read_fifo(): read data from the receiving FIFO.
- ps2_get_pkt(): check whether the FIFO is empty and read data.
- ps2_rm_pkt(): remove a packet from the head of the receiving FIFO.
- ps2_flush_fifo(): flush all packets from the receiving FIFO.

These functions are shown in Listing 16.6.

Listing 16.6

```
int ps2_is_empty(alt_u32 ps2_base)
{
  alt_u32 ctrl_reg;
  int empty_bit;

  ctrl_reg = IORD(ps2_base, CHU_PS2_CONTROL_REG);
  empty_bit = ctrl_reg & 0x00000001;
  return empty_bit;
}

alt_u8 ps2_read_fifo(alt_u32 ps2_base)
{
  alt_u32 data_reg;
  alt_u8 packet;

  data_reg = IORD(ps2_base, CHU_PS2_DATA_REG);
  packet = (alt_u8) (data_reg & 0x000000ff);
  return packet;
}

int ps2_get_pkt(alt_u32 ps2_base, alt_u8 *byte)
{
  if (!ps2_is_empty(ps2_base)) {
    *byte = ps2_read_fifo(ps2_base);
    return 1;    // got data
  }
  return 0;      // no data
}

void ps2_rm_pkt(alt_u32 ps2_base)
{
  IOWR(ps2_base, CHU_PS2_DATA_REG, 0x00000000); // write a dummy data
}

void ps2_flush_fifo(alt_u32 ps2_base)
{
  while (!ps2_is_empty(ps2_base)) {
    ps2_rm_pkt(ps2_base);
  }
}
```

Both **ps2_read_fifo()** and **ps2_get_pkt()** functions retrieve data from FIFO. Since there is no special data pattern to mark an empty FIFO, the **ps2_read_fifo()** function returns invalid data if the FIFO is empty. The **ps2_get_pkt()** function reads the FIFO and indicates whether the data is valid. The **ps2_rm_pkt()** function removes a packet from FIFO by writing the PS2 controller's data register, as discussed in Section 16.5.3. The **ps2_flush_fifo()** function removes all data from the receiving FIFO buffer.

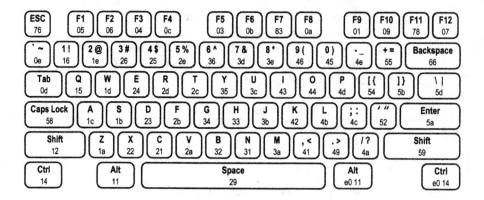

Figure 16.8   Scan code of the PS2 keyboard.

16.7   KEYBOARD DRIVER

16.7.1   Overview of the scan code

A keyboard consists of a matrix of keys and an embedded microcontroller that monitors (i.e., scans) the activities of the keys and sends the *scan code* accordingly. Three types of key activities are observed:

- When a key is pressed, the *make code* of the key is transmitted.
- When a key is held down continuously, a condition known as *typematic*, the make code is transmitted repeatedly at a specific rate. By default, a PS2 keyboard transmits the make code about every 100 ms after a key has been held down for 0.5 second.
- When a key is released, the *break code* of the key is transmitted.

The make code of the main part of a PS2 keyboard is shown in Figure 16.8. It is normally 1 byte wide and represented by two hexadecimal numbers. For example, the make code of the A key is 0x1c. This code can be conveyed by one packet when transmitted. The make codes of a handful of special-purpose keys, which are known as the *extended keys*, can have 2 to 4 bytes. A few of these keys are shown in Figure 16.8. For example, the make code of the right control key (labeled Ctrl) is 0xe0 0x14. Multiple packets are needed for the transmission. The break codes of the regular keys consist of 0xf0 followed by the make code of the key. For example, the break code of the A key is 0xf0 0x1c.

A PS2 keyboard transmits a sequence of codes according to the key activities. For example, when we press and release the A key, the keyboard first transmits its make code and then the break code:

```
0x1c 0xf0 0x1c
```

If we hold the key down for awhile before releasing it, the make code will be transmitted multiple times:

```
0x1c 0x1c 0x1c ... 0x1c 0xf0 0x1c
```

Multiple keys can be pressed at the same time. For example, we can first press the **shift** key (whose make code is 0x12) and then the A key, and release the A key and then release the `shift` key. The transmitted code sequence follows the make and break codes of the two keys:

```
0x12 0x1c 0xf0 0x1c 0xf0 0x12
```

The previous sequence is how we normally obtain an uppercase **A**. Note that there is no special code to distinguish the lowercase and uppercase keys. It is the responsibility of the host device to keep track of whether the shift key is pressed and to determine the case accordingly.

16.7.2   Interaction with host

A PS2 keyboard has an internal controller, which monitors the key's activities and generates the scan codes. At power-on, it automatically resets the parameters, performs a diagnostic self-test, and transmits a 0xaa packet to a host after passing the test. A host can send a command to the keyboard controller to inquire the status and set certain parameters, such as typematic rate. The commands are in the form of 8-bit packets. After receiving the command, the keyboard controller first transmits a 0xfa acknowledge packet and then performs the designated operation.

The power-on default setting works properly for most applications and thus we can use the keyboard without ever sending any command. However, several commands may be useful for driver development:

- The host can reset the keyboard by sending a 0xff packet. The keyboard acknowledges with a 0xfa packet and then performs the diagnostic self-test. The 0xaa packet is sent to the host after a successful test.
- The host can obtain the PS2 device id by sending a 0xf2 packet. The keyboard acknowledges with a 0xfa packet and then responds with a two-packet pattern of 0xab 0x83. This can be used to verify the existence of a PS2 keyboard.
- The host can control the three keyboard LEDs (for Caps Lock, Num Lock, and Scroll Lock) by sending two 8-bit packets. The first packet is 0xed. The second 8-bit packet is in the form of $00000cns$. The c, n, and s are three bits representing the Caps Lock, Num Lock, and Scroll Lock LEDs, respectively. The value of 1 turns on the corresponding LED. The keyboard acknowledges with a 0xfa packet.

Additional commands can be found in the references of the bibliographic section.

16.7.3   Driver routines

A keyboard is primarily used to obtain character inputs. The main task of a driver routine is to convert the scan codes to proper characters. Our development ignores the extended scan codes and assumes that these keys are not used.

In C, a character is represented by the 8-bit `char` data type. The representations are based on ASCII codes, which are 7 bits and consist of 128 code words (0x00 to 0x7f). The complete characters and their code words are shown in Table 16.1. There is no clear relationship between the scan codes and ASCII codes. A simple way to do the conversion is to define the mapping in a lookup table. In C, the lookup table can be defined as a one-dimensional constant array with the scan code as the index. The table for the lowercase characters is:

Table 16.1 ASCII codes

Code	Char	Code	Char	Code	Char	Code	Char
00	(nul)	20	(sp)	40	@	60	`
01	(soh)	21	!	41	A	61	a
02	(stx)	22	"	42	B	62	b
03	(etx)	23	#	43	C	63	c
04	(eot)	24	$	44	D	64	d
05	(enq)	25	%	45	E	65	e
06	(ack)	26	&	46	F	66	f
07	(bel)	27	'	47	G	67	g
08	(bs)	28	(	48	H	68	h
09	(ht)	29	)	49	I	69	i
0a	(nl)	2a	*	4a	J	6a	j
0b	(vt)	2b	+	4b	K	6b	k
0c	(np)	2c	,	4c	L	6c	l
0d	(cr)	2d	-	4d	M	6d	m
0e	(so)	2e	.	4e	N	6e	n
0f	(si)	2f	/	4f	O	6f	o
10	(dle)	30	0	50	P	70	p
11	(dc1)	31	1	51	Q	71	q
12	(dc2)	32	2	52	R	72	r
13	(dc3)	33	3	53	S	73	s
14	(dc4)	34	4	54	T	74	t
15	(nak)	35	5	55	U	75	u
16	(syn)	36	6	56	V	76	v
17	(etb)	37	7	57	W	77	w
18	(can)	38	8	58	X	78	x
19	(em)	39	9	59	Y	79	y
1a	(sub)	3a	:	5a	Z	7a	z
1b	(esc)	3b	;	5b	[	7b	{
1c	(fs)	3c	<	5c	\	7c	—
1d	(gs)	3d	=	5d	]	7d	}
1e	(rs)	3e	>	5e	^	7e	~
1f	(us)	3f	?	5f	_	7f	(del)

```
const char SCAN2ASCII_LO_TABLE[128]={
    0,    F9,    0,      F5,    F3,    F1,     F2,    F12,   //00
    0,    F10,   F8,     F6,    F4,    TAB,    '`',   0,     //08
    0,    0,     L_SFT,  0,     L_CTR, 'q',    '1',   0,     //10
    0,    0,     'z',    's',   'a',   'w',    '2',   0,     //18
    0,    'c',   'x',    'd',   'e',   '4',    '3',   0,     //20
    0,    ' ',   'v',    'f',   't',   'r',    '5',   0,     //28
    0,    'n',   'b',    'h',   'g',   'y',    '6',   0,     //30
    0,    0,     'm',    'j',   'u',   '7',    '8',   0,     //38
    0,    ',',   'k',    'i',   'o',   '0',    '9',   0,     //40
    0,    '.',   '/',    'l',   ';',   'p',    '-',   0,     //48
    0,    0,     '\'',   0,     '[',   '=',    0,     0,     //50
    CAPS, R_SFT, ENTER,  ']',   0,     BKSL,   0,     0,     //58
    0,    0,     0,      0,     0,     0,      BKSP,  0,     //60
    0,    '1',   0,      '4',   '7',   0,      0,     0,     //68
    0,    '.',   '2',    '5',   '6',   '8',    ESC,   NUM,   //70
    F11,  '+',   '3',    '-',   '*',   '9',    0,     0      //78
};
```

For example, we can use the expression SCAN2ASCII_LO_TABLE[21] to obtain the corresponding character of scan code 21, which is the letter c. In addition to the normal single-quoted characters, the table also contains 0's, which correspond to undefined scan codes, and uppercase constants, which are special ASCII characters and unmapped keys. The special C characters consist of

```
#define TAB     0x09    // tab
#define BKSP    0x08    // backspace
#define ENTER   0x0d    // enter (new line)
#define ESC     0x1b    // escape
#define BKSL    0x5c    // back slash
```

The other uppercase constants correspond to the keys that don't map to ASCII characters, such as function keys (F1, ⋯, F12), control key (Ctrl), etc. We can assign unused 8-bit values (0x80 to 0xff) to these keys and use them for special purposes. For example, we can display the help message when the F1 key is pressed. The complete constant assignments are shown in the Listing 16.16 at the end of the chapter.

A similar table is needed for the uppercase characters as well:

```
const char SCAN2ASCII_UP_TABLE[128] = {
    0,    F9,    0,      F5,    F3,    F1,     F2,    F12,   //00
    0,    F10,   F8,     F6,    F4,    TAB,    '~',   0,     //08
    0,    0,     L_SFT,  0,     L_CTR, 'Q',    '!',   0,     //10
    0,    0,     'Z',    'S',   'A',   'W',    '@',   0,     //18
    0,    'C',   'X',    'D',   'E',   '$',    '#',   0,     //20
    0,    '\ ',  'V',    'F',   'T',   'R',    '%',   0,     //28
    0,    'N',   'B',    'H',   'G',   'Y',    '^',   0,     //30
    0,    0,     'M',    'J',   'U',   '&',    '*',   0,     //38
    0,    '<',   'K',    'I',   'O',   ')',    '(',   0,     //40
    0,    '>',   '?',    'L',   ':',   'P',    '_',   0,     //48
    0,    0,     '\"',   0,     '{',   '+',    0,     0,     //50
    CAPS, R_SFT, ENTER,  '}',   0,     '|',    0,     0,     //58
    0,    0,     0,      0,     0,     0,      BKSP,  0,     //60
    0,    '1',   0,      '4',   '7',   0,      0,     0,     //68
```

```
    0,     '.',     '2',     '5', '6',     '8',    ESC,   NUM,  //70
    F11,   '+',     '3',     '-', '*',     '9',    0,     0     //78
};
```

With the lookup tables, we can construct a driver routine to read a character
from a keyboard. Because a PS keyboard contains many special purpose keys and
the keys can be pressed and released in an arbitrary combination (such as Ctrl-D,
Ctrl-Alt-Del, etc.), developing a robust, comprehensive routine is quite involved
and beyond the scope of this book. For our purposes, we use the keyboard to
obtain "printable" ASCII characters and digits and develop a routine accordingly.
Except for the shift keys, no other special purpose key is processed. The code of
this function, kb_get_ch(), is shown in Listing 16.7.

Listing 16.7

```
int kb_get_ch (alt_u32 ps2_base, char *ch){
  // special   characters
  #define TAB    0x09    // tab
  #define BKSP   0x08    // backspace
  #define ENTER  0x0d    // enter (new line)
  #define ESC    0x1b    // escape
  #define BKSL   0x5c    // back slash
  #define SFT_L  0x59    // left shift
  #define SFT_R  0x12    // right shift

  #define CAPS   0x80
  #define NUM    0x81
  #define CTR_L  0x82
  #define F1     0xf0
  #define F2     0xf1
  . . .

  // keyboard scan code to ascii (lower case)
  const char SCAN2ASCII_LO_TABLE[128] = {...}
  // keyboard scan code to ascii (upper case)
  const char SCAN2ASCII_UP_TABLE[128] = {...};

  static int sft_on = 0;
  alt_u8 scode;

  while (1){
    if (!ps2_get_pkt(ps2_base, &scode))          // no packet
      return(0);
    ps2_rm_pkt(ps2_base);
    switch (scode){
      case 0xf0:    // break code
        while (!ps2_get_pkt(ps2_base, &scode)); // get next
        ps2_rm_pkt(ps2_base);
        if (scode==SFT_L || scode==SFT_R)
          sft_on = 0;
        break;
      case SFT_L:    // shift key make code
      case SFT_R:
        sft_on = 1;
        break;
      default:       // normal make code: use lookup table
        if (sft_on)
          *ch = SCAN2ASCII_UP_TABLE[scode];
        else
          *ch = SCAN2ASCII_LO_TABLE[scode];
```

```
        return(1);
    }  // end switch
  }  // end while
}
```

The routine treats the two shift keys as special cases. It keeps track of whether a shift key is pressed and then uses the lowercase or uppercase lookup table accordingly. A static variable, sft_on, is used for tracking. The routine processes the received packets as follows:

- If it is the break code (i.e., beginning with a 0xf0 packet), remove two packets. If the code is for the shift key, clear sft_on to 0.
- If it is the make code of the shift key, set sft_on to 1.
- If it is the make code of the other key, obtain the character value from the proper lookup table and return the character.

Note that this routine does not process other special keys but just returns their designated codes.

Based on kb_get_ch(), we can develop a routine, kb_get_line(), to read a sequence of characters and convert them to a string. It keeps reading characters until the Enter key is pressed or the designated maximal limit, lim, is reached. The code is shown in Listing 16.8.

Listing 16.8

```
int kb_get_line(alt_u32 ps2_base, char *s, int lim)
{
  char ch;
  int i;

  i=0;
  while (1){
    while(!kb_get_ch(ps2_base, &ch));
    if ((ch=='\n')|(i==(lim - 1)))
      break;
    else
      s[i++] = ch;
  } // end while
  s[i] = '\0';
  return i;
}
```

Once obtaining a string, we can use stdio library's sscanf() function, which is similar to scanf() but gets its input from a string argument, to parse and process the string and convert it to the desired data formats.

16.8 MOUSE DRIVER

16.8.1 Overview of PS2 mouse protocol

A computer mouse is designed to detect two-dimensional motion on a surface. Its internal circuit measures the relative distance of movement. A standard PS2 mouse reports the x-axis (right/left) and y-axis (up/down) movement and the status of the left button, middle button, and right button. The amount of each movement is recorded in a mouse's internal counter. When the data is transmitted to the host, the counter is cleared to zero and restarts the counting. The content of the counter

Table 16.2 Mouse data packet format

byte 1	y_v	x_v	y_8	x_8	1	m	r	l
byte 2	x_7	x_6	x_5	x_4	x_3	x_2	x_1	x_0
byte 3	y_7	y_6	y_5	y_4	y_3	y_2	y_1	y_0

represents a 9-bit signed integer in which a positive number indicates the right or up movement and a negative number indicates the left or down movement.

The relationship between the physical distances is defined by the mouse's *resolution* parameter. The default value of resolution is four counts per millimeter. When a mouse moves continuously, the data is transmitted in a regular rate. The rate is defined by the mouse's *sampling rate* parameter. The default value of the sampling rate is 100 samples per second. If a mouse moves too fast, the amount of the movement during the sampling period may exceed the maximal range of the counter. The counter is set to the maximum magnitude in the appropriate direction. Two overflow bits are used to indicate the conditions.

The mouse reports the movement and button activities in 3 bytes, which are embedded in three PS2 packets. The detailed format of the 3-byte data is shown in Table 16.2. It contains the following information:

- $x_8, \ldots, x_0$: x-axis movement in 2's-complement format
- x_v: x-axis movement overflow
- $y_8, \ldots, y_0$: y-axis movement in 2's-complement format
- y_v: y-axis movement overflow
- l: left button status, which is 1 when the left button is pressed
- r: right button status, which is 1 when the right button is pressed
- m: optional middle button status, which is 1 when the middle button is pressed

During transmission, the **byte 1** packet is sent first and the **byte 3** packet is sent last.

A mouse has several different operation modes. The most commonly used one is the *stream mode*, in which a mouse sends the movement data when it detects movement or button activity. If the movement is continuous, the data is generated at the designated sampling rate.

16.8.2 Interaction with host

Similar to a PS2 keyboard, a mouse has an internal controller. At power-on, the controller automatically resets the parameters and performs a diagnostic self-test. If the test is successful, it transmits two packets, 0xaa and then 0x00, to a host. A host can send a command to inquire the status and set certain parameters. After receiving the command, the mouse controller first transmits a 0xfa acknowledge packet and then performs the designated operation.

Two commands are involved in driver development:

- The host can reset the mouse by sending a 0xff packet. The mouse acknowledges with a 0xfa packet and then performs the diagnostic self-test. The 0xaa and 0x00 packets are sent to the host after a successful test.

- The host can enable the stream mode by sending a 0xf4 packet. The mouse acknowledges with a 0xfa packet.

Additional commands can be found in the references of the bibliographic section.

16.8.3   Driver routines

The driver consists of a mouse initialization routine to enable the stream mode and a routine to obtain the mouse movements and button activities. A PS2 driver routine, **ps2_reset_device()**, is used to initialize a PS2 device and is discussed in this subsection as well.

The **ps2_reset_device()** function sends a command to reset a PS2 device and determines the type of the device, which can be a keyboard, mouse, or unknown, based on the response. The code is shown in Listing 16.9.

Listing 16.9

```
int ps2_reset_device(alt_u32 ps2_base)
{
  alt_u8 packet;

  ps2_flush_fifo(ps2_base);
  /* send reset 0xff */
  ps2_wr_cmd(ps2_base, 0xff);
  usleep(1000000);        // wait
  /* check 0xfa 0xaa */
  if (ps2_get_pkt(ps2_base, &packet)==0 || packet!=0xfa)
    return (0);           // no response or wrong response
  ps2_rm_pkt(ps2_base);
  if (ps2_get_pkt(ps2_base, &packet)==0 || packet!=0xaa)
    return (0);           // no response or wrong response
  ps2_rm_pkt(ps2_base);
  /* check whether 0x00 is received */
  if (ps2_get_pkt(ps2_base, &packet)==0)
    return (1);           // fifo has no more packet, device is keyboard
  ps2_rm_pkt(ps2_base);
  if (packet==0x00)
    return (2);           // mouse id
  else
    return (0);           // unknown device id
}
```

The routine flushes the FIFO buffer, issues a reset command, and examines the responses. The device is a keyboard if the received response packets are 0xfa 0xaa and a mouse if the packets are 0xfa 0xaa 0x00. Otherwise, it is an unknown device or nothing is connected to the port.

The **mouse_init()** function sets a mouse to the desired initial state. It first resets the mouse and then issues a command to enable the stream mode. The function returns 1 if the initialization is successful. The code is shown in Listing 16.10.

Listing 16.10

```
int mouse_init(alt_u32 ps2_base)
{
  alt_u8 packet;

  if (ps2_reset_device(ps2_base)!=2)
    return (0);
```

```
/* send stream mode command 0xf4 */
ps2_wr_cmd(ps2_base, 0xf4);
usleep(1000000);          // wait
/* check 0xfa (acknowledge) */
if (ps2_get_pkt(ps2_base, &packet)==0 || packet!=0xfa)
  return (0);             // no response or wrong response
/* everything is ok */
ps2_rm_pkt(ps2_base);
return(1);
}
```

The main mouse driver routine obtains the movement and button activities. We define a record for the received data:

```
typedef struct mouse_move
{
  int lbtn;      // left button
  int rbtn;      // right button
  int xmov;      // x-axis movement
  int ymov;      // x-axis movement
} mouse_mv_type;
```

The code is shown in Listing 16.11.

Listing 16.11

```
int mouse_get_activity(alt_u32 ps2_base, mouse_mv_type *mv)
{
  alt_u8 b1, b2, b3;

  alt_u32 tmp;

  /* check and retrieve 1st byte */
  if (!ps2_get_pkt(ps2_base, &b1))
    return (0);                              // no data in rx fifo buffer
  ps2_rm_pkt(ps2_base);                      // remove 1st byte
  /* wait and retrieve 2nd byte */
  while (!ps2_get_pkt(ps2_base, &b2));       // get 2nd byte
  ps2_rm_pkt(ps2_base);
  /* wait and retrieve 3rd byte */
  while (!ps2_get_pkt(ps2_base, &b3));       // get 3rd byte
  ps2_rm_pkt(ps2_base);
  /* extract button info */
  mv->lbtn = (int) (b1 & 0x01);              // extract bit 0
  mv->rbtn = (int) (b1 & 0x02)>>1;           // extract bit 1
  /* extract x movement; manually convert 9-bit 2's comp to int */
  tmp = (alt_u32) b2;
  if (b1 & 0x10)                             // check MSB (sign bit) of x movement
    tmp = tmp | 0xffffff00;                  // manual sign-extension if negative
  mv->xmov = (int) tmp;                      // data conversion
  /* extract y movement; manually convert 9-bit 2's comp to int */
  tmp = (alt_u32) b3;
  if (b1 & 0x20)                             // check MSB (sign bit) of y movement
    tmp = tmp | 0xffffff00;                  // manual sign-extension if negative
  mv->ymov = (int) tmp;                      // data conversion
  /* success */
  return(1);
}
```

The routine obtains three packets, extracts the information, and stores them to the proper fields. The sign extension is performed manually to extend the 9-bit

movement data to the 32-bit integer data type. It is done by setting 24 MSBs
to 1's if the movement is negative (i.e., MSB of the 9-bit data is 1).

A mouse is usually used in conjunction with a graphic display. An extended
graphic driver routine is discussed in Section 17.7.

16.9   TEST

To simplify the testing process, we construct a comprehensive Nios II system that
incorporates major IP cores designed in Parts III and IV of this book. The derivation is discussed in Section 16.10. This system can be used to verify the operation
of the PS2 controller IP core.

We develop a simple test program to examine the low-level functionalities of the
PS2 controller and verify the operation of the keyboard and mouse driver routines.
The program performs the following tests:

- Reset a PS2 device.
- Initialize a mouse.
- Send a command to a PS2 device.
- Display the PS2 input stream continuously.
- Display the decoded keyboard character stream continuously.
- Display the decoded mouse data stream continuously.

The complete program is shown in Listing 16.12.

Listing 16.12

```
int main(void){
  alt_u8 cmd, packet;
  int sw, btn, id;
  mouse_mv_type mv;
  char ch;
  alt_u8 ps2_msg[4]={0xff,0x0c,sseg_conv_hex(5),sseg_conv_hex(2)};

  sseg_disp_ptn(SSEG_BASE, ps2_msg);            // display " PS2"
  printf("PS2 test: \n");
  btn_clear(BTN_BASE);
  while (1){
    while (!btn_is_pressed(BTN_BASE)){ };        // wait for button
    btn=btn_read(BTN_BASE);                      // read button
    if (btn & 0x02){                             // key1 pressed
      sw=pio_read(SWITCH_BASE);                  // read switch
      printf("key/sw: %d/%d\n", btn, sw);
    }
    btn_clear(BTN_BASE);
    switch (sw){
      case 0:  // reset
        id=ps2_reset_device(PS2_BASE);
        printf("PS2 device type: %d (0/1/2: unknown/keyboard/mouse)\n", id);
        break;
      case 1:  // initialize mouse to stream mode
        id=mouse_init(PS2_BASE);
        printf("Mouse initialization status: %d (0/1: fail/succeed)\n", id);
        break;
      case 2:  // issue a ps2 command
        printf("Enter ps2 command in 2-digit hex format:\n");
        scanf("%x", &cmd);
        ps2_wr_cmd(PS2_BASE, cmd);
        printf("PS2 command 0x%02x issued.\n", cmd);
```

```
        usleep(500000);          // wait for 200 ms
        printf("PS2 response: ");
        while (ps2_get_pkt(PS2_BASE, &packet)){  // get all packets
          printf("0x%02x  ", packet);
          ps2_rm_pkt(PS2_BASE);
        }
        printf("\n");
        break;
      case 3:  // display ps2 stream
        printf("PS2 packet stream:\n");
        while (!btn_is_pressed(BTN_BASE)){
          if (ps2_get_pkt(PS2_BASE, &packet)){   // get one packet
            printf("0x%02x ", packet);
            ps2_rm_pkt(PS2_BASE);
          }
        }
        printf("\n");
        break;
      case 4:  // display decoded keyboard input stream
        printf("Keyboard char stream: \n");
        while (!btn_is_pressed(BTN_BASE)){
          if (kb_get_ch(PS2_BASE, &ch))
              printf("%c", ch);
        }
        printf("\n");
        break;
      case 5:  // display decoded mouse data stream
        printf("Mouse data stream: (left button, right button, "
               "x-axis move, y-axis move)\n");
        while (!btn_is_pressed(BTN_BASE)){
          if (mouse_get_activity(PS2_BASE, &mv)){
            printf("(%d,%d,%d,%d) ", mv.lbtn, mv.rbtn, mv.xmov, mv.ymov);
            mouse_led(LED_BASE, &mv);
          }
        }
        printf("\n");
        break;
    }  //end switch
  } // end while
}
```

The main part of the program is an infinite loop. To perform a test, we use the slide switches to select one of the five tests and press the pushbutton switch 1 (labeled **key1** on the DE1 board) to start the operation. The last three tests run continuously until a pushbutton switch is pressed again. Either a mouse or a keyboard can be connected to the DE1 board's PS2 port.

There is a function, **mouse_led()**, used in test 5. It turns on one LED of the discrete LED array according to the horizontal movement of the mouse. The code is shown in Listing 16.13.

Listing 16.13

```
void mouse_led(alt_u32 led_base, mouse_mv_type *mv){
  static int count=0;
  int pos;
  alt_u32 led_ptn;

  if(mv->lbtn)
    count = 0;
  else if (mv->rbtn)
```

```
    count = 255;
  else{
    count = count + mv->xmov;
    if (count >255)
      count = 255;
    if (count <0)
      count = 0;
  }
  pos = (count >> 5);            // get 3 MSBs
  led_ptn = (0x00000080) >> pos; // 0b10000000
  pio_write(led_base, led_ptn);
}
```

The graphic mouse test routine is discussed in Section 17.8.

The names of the base addresses, such as BTN_BASE, are based on the names of instantiated modules of a particular Nios II configuration. The module names of the comprehensive Nios II system are shown in Figure 16.9. These names may not be the same for a different system. It is good practice to include base addresses as arguments for functions and use the constant base names only in the top-level main program.

Since our driver routines are developed in an ad hoc manner and are not HAL-compliant, as discussed in Section 11.4, the relevant header and program files need to be added to the test program's project directory manually.

16.10 USE OF BOOK'S CUSTOM IP CORES

A collection of IP cores, similar to the PS2 controller IP core discussed in this chapter, are constructed in the remaining chapters of the book. They can be used as normal SPOC IP cores and instantiated in a Nios II system. The file structure, library integration, and the construction of a comprehensive Nios II testing system are discussed in the following subsections.

16.10.1 IP core files

File organization for individual chapters   The derivation of IP cores in Parts III and IV involves three basic steps:

1. Develop hardware.
2. Develop the software driver.
3. Construct a testing system to verify the IP core's operation.

The file structure of an IP core reflects these steps. In each chapter, the files are organized as follows:

- hdl directory: it contains HDL source files and the SOPC component description file (i.e., the .tcl file) of the IP core.
- drv directory: it contains the relevant C header files and program files (i.e., the .h and .c files) of the IP core's software driver.
- test directory: it contains the hardware prototype and software routines to verify the IP core's operation. It contains three separate sub-directories:
  - hardware directory: its contains a prototyping Nios II system used to verify the IP core's operation. It usually includes the following:
    * .sopc file: the testing Nios II system

    ∗ .vhd or .v file: the top-level HDL description that instantiates the Nios II system

   – **software** directory: it contains the testing C program files, which include auxiliary testing functions and the main program.

  • build directory: it contains the files generated from the synthesis and compiling processes:

   – .sopcinfo file: the Nios II system information file to be used to generate a BSP

   – .sof file: the FPGA device configuration file

   – .elf file: the final testing software image file

   – .qar file: the archived Quartus project file that contains the relevant files to regenerate the prototyping Nios II system

Note that the files in this directory are constructed for the DE1 board and thus cannot be used for other boards.

File organization for all IP cores For convenience, the relevant IP core files of Parts III and IV are gathered in two compressed files, which are chu_ip_vhdl.zip and chu_ip_drv.zip. The former contains the hardware files (i.e., HDL and .tcl files) and the latter contains the software drivers. They can be downloaded from the companion web site.

SOPC library integration The hardware IP core files can be integrated into the SOPC Builder's library. The procedure is:

1. Create a directory, say chu_ip, on the hard disk.
2. Uncompress the chu_ip_vhdl.zip file and store the uncompressed files under the chu_ip directory.
3. In the SOPC Builder window, select Tools ≻ Options... to invoke the Options dialog.
4. In the dialog, select the IP Search Path page, click the Add... button, navigate to the chu_ip directory, and click the Open button to add the directory to search path. The resulting page is shown Figure 10.11.
5. Click the Finish button and SOPC Builder searches the paths and adds the found IP cores to the left Library panel.
6. In the Block Type field, select Auto.
7. A new category, chu_ip, should appear. Expand the category and the IP cores should be listed under this category, as shown in the left panel of Figure 16.9.

A core can be added to the Nios II system, just like other normal IP cores.

Since the software drivers are constructed on an ad hoc fashion and are not HAL-compatible, they need to be copied to the software application directory as needed.

16.10.2 Comprehensive Nios II testing system

Nios II system A collection of IP cores similar to the PS2 controller is introduced in the remaining chapters of the book. Instead of creating a Nios II testing system repetitively for each individual core, we construct a comprehensive testing system that incorporates all the main IP cores. The generated FPGA configuration file (i.e., .sop file) can be used in the remaining chapters of the book. The system uses the SDRAM for the main memory and contains I/O cores to exercise most I/O peripherals on the DE1 board.

The procedure to create this testing system is:

1. Follow the procedure from the previous subsection to integrate the IP cores into the SOPC Builder's library.
2. Follow the procedure in Section 15.7 to add an ALTPLL module and rename it `pll`. Rename the the external clock `clk_50M` and rename two PLL clocks `clk_sys` and `clk_sdram`. Use `clk_sys` as the clock source for all other SOPC modules.
3. Add a Nios II/f processor, configure it with 4 KB instruction cache and 4 KB data cache, and rename it `cpu_f`. The fast processor is needed to accommodate several computation intensive applications.
4. Follow the procedure in Section 15.7 to add an SDRAM controller module and rename it `sdram`.
5. Select the `sdram` module for Nios II's reset and exception vectors.
6. Add and configure a JTAG UART module and rename it `jtag_uart`.
7. Add a PIO module, configure it as a 10-bit input port, and rename it `switch`. This is for the ten slide switches
8. Add a PIO module, configure it as a 4-bit input port, follow the procedure in Section 10.5.1 to enable the edge capture and interrupt request, and rename it `btn`. This is for the four pushbutton switches.
9. Add a PIO module, configure it as a 18-bit output port, and rename it `led`. This is for the discrete red and green LEDs on the DE1 board.
10. Add a PIO module, configure it as a 32-bit output port, and rename it `sseg`. This is for the four seven-segment LED displays on the DE1 board.
11. Add two timer modules, and rename them `sys_timer` and `stamp_timer`. As their names indicate, one timer is used to facilitate system function and the other is used to maintain a "time stamp."
12. Add a system id module and rename it `sysid`.
13. Add a chu_avalon_ps2 module from the chu_ip category, set `W_SZIE` to 2, and rename it `ps2`. This is a PS2 controller with a 4-byte FIFO buffer.
14. Add a chu_avalon_vga module from the chu_ip category and rename it `vram`. This is an SRAM-based VGA video controller discussed in Chapter 17.
15. Add a chu_avalon_audio module from the chu_ip category, set `FIFO_SIZE` to 3, and rename it `audio`. This is an audio codec controller with 8-byte FIFO buffers discussed in Chapter 18.
16. Add a chu_avalon_sd module from the chu_ip category and rename it `sdc`. This is an SD card controller discussed in Chapter 19.
17. Add a chu_avalon_gcd module from the chu_ip category and rename it `g_engine`. This is a GCD (greatest common divisor) hardware accelerator discussed in Chapter 20.
18. Add a chu_avalon_frac module from the chu_ip category and rename it `f_engine`. This is a Mandelbrot set fractal hardware accelerator discussed in Chapter 21.
19. Add a chu_avalon_ddfs module from the chu_ip category and rename it `d_engine`. This is a DDFS (direct digital frequency synthesis) hardware accelerator discussed in Chapter 22.
20. Adjust interrupt request priorities in the IRQ column.
21. Generate HDL and information files.

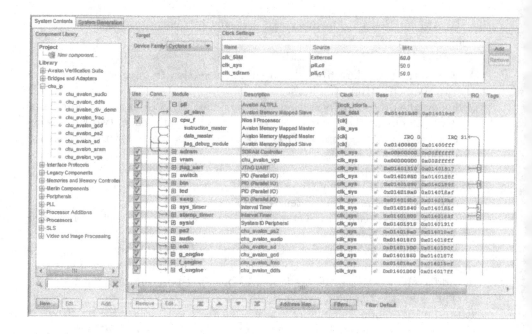

Figure 16.9   Nios II configuration for Parts III and IV.

The completed configuration screen is shown in Figure 16.9.

Top-level HDL file   After the HDL files are generated, we can create a top-level module that incorporates the Nios II system.   The HDL code is shown in Listing 16.14.

Listing 16.14   Top-level comprehensive Nios II testing circuit

```
library ieee;
use ieee.std_logic_1164.all;
entity p34_top is
  port(
5     clk: in std_logic;
      -- switch/LEDs
      sw: in std_logic_vector(9 downto 0);
      key: in std_logic_vector(3 downto 0);
      ledg: out std_logic_vector(7 downto 0);
10    ledr: out std_logic_vector(9 downto 0);
      hex3, hex2, hex1, hex0: out std_logic_vector(6 downto 0);
      -- to/from ps2
      ps2c, ps2d: inout std_logic;
      -- to VGA
15    vsync, hsync: out std_logic;
      rgb: out std_logic_vector (11 downto 0);
      -- to/from audiocodec
      m_clk, b_clk, dac_lr_clk, adc_lr_clk: out std_logic;
      dacdat: out std_logic;
20    adcdat: in std_logic;
      i2c_sclk: out std_logic;
      i2c_sdat: inout std_logic;
      -- to/from SD card
      sd_clk: out std_logic;
```

```
25        sd_di: out std_logic;
          sd_do: in std_logic;
          sd_cs: out std_logic;
          —— to/from SRAM
          sram_addr: out std_logic_vector (17 downto 0);
30        sram_dq: inout std_logic_vector (15 downto 0);
          sram_ce_n: out std_logic;
          sram_lb_n: out std_logic;
          sram_oe_n: out std_logic;
          sram_ub_n: out std_logic;
35        sram_we_n: out std_logic;
          —— to/from SDRAM
          dram_clk: out std_logic;
          dram_cs_n , dram_cke: out std_logic;
          dram_ldqm , dram_udqm: out std_logic;
40        dram_cas_n , dram_ras_n , dram_we_n: out std_logic;
          dram_addr: out std_logic_vector (11 downto 0);
          dram_ba_0 , dram_ba_1: out std_logic;
          dram_dq: inout std_logic_vector (15 downto 0)
       );
45 end p34_top;

   architecture structure of p34_top is
      component nios_p34 is
      port (
50        —— clock and reset
          signal clk_50m: in std_logic;
          signal clk_sdram: out std_logic;
          signal clk_sys: out std_logic;
          signal reset_n: in std_logic;
55        —— SDRAM
          signal zs_addr_from_the_sdram: out std_logic_vector (11 downto 0);
          signal zs_ba_from_the_sdram: out std_logic_vector (1 downto 0);
          signal zs_dq_to_and_from_the_sdram:
                inout std_logic_vector (15 downto 0);
60        signal zs_cas_n_from_the_sdram: out std_logic;
          signal zs_cke_from_the_sdram: out std_logic;
          signal zs_cs_n_from_the_sdram: out std_logic;
          signal zs_ras_n_from_the_sdram: out std_logic;
          signal zs_we_n_from_the_sdram: out std_logic;
65        signal zs_dqm_from_the_sdram: out std_logic_vector (1 downto 0);
          —— siwtch and leds
          signal in_port_to_the_switch: in std_logic_vector (9 downto 0);
          signal in_port_to_the_btn: in std_logic_vector (3 downto 0);
          signal out_port_from_the_led: out std_logic_vector (17 downto 0);
70        signal out_port_from_the_sseg: out std_logic_vector (31 downto 0);
          —— PS2
          signal ps2c_to_and_from_the_ps2: inout std_logic;
          signal ps2d_to_and_from_the_ps2: inout std_logic;
          —— video RAM (VGA)
75        signal hsync_from_the_vram: out std_logic;
          signal vsync_from_the_vram: out std_logic;
          signal rgb_from_the_vram: out std_logic_vector (11 downto 0);
          —— video RAM (SRAM)
          signal sram_addr_from_the_vram:
80                out std_logic_vector (17 downto 0);
          signal sram_dq_to_and_from_the_vram:
                inout std_logic_vector (15 downto 0);
          signal sram_ce_n_from_the_vram:out std_logic;
          signal sram_lb_n_from_the_vram: out std_logic;
85        signal sram_oe_n_from_the_vram: out std_logic;
```

```
        signal sram_ub_n_from_the_vram: out std_logic;
        signal sram_we_n_from_the_vram: out std_logic;
        -- audio codec
        signal m_clk_from_the_audio: out std_logic;
90      signal b_clk_from_the_audio: out std_logic;
        signal adc_lr_clk_from_the_audio: out std_logic;
        signal adcdat_to_the_audio: in std_logic;
        signal dac_lr_clk_from_the_audio: out std_logic;
        signal dacdat_from_the_audio: out std_logic;
95      signal codec_adc_data_out_from_the_audio:
                out std_logic_vector (31 downto 0);
        signal codec_dac_data_in_to_the_audio:
                in std_logic_vector (31 downto 0);
        signal codec_adc_rd_to_the_audio: in std_logic;
100     signal codec_dac_wr_to_the_audio: in std_logic;
        signal codec_sample_tick_from_the_audio: out std_logic;
        signal i2c_sclk_from_the_audio: out std_logic;
        signal i2c_sdat_to_and_from_the_audio: inout std_logic;
        -- SD card
105     signal sd_clk_from_the_sdc: out std_logic;
        signal sd_cs_from_the_sdc: out std_logic;
        signal sd_di_from_the_sdc: out std_logic;
        signal sd_do_to_the_sdc: in std_logic;
        -- DDFS
110     signal ddfs_data_out_from_the_d_engine:
                out std_logic_vector (15 downto 0)
        );
    end component nios_p34;
    signal sseg4: std_logic_vector(31 downto 0);
115 signal led: std_logic_vector(17 downto 0);
    signal ddfs_data: std_logic_vector(15 downto 0);
    signal dac_load_tick: std_logic;

  begin
120 nios: nios_p34
    port map(
        clk_50M=>clk,
        clk_sdram=>dram_clk,
        clk_sys=>open,
125     reset_n=>key(3),
        in_port_to_the_btn=> '0'&key(2 downto 0),
        in_port_to_the_switch=>sw,
        out_port_from_the_led=>led,
        out_port_from_the_sseg=>sseg4,
130     -- SDRAM
        zs_addr_from_the_sdram=>dram_addr,
        zs_ba_from_the_sdram(1)=>dram_ba_1,
        zs_ba_from_the_sdram(0)=>dram_ba_0,
        zs_cas_n_from_the_sdram=>dram_cas_n,
135     zs_cke_from_the_sdram=>dram_cke,
        zs_cs_n_from_the_sdram=>dram_cs_n,
        zs_dq_to_and_from_the_sdram=>dram_dq,
        zs_ras_n_from_the_sdram=>dram_ras_n,
        zs_we_n_from_the_sdram=>dram_we_n,
140     zs_dqm_from_the_sdram(1)=> dram_udqm,
        zs_dqm_from_the_sdram(0)=> dram_ldqm,
        -- PS2
        ps2c_to_and_from_the_ps2=>ps2c,
        ps2d_to_and_from_the_ps2=>ps2d,
145     -- video ram
        hsync_from_the_vram=>hsync,
```

```
        vsync_from_the_vram => vsync,
        rgb_from_the_vram=>rgb,
        sram_addr_from_the_vram => sram_addr,
150     sram_ce_n_from_the_vram => sram_ce_n,
        sram_dq_to_and_from_the_vram => sram_dq,
        sram_lb_n_from_the_vram => sram_lb_n,
        sram_oe_n_from_the_vram => sram_oe_n,
        sram_ub_n_from_the_vram => sram_ub_n,
155     sram_we_n_from_the_vram => sram_we_n,
        --- audio codec
        b_clk_from_the_audio=>b_clk,
        m_clk_from_the_audio=>m_clk,
        dac_lr_clk_from_the_audio=>dac_lr_clk,
160     adc_lr_clk_from_the_audio=>adc_lr_clk,
        dacdat_from_the_audio=>dacdat,
        adcdat_to_the_audio=>adcdat,
        i2c_sclk_from_the_audio=>i2c_sclk,
        i2c_sdat_to_and_from_the_audio=>i2c_sdat,
165     codec_adc_data_out_from_the_audio=>open,
        codec_adc_rd_to_the_audio=>dac_load_tick,
        codec_dac_data_in_to_the_audio=>(ddfs_data & ddfs_data),
        codec_dac_wr_to_the_audio=>dac_load_tick,
        codec_sample_tick_from_the_audio=>dac_load_tick,
170     --- SD card
        sd_clk_from_the_sdc=>sd_clk,
        sd_do_to_the_sdc=>sd_do,
        sd_di_from_the_sdc=>sd_di,
        sd_cs_from_the_sdc=>sd_cs,
175     --- DDFS
        ddfs_data_out_from_the_d_engine=>ddfs_data
     );
     ledr <= led(17 downto 8);
     ledg <= led(7 downto 0);
180  hex3 <= sseg4(30 downto 24);
     hex2 <= sseg4(22 downto 16);
     hex1 <= sseg4(14 downto 8);
     hex0 <= sseg4(6 downto 0);
  end structure;
```

We can follow the previous procedure to synthesize the HDL file and program the FPGA device. There are two caveats about this system. First, the DE1 board uses a dual-purpose I/O pin of the EP2C20 device for SD card's **sd_do** signal. The pin needs to be configured as a "normal I/O pin," as discussed in Section 3.5.1. Second, the Quartus II Web edition only provides a "time-limited license" for the Nios II/f core. To use the core, the Quartus Programmer window must be remained open and the USB cable must be kept connected all the time.

The synthesis report indicates that the system uses about 7000 LEs (37% of EP2C20 device), 27 M4K internal memory modules (52% of EP2C20 device), and 15 18-bit multipliers (58% of EP2C20 device).

BSP   Based on the .sopcinfo file, we can follow the previous procedure to create a BSP package and name it **p34_bsp** (for BSP of Parts III and IV). To facilitate subsequent testing application programs, we need to make two adjustments. First, in the Main tab page of BSP Editor, select sys_timer in the systm_clk_timer field and select stamp_timer in the timestamp_timer field, as shown in Figure 16.10. Second, in the Software Packages tab page of BSP Editor, check the altera_hostfs field, as

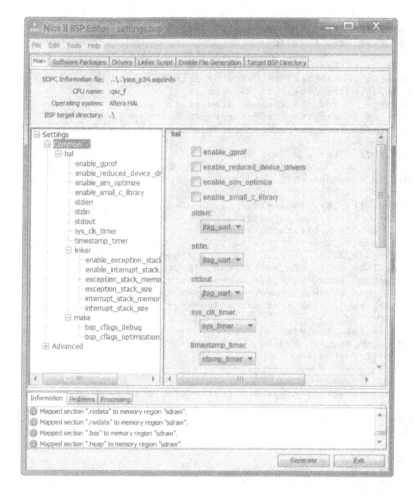

Figure 16.10   Main tab page of BSP Editor.

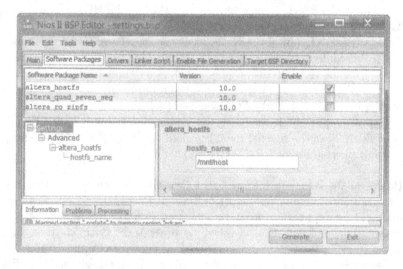

Figure 16.11   Software Packages tab page of BSP Editor.

shown in Figure 16.11. This enables the "host-based file system," which is used in Chapters 17 and 18.

16.11 BIBLIOGRAPHIC NOTES

Three articles, "PS/2 Mouse/Keyboard Protocol," "PS/2 Keyboard Interface," and "PS/2 Mouse Interface," by Adam Chapweske, provide detailed information on the PS2 keyboard and mouse interface. They can be found at the http://www.computer-engineering.org site. *Rapid Prototyping of Digital Systems: Quartus II Edition* by James O. Hamblen et al. also contains a chapter on the PS2 port and the keyboard and mouse protocols. Altera University Program IP cores consist of an alterative PS2 design. Its driver routines are more comprehensive and are integrated to the HAL framework.

16.12 SUGGESTED EXPERIMENTS

The mouse is used mainly with a graphic video interface, which is discussed in Chapter 17. Many additional experiments and projects can be found in that chapter.

16.12.1 PS2 receiving subsystem with watchdog timer

There is no error-handling capability in the PS2 receiving subsystem in Section 16.2. The potential noise and glitches in the `ps2c` signal may cause the FSMD to be stuck in an incorrect state. One way to deal with this problem is to add a watchdog timer. The timer is initiated every time the `fall_edge_tick` signal is asserted in the `dps` state. The `time_out` signal is asserted if no subsequent falling edge arrives in the next 20 μs, and the FSMD returns to the `idle` state. Design the modified receiving subsystem, derive a testbench, and use simulation to verify its operation.

16.12.2 Software receiving FIFO

In Section 16.4, a hardware FIFO buffer is attached to the receiving subsystem. Instead of using a hardware FIFO, we can develop software to perform the desired buffer functionality by using an array to mimic the operation of a FIFO buffer. One possible method is to add an interrupt request signal to the receiving subsystem. The corresponding ISR reads the packet and inserts it to the FIFO buffer each time the interrupt is asserted. The driver routines retrieve data from this buffer. Modify the PS2 top-level system and wrapping circuit, derive a new Nios II system, develop the ISR and driver routines, and verify the operation.

16.12.3 Software PS2 controller

Since the PS2's `ps2c` and `ps2d` signals are slow, we can use software and the existing timer to detect the change and to generate the desired transitions. In other words, we can eliminate receiving and transmitting subsystems and connect the two signals to a 2-bit bidirectional PIO core. Properly configure the PIO core, derive the new

Nios II system, develop the receiving and transmitting driver routines, and verify the operation.

16.12.4 Keyboard-controlled LED flashing circuit

Consider the LED flashing circuit discussed in Chapter 11. Instead of switch and buttons, we can use the keyboard to send commands:

- Use the P (for "pause") key to pause and resume the flashing operation.
- Use the following key sequence to enter the desired flashing period: F1 and then three digit keys (i.e., 000 to 999).
- All other keys or illegal sequences will be ignored.

Derive the revised program, and verify the operation.

16.12.5 Enhanced keyboard driver routine I

The kb_get_ch() function in Section 16.7 processes only the shift keys. Many additional functionalities can be added:

- Use the Caps Lock key to toggle between the lowercase or uppercase mode.
- Use the Caps Lock LED to indicate the status of the Caps Lock key.
- Use the Ctrl key for special functions (e.g., return a special Ctrl-C code when both the Ctrl and C keys are pressed).

Derive the new driver function and verify the operation.

16.12.6 Enhanced keyboard driver routine II

The kb_get_ch() function in Section 16.7 covers only the standard scan codes. A current keyboard usually contains extended scan codes for additional keys and these codes can be found in the references of the bibliographic section. Derive a new function to cover the extended scan codes and verify the operation.

16.12.7 Remote-mode mouse driver

An alternative to a mouse's stream mode is the *remote mode*, in which the mouse only transmits data packets after receiving a read data command from the host. More detailed information can be found in the references of the bibliographic section. Derive new mouse driver routines using this mode and verify the operation.

16.12.8 Scroll-wheel mouse driver

In addition to the x- and y-axis activities, newer mice add a third dimension, which corresponds to the movement of the scroll wheel. More detailed information can be found in the references of the bibliographic section. Derive new mouse driver routines for this type of mice and verify the operation.

16.13 COMPLETE PROGRAM LISTING

Listing 16.15   chu_avalon_ps2.h

```
/*********************************************************************
 *
 * Module:   PS2, keyboard, and mouse driver header
 * File:     chu_avalon_ps2.h
 * Purpose:  Routines to access PS2 port and get keyboard scan codes and
 *           mouse activities
 *
 *********************************************************************
 * Register map
 *********************************************************************
 * Read (data to cpu):
 *   offset 0
 *     * bit 7-0: 8-bit ps2 packet
 *   offset 1
 *     * bit 0: ps2 receiving fifo empty
 *     * bit 1: ps2 transmitter idle
 * Write (data from cpu):
 *   offset 0:  dummy data, used to generate a pulse
 *   offset 2:
 *     * bit 7-0: 8-bit ps2 command data
 *
 *********************************************************************/
/* file inclusion */
#include <alt_types.h>

/*********************************************************************
 * constant definitions
 *********************************************************************/
#define CHU_PS2_DATA_REG      0
#define CHU_PS2_CONTROL_REG   1
#define CHU_PS2_WR_DATA_REG   2

/*********************************************************************
 * Data type definitions
 *********************************************************************/
/* data type for mouse activities */
typedef struct mouse_move
{
  int lbtn;      // left button
  int rbtn;      // right button
  int xmov;      // x-axis movement
  int ymov;      // x-axis movement
} mouse_mv_type;
```

```
/***********************************************************************
 * Function prototypes
 ***********************************************************************/
/* PS2 functions */
int ps2_tx_is_idle(alt_u32 ps2_base);
void ps2_wr_cmd(alt_u32 ps2_base, alt_u8 cmd);
int ps2_is_empty(alt_u32 ps2_base);
alt_u8 ps2_read_fifo(alt_u32 ps2_base);
void ps2_rm_pkt(alt_u32 ps2_base);
int ps2_get_pkt(alt_u32 ps2_base, alt_u8 *byte);
void ps2_flush_fifo(alt_u32 ps2_base);
int ps2_reset_device(alt_u32 ps2_base);

//* Keyboard functions */
int kb_get_ch (alt_u32 ps2_base, char *ch);
int kb_get_line(alt_u32 ps2_base, char *s, int lim);

/* Mouse functions */
int mouse_init(alt_u32 ps2_base);
int mouse_get_activity(alt_u32 ps2_base, mouse_mv_type *mv);
```

Listing 16.16   chu_avalon_ps2.c

```
/*****************************************************************
 *
 * Module:   PS2, keyboard, and mouse driver function prototypes
 * File:     chu_avalon_ps2.c
 * Purpose:  Routines to access PS2 port and get keyboard scan codes and
 *           mouse activities
 *
 *****************************************************************
/* file inclusion */
#include <io.h>
#include <unistd.h>           // to use usleep
#include "chu_avalon_ps2.h"



/*****************************************************************
 * PS2 functions
 *****************************************************************
 *
 * Purpose: utility routines to read PS2 packets and send command
 *
 * Note:
 *   - a fifo is used in receiver but not in transmitter
 *   - reading (first item of) fifo and removing the first item are
 *     separated into two operations for flexibility
 *
 *****************************************************************/

/*****************************************************************
 * function: ps2_tx_is_idle()
 * purpose:  check whether the ps2 transmitter is idle
 * argument:
 *   ps2_base: base address of ps2 controller
 * return: 1 if idle; 0 otherwise
 * note:
 *****************************************************************/
int ps2_tx_is_idle(alt_u32 ps2_base)
{
  alt_u32 ctrl_reg;
  int idle_bit;

  ctrl_reg = IORD(ps2_base, CHU_PS2_CONTROL_REG);
  idle_bit = (ctrl_reg & 0x00000002) >> 1;
  return idle_bit;
}
```

```
/************************************************************************
* function: ps2_wr_cmd()
* purpose:  send an 8-bit command to ps2
* argument:
*   ps2_base: base address of ps2 controller
*   cmd: 8-bit command
* return:
* note:
*   - the code does not check whether ps2 is busy;
*     the calling function should not issue the command at a rapid rate
************************************************************************/
void ps2_wr_cmd(alt_u32 ps2_base, alt_u8 cmd)
{
  IOWR(ps2_base, CHU_PS2_WR_DATA_REG, (alt_u32) cmd);
}


/************************************************************************
* function: ps2_is_empty()
* purpose:  check whether the ps2 receiver fifo is empty
* argument:
*   ps2_base: base address of ps2 controller
* return: 1 if empty; 0 otherwise
* note:
************************************************************************/
int ps2_is_empty(alt_u32 ps2_base)
{
  alt_u32 ctrl_reg;
  int empty_bit;

  ctrl_reg = IORD(ps2_base, CHU_PS2_CONTROL_REG);
  empty_bit = ctrl_reg & 0x00000001;
  return empty_bit;
}


/************************************************************************
* function: ps2_read_fifo()
* purpose:  retrieve the data from the head of receiver fifo
* argument:
*   ps2_base: base address of ps2 controller
* return: data from the head of fifo
* note:
*   - the data remain in fifo after read
*   - invalid data returned if fifo is empty
************************************************************************/
alt_u8 ps2_read_fifo(alt_u32 ps2_base)
{
  alt_u32 data_reg;
  alt_u8 packet;

  data_reg = IORD(ps2_base, CHU_PS2_DATA_REG);
  packet = (alt_u8) (data_reg & 0x000000ff);
  return packet;
}
```

```
/*************************************************************
* function: ps2_rm_pkt()
* purpose:  remove data from the head of receiver fifo
* argument:
*   ps2_base: base address of ps2 controller
* return:
* note:
*************************************************************/
void ps2_rm_pkt(alt_u32 ps2_base)
{
  IOWR(ps2_base, CHU_PS2_DATA_REG, 0x00000000); // write a dummy data
}


/*************************************************************
* function: ps2_get_pkt()
* purpose:  check ps2 fifo and, if not empty, read data
* argument:
*   ps2_base: base address of ps2 controller
*   byte: pointer to the retrieved data
* return:
*   1 if data available; 0 otherwise
*   byte updated if data available
* note:
*************************************************************/
int ps2_get_pkt(alt_u32 ps2_base, alt_u8 *byte)
{
  if (!ps2_is_empty(ps2_base)) {
    *byte = ps2_read_fifo(ps2_base);
    return 1;   // got data
  }
  return 0;     // no data
}


/*************************************************************
* function: ps2_flush_fifo()
* purpose:  flush all packets from fifo
* argument:
*   ps2_base: base address of ps2 controller
* return:
* note:
*************************************************************/
void ps2_flush_fifo(alt_u32 ps2_base)
{
  while (!ps2_is_empty(ps2_base)) {
    ps2_rm_pkt(ps2_base);
  }
}
```

```
/*************************************************************************
 * function: ps2_reset_device()
 * purpose:  reset and identify the type of ps2 device (mouse or keyboard)
 * argument:
 *   ps2_base: base address of ps2 controller
 * return:
 *   0: reset fails or unknown device; 1: keyboard; 2: mouse
 * note:
 *  - procedure:
 *    1. flush ps2 receiver fifo
 *    2. host sends reset command 0xff
 *    3. ps2 device acknowledges (0xfa) and performs self-test
 *    4. ps2 device responds 0xaa if test passes
 *    5. mouse sends an extra id 0x00
 *************************************************************************/
int ps2_reset_device(alt_u32 ps2_base)
{
  alt_u8 packet;

  ps2_flush_fifo(ps2_base);
  /* send reset 0xff */
  ps2_wr_cmd(ps2_base, 0xff);
  usleep(1000000);       // wait for 200 ms
  /* check 0xfa 0xaa */
  if (ps2_get_pkt(ps2_base, &packet)==0 || packet!=0xfa)
    return (0);          // no response or wrong response
  ps2_rm_pkt(ps2_base);
  if (ps2_get_pkt(ps2_base, &packet)==0 || packet!=0xaa)
    return (0);          // no response or wrong response
  ps2_rm_pkt(ps2_base);
  /* check whether 0x00 is received */
  if (ps2_get_pkt(ps2_base, &packet)==0)
    return (1);          // fifo has no more packet, device is keyboard
  ps2_rm_pkt(ps2_base);
  if (packet==0x00)
    return (2);          // mouse id
  else
    return (0);          // unknown device id
}
```

```
/*****************************************************************
 * Keyboard functions
 *****************************************************************
 *
 * Purpose: utility routines to process keyboard scan codes
 * Note:
 *
 ****************************************************************/

/*****************************************************************
 * function: kb_get_ch()
 * purpose:  get a character or special key code from keyboard
 * argument:
 *   ps2_base: base address of ps2 controller
 *   ch: pointer to the scanned charater
 * return:
 *   - 1 if there is a valid char; 0 otherwise
 *   - ch updated with character or special key code
 * note:
 *   - cannot use extended scan codes
 ****************************************************************/
int kb_get_ch (alt_u32 ps2_base, char *ch)
{
    // special characters
    #define TAB      0x09    // tab
    #define BKSP     0x08    // backspace
    #define ENTER    0x0d    // enter (new line)
    #define ESC      0x1b    // escape
    #define BKSL     0x5c    // back slash
    #define SFT_L    0x12    // left shift
    #define SFT_R    0x59    // right shift

    #define CAPS     0x80
    #define NUM      0x81
    #define CTR_L    0x82
    #define F1       0xf0
    #define F2       0xf1
    #define F3       0xf2
    #define F4       0xf3
    #define F5       0xf4
    #define F6       0xf5
    #define F7       0xf6
    #define F8       0xf7
    #define F9       0xf8
    #define F10      0xf9
    #define F11      0xfa
    #define F12      0xfb

    // keyboard scan code to ascii (lowercase)
    static const char SCAN2ASCII_LO_TABLE[128]={
        0,    F9,    0,     F5,    F3,    F1,    F2,    F12,   //00
        0,    F10,   F8,    F6,    F4,    TAB,   '`',   0,     //08
        0,    0,     SFT_L, 0,     CTR_L, 'q',   '1',   0,     //10
        0,    0,     'z',   's',   'a',   'w',   '2',   0,     //18
        0,    'c',   'x',   'd',   'e',   '4',   '3',   0,     //20
        0,    ' ',   'v',   'f',   't',   'r',   '5',   0,     //28
        0,    'n',   'b',   'h',   'g',   'y',   '6',   0,     //30
        0,    0,     'm',   'j',   'u',   '7',   '8',   0,     //38
        0,    ',',   'k',   'i',   'o',   '0',   '9',   0,     //40
        0,    '.',   '/',   'l',   ';',   'p',   '-',   0,     //48
        0,    0,     '\'',  0,     '[',   '=',   0,     0,     //50
```

```
  CAPS, SFT_R, ENTER, ']',  0,      BKSL, 0,    0,    //58
  0,    0,    0,    0,    0,      0,    BKSP, 0,    //60
  0,    '1',  0,    '4',  '7',    0,    0,    0,    //68
  0,    '.',  '2',  '5',  '6',    '8',  ESC,  NUM,  //70
  F11,  '+',  '3',  '-',  '*',    '9',  0,    0     //78
};
// keyboard scan code to ascii (uppercase)
static const char SCAN2ASCII_UP_TABLE[128] = {
  0,    F9,   0,    F5,   F3,     F1,   F2,   F12,  //00
  0,    F10,  F8,   F6,   F4,     TAB,  '~',  0,    //08
  0,    0,    SFT_L, 0,   CTR_L,  'Q',  '!',  0,    //10
  0,    0,    'Z',  'S',  'A',    'W',  '@',  0,    //18
  0,    'C',  'X',  'D',  'E',    '$',  '#',  0,    //20
  0,    ' ',  'V',  'F',  'T',    'R',  '%',  0,    //28
  0,    'N',  'B',  'H',  'G',    'Y',  '^',  0,    //30
  0,    0,    'M',  'J',  'U',    '&',  '*',  0,    //38
  0,    '<',  'K',  'I',  'O',    ')',  '(',  0,    //40
  0,    '>',  '?',  'L',  ':',    'P',  '_',  0,    //48
  0,    0,    '\"', 0,    '{',    '+',  0,    0,    //50
  CAPS, SFT_R, ENTER, '}',  0,      '|',  0,    0,    //58
  0,    0,    0,    0,    0,      0,    BKSP, 0,    //60
  0,    '1',  0,    '4',  '7',    0,    0,    0,    //68
  0,    '.',  '2',  '5',  '6',    '8',  ESC,  NUM,  //70
  F11,  '+',  '3',  '-',  '*',    '9',  0,    0     //78
};

static int sft_on = 0;
alt_u8 scode;

while (1){
  if (!ps2_get_pkt(ps2_base, &scode))              // no packet
    return(0);
  ps2_rm_pkt(ps2_base);
  switch (scode){
    case 0xf0:   // break code
      while (!ps2_get_pkt(ps2_base, &scode)); // get next
      ps2_rm_pkt(ps2_base);
    if (scode==SFT_L || scode==SFT_R)
      sft_on = 0;
      break;
    case SFT_L:                                    // shift key make code
    case SFT_R:
    sft_on = 1;
      break;
    default:                                        // normal make code
      if (sft_on)
        *ch = SCAN2ASCII_UP_TABLE[scode];
      else
        *ch = SCAN2ASCII_LO_TABLE[scode];
      return(1);
  }  // end switch
}  // end while
}
```

```c
/**************************************************************************
* function: kb_get_line()
* purpose:  get a line from keyboard
*   ps2_base: base address of ps2 controller
*   s: pointer to the returned string
*   lim: max number in the string
* return: number of chars in the string
* note:
*   - procedure: read string until \n or max number reached
**************************************************************************/
int kb_get_line(alt_u32 ps2_base, char *s, int lim)
{
  char ch;
  int i;

  i=0;
  while (1){
    while(!kb_get_ch(ps2_base, &ch));
    if ((ch=='\n')|(i==(lim - 1)))
      break;
    else
      s[i++] = ch;
  } // end while
  s[i] = '\0';
  return i;
}
```

```
/******************************************************************
 * Mouse functions
 ******************************************************************
 *
 * Purpose: utility routines to process mouse command
 *
 * Note:
 *   - no timeout for error checking
 *
 ******************************************************************/

/******************************************************************
 * function: mouse_init()
 * purpose:  initialize mouse in stream mode
 * argument:
 *   ps2_base: base address of ps2 controller
 * return:
 *   1 if successful; 0 otherwise
 * note:
 *   - sequence:
 *     1. reset mouse
 *     3. host sends entering stream mode command 0xf4
 *     4. mouse acknowledge with 0xfa
 ******************************************************************/
int mouse_init(alt_u32 ps2_base)
{
  alt_u8 packet;

  if (ps2_reset_device(ps2_base)!=2)
    return (0);
  /* send stream mode command 0xf4 */
  ps2_wr_cmd(ps2_base, 0xf4);
  usleep(1000000);        // wait
  /* check 0xfa (acknowledge) */
  if (ps2_get_pkt(ps2_base, &packet)==0 || packet!=0xfa)
    return (0);           // no response or wrong response
  /* everything is ok */
  ps2_rm_pkt(ps2_base);
  return(1);
}
```

```
/*****************************************************************************
 * function: mouse_get_activity()
 * purpose:  retrieve 3 packets and decode mouse movement info
 * argument:
 *   ps2_base: base address of ps2 controller
 *   mv: pointer to the mouse movement data
 * return:
 *   * 1 if there is mouse activity; 0 otherwise
 *   * mv updated if there is mouse activity
 * note:
 *   - ignore middle button
 *   - ignore x, y overflow
 *   - manually performing 9-bit signed to 32-bit integer
 *****************************************************************************/
int mouse_get_activity(alt_u32 ps2_base, mouse_mv_type *mv)
{
  alt_u8 b1, b2, b3;

  alt_u32 tmp;

  /* check and retrieve 1st byte */
  if (!ps2_get_pkt(ps2_base, &b1))
    return (0);                         // no data in rx fifo buffer
  ps2_rm_pkt(ps2_base);                 // remove 1st byte
  /* wait and retrieve 2nd byte */
  while (!ps2_get_pkt(ps2_base, &b2));
  ps2_rm_pkt(ps2_base);
  /* wait and retrieve 3rd byte */
  while (!ps2_get_pkt(ps2_base, &b3));
  ps2_rm_pkt(ps2_base);
  /* extract button info */
  mv->lbtn = (int) (b1 & 0x01);         // extract bit 0
  mv->rbtn = (int) (b1 & 0x02)>>1;      // extract bit 1
  /* extract x movement; manually convert 9-bit 2's comp to int */
  tmp = (alt_u32) b2;
  if (b1 & 0x10)                        // check MSB (sign bit) of x movement
    tmp = tmp | 0xffffff00;             // manual sign-extension if negative
  mv->xmov = (int) tmp;                 // data conversion
  /* extract y movement; manually convert 9-bit 2's comp to int */
  tmp = (alt_u32) b3;
  if (b1 & 0x20)                        // check MSB (sign bit) of y movement
    tmp = tmp | 0xffffff00;             // manual sign-extension if negative
  mv->ymov = (int) tmp;                 // data conversion
  /* success */
  return(1);
}
```

Listing 16.17 chu_main_ps2_test.c

```c
/****************************************************************************
*
* Module:  PS2, keyboard, and mouse test
* File:    chu_main_ps2_test.c
* Purpose: Test PS2, keyboard and mouse driver routines
* IP core base addresses:
*    - SWITCH_BASE: slide switch
*    - BTN_BASE: pushbutton
*    - LED_BASE: discrete LEDs
*    - SSEG_BASE: 7-segment LED
*    - PS2_BASE: PS2 controller
*
****************************************************************************/
/* file inclusion */
#include <stdio.h>
#include <unistd.h>
#include "system.h"
#include "chu_avalon_gpio.h"
#include "chu_avalon_ps2.h"


/****************************************************************************
* function: mouse_led()
* purpose:  use mouse x-axis movement to control 8 leds
* argument:
*   led_base: base address of discrete LEDs
*   mv: pointer to mouse movement data
* return:
* note:
****************************************************************************/
void mouse_led(alt_u32 led_base, mouse_mv_type *mv){
  static int count=0;
  int pos;
  alt_u32 led_ptn;

  if(mv->lbtn)
    count = 0;
  else if (mv->rbtn)
    count = 255;
  else{
    count = count + mv->xmov;
    if (count>255)
      count = 255;
    if (count<0)
      count = 0;
  }
  pos = (count >> 5);              // get 3 MSBs
  led_ptn = (0x00000080) >> pos;  // 0b10000000
  pio_write(led_base, led_ptn);
}
```

```
/******************************************************************
 * function: main()
 * purpose:  test PS2, keyboard, and mouse operations
 * note:
 *   - keyboard/mouse operation assumes that the fifo is empty initially;
 *     perform "display ps2 stream" first to flush the fifo buffer
 ******************************************************************/
int main(void){
  alt_u8 cmd, packet;
  int sw, btn, id;
  mouse_mv_type mv;
  char ch;
  alt_u8 ps2_msg[4]={0xff,0x0c,sseg_conv_hex(5),sseg_conv_hex(2)};

  sseg_disp_ptn(SSEG_BASE, ps2_msg);               // display " PS2"
  printf("PS2 test: \n");
  btn_clear(BTN_BASE);
  while (1){
    while (!btn_is_pressed(BTN_BASE)){ };           // wait for button
    btn=btn_read(BTN_BASE);                         // read button
    if (btn & 0x02){                                // key1 pressed
      sw=pio_read(SWITCH_BASE);                     // read switch
      printf("key/sw: %d/%d\n", btn, sw);
    }
    btn_clear(BTN_BASE);
    switch (sw){
      case 0:   // reset
        id=ps2_reset_device(PS2_BASE);
        printf("PS2 device type: %d (0/1/2: unknown/keyboard/mouse)\n", id);
        break;
      case 1:   // initialize mouse to stream mode
        id=mouse_init(PS2_BASE);
        printf("Mouse initialization status: %d (0/1: fail/succeed)\n", id);
        break;
      case 2:   // issue a ps2 command
        printf("Enter ps2 command in 2-digit hex format:\n");
        scanf("%x", &cmd);
        ps2_wr_cmd(PS2_BASE, cmd);
        printf("PS2 command 0x%02x issued.\n", cmd);
        usleep(500000);       // wait
        printf("PS2 response: ");
        while (ps2_get_pkt(PS2_BASE, &packet)){  // get all packets
          printf("0x%02x  ", packet);
          ps2_rm_pkt(PS2_BASE);
        }
        printf("\n");
        break;
      case 3:   // display ps2 stream
        printf("PS2 packet stream:\n");
        while (!btn_is_pressed(BTN_BASE)){
          if (ps2_get_pkt(PS2_BASE, &packet)){   // get one packet
            printf("0x%02x ", packet);
            ps2_rm_pkt(PS2_BASE);
          }
        }
        printf("\n");
        break;
      case 4:   // display decoded keyboard input stream
        printf("Keyboard char stream: \n");
        while (!btn_is_pressed(BTN_BASE)){
          if (kb_get_ch(PS2_BASE, &ch))
```

```
                    printf("%c", ch);
          }
          printf("\n");
          break;
      case 5:     // display decoded mouse data stream
          printf("Mouse data stream: (left button, right button, "
                 "x-axis move, y-axis move)\n");
          while (!btn_is_pressed(BTN_BASE)){
            if (mouse_get_activity(PS2_BASE, &mv)){
              printf("(%d,%d,%d,%d) ", mv.lbtn, mv.rbtn, mv.xmov, mv.ymov);
              mouse_led(LED_BASE, &mv);
            }
          }
          printf("\n");
          break;
    }  //end switch
  } // end while
}
```

CHAPTER 17

VGA CONTROLLER

VGA (video graphics array) is a video display standard. A VGA controller generates the synchronization signals and outputs data pixels serially. In this chapter, we discuss the development of a 640-by-480 resolution VGA controller and software driver. The hardware portion includes a VGA synchronization circuit, an SRAM based dual-port video memory controller, and an Avalon interface wrapping circuit. The software portion includes the driver routines to read and write the video memory, to plot simple lines, and to display and erase texts and bitmap images as well as basic routines to retrieve an image from a file in BMP format.

17.1 INTRODUCTION

VGA is a video display standard introduced in the late 1980s in IBM PCs and is widely supported by PC graphics hardware and monitors. While it was designed for CRT (cathode ray tube) monitors, most modern LCD (liquid crystal display) monitors usually include a VGA port. Thus, the circuit developed in this chapter can be used in conjunction with LCD monitors as well.

17.1.1 Basic operation of a CRT

The conceptual sketch of a monochrome CRT monitor is shown in Figure 17.1. The electron gun (cathode) generates a focused electron beam, which traverses a vacuum tube and eventually hits the phosphorescent screen. Light is emitted at

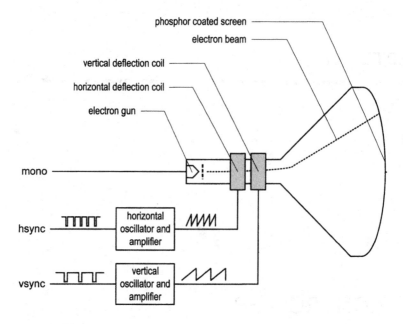

Figure 17.1   Conceptual diagram of a CRT monitor.

the instant that electrons hit a phosphor dot on the screen. The intensity of the electron beam and the brightness of the dot are determined by the voltage level of the external video input signal, labeled mono in Figure 17.1. The mono signal is an analog signal whose voltage level is between 0 and 0.7 V.

A vertical deflection coil and a horizontal deflection coil outside the tube produce magnetic fields to control how the electron beam travels and to determine where on the screen the electrons hit. In today's monitors, the electron beam traverses (i.e., scans) the screen systematically in a fixed pattern, from left to right and from top to bottom, as shown in Figure 17.2.

The monitor's internal oscillators and amplifiers generate sawtooth waveforms to control the two deflection coils. For example, the electron beam moves from the left edge to the right edge as the voltage applied to the horizontal deflection coil gradually increases. After reaching the right edge, the beam returns rapidly to the left edge (i.e., *retraces*) when the voltage changes to 0. The relationship between the sawtooth waveform and the scan is shown in Figure 17.4. Two external synchronization signals, hsync and vsync, control generation of the sawtooth waveforms. These signals are digital signals. The relationship between the hsync signal and the horizontal sawtooth is also shown in Figure 17.4. Note that the "1" and "0" periods of the hsync signal correspond to the rising and falling ramps of the sawtooth waveform.

The basic operation of a color CRT is similar except that it has three electron beams, which are projected to the red, green, and blue phosphor dots on the screen. The three dots are combined to form a pixel. We can adjust the voltage levels of the three video input signals to obtain the desired pixel color.

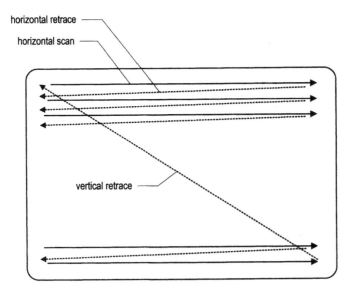

Figure 17.2   CRT scanning pattern.

Table 17.1   Three-bit VGA color combinations

Red	Green	Blue	Resulting color
0000	0000	0000	black
0000	0000	1111	blue
0000	1111	0000	green
0000	1111	1111	cyan
1111	0000	0000	red
1111	0000	1111	magenta
1111	1111	0000	yellow
1111	1111	1111	white

17.1.2   VGA port of the DE1 board

The VGA port has five active signals, including the horizontal and vertical synchronization signals, hsync and vsync, and three video signals for the red, green, and blue beams. It is physically connected to a 15-pin D-subminiature connector. A video signal is an analog signal and the video controller uses a digital-to-analog converter to convert the digital output to the desired analog level. If a video signal is represented by an N-bit word, it can be converted to 2^N analog levels. The three video signals can generate 2^{3N} different colors. This is also known as *3N-bit color* or *color depth of 3N* since a pixel is defined by $3N$ bits. In the DE1 board, a 4-bit word is used for each beam, and thus the board can support 12-bit color, which leads to 4096 (i.e., 2^{12}) possible colors. The primary color combinations are shown in Table 17.1.

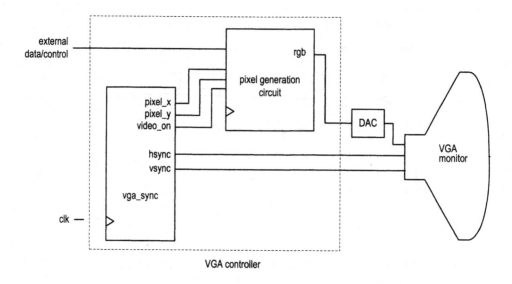

Figure 17.3 Simplified block diagram of a VGA controller.

17.1.3 Video controller

A video controller generates the synchronization signals and outputs data pixels serially. A simplified block diagram of a VGA controller is shown in Figure 17.3. It contains a synchronization circuit, labeled **vga_sync**, and a pixel generation circuit.

Synchronization circuit   The **vga_sync** circuit generates the timing and synchronization signals. The **hsync** and **vsync** signals are connected to the VGA port to control the horizontal and vertical scans of the monitor. The two signals are decoded from the internal counters, whose outputs are the **pixel_x** and **pixel_y** signals. The **pixel_x** and **pixel_y** signals indicate the relative positions of the scans and essentially specify the location of the current pixel. The **vga_sync** circuit also generates the **video_on** signal to indicate whether to enable or disable the display. The design of this circuit is discussed in Section 17.2.

Pixel generation circuit   The pixel generation circuit generates the three video signals, which are collectively referred to as the **rgb** signal. A color value is obtained according to the current coordinates of the pixel (the **pixel_x** and **pixel_y** signals) and the external control and data signals. In the Nios II framework, these signals are connected to the Avalon interconnect and the video controller is treated as an I/O device by the processor.

Generating 12-bit color demands a large amount of hardware resources and these many colors may not be necessary for many applications. The pixel generation circuit can use fewer bits internally and then map the signal to the desired 12-bit pattern. For example, we can use three bits internally and map the eight possible combinations to the patterns in Table 17.1. This leads to a 3-bit color video. Similarly, we can use one bit internally and map the two possible values to 000000000000 and 111111111111. This make the monitor function as a black-and-white monochrome monitor.

For our discussion purposes, we divided this circuit into three broad categories:

- Bit-mapped scheme
- Tile-mapped scheme
- Object-mapped scheme

In a *bit-mapped scheme*, a *video memory* is used to store the data to be displayed on the screen. Each pixel of the screen is mapped directly to a memory word, and the `pixel_x` and `pixel_y` signals form the address. A graphics processing circuit continuously updates the screen and writes relevant data to the video memory. A retrieval circuit continuously reads the video memory and routes the data to the `rgb` signal. This is the scheme used in today's high-performance video controllers. For 640-by-480 resolution, there are about 310 k (i.e., 640*480) pixels on a screen. This translates to 38.4 kB (i.e., 640*480*1/8) memory for a monochrome display, 310 kB memory (i.e., 640*480*8/8) for a 8-bit color display, and 461 kB (i.e., 640*480*12/8) memory for a 12-bit color display.

To reduce the memory requirement, one alternative is to use a *tile-mapped scheme*. In this scheme, we group a collection of bits to form a *tile* and treat each tile as a display unit. For example, we can define an 8-by-8 square of pixels (i.e., 64 pixels) as a tile. The 640-by-480 pixel-oriented screen becomes an 80-by-60 tile-oriented screen. Only 4800 (i.e., 80*60) words are needed for the *tile memory*. The number of bits in a word depends on the number of tile patterns. For example, if there are 256 tile patterns, each word should contain 8 bits, and the size of the tile memory is about 4.8 kB (i.e., 4800*8/8). The tile-mapped scheme usually requires a ROM to store the tile patterns. We call it *pattern memory*. Assume that 8-bit color patterns are used in the previous example. Each 8-by-8 tile pattern requires 64 bytes, and the entire 256 patterns need 16 KB. The overall memory requirement is about 21 KB, which is much smaller than the 310 kB of the bit-mapped scheme.

For some applications, the video display can be very simple and contains only a few objects. Instead of wasting memory to store a mostly blank screen, we can generate these objects using simple object generation circuits. We call this approach an *object-mapped scheme*.

The three schemes can be mixed together to generate a full screen. For example, we can use a bit-mapped scheme to generate the background and use an object-mapped scheme to produce the main objects. We can also use a bit-mapped scheme for one portion of a screen and tile-mapped text for another portion of the screen.

We discuss a graphic display based on the bit-mapped scheme in this chapter. References for two other schemes can be found in the bibliographic section.

17.2 VGA SYNCHRONIZATION

The video synchronization circuit generates the `hsync` signal, which specifies the required time to traverse (scan) a row, and the `vsync` signal, which specifies the required time to traverse (scan) the entire screen. Subsequent discussions are based on a 640-by-480 VGA screen with a 25-MHz *pixel rate*, which means that 25M pixels are processed in a second. Note that this resolution is also known as the *VGA mode*.

The screen of a CRT monitor usually includes a small black border, as shown at the top of Figure 17.4. The middle rectangle is the visible portion. Note that the coordinate of the vertical axis increases downward. The coordinates of the top-left and bottom-right corners are (0,0) and (639,479), respectively.

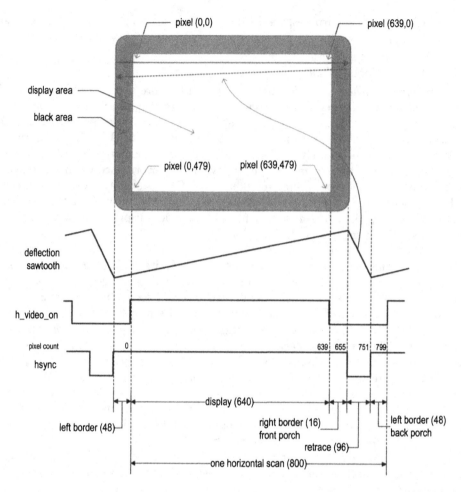

Figure 17.4 Timing diagram of a horizontal scan.

17.2.1 Horizontal synchronization

A detailed timing diagram of one horizontal scan is shown in Figure 17.4. A period of the `hsync` signal contains 800 pixels and can be divided into four regions:

- *Display*: region where the pixels are actually displayed on the screen. The length of this region is 640 pixels.
- *Retrace*: region in which the electron beams return to the left edge. The video signal should be disabled (i.e., black), and the length of this region is 96 pixels.
- *Right border*: region that forms the right border of the display region. It is also known as the *front porch* (i.e., porch before retrace). The video signal should be disabled, and the length of this region is 16 pixels.
- *Left border*: region that forms the left border of the display region. It is also known as the *back porch* (i.e., porch after retrace). The video signal should be disabled, and the length of this region is 48 pixels.

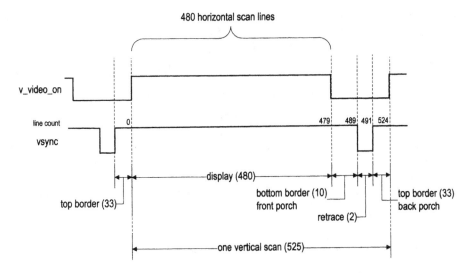

Figure 17.5 Timing diagram of a vertical scan.

Note that the lengths of the right and left borders may vary for different brands of monitors.

The **hsync** signal can be obtained by a special mod-800 counter and a decoding circuit. The counts are marked on the top of the **hsync** signal in Figure 17.4. We intentionally start the counting from the beginning of the display region. This allows us to use the counter output as the horizontal (x-axis) coordinate. This output constitutes the **pixel_x** signal. The **hsync** signal goes low when the counter's output is between 656 and 751.

Note that the CRT monitor should be black in the right and left borders and during retrace. We use the **h_video_on** signal to indicate whether the current horizontal coordinate is in the displayable region. It is asserted only when the pixel count is smaller than 640.

17.2.2 Vertical synchronization

During the vertical scan, the electron beams move gradually from top to bottom and then return to the top. This corresponds to the time required to refresh the entire screen. The format of the **vsync** signal is similar to that of the **hsync** signal, as shown in Figure 17.5. The time unit of the movement is represented in terms of horizontal scan lines. A period of the **vsync** signal is 525 lines and can be divided into four regions:

- *Display*: region where the horizontal lines are actually displayed on the screen. The length of this region is 480 lines.
- *Retrace*: region that the electron beams return to the top of the screen. The video signal should be disabled, and the length of this region is 2 lines.
- *Bottom border*: region that forms the bottom border of the display region. It is also known as the *front porch* (i.e., porch before retrace). The video signal should be disabled, and the length of this region is 10 lines.

- *Top border*: region that forms the top border of the display region. It is also known as the *back porch* (i.e., porch after retrace). The video signal should be disabled, and the length of this region is 33 lines.

As in the horizontal scan, the lengths of the top and bottom borders may vary for different brands of monitors.

The vsync signal can be obtained by a special mod-525 counter and a decoding circuit. Again, we intentionally start counting from the beginning of the display region. This allows us to use the counter output as the vertical (y-axis) coordinate. This output constitutes the pixel_y signal. The vsync signal goes low when the line count is 490 or 491.

As in the horizontal scan, we use the v_video_on signal to indicate whether the current vertical coordinate is in the displayable region. It is asserted only when the line count is smaller than 480.

17.2.3   Timing calculation of VGA synchronization signals

As mentioned earlier, we assume that the pixel rate is 25 MHz. It is determined by three parameters:

- p: the number of pixels in a horizontal scan line. For 640-by-480 resolution, it is

$$p = 800 \, \frac{pixels}{line}$$

- l: the number of lines in a screen (i.e., a vertical scan). For 640-by-480 resolution, it is

$$l = 525 \, \frac{lines}{screen}$$

- s: the number of screens per second. For flicker-free operation, we can set it to

$$s = 60 \, \frac{screens}{second}$$

The s parameter specifies how fast the screen should be refreshed. For a human eye, the refresh rate must be at least 30 screens per second to make the motion appear to be continuous. To reduce flickering, the monitor usually has a much higher rate, such as the 60 screens per second specification above. The pixel rate can be calculated by the three parameters:

$$\text{pixel rate} = p * l * s \approx 25M \, \frac{pixels}{second}$$

The pixel rate for other resolutions and refresh rates can be calculated in a similar fashion. Clearly, the rate increases as the resolution and refresh rate grow.

17.2.4   HDL implementation

The function of the vga_sync circuit is discussed in Section 17.1.3. Ideally the clock rate of a synchronization circuit should be the same as the pixel rate, which is 25 MHz for a VGA monitor with 640-by-480 resolution. If this is the case, the synchronization circuit can be implemented by two special counters: a mod-800 counter to keep track of the horizontal scan and a mod-525 counter to keep track of the vertical scan.

If the system clock rate and pixel rate are different, we usually need to create a separate clock domain for the video system. This can be done by using the PLL circuit discussed in Section 15.2.2. Since our designs generally use the 50-MHz oscillator of the prototyping board, the system clock rate is twice the pixel rate. Instead of creating a separate 25-MHz clock domain, which complicates the timing, we can generate a 25-MHz enable tick to enable or pause the counting. The tick is also routed to the p_tick port as an output signal to coordinate operation of the pixel generation circuit.

The HDL code is shown in Listing 17.1. It consists of a mod-2 counter to generate the 25-MHz enable tick and two counters for the horizontal and vertical scans. We use two status signals, h_end and v_end, to indicate completion of the horizontal and vertical scans. The values of various regions of the horizontal and vertical scans are defined as constants. They can be easily modified if a different resolution or refresh rate is used. Since the video synchronization signals may be buffered with additional registers when it is incorporated into the main system, we add suffix _i (for internal) in these signals.

Listing 17.1   VGA synchronization circuit

```
library ieee;
use ieee.std_logic_1164.all;
use ieee.numeric_std.all;
entity vga_sync is
5    port(
        clk, reset: in std_logic;
        hsync_i, vsync_i: out std_logic;
        video_on_i: out std_logic;
        p_tick: out std_logic;
10        pixel_x, pixel_y: out std_logic_vector (9 downto 0)
    );
end vga_sync;

architecture arch of vga_sync is
15    -- VGA 640-by-480 sync parameters
    constant HD: integer:=640; -- horizontal display area
    constant HF: integer:=16 ; -- h. front porch
    constant HB: integer:=48 ; -- h. back porch
    constant HR: integer:=96 ; -- h. retrace
20    constant VD: integer:=480; -- vertical display area
    constant VF: integer:=10; -- v. front porch
    constant VB: integer:=33; -- v. back porch
    constant VR: integer:=2; -- v. retrace
    -- mod-2 counter
25    signal mod2_reg, mod2_next: std_logic;
    -- sync counters
    signal v_count_reg, v_count_next: unsigned(9 downto 0);
    signal h_count_reg, h_count_next: unsigned(9 downto 0);
    -- status signal
30    signal h_end, v_end, pixel_tick: std_logic;
begin
    -- registers
    process (clk,reset)
    begin
35        if reset='1' then
            mod2_reg <= '0';
            v_count_reg <= (others=>'0');
            h_count_reg <= (others=>'0');
        elsif (clk'event and clk='1') then
```

```
40          mod2_reg <= mod2_next;
            v_count_reg <= v_count_next;
            h_count_reg <= h_count_next;
        end if;
    end process;
45  -- mod-2 circuit to generate 25 MHz enable tick
    mod2_next <= not mod2_reg;
    -- 25 MHz pixel tick
    pixel_tick <= '1' when mod2_reg='1' else '0';
    -- end of horizontal/vertical counter status
50  h_end <= '1' when h_count_reg=(HD+HF+HB+HR-1) else '0'; -- 799
    v_end <= '1' when v_count_reg=(VD+VF+VB+VR-1) else '0'; -- 524
    -- next-state logic of mod-800 horizontal sync counter
    process (h_count_reg,h_end,pixel_tick)
    begin
55      if pixel_tick='1' then  -- 25 MHz tick
            if h_end='1' then
                h_count_next <= (others=>'0');
            else
                h_count_next <= h_count_reg + 1;
60          end if;
        else
            h_count_next <= h_count_reg;
        end if;
    end process;
65  -- next-state logic of mod-525 vertical sync counter
    process (v_count_reg,h_end,v_end,pixel_tick)
    begin
        if pixel_tick='1' and h_end='1' then
            if (v_end='1') then
70              v_count_next <= (others=>'0');
            else
                v_count_next <= v_count_reg + 1;
            end if;
        else
75          v_count_next <= v_count_reg;
        end if;
    end process;
    -- horizontal sync
    hsync_i <=
80      '0' when h_count_reg>=(HD+HF) and h_count_reg<=(HD+HF+HR-1) else
        '1';
    -- vertical sync
    vsync_i <=
        '0' when v_count_reg>=(VD+VF) and v_count_reg<=(VD+VF+VR-1) else
85      '1';
    -- video on/off
    video_on_i <=
        '1' when (h_count_reg<HD) and (v_count_reg<VD) else
        '0';
90  pixel_x <= std_logic_vector(h_count_reg);
    pixel_y <= std_logic_vector(v_count_reg);
    p_tick <= pixel_tick;
    end arch;
```

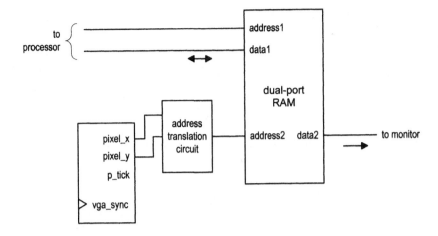

Figure 17.6   Conceptual diagram of video RAM.

17.3   SRAM-BASED VIDEO RAM CONTROLLER

17.3.1   Overview of video memory

In the bit-mapped scheme, each pixel of the screen is mapped directly to a memory word and the pixel_x and pixel_y signals form the address. The system consists of a video memory and the conceptual diagram is shown in Figure 17.6. The video memory has two ports. The "VGA port," labeled address2 and data2, is for the VGA operation. It is read continuously. The address is derived from the pixel_x and pixel_y signals, which specify the current coordinate in the screen, and the retrieved data is the color information of the corresponding pixel. The "CPU port," labeled address1 and data1, is connected to a processor. The processor writes the pixel information to the memory and updates the displayed graphic. It may need to perform a read operation occasionally when an overlay operation is involved.

The actual implementation of video memory depends on the type of physical memory devices used in the system. We can either use real dual-port memory chips or use single-port memory chips and a multiplexing circuit to mimic the dual-port access. Because of the cost and availability of dual-port memory devices, most systems utilize the latter approach.

In a single-port implementation, the VGA and CPU ports access the same address and data lines of the memory chip. Additional multiplexing and routing circuits are needed to coordinate the operation. To avoid glitches and noises on the graphic display, the VGA port usually has the priority and the processor can perform a write operation when the video display is off (i.e., when scanning is done in the black border area and retrace). An alternative scheme is the *double-buffering* scheme, in which two memory banks are used, one for the VGA port to retrieve data and one for the processor to write data. The two banks operate concurrently and switch their roles when a bank is filled with new data. This scheme essentially doubles the memory bandwidth and acts like a true dual-port memory. Other ad hoc methods can also be used to mimic the dual-port operation.

Another issue involved in the physical implementation is the mismatch between the color depth and the memory's data width. For example, we may need to use a physical memory with 8-bit data width to implement a 3-bit color video controller. Additional routing circuits and buffers are needed to pack and unpack the pixel data for memory access.

17.3.2   Memory consideration of DE1 board

The design of video memory depends on the type and characteristics of physical memory devices available. Recall that 39 kB and 310 kB are needed to support 1-bit and 8-bit colors in 640-by-480 resolution. On the DE1 board, three types of memory devices available: 26 KB FPGA on-chip memory, an external 512 KB SRAM device, and an external 8 MB SDRAM device. The on-chip memory is easy to use and supports dual-port operation but it is too small for this purpose. The SDRAM has adequate capacity. However, it is used to store executable code in our applications and its timing is more involved. Thus, the SRAM is the most reasonable choice. For clarity, we assume that 8-bit color is used in the system. Other color depths can be obtained by modifying the 8-bit color scheme.

17.3.3   Ad hoc SRAM controller

The SRAM chip on the DE1 board is a single-port memory and additional circuitry is needed to coordinate the dual-port access. One possibility is to follow the scheme discussed in Section 17.3.1. However, based on the timing characteristics of the SRAM and VGA, we can derive a more effective ad hoc design. The SRAM chip on the DE1 board is a fast device and the data sheet shows that both the read cycle and write cycle are 10 ns. In comparison, the system clock used on the DE1 board is 50 MHz (i.e., 20-ns period) and the pixel generation rate of the video is only 25 MHz (i.e., 40-ns period), which constitutes two system clock cycles. It is possible to derive a "time-multiplexing" scheme to access the SRAM and to emulate the dual-port operation. In this scheme, we can use one system cycle of the 40-ns pixel generation period for the VGA port access and the other system cycle for the CPU port access.

The p_tick signal in the synchronization circuit of Listing 17.2.4 is the output of the mod-2 circuit and can be used for the multiplexing purpose. For clarity, we rename it vga_cycle. The conceptual timing diagram is shown in the top three signals of Figure 17.9(a). The SRAM access is allocated to the VGA port when vga_cycle is 1 and to the CPU port when it is 0. The conceptual top-level diagram is shown in Figure 17.7. Note that the SRAM is treated as a 512K-by-8 device. For simplicity, we assume that 8-bit color is used in the system.

VGA port design and operation   During the VGA cycle (i.e., when vga_cycle is 1), the memory controller performs a read operation and uses the address obtained from the current pixel coordinate to retrieve the color bits.

The signal paths are shown as thick gray lines in Figure 17.7 and the timing diagram is shown in Figure 17.9(a). The vga_cycle signal controls the multiplexer and routes the corresponding pixel address to the SRAM address bus, sram_addr. The vga_cycle signal also enables the vga_rd_data register. At the rising edge of the next clock cycle, the retrieved color data is stored into the register and the

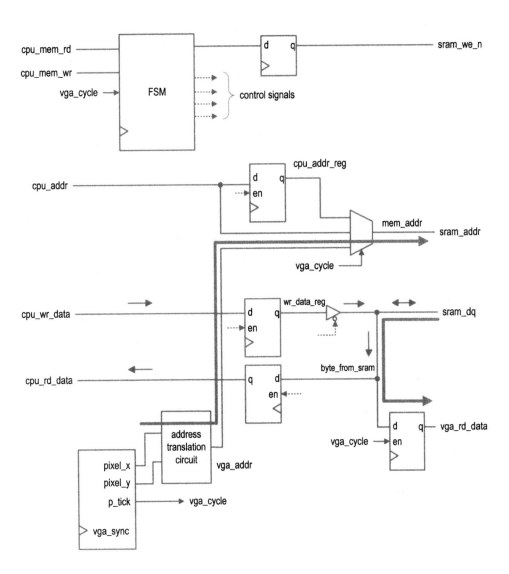

Figure 17.7   Conceptual diagram of SRAM-based video RAM controller.

register's output, vga_rd_data, is routed to the VGA display. The color data in the register remains stable for two clock cycles (i.e., 40 ns) and is not effected by the operation of the CPU cycle (i.e., when vga_cycle is 0).

The VGA's pixel location is represented by a two-dimensional coordinate, which is specified by the pixel_x and pixel_y signals of the video synchronization circuit. Since SRAM is constructed as a one-dimensional array, the two-dimensional coordinate must be converted to a one-dimensional memory address. Let the coordinate be (x, y). For the 640-by-480 resolution, one way to obtain the one-dimensional offset is

$$offset = 640 * y + x$$

Despite its simplicity, the $*$ operator implies an expensive hardware multiplier. Closer examination shows that

$$offset = 640 * y + x = 512 * y + 128 * y + x = y \ll 9 + y \ll 7 + x$$

where $\ll$ is the shift left operator (recall that in the binary system $y \ll n$ corresponds to $y * 2^n$). This indicates that the multiplier can be replaced by an adder and two fixed-amount shifters. This implementation is much more efficient since shifting a fixed amount just involves reconnection of the input and output wires. The address translation circuit in Figure 17.7 performs this task.

CPU port design and operation During the CPU cycle (i.e., when vga_cycle is 0), the memory controller performs a transaction between the processor and the SRAM device. The processor mainly writes the memory to update the display graphics but occasionally needs to read the memory for overlay operation (the overlay operation is explained in Section 17.6.3).

During the CPU cycle, the SRAM is treated as a normal memory module. The design of this type of controller is discussed in Section 15.3. The ad hoc video memory controller performs similar functions except that these operations must be done during the CPU cycle. Since the processor is not aware of the current phase of operation, it can initiate a read or write operation either in a VGA cycle or CPU cycle. The memory controller must take this into consideration and accommodate both scenarios.

The conceptual block diagram is shown in Figure 17.7, which contains several registers and an FSM. The FSM controls and coordinates the overall operation and its ASMD chart is shown in Figure 17.8. In the idle state, the FSM checks the read and write commands, cpu_mem_rd and cpu_mem_wr, and determines the type of operation. The cpu_mem_rd and cpu_mem_wr signals are later mapped to the read and write signals of the Avalon MM interface.

The assertion of cpu_mem_rd initiates the read operation. It can occur in either a VGA cycle or CPU cycle. If the current cycle is a VGA cycle, the FSM moves to the rd state and samples and stores data to a register in that state. The registered data becomes available in the fetch state and is sampled and retrieved by the Avalon MM master. The timing diagram is shown in the first part of Figure 17.9(b). At t_0, FSM detects the assertion of cpu_mem_rd and moves to the rd state. At t_1, it samples and stores the data to the cpu_rd_data_reg register. The register data becomes available after a small delay. At t_2, the Avalon MM master reads the data and deactivates the cpu_mem_rd signal. Since the data is available after three clock cycles, the cpu_mem_rd signal must remain active in this interval. When creating

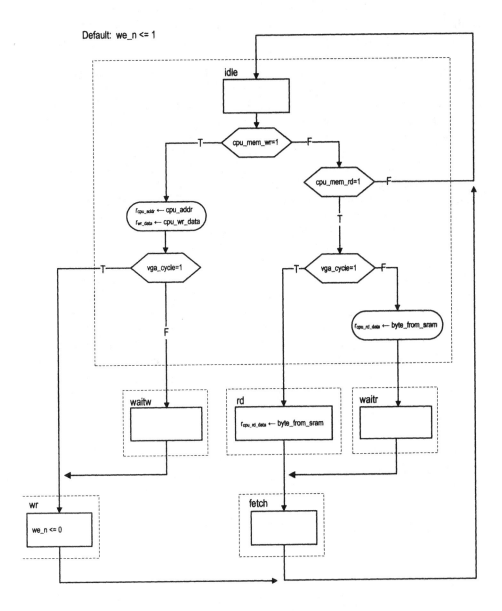

Figure 17.8   ASMD of ad hoc video SRAM controller.

the core in SOPC Component Editor, we must add a proper waiting time. This can be achieved by setting the Read Wait field to 2.

If the assertion occurs in the VGA cycle, the FSM can simply read the data from the SRAM data bus and store it to a register. To accommodate the previous Avalon MM interface read timing constraint, we add a wait state, `waitr`, to increase the reading operation to three clock cycles. The timing diagram is shown in the second part of Figure 17.9(b). Note that the Avalon MM master retrieves the data at t_3.

The assertion of `cpu_mem_wr` initiates the write operation. The timing requirement for write is more stringent. As discussed in Section 15.3, the data and write enable signals must be stable for a specific interval and the enable signal must be deasserted properly to latch the data. To achieve this, we use a dedicated state, `wr`, in FSM for this purpose. When `cpu_mem_wr` is asserted in the `idle` state, the FSM stores the data and address to registers and then examines the type of current cycle. If it is a VGA cycle, the FSM moves to the `wr` state, in which the write enable signal is asserted, which turns on the tristate buffer to route the data to SRAM's data bus and enables the SRAM's write enable signal, `we_n`. The timing diagram is shown in the first part of Figure 17.9(c). At t_0, the FSM detects the assertion of `cpu_mem_wr` and moves to the `wr` state. In this state, the registered data and address are placed on the SRAM bus and `we_n` is asserted. At t_1, the FSM returns to the `idle` state. After a short delay, `we_n` is deasserted and the data is latched to SRAM at t_2.

If the assertion occurs at the VGA cycle, the FSM does not perform any write operation so that potential glitches and timing complication can be avoided. It moves to the `waitw` state to wait for one clock cycle (i.e., the VGA cycle) and then moves to the `wr` state to write the SRAM. The write operation may take up to three clock cycles. However, since the data and address are stored to the registers when the write operation is initiated in the first clock cycle, there is no need to insert additional wait time in the Avalon MM interface. The write can be performed properly as long as the processor does not issue another memory access within the next three clock cycles.

17.3.4 HDL code

The HDL code follows the block diagram and the ASMD chart and is shown in Listing 17.2.

Listing 17.2   SRAM-based video memory controller

```
  library ieee;
2 use ieee.std_logic_1164.all;
  use ieee.numeric_std.all;
  entity vram_ctrl is
     port (
        clk, reset: in std_logic;
7       -- from video sync
        pixel_x, pixel_y: in std_logic_vector(9 downto 0);
        p_tick: in std_logic;
        -- memory interface to vga read
        vga_rd_data: out std_logic_vector(7 downto 0);
12      -- memory interface to cpu
        cpu_rd_data: out std_logic_vector(7 downto 0);
        cpu_wr_data: in std_logic_vector(7 downto 0);
        cpu_addr: in std_logic_vector(18 downto 0);
```

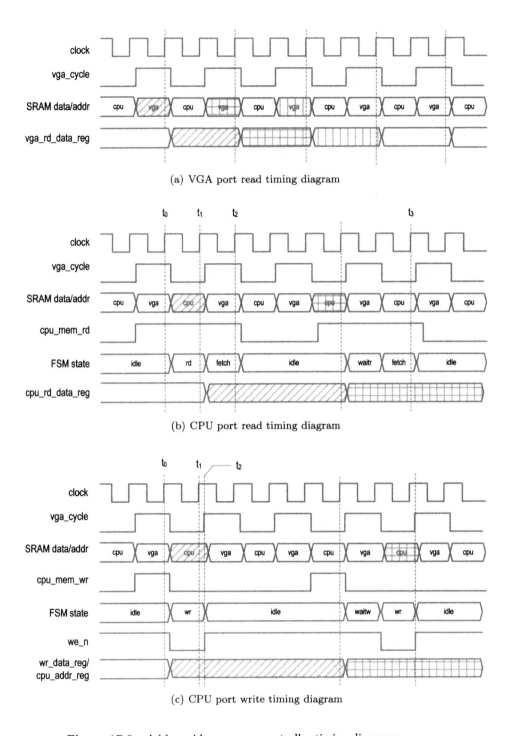

(a) VGA port read timing diagram

(b) CPU port read timing diagram

(c) CPU port write timing diagram

Figure 17.9   Ad hoc video memory controller timing diagrams.

```vhdl
            cpu_mem_wr: in std_logic;
17          cpu_mem_rd: in std_logic;
            -- to/from SRAM chip
            sram_addr: out std_logic_vector(17 downto 0);
            sram_dq: inout std_logic_vector(15 downto 0);
            sram_we_n, sram_oe_n: out std_logic;
22          sram_ce_n, sram_ub_n, sram_lb_n: out std_logic
        );
     end vram_ctrl;

   architecture arch of vram_ctrl is
27     type wr_state_type is (idle, waitr, rd, fetch, waitw, wr);
       signal state_reg, state_next: wr_state_type;
       signal vga_rd_data_reg: std_logic_vector(7 downto 0);
       signal cpu_addr_reg, cpu_addr_next: std_logic_vector(18 downto 0);
       signal wr_data_reg, wr_data_next: std_logic_vector(7 downto 0);
32     signal cpu_rd_data_reg, cpu_rd_data_next:
              std_logic_vector(7 downto 0);
       signal we_n_reg, we_n_next: std_logic;
       signal y_offset: unsigned(18 downto 0);
       signal vga_addr, mem_addr: std_logic_vector(18 downto 0);
37     signal byte_from_sram: std_logic_vector(7 downto 0);
       signal vga_cycle: std_logic;
     begin
       -- p_tick asserted every 2 clock cycles;
       vga_cycle <= p_tick;
42     --===============================================
       -- VGA port SRAM read operation
       --===============================================
       -- VGA port reads SRAM continuously
       -- read registers
47     process (clk)
       begin
          if (clk'event and clk='1') then
             if (vga_cycle='1') then
                vga_rd_data_reg <= byte_from_sram;
52           end if;
          end if;
       end process;
       -- VGA port address offset = 640*y + x = 512*y + 128*y + x
       y_offset <= unsigned('0'& pixel_y(8 downto 0) & "000000000") +
57                 unsigned("000" & pixel_y(8 downto 0) & "0000000");
       vga_addr <= std_logic_vector(y_offset + unsigned(pixel_x));
       vga_rd_data <= vga_rd_data_reg;
       --===============================================
62     -- CPU port SRAM read/write operation
       --===============================================
       -- FSMD state & data registers
       process(clk,reset)
       begin
          if (reset='1') then
67           state_reg <= idle;
             cpu_addr_reg <= (others=>'0');
             wr_data_reg <= (others=>'0');
             cpu_rd_data_reg <= (others=>'0');
             we_n_reg <= '1';
72        elsif (clk'event and clk='1') then
             state_reg <= state_next;
             cpu_addr_reg <= cpu_addr_next;
             wr_data_reg <= wr_data_next;
             cpu_rd_data_reg <= cpu_rd_data_next;
```

```
77          we_n_reg <= we_n_next;
         end if;
      end process;
      cpu_rd_data <= cpu_rd_data_reg;
      -- next-state logic
82    process(state_reg,cpu_addr_reg,wr_data_reg,byte_from_sram,
               cpu_rd_data_reg,cpu_mem_wr,cpu_mem_rd,vga_cycle,
               cpu_addr,cpu_wr_data)
      begin
         state_next <= state_reg;
87       cpu_addr_next <= cpu_addr_reg;
         wr_data_next <= wr_data_reg;
         cpu_rd_data_next <= cpu_rd_data_reg;
         case state_reg is
            when idle =>
92             if cpu_mem_wr='1' then
                  cpu_addr_next <= cpu_addr;
                  wr_data_next <= cpu_wr_data;
                  if vga_cycle = '1' then
                     state_next <= wr;
97                else
                     state_next <= waitw;
                  end if;
               elsif cpu_mem_rd='1' then
                  if vga_cycle = '1' then
102                  state_next <= rd;
                  else
                     cpu_rd_data_next <= byte_from_sram;
                     state_next <= waitr;
                  end if;
107            end if;
            when rd =>
               cpu_rd_data_next <= byte_from_sram;
               state_next <= fetch;
            when waitr =>
112            state_next <= fetch;
            when fetch =>
               state_next <= idle;
            when waitw =>
               state_next <= wr;
117         when wr =>
               state_next <= idle;
         end case;
      end process;
      -- look-ahead output
122   we_n_next <= '0' when state_next=wr else '1';

      ========================================================
      -- SRAM interface signals
      ========================================================

      -- configure SRAM as 512K-by-8
127   mem_addr <= vga_addr when vga_cycle='1' else
                  cpu_addr_reg when we_n_reg='0' else
                  cpu_addr;
      sram_addr <= mem_addr(18 downto 1);
      sram_lb_n <= '0' when mem_addr(0)='0' else '1';
132   sram_ub_n <= '0' when mem_addr(0)='1' else '1';
      sram_ce_n <= '0';
      sram_oe_n <= '0';
      sram_we_n <= we_n_reg;
      sram_dq <= wr_data_reg & wr_data_reg when we_n_reg='0' else
137               (others=>'Z');
```

```
-- LSB control lb ub
byte_from_sram <= sram_dq(15 downto 8) when mem_addr(0)='1' else
                  sram_dq(7 downto 0);
end arch;
```

In addition to the basic FSMD, there are several subtle design issues in the
code. First, we use the register address, cpu_addr_reg, for write operation but use
an unregistered address, cpu_addr, for read operation, as in

```
mem_addr <= vga_addr when vga_cycle='1' else
            cpu_addr_reg when we_n_reg='0' else
            cpu_addr;
```

This is due to the fact that cpu_mem_rd is asserted for three clocks and the address
in the Avalon MM interface is stable during this interval. On the other hand,
cpu_mem_wr is only asserted for one clock cycle and the address must be stored in
a register for later use.

Second, we use a special "lookahead buffer" for the we_n signal. The we_n signal
is asserted in the wr state and is usually coded as

```
we_n <= '0' when state_reg=wr else '1';
```

We can register we_n to make it fast and clean. To avoid the one-clock-cycle delay,
the decoding can be done by using the state's future value (i.e., state_next):

```
we_n_next <= '0' when state_next=wr else '1';
process(clk,reset)
begin
   ...
   if (clk'event and clk='1') then
      we_n_reg <= we_n_next;
      ...
   end if;
end process;
```

Finally, the SRAM on the FPGA board is a 256K-by-16 (i.e., 2^{18}-by-16) device.
However, it has two additional control signals, lb_n and ub_n, to enable the lower
byte and upper byte individually. We can configure it as a 512K-by-8 (i.e., 2^{19}-by-8)
device by using the LSB to control these signals:

```
sram_addr <= mem_addr(18 downto 1);
sram_lb_n <= '0' when mem_addr(0)='0' else '1';
sram_ub_n <= '0' when mem_addr(0)='1' else '1';
```

17.4 PALETTE CIRCUIT

Our design uses 8 bits to represent the color of a pixel, with 3 bits for the red
beam, 3 bits for the green beam, and 2 bits for the blue beam. Assume that the
data is $d_7 d_6 \cdots d_1 d_0$. The red, green, and blue colors are the $d_7 d_6 d_5$, $d_4 d_3 d_2$, and
$d_1 d_0$ fields, respectively. The DE1 board requires 4 bits for each color beam and
the three color beams constitute 12-bit color depth. We use a "palette circuit"
to convert the 8-bit color data to 12-bit color output. The simplest approach is
to treat the bits of the 8-bit color data as the MSBs of 12-bit color output and
duplicate the LSBs. The corresponding HDL code is shown in Listing 17.3.

Listing 17.3   Palette circuit

```
library ieee;
use ieee.std_logic_1164.all;
entity palette is
    port (
        color_in: in  std_logic_vector(7 downto 0);
        color_out: out std_logic_vector(11 downto 0)
    );
end palette;

architecture arch of palette is
begin
    color_out <=
        color_in(7 downto 5) & color_in(5) &
        color_in(4 downto 2) & color_in(2) &
        color_in(1 downto 0) & color_in(0) & color_in(0);
end arch;
```

The code converts 3-bit red to 4-bit red, 3-bit green to 4-bit green, and 2-bit blue to 4-bit blue. We use this scheme in the chapter.

A more sophisticated alternative is to use a color lookup table. In this scheme, the 8-bit input color data serves as the address of the table and each entry contains a 12-bit color. The 8-bit address leads to 256 (i.e., 2^8) entries in the table and thus up to 256 12-bit colors can be displayed. In other words, we can use 12-bit color in a graphic image but only 256 "simultaneous" colors from a total of 4096 (i.e., 2^{12}) possible colors can be used. The size of the lookup table is 256-by-12 (i.e., 3072) bits and can be accommodated by one internal M4K memory module of the Cyclone II device. Many image file formats contain an internal palette lookup table, which can be downloaded to the palette lookup table as needed.

17.5   VIDEO CONTROLLER IP CORE DEVELOPMENT

17.5.1   Complete video controller

The complete graphic video control system consists of the synchronization circuit, video memory controller, and palette circuit. The block diagram is shown in Figure 17.10. The vsync_i, hsync_i, and video_on_i signals are buffered to match the registered delay of memory access. In addition, the buffers can also remove potential glitches.

17.5.2   Avalon interfaces

We can add a wrapping circuit for the video controller and create an SOPC component. It includes an Avalon MM slave interface for the CPU port to interact with the host, a clock input interface for system clock, and a conduit interface for SRAM's I/O signals.

17.5.3   Register map

The VGA port of the video memory is transparent to the processor. From a Nios II processor's point of view, the video RAM is simply a single-port 512K-by-8 memory module and normal read and write is performed accordingly. To increase flexibility,

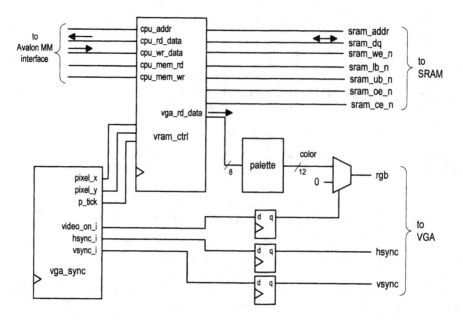

Figure 17.10   Block diagram video control system.

we increases the address space so that the values of horizontal and vertical counters of the VGA synchronization circuit can be retrieved (i.e., read). The address offsets and fields are:

- Read addresses (data to cpu)
  - offset 0x0000 to 0x7fff (normal memory)
    * bits 7 to 0: 8-bit data
  - offset 0x8000 (horizontal and vertical counters)
    * bits 19 to 10: value of vertical counter
    * bits 9 to 0: value of horizontal counter
- Write addresses (data from cpu)
  - offset 0x0000 to 0x7fff (normal memory)
    * bits 7 to 0: 8-bit data

17.5.4   Wrapped video controller

The wrapped video controller includes additional logic to decode address, align and multiplex data, and generate proper enable signals. The HDL code is shown in Listing 17.4. It basically follows the block diagram in Figure 17.10 and adds necessary segments for the wrapping circuit.

Listing 17.4   Wrapped video controller

```
library ieee;
use ieee.std_logic_1164.all;
use ieee.numeric_std.all;
entity chu_avalon_vga_graf is
    port (
        clk, reset: in  std_logic;
```

```vhdl
              -- to vga monitor
              hsync, vsync: out std_logic;
 9            rgb: out std_logic_vector(11 downto 0);
              -- to/from SRAM chip
              sram_addr: out std_logic_vector(17 downto 0);
              sram_dq: inout std_logic_vector(15 downto 0);
              sram_we_n, sram_oe_n: out std_logic;
14            sram_ce_n, sram_ub_n, sram_lb_n: out std_logic;
              -- avalon interface
              vga_address: in  std_logic_vector(19 downto 0);
              vga_chipselect: in  std_logic;
              vga_write: in std_logic;
19            vga_read: in std_logic;
              vga_writedata: in std_logic_vector(31 downto 0);
              vga_readdata: out std_logic_vector(31 downto 0)
     );
   end chu_avalon_vga_graf;
24
   architecture arch of chu_avalon_vga_graf is
       signal video_on_reg,video_on_i: std_logic;
       signal vsync_i, hsync_i: std_logic;
       signal pixel_x, pixel_y: std_logic_vector(9 downto 0);
29     signal p_tick: std_logic;
       signal cpu_rd_data: std_logic_vector(7 downto 0);
       signal vga_rd_data: std_logic_vector(7 downto 0);
       signal color: std_logic_vector(11 downto 0);
       signal wr_vram, rd_vram: std_logic;
34 begin
       -- instantiate VGA sync circuit
       vga_sync_unit: entity work.vga_sync
          port map(clk=>clk, reset=>reset,
                   hsync_i=>hsync_i, vsync_i=>vsync_i,
39                 video_on_i=>video_on_i, p_tick=>p_tick,
                   pixel_x=>pixel_x, pixel_y=>pixel_y);
       -- instantiate video SRAM control
       vram_unit: entity work.vram_ctrl
       port map (
44        clk=>clk, reset=>reset,
          -- from video sync
          pixel_x=>pixel_x, pixel_y=>pixel_y,
          p_tick=>p_tick,
          -- avalon bus interface
49        vga_rd_data=>vga_rd_data,
          cpu_rd_data=>cpu_rd_data,
          cpu_wr_data=>vga_writedata(7 downto 0),
          cpu_addr=>vga_address(18 downto 0),
          cpu_mem_wr=>wr_vram,
54        cpu_mem_rd=>rd_vram,
          -- to/from SRAM chip
          sram_addr=>sram_addr,sram_dq=>sram_dq,
          sram_we_n=>sram_we_n, sram_oe_n=>sram_oe_n,
          sram_ce_n=>sram_ce_n, sram_ub_n=>sram_ub_n,
59        sram_lb_n=>sram_lb_n
       );
       -- instantiate palette table (8-bit to 12-bit conversion)
       palet_unit: entity work.palette
       port map (
64        color_in=>vga_rd_data,
          color_out=>color);
       -- delay vga sync to accomodate memory access
       process (clk)
```

```
        begin
69          if (clk'event and clk='1') then
                if (p_tick='1') then
                    vsync <= vsync_i;
                    hsync <= hsync_i;
                    video_on_reg <= video_on_i;
74              end if;
            end if;
        end process;
        -- memory read/write decoding
        wr_vram <=
79          '1' when vga_write='1' and vga_chipselect='1' and
                    vga_address(19)='0' else
            '0';
        rd_vram <=
            '1' when vga_read='1' and vga_chipselect='1' and
84                  vga_address(19)='0' else
            '0';
        -- input data mux
        vga_readdata <= x"000000" & cpu_rd_data when vga_address(19)='0' else
                        x"000" & pixel_y & pixel_x;
89      rgb <= (others=>'0') when video_on_reg='0' else color;
    end arch;
```

17.5.5 SOPC component creation

After developing the top-level design and an Avalon wrapping circuit, we can create a new SOPC component in Component Editor following the procedure outlined in Section 14.5.4. Recall that the memory controller requires three clock cycles to complete a read operation, as discussed in Section 17.3.3. We must insert two extra wait cycles in the memory interface. This can be done by setting the **Read Wait** field to 2 in the Avalon MM slave interface of Component Editor.

17.6 VIDEO DRIVER

The basic function of a graphic video driver is to put and retrieve a pixel at the designated coordinate. We then can add additional routines to generate and process various types of graphic objects, which can be geometrical models, bitmap images, or text. The following subsections discuss these routines.

We define two symbolic constants for the VGA resolution:

```
#define DISP_GRF_X_MAX   640      // 640 columns  (0 to 639)
#define DISP_GRF_Y_MAX   480      // 480 rows     (0 to 479)
```

17.6.1 Video memory access routines

The main function of video memory access routines is to read and write a pixel at the designated location. The basic routines are shown in Listing 17.5.

Listing 17.5

```
alt_u32 vga_calc_sram_addr(int x, int y)
{
  alt_u32 addr;

  addr = (alt_u32) (DISP_GRF_X_MAX*y + x);
  return addr;
}

alt_u8 vga_rd_pix(alt_u32 vga_base, int x, int y)
{
  alt_u32 offset;
  alt_u8 color;

  offset = vga_calc_sram_addr(x,y);        // form address offset
  color = (alt_u8) IORD(vga_base, offset); // read video memory
  return color;
}

void vga_wr_pix(alt_u32 vga_base, int x, int y, alt_u8 color)
{
  alt_u32 offset;

  offset = vga_calc_sram_addr(x,y);        // form address offset
  IOWR(vga_base, offset, color);           // write video SRAM
  return;
}

void vga_rd_xy(alt_u32 vga_base, int *x, int *y)
{
  alt_u32 data;
  const alt_u32 XY_ADDR=0x00080000;

  data = IORD(vga_base, XY_ADDR);     // read video SRAM
  *x = 0x000003ff & data;             // 10 LSBs for horizontal counter
  *y = 0x000003ff & (data >> 10);     // next 10 LSBs for vertical counter
  return;
}

void vga_clr_screen(alt_u32 vga_base, alt_u8 color)
{
  int x, y;

  for(x=0; x<DISP_GRF_X_MAX; x++)
    for(y=0; y<DISP_GRF_Y_MAX; y++)
      vga_wr_pix(vga_base, x, y, color);
}
```

The vga_calc_sram_addr() function calculates the corresponding SRAM offset address for a pixel located at (x, y) by using the equation

$$offset = 640 * y + x$$

The calculation involves a multiplication operation. It is not a problem for the Nios II/s and Nios II/f configurations since their implementation includes a hardware multiplier. For the Nios II/e configuration, which uses a software routine to perform multiplication and can be slow, we can follow the discussion in Section 17.3.3 and substitute it with two shift operations:

```
addr = (alt_u32) (y<<9 + y<<7 +x);
```

The `vga_rd_pix()` and `vga_wr_pix()` functions use this function to obtain the SRAM address and then perform the read and write operations accordingly. If better performance is desired, we can move the offset calculation to hardware.

The `vga_rd_xy()` function performs the read operation and retrieves the current coordinate of the scan. It retrieves the data from the designated address and unpacks the word to obtain the x-axis and y-axis coordinates. These numbers can be used to determine whether the scan is in the retrace area (i.e., video is off) and can be used as a 60-Hz clock tick (since the counters reset to zero every $\frac{1}{60}$ second). Note that the retrieved numbers may be off somewhat since the execution of the routine may take several clock cycles.

The last function in this category, `vga_clr_screen()`, clears the screen by writing the entire video memory with a designated color.

17.6.2 Geometrical model routine

A geometrical model is generated by the mathematical description of the object, which sometimes is referred to as a *vector graphic*. For example, we can obtain a line segment from two given points, (x_1, y_1) and (x_2, y_2), by generating a series of pixels based on the equation

$$\frac{y - y_1}{x - x_1} = \frac{y_2 - y_1}{x_2 - x_1}$$

We can obtain the pixel coordinates by using the x as the independent variable and then calculating y:

$$y = \frac{y_2 - y_1}{x_2 - x_1} * (x - x_1) + y_1$$

However, for a steeper line (i.e., a line with a large slope), there are only a few dots that can be plotted within the given range and the line appears to be disconnected. It is better to use y as the independent variable to obtain more points:

$$x = \frac{x_2 - x_1}{y_2 - y_1} * (y - y_1) + x_1$$

The basic line plotting function is shown in Listing 17.6.

Listing 17.6

```
void vga_plot_line(alt_u32 vga_base, int x1, int y1, int x2, int y2,
                   alt_u8 color)
{
  int horiz, step, x, y;
  float slope;

  if ((y1==y2) && (x1==x2)){    // special case of x1==x2 and y1==y2
    vga_wr_pix(vga_base, x1, y1, color);
    return;
  }
  horiz = (abs(x2-x1)>abs(y2-y1)) ? 1 : 0;
  if (horiz){    // line is more horizontal and x2!=x1
    slope=(float)(y2-y1) / (float)(x2-x1);
    step = ((x2-x1)>1) ? 1 : -1;
    for(x=x1;x!=x2;x=x+step){
```

Figure 17.11   Bitmap of mouse pointer.

```
    y = slope*(x-x1) + y1;
    vga_wr_pix(vga_base, x, y, color);
  } // end for
} else {          // line is more vertical
  slope=(float)(x2-x1)/ (float)(y2-y1);
  step = ((y2-y1)>1) ? 1 : -1;
  for(y=y1;y!=y2;y=y+step){
    x = slope*(y-y1) + x1;
    vga_wr_pix(vga_base, x, y, color);
  } // end for
}
return;
}
```

It first checks the type of slope by comparing the distances, $|x_2 - x_1|$ and $|y_2 - y_1|$, and then selects the proper axis as the independent variable to generate the pixels.

The geometric model itself is a separate discipline and involves many sophisticated techniques and algorithms. For example, because of the floating-point operations, the previous `vga_plot_line()` function is not very efficient. A better alternative is to use the Bresenham algorithm, which uses integer arithmetic exclusively. Deriving even a simple set of driver routines is beyond the scope of the book. The line plotting function just gives us a taste of this type of program and additional information can be found in the bibliographic section.

17.6.3   Bitmap processing routines

Many graphic objects are not regular and cannot be described mathematically. We can explicitly draw the graph in rectangular grid of pixels. It is known as *raster graphics* or *bitmap*. The 12-by-20 bitmap of a mouse pointer is shown in Figure 17.11. A bitmap is usually stored in a file or an internal data structure and copied to the video memory when needed. A bitmap is frequently placed on top of a larger existing graphic and becomes another layer of the graphics, which is sometimes referred to as an *overlay*.

An overlay can be best explained by the operation of a mouse pointer over a graphic. The mouse pointer bitmap is the top layer and the original graphic is the bottom layer. When a mouse moves to a location, the pointer bitmap covers a portion of the original graphic and the pointer is shown on the display. However, the content below the pointer bitmap remains intact and will be restored when the mouse moves away. In the actual implementation, the original graphics are stored in the video memory permanently and the bitmaps are copied to the memory as needed. The *below area* may need to be retrieved to a temporary buffer so that it can be restored later. Because of the overlay operation, both write and read are needed for the CPU port of the video memory. In our program, we define a structure data type for a bitmap:

```
typedef struct tag_bmp
{
  int width;
  int height;
  alt_u8 *pdata;    // pointer to pixel array
} bmp_type;
```

In the definition, **width** and **height** specify the width and height of the bitmap and **pdata** is a pointer pointing to an array that stores the pixel data. The size of the array should be at least **width*height**. The element's data type, **alt_u8**, reflects the 8-bit color system. With this structure, the pointer bitmap of Figure 17.11 is shown in Listing 17.7.

Listing 17.7

```
alt_u8 MOUSE_DATA[]=
  {
  0x00, 0x00, 0x00, 0x00, 0x00, 0x00, 0x00, 0x00, 0x00, 0x00, 0x00, 0x00,
  0x00, 0xff, 0x00, 0x00, 0x00, 0x00, 0x00, 0x00, 0x00, 0x00, 0x00, 0x00,
  0x00, 0xff, 0x6d, 0x00, 0x00, 0x00, 0x00, 0x00, 0x00, 0x00, 0x00, 0x00,
  0x00, 0xff, 0x92, 0x6d, 0x00, 0x00, 0x00, 0x00, 0x00, 0x00, 0x00, 0x00,
  0x00, 0xff, 0x92, 0x92, 0x6d, 0x00, 0x00, 0x00, 0x00, 0x00, 0x00, 0x00,
  0x00, 0xff, 0x92, 0x92, 0x92, 0x6d, 0x00, 0x00, 0x00, 0x00, 0x00, 0x00,
  0x00, 0xff, 0x92, 0x92, 0x92, 0x92, 0x6d, 0x00, 0x00, 0x00, 0x00, 0x00,
  0x00, 0xff, 0x92, 0x92, 0x92, 0x92, 0x92, 0x6d, 0x00, 0x00, 0x00, 0x00,
  0x00, 0xff, 0x92, 0x92, 0x92, 0x92, 0x92, 0x92, 0x6d, 0x00, 0x00, 0x00,
  0x00, 0xff, 0x92, 0x92, 0x92, 0x92, 0x92, 0x92, 0x92, 0x6d, 0x00, 0x00,
  0x00, 0xff, 0x92, 0x92, 0x92, 0x92, 0x92, 0x92, 0x92, 0x92, 0x6d, 0x00,
  0x00, 0xff, 0x92, 0x92, 0x92, 0x92, 0x6d, 0x6d, 0x6d, 0x6d, 0x6d, 0x6d,
  0x00, 0xff, 0x92, 0x92, 0x6d, 0x92, 0x92, 0x6d, 0x00, 0x00, 0x00, 0x00,
  0x00, 0xff, 0x92, 0x92, 0x00, 0x6d, 0x92, 0x92, 0x6d, 0x00, 0x00, 0x00,
  0x00, 0xff, 0x92, 0x00, 0x00, 0x6d, 0x92, 0x92, 0x6d, 0x00, 0x00, 0x00,
  0x00, 0xff, 0x00, 0x00, 0x00, 0x00, 0x6d, 0x92, 0x92, 0x6d, 0x00, 0x00,
  0x00, 0x00, 0x00, 0x00, 0x00, 0x00, 0x6d, 0x92, 0x92, 0x6d, 0x00, 0x00,
  0x00, 0x00, 0x00, 0x00, 0x00, 0x00, 0x00, 0x6d, 0x92, 0x92, 0x6d, 0x00,
  0x00, 0x00, 0x00, 0x00, 0x00, 0x00, 0x00, 0x6d, 0x92, 0x92, 0x6d, 0x00,
  0x00, 0x00, 0x00, 0x00, 0x00, 0x00, 0x00, 0x00, 0x6d, 0x6d, 0x00, 0x00
  };

bmp_type MOUSE_BMP={
  12,             // width
  20,             // height
  MOUSE_DATA      // bitmap array
};
```

The video driver consists of three bitmap processing functions, which retrieve (i.e., read) an area from the video memory, write a bitmap to the video memory, and move a bitmap from one region of the video memory to another. The three functions are shown in Listing 17.8.

Listing 17.8

```
void vga_rd_bitmap(alt_u32 vga_base, int x, int y, bmp_type *bmp)
{
  int i,j;
  alt_u8 color;

  for(j=0;j<bmp->height;j++){
    for(i=0;i<bmp->width;i++){
      color = vga_rd_pix(vga_base, i+x, j+y);
      bmp->pdata[(j*bmp->width) + i] = color;
    } // end for loop i
  } // end for loop j
}

void vga_wr_bitmap(alt_u32 vga_base, int x, int y, bmp_type *bmp,
                   int tran)
{
  int i,j;
  alt_u8 color;

  for(j=0;j<bmp->height;j++){
    for(i=0;i<bmp->width;i++){
      color = bmp->pdata[(j*bmp->width) + i];
      if (tran==0 || color!=0)
        vga_wr_pix(vga_base, i+x, j+y, color);
    } // end for loop i
  } // end for loop j
}

void vga_move_bitmap(alt_u32 vga_base,
                     int xold, int yold, bmp_type *below,
                     int xnew, int ynew, bmp_type *bmp)
{
  /* restore the hidden pixels at (xold, yold) */
  vga_wr_bitmap(vga_base, xold, yold, below, 0);
  /* read the hidden pixels at (xnew, ynew) */
  vga_rd_bitmap(vga_base, xnew, ynew, below);
  /* write the top bitmap at (xnew, ynew) */
  vga_wr_bitmap(vga_base, xnew, ynew, bmp, 1);
}
```

The vga_rd_bitmap() function reads a rectangular area from the video memory. The tip (top-left corner) of the area is specified by x and y and its size is obtained from the width and height fields of bmp. The retrieved pixel data is stored to the array pointed by the pdata field of bmp.

The vga_wr_bitmap() function writes a bitmap to the video memory. Its basic program structure is similar to that of vga_rd_bitmap(). It includes one additional argument, tran, to indicate whether "transparency" operation is desired. A bitmap usually contains an irregular object and only this portion should be displayed when it is overlaid over a graphic. For example, only the arrow-shaped object in the mouse pointer bitmap, rather than the entire rectangle, should be displayed on a graphic. To achieve this, we can paint unused pixels with a special color, such as black (i.e., 0x00), and treat them as "transparent" pixels, which should not overwrite the corresponding underlying pixels when the bitmap is overlaid over a graphic. The tran argument indicates whether the black pixels should be treated as transparent pixels or normal black pixels.

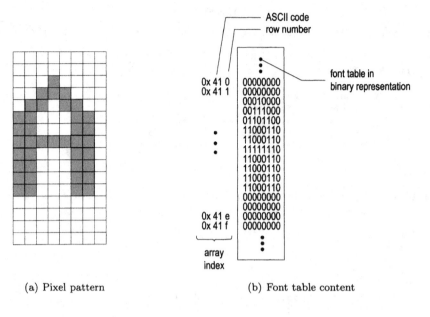

(a) Pixel pattern   (b) Font table content

Figure 17.12   Font pattern for the letter A.

The `vga_move_bitmap()` function moves a bitmap, pointed by `bmp`, from one location to another and restores and archives underlying pixels. The `below` pointer points to a buffer for the underlying pixels. This function can be best explained by mouse pointer movement. The mouse pointer bitmap is stored in an array pointed by `bmp` and the mouse is currently located at (`xold`, `yold`). The corresponding pixels of the underlying graphics below the mouse bitmap are stored in the `below` buffer. When we move the mouse pointer to a new location, (`xnew`, `ynew`), the required operations are:

- restore the graphics at (`xold`, `yold`) by writing back the stored pixels in the `below` buffer.
- read the area at (`xnew`, `ynew`) and store it to the `below` buffer.
- write the mouse pointer bitmap (pointed by `bmp`) at (`xnew`, `ynew`).

Note that the calling function must allocate memory for the `below` buffer.

17.6.4   Bit-mapped text routines

Graphic display frequently contains some texts. One way to handle text is to design a collection of bitmaps for the character set. The bitmap patterns are referred to as the *font*. The processing of a font bitmap is somewhat different from regular bitmaps and two driver routines are developed for this purpose.

Since there is no need to have multiple colors within a single character, the font bitmaps require one bit for each pixels (i.e., either "on" or "off"). We choose an 8-by-16 (i.e., 8-column-by-16-row) font similar to the one used in early IBM PCs. In this font, each character is represented as an 8-by-16 pixel pattern. The pattern for the letter "A" is shown in Figure 17.12(a). The original font set consists of 256 (2^8) patterns, including digits, uppercase and lowercase letters, punctuation symbols, and many special-purpose graphic symbols. We implement only the first half of the patterns, which correspond to the 128 (2^7) characters of the ASCII code, listed in Table 16.1.   Note that the first column of the table (ASCII codes 0x00 to 0x1f) consists of nonprintable control characters. The font uses these codes to implement special graphic symbols. For example, the 0x16 code will generate a spade pattern, ♠, on the screen. Note that the 0x00 code is reserved for a blank tile. Since each pattern requires 16 (2^4) bytes, the entire font set can be stored in a constant array of 2048 (i.e., $128 * 16$) elements. The first few lines are:

```
const alt_u8 FONT[]={
// 0x00   blank
0x00,   //
0x00,   //
0x00,   //
0x00,   //
0x00,   //
0x00,   //
0x00,   //
0x00,   //
0x00,   //
0x00,   //
0x00,   //
0x00,   //
0x00,   //
0x00,   //
0x00,   //
0x00,   //
// 0x01  smiley face
0x00,   //
0x00,   //
0x7E,   //      ******
0x81,   //     *        *
0xA5,   //     *  *    *  *
0x81,   //     *        *
0x81,   //     *        *
0xBD,   //     *  ****  *
0x99,   //     *   **   *
0x81,   //     *        *
0x81,   //     *        *
0x7E,   //      ******
0x00,   //
0x00,   //
0x00,   //
0x00,   //
```

. . .

The complete definition has 2048 rows and is stored in chu_avalon_vga_font_table.h.

Because of the 1-bit color depth, a new bitmap process function is needed. The basic code to display character ch at location (x, y) is

```
for(j=0; j<16; j++){
   ch_line_addr = 16*ch + j;   // char base address + offset
   row = FONT[ch_line_addr];
```

```
      for(i=0; i<8; i++){
        bit = row & (0x80 >> i);
        if (bit!=0)
          vga_wr_pix(vga_base, x+i, y+j, color);
      }
    }
```

The FONT[] array contains the bitmaps of 128 characters and the "base address" for character ch is 16*ch. The outer loop iterates through the 16 rows of the bitmap, starting with row 16*ch. The inner loop iterates through the row and extracts the individual bit. Note that 0x80 is 10000000 in binary format and can be treated with a 1-bit mask and the expression 0x80>>i shifts the enable bit to position i. If the corresponding bit is 1, the designated color is written to the pixel. The actual routine, vga_wr_bit_ch(), is shown in Listing 17.9.

Listing 17.9

```
void vga_wr_bit_ch(alt_u32 vga_base, int x, int y, char ch, int color,
                   int zoom)
{
  int i,j, ch_line_addr, bit;
  alt_u8 row;

  for(j=0; j<16*zoom; j++){
    /* get a row from font table */
    ch_line_addr = 16*ch + j/zoom;    // char base address + offset
    row = FONT[ch_line_addr];
    /* process bit */
    for(i=0;i<8*zoom;i++){
      bit = row & (0x80 >> (i/zoom));
      if (bit!=0)
        vga_wr_pix(vga_base, x+i, y+j, color);
    } // end for loop i
  } // end for loop j
}
```

The routine has an additional argument, zoom, which specifies the "magnification factor" of a character. For example, if zoom is 2, the font size is increased from 8-by-16 to 16-by-32. Since the resolution of the bitmap remains unchanged, the magnified fonts may have jagged edges. Note that we use ASCII code 0x7f for a special solid rectangle pattern. If we select the background color for the color argument, it can be used to erase the current character.

A relevant driver routine, vga_wr_bit_str(), writes a string from the specified coordinate. The code is shown in Listing 17.10.

Listing 17.10

```
void vga_wr_bit_str(alt_u32 vga_base, int x, int y, char *s,
                    int color, int zoom)
{
  int cx, cy;           // current x, y

  cx = x;
  cy = y;
  while (*s) {
    if (*s=='\n') {     // new line
      cx = x;
      cy = cy + 16*zoom;
```

```
      s++;
    }
    else {
      vga_wr_bit_ch(vga_base, cx, cy, *s, color, zoom);
      s++;
      cx = cx + 8*zoom;
    }  // end else
  }  // end while
}
```

The code loops through the characters until '\0' is reached. It treats the newline code (i.e., '\n') as a special character and starts a new line. This function can be used in conjunction with C's `sprintf(s,...)` function, which is similar to `print(...)` but stores the results in a string `s`, to display a formatted string, as demonstrated by the following code segment:

```
char s[11];
...
/* s store  "Index is ddd", where ddd is 3 decimal digits */
sprintf(s, "Index is %3d", n);
vga_wr_bit_str(vga_base, 200, 100, s, 0x03, 1);
```

17.7 MOUSE PROCESSING ROUTINES

A mouse is frequently used in conjunction with the graphic interface as a pointing device. The pointer bitmap, as shown in Figure 17.11, moves along with the mouse and indicates the current location on the screen. With the availability of a PS2 mouse driver and a VGA video driver, we can create two functions to support this operation. The `vga_init_mouse_ptr` function initializes the pointer bitmap and the `vga_move_mouse_ptr` function moves the pointer according to the mouse activities. The bitmap `MOUSE_BMP` defined in Section 17.6.3 is used for the mouse pointer and the two functions are shown in Listing 17.11.

Listing 17.11

```
...
#include "chu_avalon_ps2.h"
#include "chu_avalon_vga.h"
...
void vga_init_mouse_ptr(alt_u32 vga_base, alt_u32 ps2_base,
                        int x, int y, bmp_type *mouse, bmp_type *below)
{
  /* read hidden pixels */
  vga_rd_bitmap(vga_base, x, y, below);
  /* draw initial pointer */
  vga_wr_bitmap(vga_base, x, y, mouse, 1);
}

int vga_move_mouse_ptr(alt_u32 vga_base, alt_u32 ps2_base,
      int xold, int yold, bmp_type *below, int *xnew, int *ynew,
      bmp_type *mouse, mouse_mv_type *mv)
{
  if (mouse_get_activity(ps2_base, mv)==0)  // no movement
    return(0);
  /* calculate new mouse pointer position */
  *xnew = xold + mv->xmov;
```

```
   if (*xnew > (639 - mouse->width))
     *xnew = 639- mouse->width;
   if (*xnew<0)
     *xnew=0;
   *ynew = yold - mv->ymov;                     // VGA y-axis goes downward
   if (*ynew>(479 - mouse->height))
     *ynew = 479 - mouse->height;
   if (*ynew<0)
     *ynew=0;
   /* draw the updated mouse pointer, restore "underlying" area */
   vga_move_bitmap(vga_base, xold, yold, below, *xnew, *ynew, mouse);
   return(1);
}
```

The `vga_init_mouse_ptr()` function performs the initialization. It should be called when a mouse pointer is first placed on a graphic screen. The function first reads pixels below the designated location and stores it in a buffer and then writes the mouse pointer bitmap at this location. Note that the calling function must allocate memory for the buffer.

The `vga_move_mouse_ptr()` retrieves data movement information from the mouse, calculates the new coordinates, and moves the pointer bitmap to the new location. It utilizes `vga_move_bitmap()` to move the bitmap and includes the `below` buffer to store and restore the underlying pixels. The function also returns the new pointer coordinates and mouse activities data to the calling function for further processing and therefore can be treated as an extended mouse driver routine.

We can remove the pointer from the screen by writing the buffered underlying pixels to the current mouse location:

```
   vga_wr_bitmap(vga_base, x, y, &below, 0);
```

and thus there is no need for a separate driver routine for this task.

Note that the two functions are not a part of the video driver and are included in the main testing program.

17.8 TESTING PROGRAM

The VGA controller core can be instantiated and integrated to a Nios II system like a normal IP core. The system derived in Section 16.10 includes the VGA controller core and can be used for testing. We construct a program to demonstrate and verify the driver routines. The program consists of the following tests:

- Clear the screen.
- Plot a color chart that shows all colors.
- Plot random pixels.
- Plot random lines.
- Plot several functions.
- Swap two vertical strips on screen.
- Demonstrate the use of mouse pointer.
- Display bit-mapped text.

The main program is shown in Listing 17.12.

Listing 17.12

```
int main(void){
  int sw, btn;
  alt_u8 disp_msg[4]={sseg_conv_hex(13), sseg_conv_hex(1),
                      sseg_conv_hex(5), 0x0c};

  sseg_disp_ptn(SSEG_BASE, disp_msg);      // show "dISP" for display
  vga_clr_screen(VRAM_BASE,0);             // clear screen
  printf("VGA video controller test: \n\n");
  btn_clear(BTN_BASE);
  while (1){
    while (!btn_is_pressed(BTN_BASE)){ };  // wait for button
    btn=btn_read(BTN_BASE);                // read button
    if (btn & 0x02){                       // key1 pressed
      sw=pio_read(SWITCH_BASE);            // read switch
      printf("key/sw: %d/%d\n", btn, sw);
    }
    btn_clear(BTN_BASE);
    switch (sw){
      case 0:  // clear screen
        vga_clr_screen(VRAM_BASE,0);
        break;
      case 1:  // plot color chart
        plot_color_chart(VRAM_BASE);
        break;
      case 2:  // plot random pixels
        plot_random_pix(VRAM_BASE);
        break;
      case 3:  // plot random lines
        plot_random_line(VRAM_BASE);
        break;
      case 4:  // plot several functions
        plot_function(VRAM_BASE);
         break;
      case 5:  // swap two vertical strips on screen
        plot_swap(VRAM_BASE);
        break;
      case 6:  // test mouse pointer
        plot_mouse(VRAM_BASE,PS2_BASE,BTN_BASE);
        break;
      case 7:  // display bit-mapped text
        plot_text(VRAM_BASE);
        break;
    } //end switch
  } // end while
}
```

The program's basic structure is similar to that in Listing 16.12. We use the slide switches to specify the desired test and use the pushbutton switch 1 (labeled **key1** on the board) to initiate the test. The relevant functions are discussed in the following subsections. Note that only the first two tests write the entire screen and thus erase the previous image. Other tests impose some new graphics over the previous image.

17.8.1 Chart plotting routine

The chart plotting routine is shown in Listing 17.13.

Listing 17.13

```
void plot_color_chart(alt_u32 vga_base)
{
  int x, y;
  alt_u8 i;
  alt_u8 color_r, color_g, color_b, color_rgb;

  for(x=0; x<DISP_GRF_X_MAX; x++){
    for(y=0; y<DISP_GRF_Y_MAX; y++){
      if (x<240) {
        /* x < 240 */
        color_r = (alt_u8)(x/30);
        if (y<240){  // region 0
          color_g = (alt_u8)(y/30);
          color_b = 0x00;
        } else {
          color_g = (alt_u8)((y-240)/30);
          color_b = 0x02;
        }
      } else if (x<480) {
        /* 240 <= x < 480 */
        color_r = (alt_u8)((x-240)/30);
        if (y<240){  // region 2
          color_g = (alt_u8)(y/30);
          color_b = 0x01;
        } else {     // region 2
          color_g = (alt_u8)((y-240)/30);
          color_b = 0x03;
        }
      } else {
        /* 480 <= x < 640 */
        /* 3-bit color test strips */
        i = (x - 480)/20;
        if (i & 0x04 )
          color_r = 0x07;
        else
          color_r = 0x00;
        if (i & 0x02 )
          color_g = 0x07;
        else
          color_g = 0x00;
        if (i & 0x01 )
          color_b = 0x03;
        else
          color_b = 0x00;
      } // end outer if
      color_rgb = (color_r<<5) + (color_g<<2) + color_b;
      vga_wr_pix(vga_base, x, y, color_rgb);
    } // end y loop
  } // end x loop
}
```

We divide the screen into five regions. The first four regions are 240-by-240 squares, each containing 64 (i.e., 8-by-8) smaller 30-by-30 squares. Within a region, the intensities of red and green colors increase from 0 to 7 along the x-axis and y-axis, respectively. The four regions are assigned with blue color intensities of 0, 1, 2, and 3. The last region is a 160-by-480 rectangular that shows the strips of eight primary colors.

17.8.2 General plotting functions

The main program includes three routines that plots random pixels, random line segments, and a series of functions. These routines are shown in Listing 17.14.

Listing 17.14

```
void plot_random_pix(alt_u32 vga_base)
{
  int i, x, y;
  alt_u8 color;

  for(i=0;i<30000;i++){
    x=rand()%DISP_GRF_X_MAX;
    y=rand()%DISP_GRF_Y_MAX;
    color=rand()%256;
    vga_wr_pix(vga_base,x,y,color);
  }
}

void plot_random_line(alt_u32 vga_base)
{
  int i, x, y;
  alt_u8 color;

  /* test for a white dot */
  vga_plot_line(vga_base,10,10,10,10,0xff);
  /* a blue vertical line */
  vga_plot_line(vga_base,600,0,600,DISP_GRF_Y_MAX-1,0x03);
  /* a green horizontal line */
  vga_plot_line(vga_base,0,400,DISP_GRF_X_MAX-1,400,0x1c);
  /* 30 random lines from center */
  for(i=0;i<30;i++){
    x=rand()%DISP_GRF_X_MAX;
    y=rand()%DISP_GRF_Y_MAX;
    color=rand()%256;
    vga_plot_line(vga_base,DISP_GRF_X_MAX/2,DISP_GRF_Y_MAX/2,x,y,color);
  } // end for
}

void plot_function(alt_u32 vga_base){
  const float XMAX=10.0;       // max range of x-axis
  const float YMAX=10.0;       // max range of y-axis
  float x, y, step;
  int i, j;

  step = XMAX / (float)(DISP_GRF_X_MAX);
  /* red line with small slope y=0.1*x */
  x = 0.0;
  for(i=1;i<DISP_GRF_X_MAX;i++){
    x = x + step;
    y = 0.1 * x;
    if (y < YMAX){      // plot if only y is in range
      j = DISP_GRF_Y_MAX-(y/YMAX)*DISP_GRF_Y_MAX;
      vga_wr_pix(vga_base, i, j, 0xe0);
    } // end if
  } // end for
  /* blue line with 45 degree slope y=x */
  x = 0.0;
  for(i=1;i<DISP_GRF_X_MAX;i++){
    x = x + step;
    y = x;
```

```
    if (y < YMAX){      // plot if only y is in range
      j = DISP_GRF_Y_MAX-(y/YMAX)*DISP_GRF_Y_MAX;
      vga_wr_pix(vga_base, i, j, 0x003);
    } // end if
  } // end for
  /* green steep line y=10*x */
  x = 0.0;
  for(i=1;i<DISP_GRF_X_MAX;i++){
    x = x + step;
    y = 10 * x;
    if (y < YMAX){      // plot if only y is in range
      j = DISP_GRF_Y_MAX-(y/YMAX)*DISP_GRF_Y_MAX;
      vga_wr_pix(vga_base, i, j, 0x01c);
    } // end if
  } // end for
  /* y=0.2*x*x */
  x = 0.0;
  for(i=1;i<DISP_GRF_X_MAX;i++){
    x = x + step;
    y = 0.2 * x*x;
    if (y < YMAX){      // plot if only y is in range
      j = DISP_GRF_Y_MAX-(y/YMAX)*DISP_GRF_Y_MAX;
      vga_wr_pix(vga_base, i, j, 0x1f);
    } // end if
  } // end for
  /* y = 5.0 + (5.0*sin(4.0*x)-3.0*cos(4.0*x))*exp(-0.5*x) */
  x = 0.0;
  for(i=1;i<DISP_GRF_X_MAX;i++){
    x = x + step;
    y = 5.0 + (5.0*sin(4.0*x)-3.0*cos(4.0*x))*exp(-0.5*x);
    if (y < YMAX){      // plot if only y is in range
      j = DISP_GRF_Y_MAX-(y/YMAX)*DISP_GRF_Y_MAX;
      vga_wr_pix(vga_base, i, j, 0xff);
    } // end if
  } // end for
}
```

The plot_random_pix() routine plots 3000 pixels with random colors at random locations. The rand() function, which returns a random number each time it is called, is used to generate the color and x- and y-axis coordinates. The plot_random_line() routine is used to test vga_plot_line(). It plots a white dot, a vertical line, a horizontal line, and 30 random segments originating from the center of the screen. The plot_function() routine treats the display as the first quadrant of a two-dimensional plan with a range of 0 to 10 in both axes and plots the following functions:

$$y = 0.1x$$
$$y = x$$
$$y = 10x$$
$$y = 0.2x^2$$
$$y = 5 + (5\sin(4x) - 3\cos(4x))e^{-0.5x}$$

17.8.3 Strip swapping routine

The strip swapping routine treats the screen as eight 80-pixel-by-480-pixel vertical strips and randomly swaps two strips. The purpose of the routine is to test the read and write operations on a large display area. The routine is shown in Listing 17.15.

Listing 17.15

```
void plot_swap(alt_u32 vga_base)
{
  alt_u8 buf[80*480];
  alt_u8 color;
  int x, y, x1, x2;

  x1 = rand()%8;
  x2 = rand()%8;
  if (x1==x2)
    x2 = (x1+1)%8;
  x1 = 80*x1;
  x2 = 80*x2;
  /* copy area 1 from video SRAM to buffer */
  for (y=0; y<480; y++)
    for (x=0; x<80; x++)
      buf[80*y+x]=vga_rd_pix(vga_base, x+x1, y);
  /* copy area 2 to area 1 */
  for (y=0; y<480; y++)
    for (x=0; x<80; x++){
      color=vga_rd_pix(vga_base, x+x2, y);
      vga_wr_pix(vga_base, x+x1, y, color);
    }
  /* copy buffer (area 1) to area 2 of SRAM */
  for (y=0; y<480; y++)
    for (x=0; x<80; x++)
      vga_wr_pix(vga_base, x+x2, y, buf[80*y+x]);
}
```

17.8.4 Mouse demonstration routine

The mouse demonstration routine tests the mouse driver routines and indirectly verifies the operation of bitmap processing routines. Its code is shown in Listing 17.16.

Listing 17.16

```
void plot_mouse(alt_u32 vga_base, alt_u32 ps2_base, alt_u32 btn_base)
{
  mouse_mv_type mv;
  int xold, yold, xnew, ynew, act;
  static alt_u8 bdata[20*12];
  bmp_type below={12, 20, bdata};

  if (!mouse_init(ps2_base)){
    printf("Mouse initialization failed.\n");
    return;
  }
  xold = 320;
  yold = 240;
  vga_init_mouse_ptr(vga_base, ps2_base, xold, yold, &MOUSE_BMP, &below);
  /* continue until a button is pressed */
```

```
    while (!btn_is_pressed(btn_base)){
      act = vga_move_mouse_ptr(vga_base, ps2_base,
           xold, yold, &below, &xnew, &ynew, &MOUSE_BMP, &mv);
      if (act==1){
        if(mv.lbtn)
          printf("\ncurrent mouse location: %d %d", xnew, ynew);
        xold = xnew;
        yold = ynew;
      } //end if
    } // end while
    vga_wr_bitmap(vga_base, xold, yold, &below, 0);
    printf("\n");
}
```

This function keeps track of the mouse activities and displays its movements on the screen. The mouse pointer bitmap is stored in a predefined MOUSE_BMP constant in Section 17.6.3 and a static pixel array, bdata[20*12], is used to provide storage for the below buffer. In addition, the function checks the button status and displays the current mouse pointer coordinates on the console when the mouse's left button is pressed.

17.8.5   Bit-mapped text routine

The bit-mapped text demonstration function illustrates the use of bit-mapped text driver routines and is shown in Listing 17.17.

Listing 17.17

```
void plot_text(alt_u32 vga_base){

    int x, y;
    char buffer[50];
    char msg_box[]=           // a 30-by-5 message
      "******************************\n"
      "*                            *\n"
      "*        Hello World         *\n"
      "*                            *\n"
      "******************************";

    /* display four single characters at four corners */
    vga_wr_bit_ch(vga_base, 0, 0, 'a', 0xff, 1);
    vga_wr_bit_ch(vga_base, DISP_GRF_X_MAX-8, 0, 'b', 0xe0, 1);
    vga_wr_bit_ch(vga_base, 0, DISP_GRF_Y_MAX-16, 'c', 0x1c, 1);
    vga_wr_bit_ch(vga_base, DISP_GRF_X_MAX-8, DISP_GRF_Y_MAX-16,
                   'd', 0x03, 1);
    /* display a single string in three zoom factors */
    vga_wr_bit_str(vga_base, 34*8, 3*16, "Hello World", 0x1c, 1);
    vga_wr_bit_str(vga_base, 28*8, 5*16, "Hello World", 0x1c, 2);
    vga_wr_bit_str(vga_base, 23*8, 8*16, "Hello World", 0x1c, 3);
    /* display a string with multiple lines (string with \n) */
    vga_wr_bit_str(vga_base, 25*8, 16*16, msg_box, 0x1c, 1);
    /* get a formatted string by sprintf() */
    vga_rd_xy(vga_base, &x, &y);
    sprintf(buffer, "current pixel (x,y): (%3d, %3d)", x, y);
    vga_wr_bit_str(vga_base, 24*8, 24*16, buffer, 0x03, 1);
}
```

Table 17.2   24-bit BMP file header and bitmap information

Offset	Size	Purpose	value
0x00	2	magic number used to identify the BMP file	'B' 'M'
0x02	4	size of file in bytes	
0x06	4	reserved	
0x0a	4	starting address of the bitmap data	54
0x0e	4	size of the bitmap information block	40
0x12	4	bitmap width in pixels (signed integer)	
0x16	4	bitmap height in pixels (signed integer)	
0x1a	2	number of color planes used	1
0x1c	2	color depth (number of bits per pixel)	24
0x1e	4	compression method being used	0 (no compression)
0x22	4	bitmap data size	
0x26	4	horizontal resolution of the image	
0x2a	4	vertical resolution of the image	
0x2e	4	number of colors in the color palette	0 (all used)
0x32	4	number of important colors used	0 (all important)

The routine writes a, b, c, and d in four corners and writes the **Hello World** string in various formats. It also invokes the **vga_rd_xy()** function and shows the use of **sprint()** function by displaying the formatted scan counter values on screen.

17.9   BITMAP FILE PROCESSING

Although technically a bitmap image can be specified as an array in C, as the mouse pointer demonstrated in Section 17.6.3, it is too cumbersome for practical purposes. A better alternative is to draw the images in a graphic program, save them in a file, and then retrieve the bitmaps as needed. In this section, we introduce a minimal procedure to perform this task. The procedure is not comprehensive or efficient but just helps us get started.

17.9.1   BMP format overview

There are many image file formats and some involve complex color translation and data compression schemes, which are beyond the scope of this book. The BMP format is widely use in Windows. It supports various color depths and can be either compressed or uncompressed. We choose the 24-bit BMP file format since it is simple and uncompressed. A typical BMP file contains four blocks:
- BMP file header, which stores general information.
- Bitmap information, which stores information about the bitmap image.
- Color palette, which stores color definition used for indexed color bitmaps.
- Bitmap data.

The fields of the first two blocks are summarized in Table 17.2. Each row shows the address offset, the size (in terms of bytes), purpose, and the value used in a 24-bit color scheme.

The color palette block is an index table that converts color depths and its purpose is similar to that of the palette lookup table in Section 17.4. It is not used in 24-bit color BMP files.

Since there is no color palette block, the bitmap data starts at offset 0x36. The pixels are arranged from left to right and from bottom to top. This implies that the first pixel is from the bottom left corner of the bitmap. A pixel in 24-bit color scheme requires three bytes, representing the intensities of the red, green, and blue colors, respectively. The pixel data of the same row are packed together and the combined width must be a multiple of 32 bits (i.e., 4 bytes). Extra padding zeros may be needed to satisfy this constraint. For example, assume that a bitmap's width is 2 bits. A row requires 6 bytes in the 24-bit color scheme. At the end of each row, two "padding" bytes must added (to make the total width of 8 bytes) to satisfy the constraint.

17.9.2 Generation of BMP file

There are many graphic programs that create and edit bitmap images in a 24-bit BMP format. The *Paint* program distributed with Windows OS can be used to perform this task. The basic procedure in Windows 7's Paint program is:

- Start the Paint program.
- Select the View tab page and turn on Rulers, Gridlines, and Status bar boxes.
- Drag the right bottom corner of the image to expand it the desired size.
- Zoom in the image to show the individual pixels.
- Create a new image or copy an existing image and edit it.
- When completed, select the File menu and then Save As.... A subwindow appears. Specify the file name and select 24-bit Bitmap (*.bmp,*dib) in the Save as type: field.

A new file with an extension .bmp should be created. Because the file is uncompressed, its size can be much larger than a normal image file. The screenshot for the mouse pointer image is shown in Figure 17.13.

17.9.3 Sprite-based design

One common way to construct a two-dimensional scene is to first load a background bitmap and then overlay it with bitmaps of smaller objects. The latter are commonly known as *sprites*. Sprite-based designs are widely used in simple two-dimensional video games. The images of various objects are first constructed and grouped in a few "sprite sheets" and then used in the program. The sprite sheet of a game entitled *1945* is shown in Figure 17.14. It is an airplane shooting game and thus the sprites consists of various images of airplanes, ships, explosions, etc. A sample game screen capture is shown in Figure 17.15.

An object sometimes can have multiple frames to describe the various phases of action. We can create an animation effect by redrawing the frames or moving the object at a proper rate. For example, nine frames of the Earth are shown in Figure 17.16, each representing a specific view. If we load the frames in a sequential order, one at a time, in a specific interval, the Earth appears to rotate at a constant rate.

The sprite concept can be used in other applications. For example, we can create a virtual instrument panel with different types of meters, displays, switches, knobs,

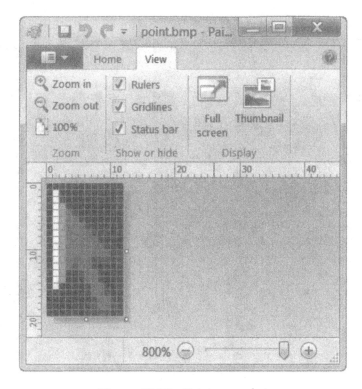

Figure 17.13   Paint screenshot.

etc. The sprite can be used for these components as well as simple animation (such as flashing LEDs and rotating knobs).

17.9.4   BMP file access

The DE1 board does not have an inherent file system. However, there are several ad hoc methods to extract and import the bitmap images:

1. C constant array
2. SD card
3. Flash-based read-only zip file
4. GDB host-based file system

The first method is to use a separate program that reads the BMP file, reformats the pixel data as a C constant array, and writes to a text file. The content of the text file looks like that of Listing 17.7. We can copy the text to a header file and include it in a C program. This method avoids file access in the C program and works fine with small images. The second method uses the board's SD card interface and reads the files from an SD card, which is discussed in Chapter 19.

The next two methods are integrated within the HAL framework. The third method groups the relevant files and converts them into a single zip file. The zip file is then downloaded to the board's flash memory and can be read by C's standard file access function. The fourth method utilizes GDB (GNU project debugger) to access the host computer's file system. This allows the Nios II system to use C's standard I/O functions to manipulate files in the PC. However, this can only be

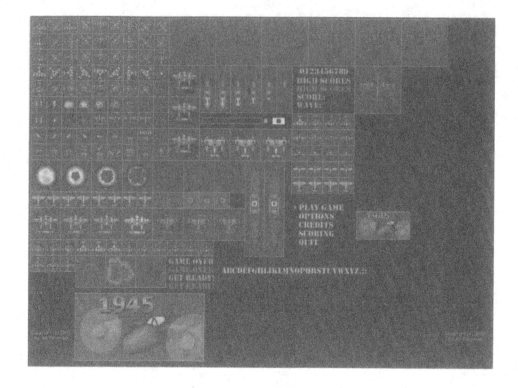

Figure 17.14 Sample sprite page (Courtesy of Ari Feldman).

done in the debugging mode and thus is only feasible in the developing process. Furthermore, the access is done via the board's JTAG link and is slow. Since the file access in this section is only for demonstration purposes, we use the last method.

17.9.5 Host-based file system

When we create the BSP package of the comprehensive Nios II testing system in Section 16.10.2, we enable and include the host-based file system in the **Software Packages** tab page, as shown in Figure 16.11, and thus this feature can be used with this BSP. Note that the page contains a **Mount-point** field, which specifies the symbolic path name, and its default name, /mnt/host, is kept. While constructing an application based on this BSP, we can access the files on a PC with normal C file I/O functions. The mount point corresponds to the physical directory where the main program resides. For example, if a file named `myfile.dat` resides on the directory of the main program, it can be open by using

```
fp = fopen("/mnt/host/myfile.dat", ...);
```

In order to use the host-based file system, we must use the "debug mode" when creating and downloading the software image. The procedure is:

1. Create an application project with a proper BSP.
2. Right-click the project and select **Debug As** and then **Nios II Hardware**.

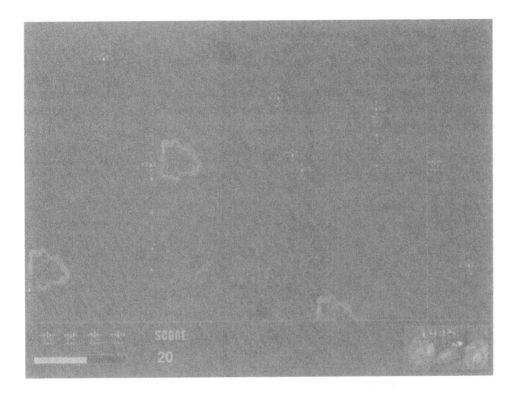

Figure 17.15   Game screenshot.

Figure 17.16   Bitmap of Earth.

3. After linking and downloading, Eclipse GUI switches to the *Debug perspective*. The execution is paused in the first line of the main program. Select the Run menu and then Resume to resume the normal execution.

17.9.6 Bitmap file retrieval routines

Basic file processing routines Although all C's standard file processing functions are available, we just use the simple `fgetc()` function. This approach allows us to access other types of medias without recreating the entire C library. A set of routines based on `fgetc()` are created to skip and retrieve 8-, 16- or 32-bit data, as shown in Listing 17.18.

Listing 17.18

```
void fskip(FILE *fp, int nbyte)
{
  int i;
  for (i=0; i<nbyte; i++)
    fgetc(fp);
}

alt_u8 fget8(FILE *fp)
{
  return( (alt_u8)fgetc(fp));
}

alt_u16 fget16(FILE *fp)
{
  alt_u16 b0, b1, r;

  b0 = (alt_u16) fgetc(fp);
  b1= (alt_u16) fgetc(fp);
  r = (b1 << 8) + b0;
  return(r);
}

alt_u32 fget32(FILE *fp)
{
  alt_u32 b0, b1, b2, b3, r;

  b0 = (alt_u32) fgetc(fp);
  b1= (alt_u32) fgetc(fp);
  b2= (alt_u32) fgetc(fp);
  b3= (alt_u32) fgetc(fp);
  r = (b3<<24) + (b2<<16) + (b1<<8) + b0;
  return(r);
}
```

The header and code of these functions are stored in files named chu_avalon_file.h and chu_avalon_file.c.

Bitmap retrieval function We can retrieve the bitmap image by using the previous file access routines. The function reads the input file, verifies the file type, extracts the pixel data, and stores the data to a buffer. The code is shown in Listing 17.19.

Listing 17.19

```
int read_bmp_file(char *file_name, bmp_type *buf)
{
  FILE *fp;
  int color_bit, x, y, slack;
  alt_u8 pixr, pixg, pixb, pix;

  /* open the file */
  fp = fopen(file_name,"rb");
  if (fp==NULL){
    printf("Error: %s fails to open. \n", file_name);
    return(-1);
  }
  /* check "magic number"; should be BM */
  if (fgetc(fp)!='B' || fgetc(fp)!='M'){
    printf("Error: %s is not a .bmp file.\n", file_name);
    fclose(fp);
    return(-1);
  }
  fskip(fp,16);                          // offset 2
  /* get height and width */
  buf->width  = (int) fget32(fp);        // offset 18
  buf->height = (int) fget32(fp);        // offset 22
  fskip(fp,2);                           // offset 26
  /* check color depth; should be 24 */
  color_bit = (int) fget16(fp);          // offset 28
  if (color_bit != 24){
    fclose(fp);
    printf("Error: color depth is not 24. \n");
    return(-2);
  }
  /* 24-bit BMP file format confirmed */
  printf("File opened. Bitmap width: %d, height: %d\n",
          buf->width, buf->height);
  fskip(fp,24);                          // offset 30
  printf("Reading in progress: \n");
  /* loop through pixel data, staring with the bottom row */
  for (y=buf->height-1; y >= 0; y--){ // offset 54
    printf(".");
    for (x=0; x< buf->width; x++) {
      /* get 24-bit color */
      pixb = (alt_u8) fgetc(fp);
      pixg = (alt_u8) fgetc(fp);
      pixr = (alt_u8) fgetc(fp);
      /* construct 8-bit color using MSBs */
      pix = (pixr & 0xe0) + ((pixg & 0xe0)>>3) + ((pixb & 0xc0)>>6);
      buf->pdata[y*(buf->width)+x]=pix;
    }  //end for x
    /* skip padding bytes, if exist */
    slack = (buf->width * 3) % 4;
    if (slack!=0)
      fskip(fp,4-slack);
  } //end for y
  fclose(fp);
  printf("\nFile loaded.\n");
  return(0);
}
```

The first portion of the code goes through the fields of the BMP file header and bitmap information. It extracts the height and width of the bitmap and checks the "magic number," which should be BM, and the color depth, which should be 24, to confirm that the file is in 24-bit BMP format.

The second part reads the pixel data and stores them to a buffer. The calling routine should allocate the buffer space and pass the buffer's pointer. Since the BMP file's pixel data format is different from that of the bmp_type data type, several tasks must be performed during the retrieval process. First, the BMP file arranges the pixels from the bottom row to the top row of the image but the bmp_type data type stores its pixels from the top row to the bottom row (recall that the y-axis of the VGA display increases downward). This can be compensated for by decrementing the outer loop:

```
for (y=buf->height-1; y >= 0; y--){
  . . .
}
```

Second, padding 0's of the BMP file must be removed. Recall that each row must be a multiple of 32 bits (i.e., 4 bytes) and each pixel in a 24-bit color scheme requires 3 bytes. The following statements check the condition and skip padding bytes:

```
slack = (buf->width * 3) % 4;
if (slack!=0)
  fskip(fp,4-slack);
```

Finally, the color depth must be reduced from 24 bits to 8 bits. This is done by extracting the first three MSBs of the red and green colors and the first two MSBs of the blue color. The reduction may lead to loss of subtle detail and sudden shifts of some colors.

Testing program   We construct a simple program to test the bitmap retrieval function and demonstrate the concept of sprite and animation. The bitmap of nine frames of the Earth is shown in Figure 17.16. The size of each frame is 100 pixels by 100 pixels and the size of the overall bitmap is 900 pixels by 100 pixels. The name of the file is **earth.bmp** and it is stored in same directory of the main program. The code is shown in Listing 17.20.

Listing 17.20

```
#define  BMP_FILE_NAME   "/mnt/host/earth9.bmp"   // path and file name
#define  NF 9                                      // # of frames
#define  NW 100                                    // width of a frame
#define  NH 100                                    // height of a frame

int main(void)
{
  alt_u8 iarray[NF*NH*NW];          // pixel data array for file buffer
  alt_u8 farray[NF][NH*NW];         // pixel data array for frames
  bmp_type img={NF*NW, NH, iarray};
  bmp_type frame[NF];
  int sw, status;
  int f, x, y;

  printf("BMP file retrieval test. \n\n");
  vga_clr_screen(VRAM_BASE, 0x08);
  status = read_bmp_file(BMP_FILE_NAME, &img);
  if (status!=0){
    printf("BMP file fails to load.  Exit program. \n");
    exit(1);
```

```
}
/* split bitmaps to 9 frames */
for (f=0; f<NF; f++){
  frame[f].height = NH;
  frame[f].width = NW;
  frame[f].pdata = farray[f];
  for (y=0; y<NH; y++) {
    for (x=0; x<NW; x++) {
      frame[f].pdata[y*NW+x]=img.pdata[y*NW*NF+f*NW+x];
    } // end x loop
  } // end y loop
} //end f loop

/* display individual frames in two rows */
for (f=0; f<5; f++)
  vga_wr_bitmap(VRAM_BASE, 30+f*(20+NW), 10, &frame[f], 1);
for (f=5; f<9; f++)
  vga_wr_bitmap(VRAM_BASE, 80+(f-5)*(20+NW), 120, &frame[f], 1);

/* animation of rotating earth */
while(1){
  for (f=0; f<NF; f++){
  sw = pio_read(SWITCH_BASE); ;
    if (sw!=0){                    // freeze if sw=0
      usleep(1000000/sw);   // sw frames per second
      vga_wr_bitmap(VRAM_BASE, 270, 250, &frame[f],1);
    } // end if
  } // end for
} // end while
}
```

The function retrieves the bitmap from the file, splits it into nine frames (which are smaller bitmaps), and displays these frames on the VGA monitor. The display is divided into two parts. The top part shows the nine frames statically in two rows. The bottom part loads the frames dynamically at a fixed interval and the Earth appears animated and rotating toward the east. The interval is $\frac{1}{sw}$ second, where sw is the value obtained from the switches. It corresponds to a rate of sw frames per second. Since the frames are loaded at the same locations, the underlying region can be overwritten. If the frames are loaded to different locations (such as a running dog), we can use vga_move_bitmap() to replace vga_wr_bitmap() and achieve animated movement.

17.10 BIBLIOGRAPHIC NOTES

The VGA standard is quite old. Its original specification can be found on the Wikipedia web site (searching by the keyword "VGA"). Altera University Program IP cores consist of an alternative video system design, which utilizes a frame buffer and Avalon's stream interface. The author's other book, *FPGA Prototyping by VHDL Examples*, contains examples for the tile-mapped and objected mapped schemes.

A wide variety of references are available in the areas of computer graphics and games. These books are either based on an existing API (such as OpenGL or DirectX) or primitive routines. *Fundamentals of Computer Graphics, second ed.* by P. Shirley and S. Marschner provides a comprehensive coverage of general computer graphics and *Designing Arcade Computer Game Graphics* by Ari Feldman discusses the development of a two-dimensional video game. The latter is out of print but an electronic version (along with sprite sheets) is available on line. The BMP file format has many options. More detailed descriptions can be found on the Wikipedia web site (searching by the keyword "bmp").

17.11 SUGGESTED EXPERIMENTS

17.11.1 PLL-based VGA controller

Our VGA controller must use a 50-MHz system clock. A more flexible alternative is to construct a subsystem based on the native 25-MHz video subsystem and use a PLL circuit to obtain the 25-MHz clock signal from the system clock. Reconstruct the VGA controller using this approach and verify its operation.

17.11.2 VGA controller with 16-bit memory configuration

The SRAM device in our VGA implementation is configured as a 512K-by-8 memory. Alternatively, we can use its native 256K-by-16 configuration and access and manipulate two pixel data at the time. Reconstruct the VGA controller using this approach and verify its operation. Compare the performance of the two approaches.

17.11.3 VGA controller with 3-bit color depth

Three-bit color displays the eight primary colors shown in Table 17.1. To simplify the design process, it can be aligned to a 4-bit boundary. The SRAM device can be configured as a 256K-by-16 memory and stores four pixels per word. Reconstruct the VGA controller using this approach and modify the driver routines and test program to verify its operation.

17.11.4 VGA controller with 1-bit color depth

Repeat experiment 17.11.3 for 1-bit color.

17.11.5 VGA controller with double buffering

For the 3-bit color scheme discussed in experiment 17.11.3, the 512KB SRAM is large enough to accommodate two VGA screens. Implement the double buffering scheme discussed in Section 17.3.1 and verify its operation.

17.11.6 VGA controller with 320-by-240 resolution

Our VGA controller has 640-by-480 resolution and 8-bit color depth. The implementation can be modified to support a configuration with 320-by-240 resolution and 12-bit color depth. Reconstruct the VGA controller for this configuration and modify the driver routines and testing program to verify its operation.

17.11.7   VGA controller with vertical mode operation

Some applications are better suited to a "vertical" screen. This can be done by turning the VGA monitor 90 degrees and treating it as a 480-by-640 display. Modify the driver routines and testing program to verify its operation.

17.11.8   Geometrical model functions

In Section 17.6.2, only one simple line plotting function is provided. Many additional geometrical model routines can be added:

- Function to plot a line by using the Bresenham algorithm.
- Function to draw a square or rectangle.
- Function to draw a polygon.
- Function to draw a circle or oval.
- Function to fill a closed shape with a specific color.

Derive these functions.

17.11.9   Bitmap manipulation functions

The basic bitmap processing functions are provided in Section 17.6.3. Several additional routines will be useful:

- Function to scale (i.e., enlarge or reduce) a bitmap.
- Function to obtain a horizontal or vertical mirror image.
- Function to rotate a bitmap to a specific degree.

Derive these functions.

17.11.10   Simulated "Etch A Sketch" toy

We can implement a simulated "Etch A Sketch" toy with a mouse and VGA monitor. It functions as follows:

- The mouse pointer can be moved to the desired location.
- Whenever the left button is pressed, the system records the mouse movement and shows the trace on the monitor.
- When the right button is pressed, the system erases the screen.

Derive the code and verify its operation.

17.11.11   Palette lookup table circuit

The BMP file format has an option for 8-bit color depth. In this depth, a palette indexing table is included in the file. One way to handle this is to implement a lookup table in the palette circuit, as discussed in Section 17.4, and load the palette information from the BMP file. Modify the VGA controller and the bitmap retrieval function to include this feature and verify its operation.

17.11.12   Virtual LED flashing system panel

For the flashing system discussed in Part II, we can use a virtual graphic panel that mirrors the condition of the FPGA board. The panel should consist of ten slide switches, one pushbutton switch, two discrete LEDs, and four seven-segment LED displays. The positions of the switches and the states of LEDs should be the same as those on the board. Design the graphic panel and integrate the codes to the previous LED flashing routines.

17.11.13   Virtual analog wall clock

We wish to implement an analog wall clock on the VGA monitor. The clock should have rotating hour, minute, and second hands. Derive the code and verify its operation.

17.12   SUGGESTED PROJECTS

17.12.1   Configurable VGA controller

To increase flexibility, we can make the VGA controller configurable. The new design should support the following features:
- Selection of resolutions of 640-by-480 and 320-by-240.
- Selection of color depths of 1 bit, 3 bits, 8 bits, and 12 bits (320-by-240 only).
- Selection of horizontal or vertical mode.
- Double buffering when feasible.

Design the hardware and software driver and verify the operation.

17.12.2   VGA controller using system SDRAM

Our VGA controller utilizes a dedicated SRAM device as the video memory. An alternative is to allocate part of the system's SDRAM as the frame buffer and let the VGA controller obtain the pixel data from the SDRAM device. We can use two line buffers, each storing pixel data of a single row, in the video controller, and perform the double buffering scheme on a row-by-row basis. This scheme imposes a tight timing constraint and special hardware, such as a DMA controller and a stream interface, may be needed to facilitate the data transfer between the SDRAM device and the video controller. Design the new video controller and verify its operation.

17.12.3   Paint program

We can design a graphic editing program similar to Windows's Paint. It should include the following features:
- Selection of an array of colors.
- Selection of line thickness.
- Drawing of line segment.
- Drawing of rectangle.
- Drawing of polygon.
- Drawing of circle and oval.
- Filling of a closed object with a specific color.
- Erasing.

Derive the code and verify its operation.

17.12.4   Video game

With the VGA and PS2 controllers, a variety of video games can be constructed. The reference in the bibliographic section serves as a good starting point for this project.

17.13   COMPLETE PROGRAM LISTING

Listing 17.21   chu_avalon_vga_font_table.h

```
/****************************************************************************
*
*  Module:   Font bitmaps
*  File:     chu_avalon_vga_font_table.h
*  Purpose:  16-by-8 128 character font bitmaps
*
****************************************************************************/
/* 16-row-by-8-column 128-char 1-bit-color font table */
const alt_u8 FONT[]={
// 0x00
0x00,  //
0x00,  //
0x00,  //
0x00,  //
0x00,  //
0x00,  //
0x00,  //
0x00,  //
0x00,  //
0x00,  //
0x00,  //
0x00,  //
0x00,  //
0x00,  //
0x00,  //
0x00,  //
// 0x01
0x00,  //
0x00,  //
0x7E,  //      ******
0x81,  //     *      *
0xA5,  //     * *  * *
0x81,  //     *      *
0x81,  //     *      *
0xBD,  //     * **** *
0x99,  //     *  **  *
0x81,  //     *      *
0x81,  //     *      *
0x7E,  //      ******
0x00,  //
0x00,  //
0x00,  //
0x00,  //
// 0x02
0x00,  //
0x00,  //
0x7E,  //      ******
0xFF,  //     ********
0xDB,  //     ** ** **
0xFF,  //     ********
0xFF,  //     ********
0xC3,  //     **    **
0xE7,  //     ***  ***
0xFF,  //     ********
0xFF,  //     ********
0x7E,  //      ******
0x00,  //
```

```
0x00,  //
0x00,  //
0x00,  //


...


// 0x7e
0x00,  //
0x00,  //
0x76,  //      *** **
0xDC,  //     ** ***
0x00,  //
0x00,  //
0x00,  //
0x00,  //
0x00,  //
0x00,  //
0x00,  //
0x00,  //
0x00,  //
0x00,  //
0x00,  //
// 0x7f
0xff,  //     ********
0xff,  //     ********
0xff,  //     ********
0xff,  //     ********
0xff,  //     ********
0xff,  //     ********
0xff,  //     ********
0xff,  //     ********
0xff,  //     ********
0xff,  //     ********
0xff,  //     ********
0xff,  //     ********
0xff,  //     ********
0xff,  //     ********
0xff,  //     ********
0xff,  //     ********
};
```

Listing 17.22 chu_avalon_vga.h

```
/************************************************************************
 *
 * Module:   VGA video driver header
 * File:     chu_avalon_vga.h
 * Purpose:  Routines to access video SRAM and display bit-mapped graphics
 *           and text
 *
 ************************************************************************
 *   Register map
 ************************************************************************
 * Read (data to cpu):
 *    offset 0x00000 to 0x7ffff
 *      * bit 7-0: 8-bit color
 *    offset 0x80000
 *      * bit 9-0:   VGA scan's current x (horizontal) position
 *      * bit 19-10: VGA scan's current y (vertical) position
 * Write (data from cpu):
 *    offset 0x00000 to 0x7ffff
 *      * bit 7-0: 8-bit color
 ***********************************************************************/
/* file inclusion */
#include <alt_types.h>

/************************************************************************
 * Data type definitions
 ***********************************************************************/
/* data type for a bitmap */
typedef struct tag_bmp
{
  int width;
  int height;
  alt_u8 *pdata;    //  pointer to pixel array
} bmp_type;

/************************************************************************
 * constant definitions
 ***********************************************************************/
#define DISP_GRF_X_MAX   640                    // 640 columns (0 to 639)
#define DISP_GRF_Y_MAX   480                    // 480 rows    (0 to 479)
```

```
/* ***********************************************************************
 * Function prototypes
 *********************************************************************** */
/* Video memory access */
alt_u32 vga_calc_sram_addr(int x, int y);
alt_u8 vga_rd_pix(alt_u32 vga_base, int x, int y);
void vga_rd_xy(alt_u32 vga_base, int *x, int *y);
void vga_wr_pix(alt_u32 vga_base, int x, int y, alt_u8 color);
/* Plotting and clear */
void vga_plot_line(alt_u32 vga_base, int x1, int y1,
                   int x2, int y2, alt_u8 color);
void vga_clr_screen(alt_u32 vga_base, alt_u8 color);
/* Bitmap processing */
void vga_wr_bitmap(alt_u32 vga_base, int x, int y,
                   bmp_type *bmp, int tran);
void vga_rd_bitmap(alt_u32 vga_base, int x, int y, bmp_type *bmp);
void vga_move_bitmap(alt_u32 vga_base,
                     int xold, int yold, bmp_type *below,
                     int xnew, int ynew, bmp_type *bmp);
/* Bit-mapped text */
void vga_wr_bit_ch(alt_u32 vga_base, int x, int y,
                   char ch, int color, int zoom);
void vga_wr_bit_str(alt_u32 vga_base, int x, int y,
                    char *s, int color, int zoom);
```

Listing 17.23 chu_avalon_vga.c

```
/***************************************************************************
 *
 * Module:  VGA video driver function prototypes
 * File:    chu_avalon_vga.c
 * Purpose: Routines to access video SRAM and display bit-mapped graphics
 *          and text
 *
 ***************************************************************************/
/* file inclusion */
#include <stdlib.h>                         // to use abs()
#include <io.h>
#include "chu_avalon_vga.h"
#include "chu_avalon_vga_font_table.h"

/***************************************************************************
 * function: vga_calc_sram_addr()
 * purpose:  calculate the video SRAM address offset for location (x, y)
 * argument:
 *   x: x-axis coordinate, 10 LSBs used
 *   y: y-axis coordinate, 9 LSBs used
 * return:
 *   address
 * note:
 *   - for a Nios II configuration without hardware multiplier, use shift:
 *     offset= 640*y + x = 512*y + 128*y + x = y<<9 + y<<7 +x
 ***************************************************************************/
alt_u32 vga_calc_sram_addr(int x, int y)
{
   alt_u32 addr;

   addr = (alt_u32) (DISP_GRF_X_MAX*y + x);
   return addr;
}


/***************************************************************************
 * function: vga_rd_pix()
 * purpose:  read a pixel from location (x,y)
 * argument:
 *   vga_base: base address of video SRAM
 *   x: x-axis coordinate, 10 LSBs used
 *   y: y-axis coordinate, 9 LSBs used
 * return:
 *   8-bit color at (x,y)
 * note:
 ***************************************************************************/
alt_u8 vga_rd_pix(alt_u32 vga_base, int x, int y)
{
   alt_u32 offset;
   alt_u8 color;

   offset = vga_calc_sram_addr(x,y);            // form address offset
   color = (alt_u8) IORD(vga_base, offset);     // read video memory
   return color;
}
```

```
/*************************************************************************
* function: vga_rd_xy()
* purpose:  get current scan coordinate (x,y)
* argument:
*   vga_base: base address of video SRAM
*   x: pointer to current value of VGA sync horizontal counter
*   y: pointer to current value of VGA sync vertical counter
* return:
*   updated x, y
* note:
*************************************************************************/
void vga_rd_xy(alt_u32 vga_base, int *x, int *y)
{
  alt_u32 data;
  const alt_u32 XY_ADDR=0x00080000;

  data = IORD(vga_base, XY_ADDR);      // read video SRAM
  *x = 0x000003ff & data;              // 10 LSBs for horizontal counter
  *y = 0x000003ff & (data >> 10);      // next 10 LSBs for vertical counter
  return;
}

/*************************************************************************
* function: vga_wr_pix()
* purpose:  write a pixel to location (x,y)
* argument:
*   vga_base: base address of video SRAM
*   x: x-axis coordinate, 10 LSBs used
*   y: y-axis coordinate, 9 LSBs used
*   color: 8-bit color
* return:
*************************************************************************/
void vga_wr_pix(alt_u32 vga_base, int x, int y, alt_u8 color)
{
  alt_u32 offset;

  offset = vga_calc_sram_addr(x,y);    // form address offset
  IOWR(vga_base, offset, color);       // write video SRAM
  return;
}
```

```
/***********************************************************************
 * function: vga_plot_line()
 * purpose:  draw a line from (x1,y1) to (x2,y2)
 * argument:
 *   vga_base: base address of video SRAM
 *   x1, y1: starting point
 *   x2, y2: end point
 *   color: 8-bit color
 * return:
 * note:
 *   - plot increments via x-axis for "horizontal" line
 *     and via y-axis for "vertical line" (steep line)
 *   - not optimized (need floating-point multiplication for each point)
 ***********************************************************************/
void vga_plot_line(alt_u32 vga_base, int x1, int y1, int x2, int y2,
                   alt_u8 color)
{
  int horiz, step, x, y;
  float slope;

  if ((y1==y2) && (x1==x2)){    // special case of x1==x2 and y1==y2
    vga_wr_pix(vga_base, x1, y1, color);
    return;
  }
  horiz = (abs(x2-x1)>abs(y2-y1)) ? 1 : 0;
  if (horiz){    // line is more horizontal and x2!=x1
    slope=(float)(y2-y1) / (float)(x2-x1);
    step = ((x2-x1)>1) ? 1 : -1;
    for(x=x1;x!=x2;x=x+step){
      y = slope*(x-x1) + y1;
      vga_wr_pix(vga_base, x, y, color);
    } // end for
  } else {         // line is more vertical
    slope=(float)(x2-x1)/ (float)(y2-y1);
    step = ((y2-y1)>1) ? 1 : -1;
    for(y=y1;y!=y2;y=y+step){
      x = slope*(y-y1) + x1;
      vga_wr_pix(vga_base, x, y, color);
    } // end for
  }
  return;
}


/***********************************************************************
 * function: vga_clr_screen()
 * purpose:  clear the screen to a background color
 * argument:
 *   vga_base: base address of video SRAM
 *   color: background color
 * return:
 ***********************************************************************/
void vga_clr_screen(alt_u32 vga_base, alt_u8 color)
{
  int x, y;

  for(x=0; x<DISP_GRF_X_MAX; x++)
    for(y=0; y<DISP_GRF_Y_MAX; y++)
      vga_wr_pix(vga_base, x, y, color);     // write black
}
```

```
/******************************************************************
* function: vga_wr_bitmap()
* purpose:  write a bitmap to video SRAM starting at (x,y)
* argument:
*   vga_base: base address of video SRAM
*   x: x-axis coordinate, 10 LSBs used
*   y: y-axis coordinate, 9 LSBs used
*   bmp: pointer to the bitmap structure
*   tran: whether to draw transparent background
*         0: draw black pixels in the background bitmap
*         1: not draw black pixels in the background bitmap
* return:
* note:
******************************************************************/
void vga_wr_bitmap(alt_u32 vga_base, int x, int y, bmp_type *bmp,
                   int tran)
{
  int i,j;
  alt_u8 color;

  for(j=0;j<bmp->height;j++){
    for(i=0;i<bmp->width;i++){
      color = bmp->pdata[(j*bmp->width) + i];
      if (tran==0 || color!=0)
        vga_wr_pix(vga_base, i+x, j+y, color);
    } // end for loop i
  } // end for loop j
}

/******************************************************************
* function: vga_rd_bitmap()
* purpose:  read a bitmap from video SRAM starting at (x,y)
* argument:
*   vga_base: base address of video SRAM
*   x: x-axis coordinate, 10 LSBs used
*   y: y-axis coordinate, 9 LSBs used
*   bmp: pointer to the returned bitmap structure
* return: updated bmp structure
* note: the calling function must allocate memory for the retrieved
*       "bmp" bitmap structure
******************************************************************/
void vga_rd_bitmap(alt_u32 vga_base, int x, int y, bmp_type *bmp)
{
  int i,j;
  alt_u8 color;

  for(j=0;j<bmp->height;j++){
    for(i=0;i<bmp->width;i++){
      color = vga_rd_pix(vga_base, i+x, j+y);
      bmp->pdata[(j*bmp->width) + i] = color;
    } // end for loop i
  } // end for loop j
}
```

```
/******************************************************************************
 * function:  vga_move_bitmap()
 * purpose:   move an overlay bitmap from (xold,yold) to (xnew,ynew)
 * argument:
 *   vga_base: base address of video SRAM
 *   xold: current x-axis coordinate, 10 LSBs used
 *   yold: current y-axis coordinate, 9 LSBs used
 *   below: pointer to the buffer that stores the pixels below bmp
 *   xnew: new x-axis coordinate, 10 LSBs used
 *   ynew: new y-axis coordinate, 9 LSBs used
 *   bmp: pointer to bitmap
 * return:
 *   underlying pixels at (xnew,ynew) are stored into bellow buffer
 * note:
 *   - underlying pixels at (xold,yold) were stored in below buffer
 *     before the function is called
 *   - the calling functioning must allocate memory for below buffer
 ******************************************************************************/
void vga_move_bitmap(alt_u32 vga_base,
                     int xold, int yold, bmp_type *below,
                     int xnew, int ynew, bmp_type *bmp)
{
  /* restore the hidden pixels at (xold, yold) */
  vga_wr_bitmap(vga_base, xold, yold, below, 0);
  /* read the hidden pixels at (xnew, ynew) */
  vga_rd_bitmap(vga_base, xnew, ynew, below);
  /* write the top bitmap at (xnew, ynew) */
  vga_wr_bitmap(vga_base, xnew, ynew, bmp, 1);
}
```

```
/*****************************************************************************
 * function: vga_wr_bit_ch()
 * purpose:  write a char to video SRAM at (x,y); use 0x7f to erase
 * argument:
 *   vga_base: base address of video SRAM
 *   x: x-axis coordinate, 9 LSBs used
 *   y: y-axis coordinate, 8 LSBs used
 *   ch: ascii character (only 7 LSBs used)
 *   color: character color
 *   zoom: zoom factor (usually 1 to 3)
 * return:
 * note:
 *   - the font is 8 pixels wide and 16 pixels tall
 *   - zoom only magnifies 1-by-1 pixel to 2-by-2 or 3-by-3 pixels and
 *     char shows jagged edge
 *   - 0x7f fills tile with solid color and
 *     can be used to erase a char by using background color
 *****************************************************************************/
void vga_wr_bit_ch(alt_u32 vga_base, int x, int y, char ch, int color,
                   int zoom)
{
  int i,j, ch_line_addr, bit;
  alt_u8 row;

  for(j=0; j<16*zoom; j++){
    /* get a row from font table */
    ch_line_addr = 16*ch + j/zoom;    // char base address + offset
    row = FONT[ch_line_addr];
    /* process bit */
    for(i=0;i<8*zoom;i++){
      bit = row & (0x80 >> (i/zoom));
      if (bit!=0)
        vga_wr_pix(vga_base, x+i, y+j, color);
    } // end for loop i
  } // end for loop j
}
```

```
/**************************************************************************
* function: vga_wr_bit_string()
* purpose:  write a string to video SRAM at (x,y);
* argument:
*   vga_base: base address of video SRAM
*   x: x-axis coordinate, 9 LSBs used
*   y: y-axis coordinate, 8 LSBs used
*   s: pointer to the string
*   color: character color
*   zoom: zoom factor (usually 1 to 3)
* return:
* note:
**************************************************************************/
void vga_wr_bit_str(alt_u32 vga_base, int x, int y, char *s,
                    int color, int zoom)
{
  int cx, cy;         // current x, y

  cx = x;
  cy = y;
  while (*s) {
    if (*s=='\n') {    // new line
      cx = x;
      cy = cy + 16*zoom;
      s++;
    }
    else {
      vga_wr_bit_ch(vga_base, cx, cy, *s, color, zoom);
      s++;
      cx = cx + 8*zoom;
    }  // end else
  }  // end while
}
```

Listing 17.24   chu_main_vga_test.c

```
/***********************************************************************
 *
 * Module:  VGA graphic test
 * File:    chu_main_vga_test.c
 * Purpose: Test VGA graphic driver routines
 * IP core base addresses:
 *    - SWITCH_BASE: slide switch
 *    - BTN_BASE: pushbutton
 *    - LED_BASE: discrete LEDs
 *    - SSEG_BASE: 7-segment LED
 *    - PS2_BASE: PS2 controller
 *    - VRAM_BASE: video SRAM
 *
 ***********************************************************************/
/* file inclusion */
#include <stdio.h>
#include <stdlib.h>     // use rand()
#include <math.h>       // use exp(), sin(), cos(),
#include <unistd.h>
#include <io.h>
#include "system.h"
#include "chu_avalon_gpio.h"
#include "chu_avalon_ps2.h"
#include "chu_avalon_vga.h"

/* constant definition */
/* 12-row-by-20-column 8-bit-color mouse pointer bitmap array */
alt_u8 MOUSE_DATA[]=
  {
  0x00, 0x00, 0x00, 0x00, 0x00, 0x00, 0x00, 0x00, 0x00, 0x00, 0x00, 0x00,
  0x00, 0xff, 0x00, 0x00, 0x00, 0x00, 0x00, 0x00, 0x00, 0x00, 0x00, 0x00,
  0x00, 0xff, 0x6d, 0x00, 0x00, 0x00, 0x00, 0x00, 0x00, 0x00, 0x00, 0x00,
  0x00, 0xff, 0x92, 0x6d, 0x00, 0x00, 0x00, 0x00, 0x00, 0x00, 0x00, 0x00,
  0x00, 0xff, 0x92, 0x92, 0x6d, 0x00, 0x00, 0x00, 0x00, 0x00, 0x00, 0x00,
  0x00, 0xff, 0x92, 0x92, 0x92, 0x6d, 0x00, 0x00, 0x00, 0x00, 0x00, 0x00,
  0x00, 0xff, 0x92, 0x92, 0x92, 0x92, 0x6d, 0x00, 0x00, 0x00, 0x00, 0x00,
  0x00, 0xff, 0x92, 0x92, 0x92, 0x92, 0x92, 0x6d, 0x00, 0x00, 0x00, 0x00,
  0x00, 0xff, 0x92, 0x92, 0x92, 0x92, 0x92, 0x92, 0x6d, 0x00, 0x00, 0x00,
  0x00, 0xff, 0x92, 0x92, 0x92, 0x92, 0x92, 0x92, 0x92, 0x6d, 0x00, 0x00,
  0x00, 0xff, 0x92, 0x92, 0x92, 0x92, 0x92, 0x92, 0x92, 0x92, 0x6d, 0x00,
  0x00, 0xff, 0x92, 0x92, 0x92, 0x92, 0x6d, 0x6d, 0x6d, 0x6d, 0x6d, 0x6d,
  0x00, 0xff, 0x92, 0x92, 0x6d, 0x92, 0x92, 0x6d, 0x00, 0x00, 0x00, 0x00,
  0x00, 0xff, 0x92, 0x92, 0x00, 0x6d, 0x92, 0x92, 0x6d, 0x00, 0x00, 0x00,
  0x00, 0xff, 0x92, 0x00, 0x00, 0x6d, 0x92, 0x92, 0x6d, 0x00, 0x00, 0x00,
  0x00, 0xff, 0x00, 0x00, 0x00, 0x00, 0x6d, 0x92, 0x92, 0x6d, 0x00, 0x00,
  0x00, 0x00, 0x00, 0x00, 0x00, 0x00, 0x6d, 0x92, 0x92, 0x6d, 0x00, 0x00,
  0x00, 0x00, 0x00, 0x00, 0x00, 0x00, 0x00, 0x6d, 0x92, 0x92, 0x6d, 0x00,
  0x00, 0x00, 0x00, 0x00, 0x00, 0x00, 0x00, 0x6d, 0x92, 0x92, 0x6d, 0x00,
  0x00, 0x00, 0x00, 0x00, 0x00, 0x00, 0x00, 0x00, 0x6d, 0x6d, 0x00, 0x00
  };

/* 12-row-by-20-column 8-bit-color mouse pointer bitmap data structure */
bmp_type MOUSE_BMP={
  12,           // width
  20,           // height
  MOUSE_DATA    // bitmap array
};
```

```
/*************************************************************************
* function: vga_init_mouse_ptr()
* purpose:   initialize the mouse pointer by writing pointer bitmap
*            at (x,y) and saving the underlying pixels in below buffer
* argument:
*   vga_base: base address of video SRAM
*   ps2_base: base address of PS2 device
*   x: x-axis coordinate, 10 LSBs used
*   y: y-axis coordinate, 9 LSBs used
*   mouse: pointer to mouse pointer bitmap
*   below: pointer to the buffer storing pixels below mouse pointer
* return:
*   underlying pixels at (x,y) are stored into bellow buffer
* note:
*   - the calling function must allocate memory for below buffer
*************************************************************************/
void vga_init_mouse_ptr(alt_u32 vga_base,  alt_u32 ps2_base,
                        int x, int y, bmp_type *mouse, bmp_type *below)
{
  /* read hidden pixels */
  vga_rd_bitmap(vga_base, x, y, below);
  /* draw initial pointer  */
  vga_wr_bitmap(vga_base, x, y, mouse, 1);
}
```

```
/****************************************************************************
 * function: vga_move_mouse_ptr()
 * purpose:  move mouse pointer on VGA according to mouse movement
 * argument:
 *   vga_base: base address of video SRAM
 *   ps2_base: base address of PS2 device
 *   xold: current x-axis coordinate, 10 LSBs used
 *   yold: current y-axis coordinate, 9 LSBs used
 *   below: pointer to the buffer storing pixels below mouse pointer
 *   xnew: pointer to new x-axis coordinate, 10 LSBs used
 *   ynew: pointer to new y-axis coordinate, 9 LSBs used
 *   mouse: pointer to mouse pointer bitmap
 *   mv: pointer to mouse activity data
 * return:
 *   - 1 if mouse has activities; 0 otherwise
 *   - updated new mouse pointer coordinates (xnew, ynew)
 *   - updated mouse activity data
 *   - underlying pixels at (xnew,ynew) are stored into below buffer
 * note:
 *   - the calling function must allocate memory for below buffer
 ****************************************************************************/
int vga_move_mouse_ptr(alt_u32 vga_base,  alt_u32 ps2_base,
      int xold, int yold, bmp_type *below, int *xnew, int *ynew,
      bmp_type *mouse, mouse_mv_type *mv)
{
  if (mouse_get_activity(ps2_base, mv)==0)   // no movement
    return(0);
  /* calculate new mouse pointer position */
  *xnew = xold + mv->xmov;
  if (*xnew > (639 - mouse->width))
    *xnew = 639- mouse->width;
  if (*xnew<0)
    *xnew=0;
  *ynew = yold - mv->ymov;                    // VGA y-axis goes downward
  if (*ynew>(479 - mouse->height))
    *ynew = 479 - mouse->height;
  if (*ynew<0)
    *ynew=0;
  /* draw the updated mouse pointer, restore below area  */
  vga_move_bitmap(vga_base, xold, yold, below, *xnew, *ynew, mouse);
  return(1);
}
```

```
/***************************************************************************
* function: plot_color_chart()
* purpose:  show all 2^8 colors in a chart
* argument:
*   vga_base: base address of video SRAM
* return:
* note:
*   - main area includes 4 regions
*   - each region has 2^3-by-2^3 squares for red/green
*   - blue is 0, 1, 2, 3 in each region
***************************************************************************/
void plot_color_chart(alt_u32 vga_base)
{
  int x, y;
  alt_u8 i;
  alt_u8 color_r, color_g, color_b, color_rgb;

  for(x=0; x<DISP_GRF_X_MAX; x++){
    for(y=0; y<DISP_GRF_Y_MAX; y++){
      if (x<240) {
        /* x < 240 */
        color_r = (alt_u8)(x/30);
        if (y<240){  // region 0
          color_g = (alt_u8)(y/30);
          color_b = 0x00;
        } else {
          color_g = (alt_u8)((y-240)/30);
          color_b = 0x02;
        }
      } else if (x<480) {
        /* 240 <= x < 480 */
        color_r = (alt_u8)((x-240)/30);
        if (y<240){  // region 2
          color_g = (alt_u8)(y/30);
          color_b = 0x01;
        } else {     // region 2
          color_g = (alt_u8)((y-240)/30);
          color_b = 0x03;
        }
      } else {
        /* 480 <= x < 640 */
        /* 3-bit color test strips */
        i = (x - 480)/20;
        if (i & 0x04 )
          color_r = 0x07;
        else
          color_r = 0x00;
        if (i & 0x02 )
          color_g = 0x07;
        else
          color_g = 0x00;
        if (i & 0x01 )
          color_b = 0x03;
        else
          color_b = 0x00;
      } // end outer if
      color_rgb = (color_r<<5) + (color_g<<2) + color_b;
      vga_wr_pix(vga_base, x, y, color_rgb);
    } // end y loop
  } // end x loop
}
```

```
/* **************************************************************
 * function: plot_random_pix()
 * purpose:  plot 30k pixels randomly
 * argument:
 *   vga_base: base address of video SRAM
 * return:
 * note:
 * *************************************************************/
void plot_random_pix(alt_u32 vga_base)
{
  int i, x, y;
  alt_u8 color;

  for(i=0;i<30000;i++){
    x=rand()%DISP_GRF_X_MAX;
    y=rand()%DISP_GRF_Y_MAX;
    color=rand()%256;
    vga_wr_pix(vga_base,x,y,color);
  }
}

/* **************************************************************
 * function: plot_random_line()
 * purpose:  plot 30 random line segments from center
 * argument:
 *   vga_base: base address of video SRAM
 * return:
 * note:
 * *************************************************************/
void plot_random_line(alt_u32 vga_base)
{
  int i, x, y;
  alt_u8 color;

  /* test for a white dot */
  vga_plot_line(vga_base,10,10,10,10,0xff);
  /* a blue vertical line */
  vga_plot_line(vga_base,600,0,600,DISP_GRF_Y_MAX-1,0x03);
  /* a green horizontal line */
  vga_plot_line(vga_base,0,400,DISP_GRF_X_MAX-1,400,0x1c);
  /* 30 random lines from center */
  for(i=0;i<30;i++){
    x=rand()%DISP_GRF_X_MAX;
    y=rand()%DISP_GRF_Y_MAX;
    color=rand()%256;
    vga_plot_line(vga_base,DISP_GRF_X_MAX/2,DISP_GRF_Y_MAX/2,x,y,color);
  } // end for
}
```

```
/***********************************************************************
 * function: plot_function()
 * purpose: plot several functions in the first quadrant
 * argument:
 *   vga_base: base address of video SRAM
 * return:
 * note:
 ***********************************************************************/
void plot_function(alt_u32 vga_base){
  const float XMAX=10.0;       // max range of x-axis
  const float YMAX=10.0;       // max range of y-axis
  float x, y, step;
  int i, j;

  step = XMAX / (float)(DISP_GRF_X_MAX);
  /* red line with small slope y=0.1*x */
  x = 0.0;
  for(i=1;i<DISP_GRF_X_MAX;i++){
    x = x + step;
    y = 0.1 * x;
    if (y < YMAX){       // plot if only y is in range
      j = DISP_GRF_Y_MAX -(y/YMAX)*DISP_GRF_Y_MAX;
      vga_wr_pix(vga_base, i, j, 0xe0);
    } // end if
  } // end for
  /* blue line with 45 degree slope y=x */
  x = 0.0;
  for(i=1;i<DISP_GRF_X_MAX;i++){
    x = x + step;
    y = x;
    if (y < YMAX){       // plot if only y is in range
      j = DISP_GRF_Y_MAX -(y/YMAX)*DISP_GRF_Y_MAX;
      vga_wr_pix(vga_base, i, j, 0x003);
    } // end if
  } // end for
  /* green steep line y=10*x */
  x = 0.0;
  for(i=1;i<DISP_GRF_X_MAX;i++){
    x = x + step;
    y = 10 * x;
    if (y < YMAX){       // plot if only y is in range
      j = DISP_GRF_Y_MAX -(y/YMAX)*DISP_GRF_Y_MAX;
      vga_wr_pix(vga_base, i, j, 0x01c);
    } // end if
  } // end for
  /* y=0.2*x*x */
  x = 0.0;
  for(i=1;i<DISP_GRF_X_MAX;i++){
    x = x + step;
    y = 0.2 * x*x;
    if (y < YMAX){       // plot if only y is in range
      j = DISP_GRF_Y_MAX -(y/YMAX)*DISP_GRF_Y_MAX;
      vga_wr_pix(vga_base, i, j, 0x1f);
    } // end if
  } // end for
  /* y = 5.0 + (5.0*sin(4.0*x)-3.0*cos(4.0*x))*exp(-0.5*x) */
  x = 0.0;
  for(i=1;i<DISP_GRF_X_MAX;i++){
    x = x + step;
    y = 5.0 + (5.0*sin(4.0*x)-3.0*cos(4.0*x))*exp(-0.5*x);
    if (y < YMAX){       // plot if only y is in range
```

```
        j = DISP_GRF_Y_MAX-(y/YMAX)*DISP_GRF_Y_MAX;
        vga_wr_pix(vga_base, i, j, 0xff);
    } // end if
  } // end for
}


/**************************************************************************
* function: plot_swap()
* purpose:  swap two 80-bit wide vertical strips
* argument:
*   vga_base: base address of video SRAM
* return:
* note:
*   - test SRAM read operation
*   - can be done pixel-by-pixel basis; buffer is for demo purposes
**************************************************************************/
void plot_swap(alt_u32 vga_base)
{
  alt_u8 buf[80*480];
  alt_u8 color;
  int x, y, x1, x2;

  x1 = rand()%8;
  x2 = rand()%8;
  if (x1==x2)
     x2 = (x1+1)%8;
  x1 = 80*x1;
  x2 = 80*x2;
  /* copy area 1 from video SRAM to buffer */
  for (y=0; y<480; y++)
    for (x=0; x<80; x++)
       buf[80*y+x]=vga_rd_pix(vga_base, x+x1, y);
  /* copy area 2 to area 1 */
  for (y=0; y<480; y++)
    for (x=0; x<80; x++){
       color=vga_rd_pix(vga_base, x+x2, y);
       vga_wr_pix(vga_base, x+x1, y, color);
  }
  /* copy buffer (area 1) to area 2 of SRAM */
  for (y=0; y<480; y++)
    for (x=0; x<80; x++)
       vga_wr_pix(vga_base, x+x2, y, buf[80*y+x]);
}
```

```
/*****************************************************************************
 * function: plot_mouse()
 * purpose:  plot mouse pointer movement
 * argument:
 *   vga_base: base address of video SRAM
 *   ps2_base: base address of PS2 controller
 *   btn_base: base address of pushbutton switch
 * return:
 * note:
 *   - the 12-by-20 MOUSE_BMP bitmap is used for mouse pointer
 *****************************************************************************/
void plot_mouse(alt_u32 vga_base, alt_u32 ps2_base, alt_u32 btn_base)
{
  mouse_mv_type mv;
  int xold, yold, xnew, ynew, act;
  static alt_u8 bdata[20*12];
  bmp_type below={12, 20, bdata};

  if (!mouse_init(ps2_base)){
    printf("Mouse initialization failed.\n");
    return;
  }
  xold = 320;
  yold = 240;
  vga_init_mouse_ptr(vga_base, ps2_base, xold, yold, &MOUSE_BMP, &below);
  /* continue until a button is pressed */
  while (!btn_is_pressed(btn_base)){
    act = vga_move_mouse_ptr(vga_base, ps2_base,
           xold, yold, &below, &xnew, &ynew, &MOUSE_BMP, &mv);
    if (act==1){
      if(mv.lbtn)
         printf("\ncurrent mouse location: %d %d", xnew, ynew);
      xold = xnew;
      yold = ynew;
    } //end if
  } // end while
  vga_wr_bitmap(vga_base, xold, yold, &below, 0);
  printf("\n");
}
```

```
/*****************************************************************************
 * function: plot_text()
 * purpose:  test bit-mapped fonts for text
 * argument:
 *   vga_base: base address of video SRAM
 * return:
 * note: pixel (x,y) changes each time and
 *       new patterns overlay over old patterns
 *****************************************************************************/
void plot_text(alt_u32 vga_base){

  int x, y;
  char buffer[50];
  char msg_box[]=          // a 30-by-5 message
    "******************************\n"
    "*                            *\n"
    "*          Hello World       *\n"
    "*                            *\n"
    "******************************";

  /* display four single characters at four corners */
  vga_wr_bit_ch(vga_base, 0, 0, 'a', 0xff, 1);
  vga_wr_bit_ch(vga_base, DISP_GRF_X_MAX-8, 0, 'b', 0xe0, 1);
  vga_wr_bit_ch(vga_base, 0, DISP_GRF_Y_MAX-16, 'c', 0x1c, 1);
  vga_wr_bit_ch(vga_base, DISP_GRF_X_MAX-8, DISP_GRF_Y_MAX-16,
                'd', 0x03, 1);
  /* display a single string in three zoom factors */
  vga_wr_bit_str(vga_base, 34*8, 3*16, "Hello World", 0x1c, 1);
  vga_wr_bit_str(vga_base, 28*8, 5*16, "Hello World", 0x1c, 2);
  vga_wr_bit_str(vga_base, 23*8, 8*16, "Hello World", 0x1c, 3);
  /* display a string with multiple lines (string with \n) */
  vga_wr_bit_str(vga_base, 25*8, 16*16, msg_box, 0x1c, 1);
  /* get a formatted string by sprintf() */
  vga_rd_xy(vga_base, &x, &y);
  sprintf(buffer, "current pixel (x,y): (%3d, %3d)", x, y);
  vga_wr_bit_str(vga_base, 24*8, 24*16, buffer, 0x03, 1);
}
```

```
/***************************************************************************
 * function: main()
 * purpose:  test video SRAM access, mouse and bit-mapped font
 * note:
 ***************************************************************************/
int main(void){
  int sw, btn;
  alt_u8 disp_msg[4]={sseg_conv_hex(13), sseg_conv_hex(1),
                      sseg_conv_hex(5), 0x0c};

  sseg_disp_ptn(SSEG_BASE, disp_msg);         // show "dISP" for display
  vga_clr_screen(VRAM_BASE,0);                // clear screen
  printf("VGA video controller test: \n\n");
  btn_clear(BTN_BASE);
  while (1){
    while (!btn_is_pressed(BTN_BASE)){ };      // wait for button
    btn=btn_read(BTN_BASE);                     // read button
    if (btn & 0x02){                            // key1 pressed
      sw=pio_read(SWITCH_BASE);                 // read switch
      printf("key/sw: %d/%d\n", btn, sw);
    }
    btn_clear(BTN_BASE);
    switch (sw){
      case 0:  // clear screen
        vga_clr_screen(VRAM_BASE,0);
        break;
      case 1:  // plot color chart
        plot_color_chart(VRAM_BASE);
        break;
      case 2:  // plot random pixels
        plot_random_pix(VRAM_BASE);
        break;
      case 3:  // plot random lines
        plot_random_line(VRAM_BASE);
        break;
      case 4:  // plot several functions
        plot_function(VRAM_BASE);
        break;
      case 5:  // swap two vertical strips on screen
        plot_swap(VRAM_BASE);
        break;
      case 6:  // test mouse pointer
        plot_mouse(VRAM_BASE,PS2_BASE,BTN_BASE);
        break;
      case 7:  // display bit-mapped text
        plot_text(VRAM_BASE);
        break;
    } //end switch
  } // end while
}
```

Listing 17.25   chu_avalon_file.h

```
/*******************************************************************
*
* Module:   File retrieval header
* File:     chu_avalon_file.h
* Purpose:  Simple file access routines
*
*******************************************************************/
/* file inclusion */
#include <alt_types.h>
#include <stdlib.h>
#include <io.h>


/*******************************************************************
* Function prototypes
*******************************************************************/
/* basic file access routines */
void fskip(FILE *fp, int nbyte);
alt_u8 fget8(FILE *fp);
alt_u16 fget16(FILE *fp);
alt_u32 fget32(FILE *fp);
```

Listing 17.26   chu_avalon_file.c

```
/*****************************************************************
 *
 * Module:    File retrieval function prototypes
 * File:      chu_avalon_file.c
 * Purpose:   Routines to retrieve bitmap from BMP file
 *            Note: must use debug mode to use host-based file system
 *
 *****************************************************************/
/* file inclusion */
#include <stdio.h>
#include "chu_avalon_file.h"


/*****************************************************************
 * function: fskip()
 * purpose:   skip a specific number of bytes in a file
 * argument:
 *   fp: file pointer
 *   nbyte: # bytes to skip
 * return:
 * note:
 *****************************************************************/
void fskip(FILE *fp, int nbyte)
{
  int i;
  for (i=0; i<nbyte; i++)
    fgetc(fp);
}


/*****************************************************************
 * function: fget8()
 * purpose:   get 8 bits (one byte) from a file
 * argument:
 *   fp: file pointer
 * return: a byte in alt_u8 format
 * note:
 *****************************************************************/
alt_u8 fget8(FILE *fp)
{
  return( (alt_u8)fgetc(fp));
}


/*****************************************************************
 * function: fget16()
 * purpose:   get 16 bits (half word) from a file
 * argument:
 *   fp: file pointer
 * return: a half word in alt_u16 format
 * note: "little endian" byte ordering
 *****************************************************************/
alt_u16 fget16(FILE *fp)
{
  alt_u16 b0, b1, r;

  b0 = (alt_u16) fgetc(fp);
  b1= (alt_u16) fgetc(fp);
  r = (b1 << 8) + b0;
  return(r);
}
```

```
/****************************************************************************
* function: fget32()
* purpose:  get 32 bits (a word) from a file
* argument:
*   fp: file pointer
* return: a word in alt_u32 data type
* note: "little endian" byte ordering
****************************************************************************/
alt_u32 fget32(FILE *fp)
{
  alt_u32 b0, b1, b2, b3, r;

  b0 = (alt_u32) fgetc(fp);
  b1= (alt_u32) fgetc(fp);
  b2= (alt_u32) fgetc(fp);
  b3= (alt_u32) fgetc(fp);
  r = (b3<<24) + (b2<<16) + (b1<<8) + b0;
  return(r);
}
```

Listing 17.27 chu_main_bmp_file_test.c

```
/*****************************************************************
 *
 * Module:   BMP file retrieval test
 * File:     chu_main_bmp_file_test.c
 * Purpose:  Test BMP (.bmp) file access and demonstrate simple animation
 * IP core base addresses:
 *    - SWITCH_BASE: slide switch
 *    - VRAM_BASE: video SRAM
 *
 *****************************************************************/
/* file inclusion */
#include <stdio.h>
#include <unistd.h>
#include "system.h"
#include "chu_avalon_gpio.h"
#include "chu_avalon_vga.h"
#include "chu_avalon_file.h"

/* constants */
#define   BMP_FILE_NAME   "/mnt/host/earth9.bmp"    // path and file name
#define   NF 9                                       // # of frames
#define   NW 100                                     // width of a frame
#define   NH 100                                     // height of a frame

/*****************************************************************
 * function: read_bmp_file()
 * purpose:  read bitmap from BMP file and store it in a buffer
 * argument:
 *    file_name: name of the BMP file
 *    buf: pointer to bitmap buffer
 * return: 0 if successful
 * note:
 *    - calling function must allocate buffer space
 *    - only 24-bit BMP file is supported
 *    - 24-bit color is converted to 8-bit color
 *****************************************************************/
int read_bmp_file(char *file_name, bmp_type *buf)
{

  FILE *fp;
  int color_bit, x, y, slack;
  alt_u8 pixr, pixg, pixb, pix;

  /* open the file */
  fp = fopen(file_name,"rb");
  if (fp==NULL){
    printf("Error: %s fails to open. \n", file_name);
    return(-1);
  }
  /* check "magic number"; should be BM */
  if (fgetc(fp)!='B' || fgetc(fp)!='M'){
    printf("Error: %s is not a .bmp file.\n", file_name);
    fclose(fp);
    return(-1);
  }
  fskip(fp,16);                           // offset 2
  /* get height and width */
  buf->width  = (int) fget32(fp);         // offset 18
  buf->height = (int) fget32(fp);         // offset 22
  fskip(fp,2);                            // offset 26
```

```
/* check color depth; should be 24 */
color_bit = (int) fget16(fp);          // offset 28
if (color_bit != 24){
  fclose(fp);
  printf("Error: color depth is not 24. \n");
  return(-2);
}
/* 24-bit BMP file format confirmed */
printf("File opened. Bitmap width: %d, height: %d\n",
        buf->width, buf->height);
fskip(fp,24);                          // offset 30
printf("Reading in progress: \n");
/* loop through pixel data, staring with the bottom row */
for (y=buf->height-1; y >= 0; y--){ // offset 54
  printf(".");
  for (x=0; x< buf->width; x++) {
    /* get 24-bit color */
    pixb = (alt_u8) fgetc(fp);
    pixg = (alt_u8) fgetc(fp);
    pixr = (alt_u8) fgetc(fp);
    /* construct 8-bit color using MSBs */
    pix = (pixr & 0xe0) + ((pixg & 0xe0)>>3) + ((pixb & 0xc0)>>6);
    buf->pdata[y*(buf->width)+x]=pix;
  } //end for x
  /* skip padding bytes, if exist */
  slack = (buf->width * 3) % 4;
  if (slack!=0)
    fskip(fp,4-slack);
} //end for y
fclose(fp);
printf("\nFile loaded.\n");
return(0);
}
```

```
/*****************************************************************************
* function: main()
* purpose:  Read a bitmap file and demonstrate simple animation
* note:
*   - the host-based file system must be enabled in BSP editor
*   - the earth9.bmp bitmap file should be in the project directory
*   - build/load the program with "Debug As => Nios II Hardware"
*****************************************************************************/
int main(void)
{
  alt_u8 iarray[NF*NH*NW];            // pixel data array for file buffer
  alt_u8 farray[NF][NH*NW];           // pixel data array for frames
  bmp_type img={NF*NW, NH, iarray};
  bmp_type frame[NF];
  int sw, status;
  int f, x, y;

  printf("BMP file retrieval test. \n\n");
  vga_clr_screen(VRAM_BASE, 0x08);
  status = read_bmp_file(BMP_FILE_NAME, &img);
  if (status!=0){
    printf("BMP file fails to load.  Exit program. \n");
    exit(1);
  }
  /* split bitmaps to 9 frames */
  for (f=0; f<NF; f++){
    frame[f].height = NH;
    frame[f].width = NW;
    frame[f].pdata = farray[f];
    for (y=0; y<NH; y++) {
      for (x=0; x<NW; x++) {
        frame[f].pdata[y*NW+x]=img.pdata[y*NW*NF+f*NW+x];
      } // end x loop
    } // end y loop
  } //end f loop

  /* display individual frames in two rows  */
  for (f=0; f<5; f++)
    vga_wr_bitmap(VRAM_BASE, 30+f*(20+NW), 10, &frame[f], 1);
  for (f=5; f<9; f++)
    vga_wr_bitmap(VRAM_BASE, 80+(f-5)*(20+NW), 120, &frame[f], 1);

  /* animation of rotating earth  */
  while(1){
    for (f=0; f<NF; f++){
    sw = pio_read(SWITCH_BASE); ;
      if (sw!=0){                  // freeze if sw=0
        usleep(1000000/sw);   // sw frames per second
        vga_wr_bitmap(VRAM_BASE, 270, 250, &frame[f],1);
      } // end if
    } // end for
  } // end while
}
```

CHAPTER 18

AUDIO CODEC CONTROLLER

An audio codec device can digitize an analog audio signal and convert the digitized signal back to analog format. The DE1 board contains a Wolfson WM8731 codec device. In this chapter, we discuss the interface circuit for WM8731 and and the software driver. The hardware portion includes an I^{2}C controller for device configuration and a data access control circuit to process incoming and outgoing audio data streams. The software portion includes the driver routines to send an I^{2}C packet, set up WM8731, and to transmit and receive audio data as well as basic routines to process an audio file in uncompressed WAV format.

18.1 INTRODUCTION

Audio signals are in a continuous analog format. To take advantage of the capability of a digital system, an analog signal must be converted to discrete digital numbers. The digitized data then can be stored and processed by a digital system. During playback, the data is converted back to an analog signal. A specific device, known as an *audio codec*, is designed for this purpose. The DE1 board contains a codec device, WM8731, from Wolfson Microelectronics.

18.1.1 Overview of codec

An audio codec device converts (i.e., *co*de) analog signals into digital data and converts (i.e., *decode*) the data back to analog signals. It is essentially an interface from and to the analog domain. A codec contains an *ADC* (*analog-to-digital converter*) and a *DAC* (*digital-to-analog converter*). An ADC converts an input analog voltage to a discrete digital number proportional to the value of the voltage. A DAC performs the reversed operation and converts a digital number to an analog voltage level.

Two important parameters of ADC and DAC specifications are the *resolution* and *sampling rate*. The resolution of a convertor specifies the number of discrete values produced over the range of input voltage. It is represented in terms of bits in the discrete values. For example, a resolution of 16 bits indicates that an input voltage can be converted to one of the possible 2^{16} levels. Higher resolution can reduce quantization error. The sampling rate of a convertor specifies the number of analog values sampled per second. The Nyquist–Shannon sampling theorem indicates that to reconstruct a signal, it must be sampled at least at *Nyquist rate*, which is two times the bandwidth of an input signal. For example, the frequency range of an audio signal is between 20 Hz and 20 kHz and thus the minimal sampling frequency should be 40 kHz. Higher sampling rates sometimes are used to facilitate the filter design and reduce noise.

Note that the *codec* term is also used as *co*mpressor and *dec*ompressor, which concerns the data compression algorithm and is not related to an audio codec device.

18.1.2 Overview of WM8731 device

The Wolfson WM8731 device is a stereo codec device. It contains two pairs of ADCs and DACs to accommodate both the left and right channels of stereo audio. The converters support 24-bit resolution and up to 96 kHz sampling rate. In addition to ADCs and DACs, the device also contains digital filters, analog multiplexing and mixing circuits, input gain control circuits, and headphone amplifiers. Some features of WM8731, such as the headphone volume and sampling rate, are "programmable." We can configure the device and select the desired options by issuing commands to its internal configuration registers.

The simplified top-level diagram is shown in Figure 18.1. It illustrates the main data flow and the key I/O signals. The I/O signals can be divided into three groups. The first group is the analog signals. It consists of rlinein and llinein (for "right-channel line input" and "left-channel line input"), micin ("for microphone input"), and rhpout and lhpout (for "right-channel headphone output" and "left-channel headphone output"). The second group is the digitalized data signals and clocks used in the *digital audio interface*. It consists of dacdat (for "data to DAC"), adcdat (for "data from ADC"), mclk (for "master clock"), bclk (for "bit clock"), daclrc (for "DAC left-right channel clock"), and adclrc (for "ADC left-right channel clock"). The last group is signals for the *control interface* and includes sdin (for "serial data in") and sclk (for "serial clock").

The main ADC data flow is shown as thick gray line in Figure 18.1. The major components of the path are:

- Amplification circuits: adjust the input voltage level (gain and mute).
- Two ADCs: digitize the analog left- and right-channel line input signals.
- Digital filter: perform decimation (reduction of the number of samples).
- Digital audio interface: output two sampled data through a 1-bit serial data line.

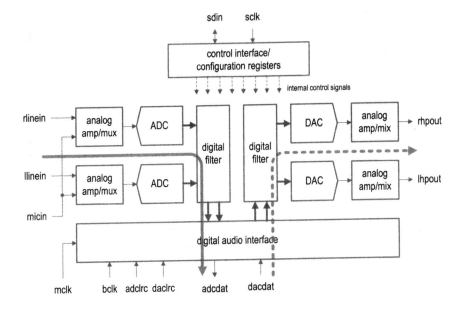

Figure 18.1   Conceptual diagram of WM8731 codec.

The microphone input can also be routed to the ADC via the analog multiplexers. The main DAC data flow is shown as dashed gray line in Figure 18.1. It mirrors the ADC flow but in the reversed order.

Some features of WM8731 can be configured by writing the desired values to its configuration registers via the control interface. The output of these registers in turn controls various internal components, as indicated by the dotted lines in Figure 18.1.

18.1.3   Registers of WM8731 device

WM8731 has 11 configuration registers. Each register contains 9 bits of data and is identified by 7 bits of address. During a write operation, the address and data bits are grouped together to form a 16-bit word and transmitted serially via the control interface. There are more than three dozen fields within these registers and the register map is shown in Figure 18.2. The complete definitions and explanations of these fields can be found in the WM8731 manual. Some key features are summarized in the following paragraphs.

Registers R0 and R1 mainly control the gains of the left and right line inputs. The fields of R0 are:

- LINVOL (for "left line input volume"): It controls the amplifier gain from the left line input. The gain is logarithmically adjustable from +12 dB to −34.5 dB in 1.5-dB steps. The default value is 10111 (defined as 0 dB).
- LINMUTE (for "left line input mute"): When it is asserted (i.e., 1), the left line input to ADC is muted.
- LRINBOTH (for "left line input controlling both channels"): When it is asserted (i.e., 1), the left line gain and mute values will be loaded to the corresponding fields of the right channel simultaneously.

register	address	data									
		B15 - B9	B8	B7	B6	B5	B4	B3	B2	B1	B0
R0	000000		LRIN BOTH	LIN MUTE	0	0	LINVOL				
R1	000001		RLIN BOTH	RIN MUTE	0	0	RINVOL				
R2	000010		LRHP BOTH	LZCEN	LHPVOL						
R3	000011		RLHP BOTH	RZCEN	RHPVOL						
R4	000100		0	SIDEATT		SIDE TONE	DAC SEL	BYPASS	INSEL	MUTE MIC	MIC BOOST
R5	000101		0	0	0	0	HPOR	DAC MU	DEEMPH		ADC HPD
R6	000110		0	PPW OFF	CLK OUTPD	OSCPD	OUTPD	DACPD	ADCPD	MICPD	LININPD
R7	000111		0	BCLK INV	MS	LR SWAP	LRP	IWL		FORMAT	
R8	001000		0	CLKO DIV2	CLKI DIV2	SR				BOSR	USB/ NORM
R9	001001		0	0	0	0	0	0	0	0	ACTIVE
R15	001111	RESET									

Figure 18.2   WM8731 register map.

The fields of R1 mirror those in R0, but for the right channel.

Registers R2 and R3 mainly control the volumes of the left and right headphone outputs. The fields of R2 are:

- LHPVOL (for "left headphone volume"): It controls the volume of the left headphone output. The gain is logarithmically adjustable from +6 dB to −73 dB in 1-dB steps. The default value is 1111001 (defined as 0 dB). The 0110000 value corresponds to −73 dB and any value smaller mutes the output.
- LZCEN (for "left zero cross detect enable"): It indicates whether to enable the zero-crossing detection circuit, which can reduce certain click noises.
- LRHPBOTH (for "left headphone controlling both channels"): When it is asserted, the left headphone volume and zero-crossing values will be loaded to the corresponding fields of the right channel simultaneously.

The fields of R3 mirror those in R2, but for the right channel.

Register R4 sets up the analog path and controls the microphone input. The ADC's input source can either from the line input or microphone input and is specified by the following field:

- INSEL (for "input select"): It is 1 for the line input and 0 for the microphone input.

The headphone amplifier has three input sources, which are DAC's output, line input (without passing through ADC), and microphone input (without passing through ADC). These sources can be summed together and fed to the amplifier. Three fields specify whether a specific source is included:

- BYPASS (for "bypass ADC"): If it is 1, the line input is included.
- DACSEL (for "DAC select"): If it is 1, the DAC output is included.
- SIDETONE (for "side tone select"): If it is 1, the microphone input (i.e., which is known as "side tone") is included.

The remaining fields control the microphone operation:

- MICBOOST (for "microphone boost"): When it is asserted, the microphone input signal is passed through a second amplifier with a fixed 20 dB gain.
- MUTEMIC (for "mute microphone"): When it is asserted, the microphone input to ADC is muted.
- SIDEATT (for "side tone attenuation"): When the microphone input is included for the headphone amplifier, it specifies whether to apply some attenuation (between −6 dB and −15 dB) to the signal.

Register R5 sets up options relevant to digital audio processing. The fields are:

- ADCHPD (for "ADC high pass filter disabled"): It specifies whether to disable or enable the ADC's high pass filter.
- DEEMP (for "de-emphasis control"): It sets up the DAC's de-emphasis filter.
- DACMU (for "DAC mute"): When it is asserted, the DAC is "digitally muted."
- HPOR (for "high-pass filter offset "): It indicates whether to store or clear the DC offset value when the DAC's high-pass filter is disabled.

Register R6 controls the power down of various circuit blocks. Since power conservation is not a concern in our application, all fields are set to 0's (i.e., power up for all circuits).

Registers R7 and R8 set up the digital audio interface and sampling rates. We design a control circuit to access the codec's data in Section 18.3. The design is based on a fixed set of parameters and the fields of R7 and R8 must be configured accordingly. The fields and the desired values in R7 are listed below and additional explanations can be found in Section 18.3.

- FORMAT (for "audio data format select"): It specifies the basic format of serial data stream and must be 01 (for "left-adjust format").
- IWL (for "input data width select"): It specifies the resolution (i.e., number of bits per sample) and must be 00 (for 16 bits).
- LRP (for "left right phase control"): WM8731 uses a signal, daclrc, to indicate the current channel (i.e., left or right) of the DAC data stream. This field indicates the "phase" interpretation of this signal. It must be 0, which means that the left channel data is transmitted when daclrc is high.
- LRSWAP (for "left right clock swap"): It should be 0 (i.e., no swap).
- MS (for "master slave mode"): WM8731 can be configured as a "master" and generates the relevant clock signals or can be configured as a "slave" and lets the external device generate clock signals. This field must be 0 (for the "slave mode").
- BCLKINV (for "bit clock invert"): It specifies whether to invert the bit clock signal and must be 0 (for "no invertion").

Register R8 specifies the codec's master clock frequency and audio data sampling rate, which are 12.288 MHz and 48 K samples per second in our design. We can achieve this by configuring R8's fields as follows:

- USB/NORMAL (for "USB or normal mode select"): It must be set to 0 to select the normal mode.
- BOSR (for "base over-sampling rate"): It must be set to 0.
- SR (for "sampling rate control"): It must be set to 0000. This value in conjunction with the previous BOSR setting specifies the sampling rate of ADC and DAC to be 48 K samples per second.
- CLKIDIV2 (for "core clock divided-by-2 select"): It must be set to 0 for not using the divided-by-2 master clock (mclk) for the system core clock.
- CLKODIV2 (for "clock out divided-by-2 select"): It must be set to 0 for not using the divided-by-2 system core clock for the codec's clockout signal.

Register R9 contains only one field, ACTIVE. It must be set to 1 to activate the digital audio interface. Register R15 is used to facilitate the device reset operation. It does not contain any field. The operation is done by writing a dummy data to the 001111 address.

18.2 I²C CONTROLLER

The WM8731 device can be configured via the control interface, which supports both 2- and 3-wire serial bus protocols. The DE1 board is constructed to support a 2-wire I^2C (*inter-integrated circuit*) bus protocol.

18.2.1 Overview of I²C interface

The I^2C protocol is a low-speed serial bus for efficient communication between devices. Its *standard mode* supports a data rate up to 100 K bits per second. The I²C bus consists of two bidirectional lines, sda (for "serial data") and scl (for "serial clock"), for data and clock, respectively. The two lines are connected to the sdin and sclk pins of WM8731. Each device on the I²C bus has a unique address and can operate as either a transmitter or receiver. During the operation, one device on the bus functions as the *master* and other devices function as *slaves*. The master generates the clock on the scl signal and also initiates and terminates the data transfer. The master and the designated slave place data on or retrieve data from the sda signal.

Electrical characteristics The physical connection of the I²C bus is shown in Figure 18.3. It uses *open-drain* technology, which means that the output stage of a device must have an open-drain structure. Both sda and scl lines are connected to the voltage source (V_{DD}) via pull-up resistors and are high when the bus is idle. A line becomes low as soon as one device's output turns to low. It thus performs the *wired-and* function.

Basic timing The basic timing diagram of a typical data transfer is shown in Figure 18.4. Both lines are high when the bus is idle. The master initiates a transfer by creating the *start* (S) condition, in which sda changes from high to low while scl is high. It then generates the clock signal on scl. Based on the type of transaction, either the master or the designated slave places data on sda. The data must be stable when scl is high and thus the change can only occur when scl is low. The definitions insure that the start and stop conditions will never be confused as data.

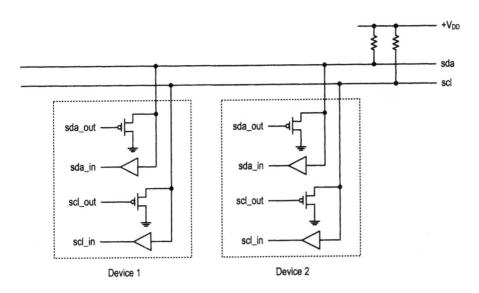

Figure 18.3   Conceptual diagram of I²C bus.

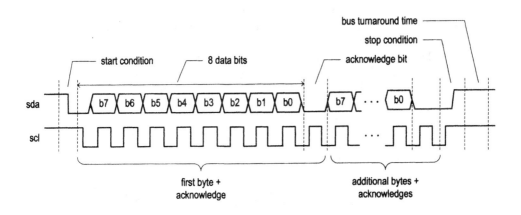

Figure 18.4   Basic timing diagram of I²C data transfer.

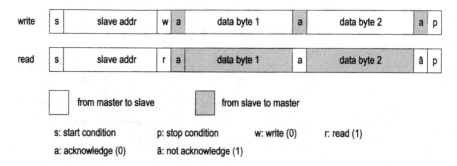

Figure 18.5   Complete sequence of reading and writing two bytes of data.

The transfer is done on a byte-by-byte basis with MSB first. Each byte is followed by an acknowledge bit in the ninth clock cycle. The number of bytes in a transfer is unrestricted. After completion, the master terminates a transfer by creating the *stop* (P) condition, in which sda changes from low to high while scl is high. After the transaction is done, the devices on the bus must wait for a small *turnaround time* before initiating another transaction.

Bus protocol   When the bus is idle, a device can initiate a data transfer and becomes the master. A typical data transfer is composed of several parts: start, slave address plus direction bit, data, and stop. The master initiates the start condition and then sends the slave's 7-bit address (i.e., device id) plus a direction bit. The address selects the desired device on the bus and the direction bit indicates the direction of data flow, in which a 1 is for read operation (i.e., from slave to master) and a 0 is for write operation (i.e., from master to slave). The master then releases the sda line and the slave with the matched address should acknowledge by pulling sda low during the ninth clock cycle. The data transfer then can proceed byte-by-byte as specified by the direction bit. The receiving device acknowledges in the ninth clock cycle. After completion, the master generates a stop condition and frees the bus. The complete sequences of writing and reading two bytes of data are shown in Figure 18.5. Note that the master should generate a "not acknowledge" (i.e., 1) in the ninth cycle of the read operation after it receives the last data byte.

The I^{2}C protocol also incorporates clock synchronization and arbitration mechanisms to support a slow slave device and accommodate multiple masters. Additional information can be found in the bibliographic section.

18.2.2   HDL implementation

The DE1 board uses I^{2}C as the interface to configure the codec device. The bus organization is very simple. It contains only two devices, the FPGA chip, which acts as a master, and the WM8731 chip, which acts as a slave. Only a write operation is performed.

Configuring a WM8731's internal register requires to send a 7-bit register id (i.e., register address) and a 9-bit register data value. The two fields are combined to form two bytes of information. Although there is only one slave on the bus, the slave address (i.e., the designated id of WM8731) and direction bit (always 0

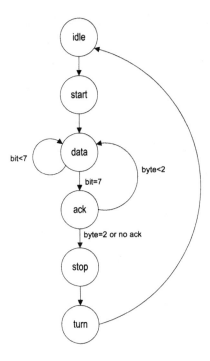

Figure 18.6   Sketch of I²C controller FSM.

for writing) are still needed. The complete transmitting sequence is similar to the writing sequence in Figure 18.5.

A custom I²C controller can be constructed for this purpose. It acts as a master and transmits three bytes of information to the designated codec slave. The controller is basically an FSMD that generates proper scl and sda signals according to the timing sequence in Figure 18.5. The sketch of the control FSM is shown in Figure 18.6. The **start**, **stop**, and **turn** states represent the I²C's start condition, stop condition, and required turnaround time between two transactions. The **data** and **ack** states are for one-bit transfer and acknowledge. Two counters, **bit** and **byte**, are used to keep track of the numbers of bits and bytes processed so far. The FSM circulates in the **data** state eight times to transfer eight bits of data and then moves to the **ack** state. If the proper acknowledge bit is received, the process repeats three times to transfer three bytes of information.

In the actual implementation, we need to divide certain "sketch states" to two or three states to meet the timing requirement. For example, the original **data** state is divided into **data1**, **data2**, and **data3** states. The division of the **data** state is shown in detail in Figure 18.7(a) and the complete division is shown in Figure 18.7(b). Note that two states, **scl_begin** and **scl_end**, are inserted to generate adequate times for the low interval.

We can derive the HDL code based on the FSM. The main input, **din**, is 24-bit (i.e., 3 bytes) information to be transferred. The two main outputs, **i2c_sclk** and **i2c_sdat**, are connected to the bidirectional **scl** and **sda** lines. In addition, the controller includes an input command, **wr_i2c**, which initiates the data transfer,

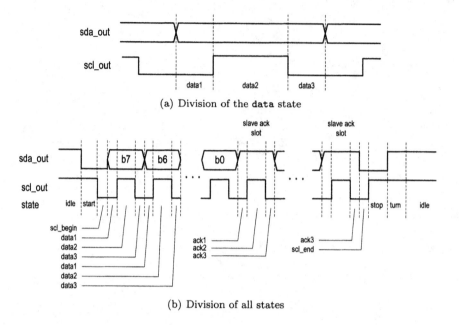

(a) Division of the **data** state

(b) Division of all states

Figure 18.7   State division of I^{2}C transactions.

and three status signals, i2c_idle, i2c_fail, and i2c_done_tick. The HDL code is shown in Listing 18.1.

Listing 18.1   I^{2}C controller

```
1 library ieee;
  use ieee.std_logic_1164.all;
  use ieee.numeric_std.all;
  entity i2c is
     port (
6        clk, reset: in  std_logic;
         din: in std_logic_vector(23 downto 0);
         wr_i2c: in std_logic;
         i2c_sclk: out std_logic;
         i2c_sdat: inout std_logic;
11       i2c_idle, i2c_fail: out std_logic;
         i2c_done_tick: out std_logic
     );
  end i2c;

16 architecture arch of i2c is
     constant HALF: integer := 249; -- 10us/20ns/2 = 250
     constant QUTR: integer := 125; -- 10us/20ns/4 = 125
     constant C_WIDTH: integer := 8;
     type statetype is (
21      idle, start, scl_begin, data1, data2, data3,
        ack1, ack2, ack3, scl_end, stop, turn);
     signal state_reg, state_next: statetype;
     signal c_reg,c_next: unsigned(C_WIDTH-1 downto 0);
     signal data_reg,data_next: std_logic_vector(23 downto 0);
26   signal bit_reg, bit_next: unsigned(2 downto 0);
     signal byte_reg, byte_next: unsigned(1 downto 0);
     signal sdat_out, sclk_out: std_logic;
```

```vhdl
   signal sdat_reg, sclk_reg: std_logic;
   signal ack_reg, ack_next: std_logic;
31 begin
   --=========================================================
   -- output
   --=========================================================
   -- buffer for sda and scl lines
36 process (clk, reset)
   begin
      if reset='1' then
         sdat_reg <= '1';
         sclk_reg <= '1';
41    elsif (clk'event and clk='1') then
         sdat_reg <= sdat_out;
         sclk_reg <= sclk_out;
      end if;
   end process;
46 -- only master drives scl line
   i2c_sclk <= sclk_reg;
   -- i2c_sdat are with pull-up resistors
   -- and becomes high when not driven
   i2c_sdat <= 'Z' when sdat_reg='1' else '0';
51 -- codec fails to acknowledge properly
   i2c_fail <= '1' when ack_reg='1' else '0';

   --=========================================================
   -- fsmd for transmitting three bytes
56 --=========================================================
   -- registers
   process (clk, reset)
   begin
      if reset='1' then
61       state_reg <= idle;
         c_reg <= (others=>'0');
         bit_reg <= (others=>'0');
         byte_reg <= (others=>'0');
         data_reg <= (others=>'0');
66       ack_reg <= '1';
      elsif (clk'event and clk='1') then
         state_reg <= state_next;
         c_reg <= c_next;
         bit_reg <= bit_next;
71       byte_reg <= byte_next;
         data_reg <= data_next;
         ack_reg <= ack_next;
      end if;
   end process;
76 -- next-state logic
   process(state_reg,bit_reg,byte_reg,data_reg,c_reg,
           din,wr_i2c,i2c_sdat)
   begin
      state_next <= state_reg;
81    sclk_out <= '1';
      sdat_out <= '1';
      c_next <= c_reg + 1;  -- timer counts continuously
      bit_next <= bit_reg;
      byte_next <= byte_reg;
86    data_next <= data_reg;
      ack_next <= ack_reg;
      i2c_done_tick <='0';
      i2c_idle <='0';
```

```vhdl
         case state_reg is
91           when idle =>
                i2c_idle <= '1';
                if wr_i2c='1' then
                   data_next <= din;
                   bit_next <= "000";
96                 byte_next <="00";
                   c_next <= (others=>'0');
                   state_next <= start;
                end if;
             when start =>          -- start condition
101             sdat_out <= '0';
                if c_reg=HALF then
                   c_next <= (others=>'0');
                   state_next <= scl_begin;
                end if;
106          when scl_begin =>   -- 1st half of scl=0
                sclk_out <= '0';
                if c_reg=QUTR then
                   c_next <= (others=>'0');
                   state_next <= data1;
111             end if;
             when data1 =>
                sdat_out <= data_reg(23);
                sclk_out <= '0';
                if c_reg=QUTR then
116                c_next <= (others=>'0');
                   state_next <= data2;
                end if;
             when data2 =>
                sdat_out <= data_reg(23);
121             if c_reg=HALF then
                   c_next <= (others=>'0');
                   state_next <= data3;
                end if;
             when data3 =>
126             sdat_out <= data_reg(23);
                sclk_out <= '0';
                if c_reg=QUTR then
                   c_next <= (others=>'0');
                   if bit_reg=7 then          -- done with 8 bits
131                   state_next <= ack1;
                   else
                      data_next <= data_reg(22 downto 0) & '0';
                      bit_next <= bit_reg + 1;
                      state_next <= data1;
136                end if;
                end if;
             when ack1 =>
                sclk_out <= '0';
                if c_reg=QUTR then
141                c_next <= (others=>'0');
                   state_next <= ack2;
                end if;
             when ack2 =>
                if c_reg=HALF then
146                c_next <= (others=>'0');
                   state_next <= ack3;
                   ack_next <= i2c_sdat;   -- read ack from slave
                end if;
             when ack3 =>
```

```
151          sclk_out <= '0';
             if c_reg=QUTR then
                 c_next <= (others=>'0');
                 if ack_reg = '1' then      -- slave fails to ack
                     state_next <= scl_end;
156              else
                     if byte_reg=2 then     -- done with 3 bytes
                         state_next <= scl_end;
                     else
                         bit_next <= "000";
161                      byte_next <= byte_reg + 1;
                         data_next <= data_reg(22 downto 0) & '0';
                         state_next <= data1;
                     end if;
                 end if;
166          end if;
         when scl_end =>                     -- 2nd half of scl=0
             sclk_out <= '0';
             sdat_out <= '0';
             if c_reg=QUTR then
171              c_next <= (others=>'0');
                 state_next <= stop;
             end if;
         when stop =>                        -- stop condition
             sdat_out <= '0';
176          if c_reg=HALF then
                 c_next <= (others=>'0');
                 state_next <= turn;
             end if;
         when turn =>                        -- turnaround time
181          if c_reg=HALF then
                 state_next <= idle;
                 i2c_done_tick <= '1';
             end if;
     end case;
186  end process;
end arch;
```

The FSMD basically follows the sequence outlined in Figure 18.6. There are several registers in the FSMD. The data_reg stores the 24-bit input data and shifts out the MSB sequentially in the data3 or ack3 state. The bit_reg and byte_reg keep track of the numbers of bits and bytes processed, respectively.

We design the I²C controller to transfer data at a rate of 100 K bits per second, which translates to 10 microseconds per bit. A state in the FSMD usually lasts a half of the interval (i.e., 5 microseconds) or a quarter of the interval (i.e., 2.5 microseconds). Since the 50-MHz (i.e., a period of 0.02 microseconds) system clock is much faster, we use the c_reg register as a timer to keep track of the amount of time spent in each state. Two constants, HALF and QUTR, are used to represent the number of system clock cycles for half and quarter of 10-microsecond intervals. The c_reg register counts continuously and is cleared to zero when the FSM exits the previous state. The FSMD can only move out from the current state when c_reg reaches the designated value.

The output from the FSMD, sclk_out and sdat_out, are passed to two buffers to remove any potential glitches and the buffered outputs are used to control the scl and sda lines, which are labeled i2c_sclk and i2c_sdat in code. Since the slave needs to send an acknowledge bit to the master, the data transfer in sda line is bidirectional. The two-way communication is achieved by the wired-and

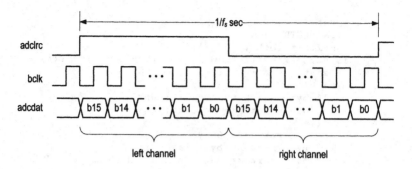

Figure 18.8   Timing diagram of ADC output in left justified mode.

operation of the open-drain bus structure. Although not supported directly, the wired-and operation can be implemented by FPGA I/O pin's tristate buffer. The corresponding HDL statement is

```
i2c_sdat <= 'Z' when sdat_reg='1' else '0';
```

In this scheme, the FPGA device turns off the tristate buffer (i.e., make the output in a high-impedance state) when a desired bus line level is 1. Since a bus line is connected to V_{DD} via a pull resister, it is driven to 1 implicitly when all devices output 1 or are in high-impedance state.

18.3   CODEC DATA ACCESS CONTROLLER

18.3.1   Overview of digital audio interface

The *digital audio interface* of WM8731 receives and transmits the digitized data. To save the physical I/O pins, the data transfer is performed serially via two pins. The adcdat pin transmits the ADC output from codec to FPGA and the dacdat pin receives the DAC input from FPGA to codec. The clock signals, bclk, daclrc, and adclrc, indicate the validity of data bits and distinguish the left and right channels. WM8731 and the data access controller (within FPGA device) must perform the needed parallel-to-serial conversion and serial-to-parallel conversion.

WM8731 can be operated in either one of the four *audio interface modes*, which specify the timing relationship between the clock signals and serial data. Our data access controller uses the *left justified mode* and a fixed 16-bit resolution. The timing diagram of the 16-bit ADC output in this mode is shown in Figure 18.8. The bclk signal functions as a bit clock and a new data bit is transmitted out in the adcdat line at every high-to-low edge. The adclrc signal is an alignment clock that indicates whether the left- or right-channel data is presented in the adcdat line. The MSB is transmitted first. For a sampling rate of f_s, there are f_s samples per second and each sampled data must be transferred in the interval of $\frac{1}{f_s}$ second. The interval becomes the clock period of adclrc, as shown in Figure 18.8.

The timing diagram of DAC output is similar except that the data is transmitted from the FPGA to the codec via the dacdat line and daclrc is used for the alignment clock.

WM8731 can be configured either to generate or to receive the bclk, adclrc, and daclrc signals, which are known as the *master clocking mode* and *slave clocking mode*, respectively. In addition to `bclk`, `adclrc`, and `daclrc`, WM8731 also needs a master reference clock, `mclk`, for its internal operation.

18.3.2   HDL implementation

WM8731 is a versatile device and many features are configurable. Our data access controller is oriented to a fixed configuration. The selected parameters are:

- Left-justified mode
- Slave clocking scheme (i.e., clock signals provided by FPGA)
- Master clock rate of 12.288 MHz
- Resolution of 16 bits
- Sampling rate of 48 K samples per second

The main tasks of the data access controller is to generate various clocking signals, perform a serial-to-parallel conversion of the incoming ADC data, and perform a parallel-to-serial conversion of the outgoing DAC data. The HDL code is shown in Listing 18.2.

Listing 18.2   Codec data access controller

```
library ieee;
use ieee.std_logic_1164.all;
use ieee.numeric_std.all;
4 entity adc_dac is
   port (
      clk, reset: in  std_logic;
      dac_data_in: in std_logic_vector(31 downto 0);
      adc_data_out: out std_logic_vector(31 downto 0);
9     m_clk, b_clk, dac_lr_clk, adc_lr_clk: out std_logic;
      dacdat: out std_logic;
      adcdat: in std_logic;
      load_done_tick: out std_logic
   );
14 end adc_dac;

   architecture arch of adc_dac is
      constant M_DVSR: integer := 2;
      constant B_DVSR: integer := 3;
19    constant LR_DVSR: integer := 5;
      signal m_reg, m_next: unsigned(M_DVSR-1 downto 0);
      signal b_reg, b_next: unsigned(B_DVSR-1 downto 0);
      signal lr_reg, lr_next: unsigned(LR_DVSR-1 downto 0);
      signal dac_buf_reg, dac_buf_next:
24       std_logic_vector(31 downto 0);
      signal adc_buf_reg, adc_buf_next:
         std_logic_vector(31 downto 0);
      signal lr_delayed_reg, b_delayed_reg: std_logic;
      signal m_12_5m_tick, load_tick: std_logic;
29    signal b_neg_tick, b_pos_tick: std_logic;
   begin
      —————————————————————————————————————————
      —— clock signals for codec digital audio interface
      —————————————————————————————————————————
34    —— registers
      process (clk, reset)
      begin
         if reset='1' then
            m_reg <= (others=>'0');
39          b_reg <= (others=>'0');
```

```
            lr_reg <= (others=>'0');
            dac_buf_reg <= (others=>'0');
            adc_buf_reg <= (others=>'0');
            b_delayed_reg <= '0';
44          lr_delayed_reg <= '0';
         elsif (clk'event and clk='1') then
            m_reg <= m_next;
            b_reg <= b_next;
            lr_reg <= lr_next;
49          dac_buf_reg <= dac_buf_next;
            adc_buf_reg <= adc_buf_next;
            b_delayed_reg <= b_reg(B_DVSR-1);
            lr_delayed_reg <= lr_reg(LR_DVSR-1);
         end if;
54    end process;
      -- codec 12.5 MHz m_clk (master clock)
      -- ideally should be 12.288 MHz
      m_next <= m_reg + 1;   -- mod-4 counter
      m_clk <= m_reg(M_DVSR-1);
59    m_12_5m_tick <= '1' when m_reg=0 else '0';
      -- b_clk (m_clk / 8 = 32*48 kHz )
      b_next <= b_reg + 1 when m_12_5m_tick='1' else
                   b_reg;   -- mod-8 counter
      b_clk <= b_reg(B_DVSR-1);
64    -- neg edge of b_clk
      b_neg_tick <= b_delayed_reg and (not b_reg(B_DVSR-1));
      -- pos edge of b_clk
      b_pos_tick <= (not b_delayed_reg) and b_reg(B_DVSR-1);
      -- adc_/dac_lr_clk (dac_lr_clk / 32 = 48 kHz )
69    lr_next <= lr_reg + 1 when b_neg_tick='1' else
                   lr_reg;   -- mod-32 counter
      dac_lr_clk <= lr_reg(LR_DVSR-1);
      adc_lr_clk <= lr_reg(LR_DVSR-1);
      -- pos edge of dac_lr_clk
74    load_tick <= (not lr_delayed_reg) and lr_reg(LR_DVSR-1);
      load_done_tick <= load_tick;

      -- DAC buffer to shift out data
      --    data shifted out at b_clk 1-to-0 edge
79
      dac_buf_next <=
         dac_data_in when load_tick='1' else
         dac_buf_reg(30 downto 0) & '0' when b_neg_tick='1' else
         dac_buf_reg;
84    dacdat <= dac_buf_reg(31);

      -- ADC buffer to shift in data
      --    data shifted out at the b_clk 1-to-0 edge from ADC
      --    use 0-to-1 edge to latch in ADC data
89
      adc_buf_next <=
         adc_buf_reg(30 downto 0) & adcdat when b_pos_tick='1'  else
         adc_buf_reg;
      adc_data_out <= adc_buf_reg;
94 end arch;
```

The first part of the code is to generate the mclk, bclk, adclrc, and daclrc clocks, which are labeled m_clk, b_clk, adc_lr_clk, and dac_lr_clk. The specification indicates that mclk frequency should be 12.288 MHz. Since DE1 board's system clock is 50 MHz, we can construct a 2-bit (i.e., mod-4) binary counter to obtain a 12.5 MHz signal (with a tolerable 1.7% error) and use the MSB of the counter for m_clk. With the dual-channel, 16-bit resolution and 48 K sampling rate configuration, 1.536 M bits (i.e., 2*16*48K bits) are transmitted per second and thus the bclk frequency should be 1.536 MHz. Note that this frequency is one eighth of the designated mclk frequency and thus we can construct a 3-bit (i.e., mod-8) binary counter to obtain b_clk from the m_clk counter. There are 32 bits in each stereo sample and we can use a 5-bit (i.e., mod-32) binary counter, lr_reg, to obtain the alignment clock. The MSB of lr_reg output is then fed to adc_lr_clk and dac_lr_clk.

Several status signals are created to facilitate the data processing. The b_neg_tick and b_pos_tick signals are asserted for one clock cycle when there is a high-to-low transition and low-to-high transition in b_clk. The load_tick signal is asserted for one clock cycle when there is a low-to-high transition in lr_reg. It indicates the completion of the processing of one stereo data sample.

The incoming serial ADC data is stored in a 32-bit shifting register, adc_buf_reg. Recall that the codec puts a new bit of data at the falling edge of bclk. We shift in the bit at the rising edge in which the data bit should be stable. The code segment is

```
adc_buf_next <=
  adc_buf_reg(30 downto 0) & adcdat when b_pos_tick='1' else
  adc_buf_reg;
```

This shifting operation performs continuously. The rising edge of the alignment clock, represented by load_tick, marks the completion of transmitting current data sample. It is connected to load_done_tick as an output status signal to inform the external circuit that a 32-bit data is ready.

The outgoing DAC data is shifted out via the 32-bit shift register, dac_buf_reg. The code segment is

```
dac_buf_next <=
    dac_data_in when load_tick='1' else
    dac_buf_reg(30 downto 0) & '0' when b_neg_tick='1' else
    dac_buf_reg;
  dacdat <= dac_buf_reg(31);
```

The load_tick signal can be interpreted as the beginning of the new data transfer and the 32-bit input data, dac_data_in, is loaded to the register. It is then shifted to the left by one position at the falling edge of bclk, as required by WM8731 specification. The MSB is connected to the serial out line, dacdat.

18.4   AUDIO CODEC CONTROLLER IP CORE DEVELOPMENT

18.4.1   Complete audio codec controller

The top-level diagram of a complete audio codec controller is shown in Figure 18.9. It consists of the I^2C controller, data access controller, and two FIFO buffers. The FIFO buffers provide cushion space for the ADC and DAC data, which are received and transmitted at a constant rate. The load_done_tick is used to write new 32-bit data to the ADC FIFO and load (and remove) new 32-bit data from the DAC FIFO. The HDL code follows the block diagram and is shown in Listing 18.3.

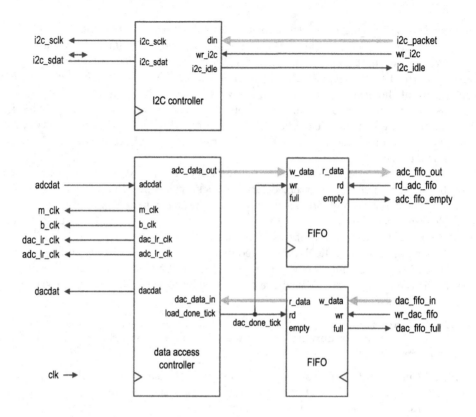

Figure 18.9 Top-level block diagram of audio codec interface.

Listing 18.3 Audio codec controller

```
library ieee;
use ieee.std_logic_1164.all;
use ieee.numeric_std.all;
entity codec_top is
5    generic(FIFO_SIZE: integer:=3);
     port (
         clk, reset: in  std_logic;
         -- to WM8731
         m_clk, b_clk, dac_lr_clk, adc_lr_clk: out std_logic;
10       dacdat: out std_logic;
         adcdat: in std_logic;
         i2c_sclk: out std_logic;
         i2c_sdat: inout std_logic;
         -- to main system
15       wr_i2c: in std_logic;
         i2c_packet: in std_logic_vector(23 downto 0);
         i2c_idle: out std_logic;
         rd_adc_fifo: in std_logic;
         adc_fifo_out: out std_logic_vector(31 downto 0);
20       adc_fifo_empty: out std_logic;
         wr_dac_fifo: in std_logic;
         dac_fifo_in: in std_logic_vector(31 downto 0);
         dac_fifo_full: out std_logic;
         sample_tick: out std_logic
25   );
```

```
   end codec_top;

   architecture arch of codec_top is
       signal dac_data_in: std_logic_vector(31 downto 0);
30     signal adc_data_out: std_logic_vector(31 downto 0);
       signal dac_done_tick: std_logic;
   begin
       sample_tick  <= dac_done_tick;
       — instantiate i2c unit
35     i2c_unit: entity work.i2c(arch)
          port map(clk=>clk, reset=>reset,
                   wr_i2c=>wr_i2c, din=>i2c_packet,
                   i2c_sclk=>i2c_sclk, i2c_sdat=>i2c_sdat,
                   i2c_idle=>i2c_idle,
40                 i2c_done_tick=>open, i2c_fail=>open);
       — instantiate codec dac/adc
       dac_adc_unit: entity work.adc_dac(arch)
          port map(clk=>clk, reset=>reset,
                   dac_data_in=>dac_data_in,
45                 adc_data_out=>adc_data_out,
                   m_clk=>m_clk, b_clk=>b_clk,
                   dac_lr_clk=>dac_lr_clk, adc_lr_clk=>adc_lr_clk,
                   dacdat=>dacdat, adcdat=>adcdat,
                   load_done_tick=>dac_done_tick);
50     — instantiate adc fifo
       fifo_adc_unit: entity work.fifo(arch)
          generic map(B=>32, W=>FIFO_SIZE)
          port map(clk=>clk, reset=>reset, rd=>rd_adc_fifo,
                   wr=>dac_done_tick, w_data=>adc_data_out,
55                 empty=>adc_fifo_empty, full=>open,
                   r_data=>adc_fifo_out);
       — instantiate dac fifo
       fifo_dac_unit: entity work.fifo(arch)
          generic map(B=>32, W=>FIFO_SIZE)
60     port map(clk=>clk, reset=>reset, rd=>dac_done_tick,
                   wr=>wr_dac_fifo, w_data=>dac_fifo_in,
                   empty=>open, full=>dac_fifo_full,
                   r_data=>dac_data_in);
   end arch;
```

18.4.2   Avalon interfaces

The wrapping circuit for the audio codec controller should include an Avalon MM slave interface to interact with the host, a clock input interface for the system clock, and a conduit interface for the external signals. One design is to use the Nios II processor to process and buffer the ADC and DAC data streams and assign the signals on the right side of the diagram of Figure 18.9 to the Avalon MM interface. However, since the ADC and DAC data streams tend to be continuous and at a fairly high rate, it is possible that the data comes from or goes to an external component rather than through the processor. For example, during the playback of an audio file, the data may be routed directly from a flash card. To increase flexibility, we add additional routing logic so that the codec controller's ADC and DAC data streams can be connected to the processor (via the Avalon MM interface) or external components. The conceptual diagram is shown in Figure 18.10. We assume that the external component is connected via a conduit interface. We use three 2-to-1 multiplexers to select control signals and DAC data streams and route

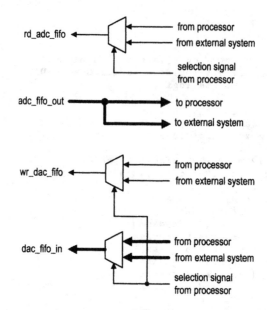

Figure 18.10   Routing for the ADC and DAC data streams and control.

the ADC data streams to both the Avalon MM interface and conduit interface. Note that the selection signals of the multiplexers are assigned to the Avalon MM interface so that the processor can specify the routing choice. This feature is used in Chapter 22.

18.4.3   Register map

A Nios II processor interacts with the audio codec controller as follows:
- check (i.e., read) various status signals.
- receive (i.e., read) 32-bit audio data from the ADC FIFO buffer.
- issue (i.e., write) a command to a configuration register via the I^2C bus.
- set up (i.e., write) the data stream selection register.
- transmit (i.e., write) 32-bit audio data to the DAC FIFO buffer.

The registers, their address offsets, and fields are:
- Read addresses (data to cpu)
  - offset 1 (status register)
    * bit 0: asserted (i.e., 1) when I^2C controller is idle
    * bit 1: asserted (i.e., 1) when DAC FIFO is full
    * bit 2: asserted (i.e., 1) when ADC FIFO is empty
  - offset 3 (ADC data register)
    * bits 31 to 0: 31-bit audio data from ADC
- Write addresses (data from cpu)
  - offset 0 (I^2C data register)
    * bits 23 to 0: 24-bit codec command packet to I^2C bus
  - offset 1 (codec data stream selection register)

* bit 0: select Avalon MM interface for codec DAC data stream when
    it is 0 and select conduit interface when it is 1.
  * bit 1: select Avalon MM interface for codec ADC data stream when
    it is 0 and select conduit interface when it is 1.
- offset 2 (DAC data register)
  * bits 31 to 0: 31-bit audio data to DAC

18.4.4   Wrapped audio codec controller

The HDL code of the wrapped audio codec controller is shown in Listing 18.4.

Listing 18.4   Wrapped audio codec controller

```
1 library ieee;
  use ieee.std_logic_1164.all;
  use ieee.numeric_std.all;
  entity chu_avalon_audio is
     generic(FIFO_SIZE: integer:=3); -- 2^FIFO_SIZE words
6    port (
         clk, reset: in  std_logic;
         -- external signals (to avalon conduit interface)
         m_clk, b_clk, dac_lr_clk, adc_lr_clk: out std_logic;
         dacdat: out std_logic;
11       adcdat: in std_logic;
         i2c_sclk: out std_logic;
         i2c_sdat: inout std_logic;
         -- avalon MM interface
         audio_address: in std_logic_vector(1 downto 0);
16       audio_chipselect: in std_logic;
         audio_write: in std_logic;
         audio_writedata: in std_logic_vector(31 downto 0);
         audio_read: in std_logic;
         audio_readdata: out std_logic_vector(31 downto 0);
21       -- avalon conduit interface
         codec_adc_rd: in std_logic;
         codec_adc_data_out: out std_logic_vector(31 downto 0);
         codec_dac_wr: in std_logic;
         codec_dac_data_in: in std_logic_vector(31 downto 0);
26       codec_sample_tick: out std_logic
     );
  end chu_avalon_audio;

  architecture arch of chu_avalon_audio is
31   signal wr_en: std_logic;
     signal wr_i2c, i2c_idle, wr_sel: std_logic;
     signal dbus_sel_reg: std_logic_vector(1 downto 0);
     signal dac_fifo_in: std_logic_vector(31 downto 0);
     signal adc_fifo_out: std_logic_vector(31 downto 0);
36   signal dac_done_tick: std_logic;
     signal wr_dac_fifo, cpu_wr_dac_fifo: std_logic;
     signal rd_adc_fifo, cpu_rd_adc_fifo: std_logic;
     signal adc_fifo_empty, dac_fifo_full: std_logic;
  begin
41   -- instantiate codec controller
     codec_unit: entity work.codec_top(arch)
        generic map(FIFO_SIZE=>FIFO_SIZE)
        port map(clk=>clk, reset=>reset,
                 i2c_sclk=>i2c_sclk, i2c_sdat=>i2c_sdat,
46               m_clk=>m_clk, b_clk=>b_clk,
```

```
                         dac_lr_clk=>dac_lr_clk, adc_lr_clk=>adc_lr_clk,
                         dacdat=>dacdat, adcdat=>adcdat,
                         wr_i2c=>wr_i2c, i2c_idle=>i2c_idle,
                         i2c_packet=>audio_writedata(23 downto 0),
51                       rd_adc_fifo=>rd_adc_fifo,
                         adc_fifo_empty=>adc_fifo_empty,
                         adc_fifo_out=>adc_fifo_out,
                         wr_dac_fifo=>wr_dac_fifo,
                         dac_fifo_full=>dac_fifo_full,
56                       dac_fifo_in=>dac_fifo_in,
                         sample_tick=>codec_sample_tick);
      -- data stream selection register
      process (clk, reset)
      begin
61        if reset='1' then
              dbus_sel_reg <= "00";
          elsif (clk'event and clk='1') then
              if wr_sel='1' then
                  dbus_sel_reg <= audio_writedata(1 downto 0);
66            end if;
          end if;
      end process;
      -- write decoding
      wr_en <= '1' when audio_write='1' and audio_chipselect='1' else
71             '0';
      wr_i2c <= '1' when audio_address="00" and wr_en='1' else
                '0';
      wr_sel <= '1' when audio_address="01" and wr_en='1' else
                '0';
76    cpu_wr_dac_fifo <= '1' when audio_address="10" and wr_en='1' else
                         '0';
      cpu_rd_adc_fifo <= '1' when audio_address="11" and audio_read='1' and
                              audio_chipselect='1' else
                         '0';
81    -- read multiplexing
      audio_readdata <=
          adc_fifo_out when audio_address="11" else
          x"0000000" & '0' & adc_fifo_empty & dac_fifo_full & i2c_idle;
      -- data stream routing & control
86    wr_dac_fifo <= cpu_wr_dac_fifo when dbus_sel_reg(0)='0' else
                         codec_dac_wr;
      dac_fifo_in <= audio_writedata when dbus_sel_reg(0)='0' else
                         codec_dac_data_in;
      rd_adc_fifo <= cpu_rd_adc_fifo when dbus_sel_reg(1)='0' else
91                       codec_adc_rd;
      codec_adc_data_out <= adc_fifo_out;
   end arch;
```

The circuit includes a 2-bit register, **dbus_sel_reg**, for the codec data stream selection signal, write decoding logic, read multiplexing logic, and data stream routing logic. The decoding logic and multiplexing logic are similar to those of the wrapped PS2 controller discussed in Section 16.5.3 except for the read ADC FIFO operation. To accommodate a faster rate, we remove the data item from the FIFO buffer during the read operation without using a separate removal operation. This is done by decoding the read instruction for the FIFO read (i.e., removal) signal:

```
      cpu_rd_adc_fifo <=
          '1' when audio_address="11" and audio_read='1' and
                  audio_chipselect='1' else
          '0';
```

The data stream selection is performed by the **dbus_sel_reg** register and the routing circuit. The desired data stream is written to the register and its output is used to select the control signals and DAC input data stream (the signals with codec_ prefix are designed for external connections):

```
wr_dac_fifo   <= cpu_wr_dac_fifo when dbus_sel_reg(0)='0' else
                 codec_dac_wr;
dac_fifo_in   <= audio_writedata when dbus_sel_reg(0)='0' else
                 codec_dac_data_in;
rd_adc_fifo   <= cpu_rd_adc_fifo when dbus_sel_reg(1)='0' else
                 codec_adc_rd;
codec_adc_data_out <= adc_fifo_out;
```

18.4.5 SOPC component creation

After constructing the wrapped audio codec controller, we can create a new SOPC component in Component Editor following the procedure outlined in Section 14.5.4. It then can be treated as a normal IP core and integrated into a Nios II system. Note that the reading and writing FIFO buffer is done in one clock cycle and thus the wait time fields in the Avalon MM interface must be cleared to 0's.

18.5 CODEC DRIVER

The audio codec driver consists of a collection of routines to configure the device, retrieve ADC data, and transmit DAC data. To make the code clear, we define the address offsets as symbolic register names in the header file:

```
#define CHU_AUD_I2C_DATA_REG    0
#define CHU_AUD_STATUS_REG      1
#define CHU_AUD_DBUS_SEL_REG    1
#define CHU_AUD_DAC_DATA_REG    2
#define CHU_AUD_ADC_DATA_REG    3
```

18.5.1 I^2C command routines

There are two basic functions related to issue a command packet:
- audio_i2c_is_idle(): check whether the I^2C controller is idle.
- audio_i2c_wr_cmd(): issue (i.e., write) a 24-bit command packet.

The two routines are shown in Listing 18.5.

Listing 18.5

```
int audio_i2c_is_idle(alt_u32 audio_base)
{
  alt_u32 status_reg;
  int i2c_idle_bit;

  status_reg = IORD(audio_base, CHU_AUD_STATUS_REG);
  i2c_idle_bit = (int) (status_reg & 0x00000001);
  return i2c_idle_bit;
}
```

```
void audio_i2c_wr_cmd(alt_u32 audio_base, alt_u8 addr, alt_u16 cmd)
{
  const alt_32 i2c_id = 0x00000034;
  alt_u32 packet;      // data written to i2c; only 24 LSBs used

  packet = i2c_id;
  packet = (packet << 7) + (addr & 0x07f);   // append 7-bit address
  packet = (packet << 9) + (cmd & 0x01ff);   // append 9-bit command
  IOWR(audio_base, CHU_AUD_I2C_DATA_REG, (alt_u32) packet);
}
```

The audio_i2c_is_idle() function extracts the idle bit from the status register. The audio_i2c_wr_cmd() function assembles the data to a 24-bit packet and sends it to the I^{2}C controller. The packet consists of four major fields. The first field is the 7-bit slave device id and the second field is the read/write bit. WM8731's id is 0011010. Since there is only one slave device on the bus and the master always performs a write operation, these fields do not change. The two fields are concatenated to form the first byte, which is 0x34 (i.e., 00110100). The third and fourth fields are a 7-bit configuration register address and 9-bit configuration data passed from the calling function. These fields are concatenated to form the final 24-bit packet and written to the I^{2}C controller. The configuration addresses and data are summarized in Figure 18.2.

18.5.2 Data source select routine

The audio_wr_src_sel() function writes the data stream selection register. The code is shown in Listing 18.6.

Listing 18.6

```
void audio_wr_src_sel(alt_u32 audio_base, int dac_sel, int adc_sel)
{
  alt_u32 sel_reg = 0x00000000;

  if (dac_sel!=0)
    sel_reg = sel_reg | 0x00000001;    // set LSB to 1
  if (adc_sel!=0)
    sel_reg = sel_reg | 0x00000002;    // set 2nd LSB to 1
  IOWR(audio_base, CHU_AUD_DBUS_SEL_REG, sel_reg);
}
```

18.5.3 Device initialization routine

The audio_init() function configures the WM8731 device to a desired initial state. It is shown in Listing 18.7.

Listing 18.7

```
void audio_init(alt_u32 audio_base)
{
  /* initial configuration values (registers R0 to R9)  */
  const alt_u16 cmds[10]={       // only 9 LSBs used
    0x0017, // R0: left line in gain 0 dB
    0x0017, // R1: right line in gain 0 dB
    0x0079, // R2: left headphone out volume 0 dB
    0x0079, // R3: right headphone out volume 0 dB
```

```
  0x0010,  // R4:  analog  path  select:  line−in  to  adc,  dac  to  line−out
  0x0000,  // R5:  digital  audio:  high−pass  filter ,  no  de−emphasis
  0x0000,  // R6:  enable  all  power
  0x0001,  // R7:  digital  interface:  left −adjust ,  16− bit  resolution
  0x0000,  // R8:  48K  sampling  rate  with  12.288MHz  master  clock
  0x0001   // R9:  activate
};
int i;

while (!audio_i2c_is_idle(audio_base)){};     // wait  until  i2c  idle
/* write a dummy data to R15 to reset the codec */
audio_i2c_wr_cmd(audio_base, 15, 0);
/* cycle through 10 commands */
for (i=0; i<10; i++){
  while (!audio_i2c_is_idle(audio_base)){}; // wait  until  i2c  idle
  audio_i2c_wr_cmd(audio_base, i, cmds[i]); // send  a  command  packet
}
audio_wr_src_sel(audio_base, 0, 0);          // dac/adc  to  Avalon  bus
}
```

A constant array, cmd[10], is used to define the data for configuration registers
R0 to R9. In the code, we first reset the codec device by writing a dummy packet
to R15 and then write configuration data to registers R0 to R9 sequentially and
then connect the codec data stream to the Avalon MM interface. Recall that the
register R7 and R8 define the digital interface format, resolution, and sampling
rate, and thus their contents should not be modified.

18.5.4 Audio data access routines

The data access routines check the statuses of FIFO buffers, retrieve audio data
from the ADC FIFO, and transmit audio data to the DAC FIFO. These functions
are shown in Listing 18.8.

Listing 18.8

```
int audio_dac_fifo_full(alt_u32 audio_base)
{
  alt_u32 status_reg;
  int dac_full_bit;

  status_reg = IORD(audio_base, CHU_AUD_STATUS_REG);
  dac_full_bit = (int)((status_reg & 0x00000002) >> 1);
  return dac_full_bit;
}

void audio_dac_wr_fifo(alt_u32 audio_base, alt_u32 data)
{
  IOWR(audio_base, CHU_AUD_DAC_DATA_REG, data);
}

int audio_adc_fifo_empty(alt_u32 audio_base)
{
  alt_u32 status_reg;
  int adc_empty_bit;

  status_reg = IORD(audio_base, CHU_AUD_STATUS_REG);
  adc_empty_bit = (int)((status_reg & 0x00000004) >> 2);
  return adc_empty_bit;
}
```

```
alt_u32 audio_adc_rd_fifo(alt_u32 audio_base)
{
  alt_u32 data_reg;

  data_reg = IORD(audio_base, CHU_AUD_ADC_DATA_REG);
  return data_reg;
}
```

18.6   TESTING PROGRAM

The audio codec controller can be instantiated and integrated into a Nios II system like a normal IP core. The system derived in Section 16.10.2 includes an audio codec controller module and can be used for testing. We construct a simple program to demonstrate and verify the driver routines. The program consists of the following tests:

- Initialize audio codec.
- Record line input and store the data to a buffer.
- Record microphone input and store the data to a buffer.
- Generate a synthetic sinusoid wave and store the data to a buffer.
- Play the buffered data.
- Issue a command to a codec configuration register.
- Adjust the headphone volume.
- Display the buffered data on a VGA monitor.
- Clear the VGA screen.

The main program is shown in Listing 18.9.

Listing 18.9

```
int main(void)
{
  const int SF=48000;        // sampling freq=48K
  const int BUF_SIZE= 5*SF;  // 5 sec @ sampling frequency SF
  alt_u32 buf[BUF_SIZE];
  int sw, btn, freq, vol,  reg, cmd, i;
  alt_u8 disp_msg[4]={sseg_conv_hex(5), 0x23, 0x2b, sseg_conv_hex(13)};

  sseg_disp_ptn(SSEG_BASE, disp_msg);      // show "Sond" for display
  vga_clr_screen(VRAM_BASE,0);             // clear screen
  audio_init(AUDIO_BASE);
  btn_clear(BTN_BASE);
  printf("Audio codec interface test \n\n");

  while (1){
    while (!btn_is_pressed(BTN_BASE)){ };  // wait for button
    btn=btn_read(BTN_BASE);                // read button
    if (btn & 0x02){                       // key1 pressed
      sw=pio_read(SWITCH_BASE);            // read switch
      printf("key/sw: %d/%d\n", btn, sw);
    }
    btn_clear(BTN_BASE);
    switch (sw){
      case 0:  // initialize codec controller
        audio_init(AUDIO_BASE);
        break;
```

```
case 1:  // record line input for 5 sec
  audio_i2c_wr_cmd(AUDIO_BASE, 4, 0x0010);        // line-in to adc
  for (i=0; i<BUF_SIZE; i++){
    while(audio_adc_fifo_empty(AUDIO_BASE)){} // wait if fifo empty
    buf[i]=audio_adc_rd_fifo(AUDIO_BASE);
  }
  break;
case 2:  // record microphone input for 5 sec
  audio_i2c_wr_cmd(AUDIO_BASE, 4, 0x0015);        // mic to adc, boost
  for(i=0; i<BUF_SIZE; i++){
    while(audio_adc_fifo_empty(AUDIO_BASE)){} // wait if fifo empty
    buf[i]=audio_adc_rd_fifo(AUDIO_BASE);
  }
  break;
case 3:  // fill buffer with sinusoidal data
  printf("enter frequency:");
  scanf("%d", &freq);
  record_sin_wave(SF, freq, BUF_SIZE, buf);
  printf("sinusoidal wave recorded. \n");
  break;
case 4:  // play buffered audio data repeatedly until a key pressed
  while (!btn_is_pressed(BTN_BASE)){
    for(i=0; i<BUF_SIZE; i++){
      while(audio_dac_fifo_full(AUDIO_BASE)){} // wait if fifo full
      audio_dac_wr_fifo(AUDIO_BASE, buf[i]);
    } // end for
  } // end while
  break;
case 5:  // issue codec command
  printf("enter codec register #:");
  scanf("%d", &reg);
  printf("enter command (in hex):");
  scanf("%x", &cmd);
  audio_i2c_wr_cmd(AUDIO_BASE, (alt_u8)reg, (alt_u16)cmd);
  printf("send 0x%x to codec register %d\n\n", cmd, reg);
  break;
case 6:  // set volume
  printf("enter volume (between 28 and 127):");
  scanf("%d", &vol);
  set_volume(AUDIO_BASE, vol);
  printf("Volume set\n");
  break;
case 7:  // plot buffered right- and left-channel data
  plot_audio_buffer(VRAM_BASE, buf);
  break;
case 8:  // clear screen
  vga_clr_screen(VRAM_BASE,0);
  break;
} //end switch
} // end while
}
```

The program uses a one-dimensional array, buf[], to store five seconds of audio data. Since the sampling rate is 48 K samples per second, the size of the buffer is 5*48000. The audio data can be obtained from the line or microphone input via ADC or manually synthesized, and then played back via DAC. We also implement an auxiliary function to display the buffered data on the VGA screen. The program's basic structure is similar to that in Listing 16.12. We use slide switches to specified the desired test and use the pushbutton switch 1 (labeled key1 on the

board) to initiate the test. The relevant functions are discussed in the following
subsections.

Sinusoidal wave generation routine   The `record_sin_wave()` function generates digitized sinusoidal data points and stores them to a buffer. The function is shown in
Listing 18.10.

Listing 18.10

```
void record_sin_wave(int sf, int freq, int size, alt_u32 *buf)
{
   const float PI=3.14159;
   const float AMP_MAX= 32767.0;        // max amplitude (2^15-1)
   float amp;
   int npoint, ncycle, i, j;
   alt_u32 left, right;

   npoint = sf/freq;                     // # of steps in one period
   /* construct 1st cycle */
   for (i=0; i<npoint; i++){
     amp = sin((float)i/(float)npoint * 2.0 * PI);
     left = (alt_u32) (amp * AMP_MAX);   // left channel; 16 LSBs used
     amp = cos((float)i/(float)npoint * 2.0 * PI);
     right = (alt_u32) (amp * AMP_MAX);  // right channel; 16 LSBs used
     buf[i] = (left << 16) + right;      // combine two channels
   }
   /* duplicate the 1st cycle for the remaining ncycle-1 cycles */
   /* no data for the last fractional cycle                     */
   ncycle = size/npoint;
   for (j=1; j<ncycle; j++){
     for (i=0; i<npoint; i++){
        buf[j*npoint+i]=buf[i];
     } // end for i
   } // end for j
}
```

The input arguments are the sampling frequency `sf`, the desired sinusoidal frequency `freq`, and the pointer to the buffer. The amplitude is set to the maximal
swing, which is defined by `AMP_MAX`. Since the audio data is defined as a 16-bit
signed number, the maximal amplitude is $2^{15} - 1$.

In the routine, we first calculate the number of data points in one cycle and
then create the audio data for the first cycle, using `sin()` for the left channel and
`cos()` for the right channel. Since the `sin()` and `cos()` functions involve many
floating-point operations and are time consuming, we fill the remaining buffer by
replicating the data from the first cycle.

Volume control routine   We can adjust the headphone volume through WM8731
device's configuration register. The volume is controlled by the 7-bit `LHPVOL` and
`RHPVOL` fields. They are logarithmically adjustable from +6 dB to −73 dB in 1-dB
steps. The value 121 (0x79) is defined to be 0 dB ("normal volume") and values 127
(0x7f) and 48 (0x30) correspond to +6 dB and −73 dB. Any value smaller than 48
mutes the output. The `set_volume()` function is shown in Listing 18.11. It first
checks the maximal volume and then writes configuration registers R2 and R3.

Listing 18.11

```
void set_volume(alt_u32 audio_base, int vol)
{
  int cmd;

  if (vol>0x7f)  // exceed maximal volume
     cmd = 0x7f;
  while (!audio_i2c_is_idle(audio_base)){};
    audio_i2c_wr_cmd(audio_base, 2, (alt_u16) cmd);
  while (!audio_i2c_is_idle(audio_base)){};
    audio_i2c_wr_cmd(audio_base, 3, (alt_u16) cmd);
}
```

Buffer plotting routine   The `plot_audio_buffer()` function displays the stored data on the VGA screen. The range of the x-axis is from 0 to $\frac{639}{48000}$ second, which correspond to the first 640 data points in the buffer, and the range of the y-axis is adjusted to accommodate the maximal amplitude swing. The code is shown in Listing 18.12.

Listing 18.12

```
void plot_audio_buffer(alt_u32 vga_grf_base, alt_u32 *buf)
{
  int i, j;
  alt_16 y_left, y_right;

  for(i=1; i<640; i++){
     y_right = (alt_16) (0x0000ffff ^ buf[i]);  // right channel data
     y_left  = (alt_16) (buf[i]>>16);           // left channel data
     j=480 - ((int)y_left/256 + 480/2);         // adjust y to -128/127
     vgag_wr_pix(vga_grf_base, i, j, 0xe0);     // plot left channel red
     j=480 - ((int)y_right/256 + 480/2);        // adjust y to -128/127
     vgag_wr_pix(vga_grf_base, i, j, 0x03);     // plot right channel blue
  }
}
```

18.7 AUDIO FILE PROCESSING

A common application of an audio codec is to play pre-recorded audio files. In this section, we introduce a minimal procedure to process audio files and retrieve data. The procedure is not comprehensive or efficient but just helps us get started.

18.7.1 WAV format overview

There are many audio file formats and most of them involve sophisticated data compression schemes, which are beyond the scope of this book. We consider the simplest form of WAV format in this section. The WAV (or WAVE) file format is a subset of RIFF (resource interchange file format) specification for storage of multimedia data. A simple WAV file frequently contains a single *WAVE chunk*. The chunk consists of an *fmt sub-chunk*, which provides information on audio format, and a *data sub-chunk*, which contains the actual audio data. For simplicity, we assume that the WAV file has been pre-processed to conform our audio codec controller specification:

Table 18.1   WAV chunk and sub-chunk information

Offset	Size	Purpose	value
0x00	4	chunk id	"RIFF"
0x04	4	chunk size (bytes)	
0x08	4	format	"WAVE"
0x0c	4	sub-chunk id	"fmt "
0x10	4	sub-chunk size (bytes)	16 (for PCM)
0x14	2	audio format	1 (for PCM)
0x16	2	number of channels	2
0x18	4	sampling rate	48000
0x1c	4	byte rate (bytes per second)	192000 ($\frac{48000*2*16}{8}$)
0x20	2	block alignment (bytes in one sample)	4 ($\frac{2*16}{8}$)
0x22	2	resolution	16
0x24	4	sub-chunk id	"data"
0x28	4	sub-chunk size (number of bytes in data)	
0x2c	*	audio data sample starting here	

- Audio data format: not compressed, which is known as *PCM* (pulse coded modulation) format.
- Number of channels: 2.
- Sampling rate: 48000 samples per second.
- Resolution: 16 bits per data sample.

A typical WAV file contains an RIFF chunk descriptor, an fmt sub-chunk, and a data sub-chunk. The fields are summarized in Table 18.1. Each row shows the address offset, the size (in terms of bytes), purpose, and the value used in our format specification.

18.7.2   Audio format conversion program

Since our demonstration code only accepts a specific format, an audio file usually needs to be preprocessed in advance. A free utility program, *SoX* (for "Sound eXchange"), can be used to perform this task. SoX is a versatile audio processing command-line utility that can convert audio files from one format to another format. The following statement converts a WAV file (`input.wav`) with arbitrary parameters to a file (`output.wav`) with our desired format:

```
sox input.wav  -r 48000 -b 16 -c 2 -e signed output.wav
```

The -r, -b, -c, and -e options represent sampling rate, bits per sample (resolution), number of channels, and encoding scheme, respectively. The `signed` option means that the encoding type is PCM and the data is stored in signed (two's complement) format. The executable file and detailed documentation of SoX can be found in http://sox.sourceforge.net/.

18.7.3 Audio data retrieval routine

The discussion and procedure used to process bitmap files in Section 17.9 can be applied to the audio WAV file as well. A similar **read_wav_file()** can be constructed and the code is shown in Listing 18.13.

Listing 18.13

```
int read_wav_file(char *file_name, alt_u32 *buf)
{
  /* note that fget32 read in little endian, but id in big endian */
  const alt_u32 RIFF_ID = 0x46464952;    //ascii for FFIR
  const alt_u32 WAVE_ID = 0x45564157;    //ascii for EVAW
  const alt_u32 FMT_ID  = 0x20746d66;    //ascii for \btmf
  const alt_u32 DATA_ID = 0x61746164;    //ascii for ATAD

  FILE *fp;
  alt_u32 r_id, w_id, f_id, d_id, srate, data_size;
  alt_u16 compression, channel, res;
  int i, s_size;

  /* open the file */
  fp = fopen(file_name,"rb");
  if (fp==NULL){
    printf("Error: cannot open file %s.\n", file_name);
    return(-1);
  }
  /* extract relevant chunk/subchunk info */
  r_id = fget32(fp);          // offset  0: "RIFF" chunk id
  fskip(fp,4);                // offset  4: chunk size
  w_id = fget32(fp);          // offset  8: "WAVE" chunk id
  f_id = fget32(fp);          // offset 12: "fmt " subchunk id
  fskip(fp,4);                // offset 16: subchunk size
  compression=fget16(fp);     // offset 20: 1 for PCM
  channel = fget16(fp);       // offset 22: 2 for stereo
  srate = fget32(fp);         // offset 24: 48K sampling rate
  fskip(fp,4);                // offset 28: byte rate
  fskip(fp,2);                // offset 32: block size
  res = fget16(fp);           // offset 34: 16 bits resolution
  d_id = fget32(fp);          // offset 36: "data" subchunk id
  data_size = fget32(fp);     // offset 40: # bytes of data subchunk
  /* check chunck/subchunk ids and paramters */
  if ((r_id!=RIFF_ID) || (w_id!=WAVE_ID) || (f_id!=FMT_ID) ||
      (d_id!=DATA_ID) || (compression!=1) || (channel!=2) ||
      (srate!=48000) || (res!=16)){
    printf("Error: incorrect wave file format\n");
    printf("Must be PCM, 2 channels, 48K rate, 16-bit resolution.\n");
    printf("RIFF/WAVE/fmt /data ids: %08x/%08x/%08x/%08x\n",
           r_id,w_id,f_id,d_id);
    printf("compression/channel/rate/res/data_szie: %d/%d/%d/%d/%d\n",
           compression, channel, srate, res, data_size);
    fclose(fp);
    return(-1);
  }
  s_size = data_size/4;
  printf("File opened.\n # audio data samples: %d\n", s_size);
  for (i=0; i<s_size; i++) {
    // get 32-bit data
    buf[i] = fget32(fp);
    if (i%1000 == 0)
      printf(".");
```

```
    }
    fclose(fp);
    printf("\nFile loaded.\n");
    return(0);
}
```

The first part of the code extracts various chunk and sub-chunk fields and verify that the file conforms to our specific format. We use four constants, RIFF_ID, WAVE_ID, FMT_ID, and DATA_ID, for the designated id string. Since fget32() constructs the 32-bit data in "little-endian" byte order, the order of the ASCII characters in these constants are revered (for example, "RIFF" is defined as "FFIR" in RIFF_ID). The second part reads the audio data and stores them to a buffer. The calling routine should allocate the buffer space and pass the buffer's pointer.

We construct a simple program to test the audio retrieval function. A SoX generated file, trumpet.wav, is read into the buffer and played repeatedly. The code is shown in Listing 18.14.

Listing 18.14

```
/* file inclusion */
#include <stdio.h>
#include <unistd.h>
#include "system.h"
#include "chu_file.h"
#include "chu_avalon_audio.h"

/* constants */
#define   WAV_FILE_NAME   "/mnt/host/trumpet.wav"   // path and file name
#define   BUF_SIZE        (6*48000)                 // 6 sec of audio data

int main(void){
  alt_u32 buf[BUF_SIZE];
  int i;

  audio_init(AUDIO_BASE);
  printf("Wave file test \n\n");
  if (read_wav_file(WAV_FILE_NAME, buf)!=0){
    printf("\n Fail to load wav file. \n");
    return (0);
  }
  //printf("\nWave File loaded \n");
  while(1){
    for(i=0; i<BUF_SIZE; i++){
      while(audio_dac_fifo_full(AUDIO_BASE)){}   // wait if dac fifo full
        audio_dac_wr_fifo(AUDIO_BASE, buf[i]);
    } // end for
  } // end while
}
```

Because no data compression is used in the file, its size can be fairly large. Each second of audio data requires a buffer space of 192 KB (i.e., $\frac{48000*32}{8}$). The calling function must take this into consideration and allocate adequate buffer space. Since the data transfer rate of the host-based file is really low, it is only feasible for testing purposes.

18.8 BIBLIOGRAPHIC NOTES

Detailed information on the I^{2}C bus can be found in *I^{2}C-Bus Specification and User Manual* published by NXP Semiconductor. It can be downloaded from the company's web site. The WM8731 manual is included in the DE1 board distribution CD and can also be downloaded from Wolfson Microelectronics's web site. The WAV file format has many options. More detailed descriptions can be found on the Wikipedia web site (searching by the keyword "wav").

Audio signal processing is a branch of DSP (digital signal processing). *Understanding Digital Signal Processing, 2nd ed.* by R. Lyons gives a comprehensive introduction to DSP and *Digital Audio Signal Processing, 2nd ed.* by U. Zolzer provides in-depth coverage on the audio-related topics. *Digital Signal Processing with Field Programmable Gate Arrays* by U. Meyer-Baese discusses the FPGA implementation of key DSP functions.

18.9 SUGGESTED EXPERIMENTS

18.9.1 Software I^{2}C controller

Instead of using custom hardware for the I^{2}C controller, we can connect the **sda** and **scl** signals to a 2-bit PIO core and use software and the existing timer functions to generate and process the transitions. Properly configure the PIO core, derive the new Nios II system, develop the driver routines, and verify its operation.

18.9.2 Hardware data access controller using master clocking mode

The data access controller in Section 18.3 assume that WM8731 is configured in slave clocking mode, in which the **bclk**, **adclrc**, and **daclrc** signals are generated by the controller (i.e., FPGA device). Redesign the data access controller using WM8731's master clocking mode, in which the codec device generates the relevant clocking signals, and verify its operation. Note that the **MS** field of configuration register R7 must be set to 1 to reflect the change.

18.9.3 Software data access controller using slave clocking mode

Instead of using custom hardware for a data access controller, we can use software plus a timer core and a PIO core to generate various clock signals and to process serial data. We also want to use WM8731's slave clocking mode in this experiment. Properly configure the timer core and PIO core, derive the new Nios II system, develop the driver routines, and verify its operation. Interrupt may be needed to accommodate the fast date rate.

18.9.4 Software data access controller using master clocking mode

Repeat Experiment 18.9.3 but use WM8731's master clocking mode.

18.9.5   Configurable data access controller

The data access controller in Section 18.3 is designed with a fixed sampling rate (48 kHz) and a fixed resolution (16 bits). WM8731 actually supports a range of sampling rates (8 to 96 kHz) and resolutions (16 to 32 bits). Redesign the controller so that the sampling rate and resolution can be dynamically reconfigured, develop the driver routines, and verify its operation.

18.9.6   Voice recorder

Use the codec to design a voice recorder, which can record input from the microphone and play back the recorded data. The system should support the following functions:

- Adjust the gain of microphone input.
- Adjust the volume of headphone.
- Play or pause the playback.
- Start and stop the recording.
- "Fast forward" and "rewind" the recorded data.

Use proper combinations of slide switches and pushbutton switches to specify the desired function. Derive the code for the system and verify its operation.

18.9.7   Real-time sinusoidal wave generator

In Section 18.6, the sinusoidal wave is stored in a buffer and then played back. An alternative is to generate the data points in real time; i.e., generate a data value every $\frac{1}{48000}$ second and write the value to the DAC FIFO. To reduce computation time, we can create a 512-entry sinusoidal function lookup table in advance and generate the desired data point by reading the table and interpolation rather than calling the computation-intensive `sin()` function. Derive the code and verify its operation.

18.9.8   Real-time audio wave display

In Section 18.6, the `plot_audio_buffer()` function displays the first 640 data points in the buffer on the VGA screen. Derive a more advanced function that can display the audio waveform continuously and in real time. The waveform can be either incoming data from ADC or outgoing data to DAC. Adjust the time scale (i.e., x-axis) and amplitude scale (i.e., y-axis) properly to obtain an observable waveform.

18.9.9   Echo effect

An echo is a delayed reflection of sound. If the original signal is $f(t)$, the echo can be expressed as $a * f(t - d)$, where a is attenuation ($a < 1$) and d is path delay, and the combined signal becomes $f(t) + a * f(t - d)$. We can digitally create an artificial echo effect with a codec. This can be done by digitizing the line input by ADC, and storing the needed delayed data in a buffer, adding the echo to the current input signal, and converting the combined signal to analog format by DAC. Derive the function and verify its operation. The a and d should be adjustable.

18.10   SUGGESTED PROJECTS

18.10.1   Full-fledged I^{2}C controller

The I^{2}C controller designed in Section 18.2 is customized as a master to write three
bytes to the designated codec device. A more general controller should contain the
following features:

- Function as either a master or a slave.
- Support standard mode (100 K bits per second) and fast mode (400 K bits
  per second) data transfer.
- Perform both read and write operations.
- Read or write an arbitrary number of bytes.
- Support clock synchronization by allowing a slow device to stretch the clock.
- Support multiple masters and arbitration.

Refer to the I^{2}C manual in the bibliographic section for detailed specifications.
Design the hardware and use simulation to verify its operation.

18.10.2   Digital equalizer

An equalizer is used to control the frequency response of an audio system. The
audible frequency spectrum is divided into several segments (usually 5 or 10). A
bandpass filter is designed for each segment to pass the designated frequency range
and to control the gain. The filtered results are then added together to form the
final waveform. We can create a digital equalizer by using a codec and a collection of
digital filters. Design the custom hardware, Avalon wrapping circuit, and software
driver, and verify its operation. The information about digital filters can be found
in the bibliographic section.

18.10.3   Digital audio oscilloscope

An oscilloscope can display the waveform of a signal and allow the user to adjust
the voltage level and time scale to obtain better visualization and measurement.
One way to design an oscilloscope is to use ADC to capture the input data and
then display the data on a monitor. Research the basic operation and layout of
an oscilloscope and use a codec and VGA controllers to design and implement a
digital oscilloscope. Note that because of the low sampling rate of an audio ADC,
the bandwidth is severely limited and thus this design can only be used to observe
low-frequency signals.

18.11 COMPLETE PROGRAM LISTING

Listing 18.15 chu_avalon_audio.h

```
/*****************************************************************************
*
* Module:   Audio codec driver header
* File:     chu_avalon_audio.h
* Purpose:  Routines to configure codec, retrieve data from ADC, and
*           send data to DAC
*
*****************************************************************************
*  Register map
*****************************************************************************
* Read (data to cpu):
*    offset 1
*      * bit 0: i2c idle
*      * bit 1: dac fifo full
*      * bit 2: adc fifo empty
*    offset 3
*      * bits 31 to 0: stereo audio data (16-bit left + 16-bit right)
* Write (data from cpu):
*    offset 0
*      * bits 23-0: 24-bit codec command to i2c bus
*    offset 1
*      * bit 0: dac data stream select
*      * bit 1: adc data stream select
*    offset 2
*      * bits 31 to 0: stereo audio data (16-bit left + 16-bit right)
*****************************************************************************/
/* file inclusion */
#include <alt_types.h>

/*****************************************************************************
* constant definitions
*****************************************************************************/
#define CHU_AUD_I2C_DATA_REG    0
#define CHU_AUD_STATUS_REG      1
#define CHU_AUD_DBUS_SEL_REG    1
#define CHU_AUD_DAC_DATA_REG    2
#define CHU_AUD_ADC_DATA_REG    3

/*****************************************************************************
* Function prototypes
*****************************************************************************/
/* I2C functions */
int audio_i2c_is_idle(alt_u32 audio_base);
void audio_i2c_wr_cmd(alt_u32 audio_base, alt_u8 addr, alt_u16 cmd);
void audio_init(alt_u32 audio_base);

/* data routing selection */
void audio_wr_src_sel(alt_u32 audio_base, int dac_sel, int adc_sel);

/* dac functions */
int audio_dac_fifo_full(alt_u32 audio_base);
void audio_dac_wr_fifo(alt_u32 audio_base, alt_u32 data);

/* adc functions */
int audio_adc_fifo_empty(alt_u32 audio_base);
alt_u32 audio_adc_rd_fifo(alt_u32 audio_base);
```

Listing 18.16   chu_avalon_audio.c

```
/********************************************************************
*
* Module:   Audio codec driver function prototypes
* File:     chu_avalon_audio.c
* Purpose:  Routines to configure codec, retrieve data from ADC, and
*           send data to DAC
*
********************************************************************/
/* file inclusion */
#include <io.h>
#include "chu_avalon_audio.h"


/********************************************************************
*
* I2C functions:  routines using I2C to send codec configuration command
*
********************************************************************/

/********************************************************************
* function: audio_i2c_is_idle
* purpose:  check whether the I2C controller is idle
* argument:
*    audio_base: base address of audio codec controller
* return: 1 if idle; 0 otherwise
* note:
********************************************************************/
int audio_i2c_is_idle(alt_u32 audio_base)
{
  alt_u32 status_reg;
  int i2c_idle_bit;

  status_reg = IORD(audio_base, CHU_AUD_STATUS_REG);
  i2c_idle_bit = (int) (status_reg & 0x00000001);
  return i2c_idle_bit;
}


/********************************************************************
* function: audio_i2c_wr_cmd
* purpose:  send a 24-bit packet via I2C bus
* argument:
*    audio_base: base address of audio codec controller
*    addr: address of codec configuration register; only 7 LSBs used
*    cmd: command data  to codec configuration register; only 9 LSBs used
* return:
* note:
*  - byte 0 is 7-bit I2C id + 1-bit r/w;
*    id of WM8731: 0011010; r/w:0 (always write for WM8731);
*    combined: 0011_0100 (0x34)
********************************************************************/
void audio_i2c_wr_cmd(alt_u32 audio_base, alt_u8 addr, alt_u16 cmd)
{
  const alt_32 i2c_id = 0x00000034;
  alt_u32 packet;      // data written to I2C; only 24 LSBs used

  packet = i2c_id;
  packet = (packet << 7) + (addr & 0x07f);  // append 7-bit address
  packet = (packet << 9) + (cmd & 0x01ff);  // append 9-bit command
  IOWR(audio_base, CHU_AUD_I2C_DATA_REG, (alt_u32) packet);
}
```

```
/*************************************************************************
 * function: audio_init
 * purpose:  initialize codec by writing configuration registers with
 *           the pre-selected values
 * argument:
 *   audio_base: base address of audio codec controller
 * return:
 * note:
 *************************************************************************/
void audio_init(alt_u32 audio_base)
{
  /* initial configuration values (registers R0 to R9)   */
  const alt_u16 cmds[10]={        // only 9 LSBs used
    0x0017, // R0: left line in gain 0 dB
    0x0017, // R1: right line in gain 0 dB
    0x0079, // R2: left headphone volume 0 dB
    0x0079, // R3: right headphone volume 0 dB
    0x0010, // R4: analog path select: line-in to adc, dac to line-out
    0x0000, // R5: digital audio: high-pass filter, no de-emphasis
    0x0000, // R6: enable all power
    0x0001, // R7: digital interface: left-adjust, 16-bit resolution
    0x0000, // R8: 48K sampling rate with 12.288MHz master clock
    0x0001  // R9: activate
  };
  int i;

  while (!audio_i2c_is_idle(audio_base)){};   // wait until I2C idle
  /* write a dummy data to R15 to reset the codec */
  audio_i2c_wr_cmd(audio_base, 15, 0);
  /* cycle through 10 commands */
  for (i=0; i<10; i++){
    while (!audio_i2c_is_idle(audio_base)){}; // wait until I2C idle
    audio_i2c_wr_cmd(audio_base, i, cmds[i]); // send a command packet
  }
  audio_wr_src_sel(audio_base, 0, 0);          // dac/adc to Avalon bus
}

/*************************************************************************
 * function: audio_wr_src_sel()
 * purpose:  set up the bus connection to codec controller
 * argument:
 *   audio_base: base address of audio codec controller
 *   dac_sel: 0 for Avalon MM interface
 *   adc_sel: 0 for Avalon MM interface
 * return:
 *************************************************************************/
void audio_wr_src_sel(alt_u32 audio_base, int dac_sel, int adc_sel)
{
  alt_u32 sel_reg = 0x00000000;

  if (dac_sel!=0)
    sel_reg = sel_reg | 0x00000001;    // set LSB to 1
  if (adc_sel!=0)
    sel_reg = sel_reg | 0x00000002;    // set 2nd LSB to 1
  IOWR(audio_base, CHU_AUD_DBUS_SEL_REG, sel_reg);
}
```

```
/************************************************************************
 *
 * Digital audio interface functions:
 *   routines to read adc fifo and write dac fifo
 *
 ************************************************************************/

/************************************************************************
 * function: audio_dac_fifo_full()
 * purpose:  check whether the codec dac fifo is full
 * argument:
 *   audio_base: base address of audio codec controller
 * return: 1 if full; 0 otherwise
 * note:
 ************************************************************************/
int audio_dac_fifo_full(alt_u32 audio_base)
{
  alt_u32 status_reg;
  int dac_full_bit;

  status_reg = IORD(audio_base, CHU_AUD_STATUS_REG);
  dac_full_bit = (int)((status_reg & 0x00000002) >> 1);
  return dac_full_bit;
}

/************************************************************************
 * function: audio_dac_wr_fifo()
 * purpose:  write data to the dac fifo
 * argument:
 *   audio_base: base address of audio codec controller
 *   data: 32-bit audio data
 * return:
 ************************************************************************/
void audio_dac_wr_fifo(alt_u32 audio_base, alt_u32 data)
{
  IOWR(audio_base, CHU_AUD_DAC_DATA_REG, data);
}

/************************************************************************
 * function: audio_adc_fifo_empty()
 * purpose:  check whether the codec adc fifo is empty
 * argument:
 *   audio_base: base address of audio codec controller
 * return: 1 if empty; 0 otherwise
 * note:
 ************************************************************************/
int audio_adc_fifo_empty(alt_u32 audio_base)
{
  alt_u32 status_reg;
  int adc_empty_bit;

  status_reg = IORD(audio_base, CHU_AUD_STATUS_REG);
  adc_empty_bit = (int)((status_reg & 0x00000004) >> 2);
  return adc_empty_bit;
}
```

```
/***************************************************************************
* function: audio_adc_rd_fifo()
* purpose:  retrieve data from the head of adc fifo
* argument:
*   audio_base: base address of audio codec controller
* return: 32-bit adc data from the head of fifo
* note:
*   - the data is removed from fifo after read
***************************************************************************/
alt_u32 audio_adc_rd_fifo(alt_u32 audio_base)
{
   alt_u32 data_reg;

   data_reg = IORD(audio_base, CHU_AUD_ADC_DATA_REG);
   return data_reg;
}
```

Listing 18.17   chu_main_audio_test.c

```
/***********************************************************************
 *
 * Module:   Audio codec test
 * File:     chu_main_audio_test.c
 * Purpose: Test audio codec driver functions
 * IP core base addresses:
 *   - SWITCH_BASE: slide switch
 *   - BTN_BASE: pushbutton
 *   - SSEG_BASE: 7-segment LED
 *   - VRAM_BASE: video SRAM
 *   - AUDIO_BASE: audio codec controller
 *
 ***********************************************************************/
/* file inclusion */
#include <stdio.h>
#include <math.h>          // for sin()
#include <unistd.h>
#include "system.h"
#include "chu_avalon_gpio.h"
#include "chu_avalon_audio.h"
#include "chu_avalon_vga.h"


/***********************************************************************
 * function: record_sin_wave()
 * purpose:  generate sin wave and record data points in buffer
 * argument:
 *   sf: sampling frequency
 *   freq: sin wave frequency
 *   size: size (# of data points) of the buffer
 *   buf: pointer to the buffer
 * return: updated buf with sin/cos data
 * note:
 *   - calling routine needs to allocate space for buffer
 ***********************************************************************/
void record_sin_wave(int sf, int freq, int size, alt_u32 *buf)
{
  const float PI=3.14159;
  const float AMP_MAX= 32767.0;         // max amplitude (2^15-1)
  float amp;
  int npoint, ncycle, i, j;
  alt_u32 left, right;

  npoint = sf/freq;                     // # of steps in one period
  /* construct 1st cycle */
  for (i=0; i<npoint; i++){
    amp = sin((float)i/(float)npoint * 2.0 * PI);
    left = (alt_u32) (amp * AMP_MAX);   // left channel; 16 LSBs used
    amp = cos((float)i/(float)npoint * 2.0 * PI);
    right = (alt_u32) (amp * AMP_MAX);  // right channel; 16 LSBs used
    buf[i] = (left << 16) + right;      // combine two channels
  }
  /* duplicate the 1st cycle for the remaining ncycle-1 cycles */
  /* no data for the last fractional cycle                     */
  ncycle = size/npoint;
  for (j=1; j<ncycle; j++){
    for (i=0; i<npoint; i++){
      buf[j*npoint+i]=buf[i];
    } // end for i
  } // end for j
}
```

```
/******************************************************************
 * function: set_volume
 * purpose:  set headphone volume
 * argument:
 *   audio_base: base address of audio codec controller
 *   vol: volume
 * return:
 * note:
 *   - codec: 7-bit headphone volume control; 80 levels in 1-dB step
 *     0x7f: +6 dB;  0x79: 0 dB;  0x30: -73 dB;  <0x30: mute
 ******************************************************************/
void set_volume(alt_u32 audio_base, int vol)
{
  alt_u16 cmd;

  if (vol>0x7f)                              // exceed maximal volume
     cmd = 0x007f;
  else
     cmd = (alt_u16) vol;
  while (!audio_i2c_is_idle(audio_base)){};
    audio_i2c_wr_cmd(audio_base, 2, cmd);  // left headphone out
  while (!audio_i2c_is_idle(audio_base)){};
    audio_i2c_wr_cmd(audio_base, 3, cmd);  // right headphone out
}

/******************************************************************
 * function: plot_audio_buffer()
 * purpose:  plot buffered audio data on screen
 * argument:
 *   vga_base: base address of VGA video SRAM
 *   buf: pointer to the buffer
 * return:
 * note:
 *   - plot only the first 640 data points
 *   - magnitude range is from -128 to +127
 ******************************************************************/
void plot_audio_buffer(alt_u32 vga_base, alt_u32 *buf)
{
  int i, j;
  alt_16 y_left, y_right;

  for(i=1; i<640; i++){
    y_right = (alt_16) (0x0000ffff ^ buf[i]); // right channel data
    y_left  =  (alt_16) (buf[i]>>16);         // left channel data
    j=480 - ((int)y_left/256 + 480/2);        // adjust y to -128/127
    vga_wr_pix(vga_base, i, j, 0xe0);         // plot left channel red
    j=480 - ((int)y_right/256 + 480/2);       // adjust y to -128/127
    vga_wr_pix(vga_base, i, j, 0x03);         // plot right channel blue
  }
}
```

```
/***************************************************************************
 * function: main()
 * purpose:  test audio codec
 * note:
 ***************************************************************************/
int main(void)
{
  const int SF=48000;       // sampling freq=48K
  const int BUF_SIZE= 5*SF; // 5 sec @ sampling frequency SF
  alt_u32 buf[BUF_SIZE];
  int sw, btn, freq, vol, reg, cmd, i;
  alt_u8 disp_msg[4]={sseg_conv_hex(5), 0x23, 0x2b, sseg_conv_hex(13)};

  sseg_disp_ptn(SSEG_BASE, disp_msg);      // show "Sond" for display
  vga_clr_screen(VRAM_BASE,0);             // clear screen
  audio_init(AUDIO_BASE);
  btn_clear(BTN_BASE);
  printf("Audio codec interface test \n\n");

  while (1){
    while (!btn_is_pressed(BTN_BASE)){ };      // wait for button
    btn=btn_read(BTN_BASE);                    // read button
    if (btn & 0x02){                           // key1 pressed
      sw=pio_read(SWITCH_BASE);                // read switch
      printf("key/sw: %d/%d\n", btn, sw);
    }
    btn_clear(BTN_BASE);
    switch (sw){
      case 0:  // initialize codec controller
        audio_init(AUDIO_BASE);
        break;
      case 1:  // record line input for 5 sec
        audio_i2c_wr_cmd(AUDIO_BASE, 4, 0x0010);     // line-in to adc
        for (i=0; i<BUF_SIZE; i++){
          while(audio_adc_fifo_empty(AUDIO_BASE)){} // wait if fifo empty
          buf[i]=audio_adc_rd_fifo(AUDIO_BASE);
        }
        break;
      case 2: // record microphone input for 5 sec
        audio_i2c_wr_cmd(AUDIO_BASE, 4, 0x0015);     // mic to adc, boost
        for(i=0; i<BUF_SIZE; i++){
          while(audio_adc_fifo_empty(AUDIO_BASE)){} // wait if fifo empty
          buf[i]=audio_adc_rd_fifo(AUDIO_BASE);
        }
        break;
      case 3:  // fill buffer with sinusoidal data
        printf("enter frequency:");
        scanf("%d", &freq);
        record_sin_wave(SF, freq, BUF_SIZE, buf);
        printf("sinusoidal wave recorded. \n");
        break;
      case 4: // play buffered audio data repeatedly until a key pressed
        while (!btn_is_pressed(BTN_BASE)){
          for(i=0; i<BUF_SIZE; i++){
            while(audio_dac_fifo_full(AUDIO_BASE)){} // wait if fifo full
            audio_dac_wr_fifo(AUDIO_BASE, buf[i]);
          } // end for
        } // end while
        break;
      case 5:  // issue codec command
        printf("enter codec register #:");
```

```
        scanf("%d", &reg);
        printf("enter command (in hex):");
        scanf("%x", &cmd);
        audio_i2c_wr_cmd(AUDIO_BASE, (alt_u8)reg, (alt_u16)cmd);
        printf("send 0x%x to codec register %d\n\n", cmd, reg);
        break;
    case 6:  // set volume
        printf("enter volume (between 28 and 127):");
        scanf("%d", &vol);
        set_volume(AUDIO_BASE, vol);
        printf("Volume set\n");
        break;
    case 7:  // plot buffered right- and left-channel data
        plot_audio_buffer(VRAM_BASE, buf);
        break;
    case 8:  // clear screen
        vga_clr_screen(VRAM_BASE,0);
        break;
    } //end switch
  } // end while
}
```

Listing 18.18   chu_main_wav_file_test.c

```
/******************************************************************
* Module:  WAV file retrieval test
* File:    chu_main_wav_file_test.c
* Purpose: Test wave (.wav) file access
* IP core base addresses:
*   - AUDIO_BASE: audio codec controller
******************************************************************/
/* file inclusion */
#include <stdio.h>
#include <unistd.h>
#include "system.h"
#include "chu_avalon_file.h"
#include "chu_avalon_audio.h"

/* constants */
#define  WAV_FILE_NAME  "/mnt/host/trumpet.wav" // path and file name
#define  BUF_SIZE       (6*48000)               // 6 sec of audio data

/******************************************************************
* function: read_wav_file()
* purpose:  read audio data from wave file and store it in a buffer
* argument:
*   file_name: name of the WAV file
*   buf: pointer to bitmap buffer
* return:
*   - 0: if successful
*   - buf updated with audio data
* note:
*   - calling function must allocate buffer space
*   - only support the following format:
*     - compression: 1 (PCM)
*     - channel: 2
*     - sampling rate: 48K
*     - resolution: 16 bits
******************************************************************/
int read_wav_file(char *file_name, alt_u32 *buf)
{
  /* note that fget32 read in little endian, but id in big endian */
  const alt_u32 RIFF_ID = 0x46464952;   //ascii for FFIR
  const alt_u32 WAVE_ID = 0x45564157;   //ascii for EVAW
  const alt_u32 FMT_ID  = 0x20746d66;   //ascii for \btmf
  const alt_u32 DATA_ID = 0x61746164;   //ascii for ATAD

  FILE *fp;
  alt_u32 r_id, w_id, f_id, d_id, srate, data_size;
  alt_u16 compression, channel, res;
  int i, s_size;

  /* open the file */
  fp = fopen(file_name,"rb");
  if (fp==NULL){
    printf("Error: cannot open file %s.\n", file_name);
    return(-1);
  }
  /* extract relevant chunk/subchunk info */
  r_id = fget32(fp);          // offset  0: "RIFF" chunk id
  fskip(fp,4);                // offset  4: chunk size
  w_id = fget32(fp);          // offset  8: "WAVE" chunk id
  f_id = fget32(fp);          // offset 12: "fmt " subchunk id
  fskip(fp,4);                // offset 16: subchunk size
```

```
  compression=fget16(fp);      // offset 20: 1 for PCM
  channel = fget16(fp);        // offset 22: 2 for stereo
  srate = fget32(fp);          // offset 24: 48K sampling rate
  fskip(fp,4);                 // offset 28: byte rate
  fskip(fp,2);                 // offset 32: block size
  res = fget16(fp);            // offset 34: 16 bits resolution
  d_id = fget32(fp);           // offset 36: "data" subchunk id
  data_size = fget32(fp);      // offset 40: # bytes of data subchunk
  /* check chunk/subchunk ids and parameters */
  if ((r_id!=RIFF_ID) || (w_id!=WAVE_ID)  || (f_id!=FMT_ID) ||
      (d_id!=DATA_ID) || (compression!=1) || (channel!=2)   ||
      (srate!=48000)  || (res!=16)){
    printf("Error: incorrect wave file format\n");
    printf("Must be PCM, 2 channels, 48K rate, 16-bit resolution.\n");
    printf("RIFF/WAVE/fmt/data ids: %08x/%08x/%08x/%08x\n",
           r_id, w_id, f_id, d_id);
    printf("compression/channel/srate/res/d_szie: %d/%d/%d/%d/%d\n",
           compression, channel, srate, res, data_size);
    fclose(fp);
    return(-1);
  }
  s_size = data_size/4;
  printf("File opened.\n # audio data samples: %d\n", s_size);
  /*if (s_size > BUF_SIZE);
     s_size = BUF_SIZE; */
  for (i=0; i<s_size; i++) {
    // get 32-bit data
    buf[i] = fget32(fp);
    if (i%1000 == 0)
      printf(".");
  }
  fclose(fp);
  printf("\nFile loaded.\n");
  return(0);
}


/******************************************************************************
 * function: main()
 * purpose:  Read a wave file and play it back
 * note:
 *    - the host-based file system must be enabled in BSP editor
 *    - the trumpet.wav sound file should be in the project directory
 *    - build/load the program with "Debug As => Nios II Hardware"
 ******************************************************************************/
int main(void){
  alt_u32 buf[BUF_SIZE];
  int i;

  audio_init(AUDIO_BASE);
  printf("Wave file test \n\n");
  if (read_wav_file(WAV_FILE_NAME, buf)!=0){
    printf("\n Fail to load wav file. \n");
    return (0);
  }
  while(1){
    for(i=0; i<BUF_SIZE; i++){
      while(audio_dac_fifo_full(AUDIO_BASE)){}  // wait if dac fifo full
        audio_dac_wr_fifo(AUDIO_BASE, buf[i]);
    } // end for
  } // end while
}
```

CHAPTER 19

SD CARD CONTROLLER

A flash memory card can be used as massive storage for an embedded system. The DE1 board has a slot for an SD (secure digital) card. In this chapter, we develop hardware and software routines to retrieve data from an SD card. The hardware portion is an SPI (serial peripheral interface) controller, which sends commands and transfers data between the SD card and processor. The software portion includes the SPI driver routines to transmit and receive data over the SPI bus, the SD card driver routines to send a command and read and write a sector, and basic routines to read a file from a FAT16 file system.

19.1 OVERVIEW OF SD CARD

The FPGA's internal memory, external SRAM, and SDRAM discussed in Chapter 15 are all *volatile memory*, which requires power to maintain stored data. *Non-volatile* memory, on the other hand, can keep data after power-off. EEPROM (electrically erasable programmable read-only memory) device is a type of non-volatile memory. Its content can be electrically erased and written again. *Flash memory* is a special type of EEPROM in which a large block of data is erased and written at a time and costs much less than a byte-programmable device. To write a single word, a flash memory device needs to read the entire block containing the word, modify the specific word, erase the entire block, and then write the entire block back. Because of the high overhead, it is more feasible to transfer a large amount

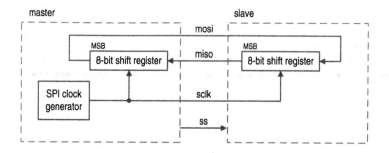

master   slave

Figure 19.1   Conceptual diagram of SPI bus.

of data at a time. This behaves more like a hard disk and thus flash memory is usually used for massive storage rather than "random-access" memory.

A *flash memory card* is a device that contains flash memory and a controller. Because of flash memory's relative low per-byte cost, the memory card is frequently used as an external massive storage for embedded applications. An *SD* (*secure digital*) card is a widely used memory card format. The SD card format is the successor to the MMC (*multimedia card*) format and is designed by the SD Card Association. The card's form factor, electrical interface, and protocol are all part of the specification. The capacity of a standard SD card ranges from a few megabytes up to a maximum of two gigabytes. The *SDHC* (*secure digital high capacity*) card is an extension of the SD card standard and supports storage capacity between 4 gigabytes and 32 gigabytes. SHDC cards have the same physical shape and form factor but with slightly different protocols. In this chapter, we use the generic term "SD card" to refer to both types of cards. When the distinction is important, we use more specific terms "standard capacity SD card" and "high capacity SDHC card" to emphasize a particular type of card.

An SD card supports two operation modes, which are *SD mode* and *SPI (serial peripheral interface) mode*. The SD mode is a proprietary format and uses four lines for data transfer. SPI is an open standard for serial interfaces and is used widely in embedded applications. The SD card socket on the DE1 board is wired to support the SPI mode and our discussion in this chapter is limited to this mode.

19.2   SPI CONTROLLER

19.2.1   Overview of SPI interface

Basic protocol   The *SPI (serial peripheral interface)* bus is a synchronous serial data link standard originally developed by Motorola. It uses three lines for communication, including one for the clock, one for serial data input, and one for serial data output. An SPI bus contains one master device and one or more slave devices. The master generates the clock signal and initiates the data transfer. The conceptual diagram of an SPI bus with two devices is shown in Figure 19.1. There are two shift registers, one in the master and one in the slave. The two shift registers are connected as a ring via the `mosi` (for "master-out-slave-in") and `miso` (for "master-in-slave-out") lines and their operation is controlled by the same clock signal, `sclk`. We assume that both registers are eight bits wide and data transfer is done on a

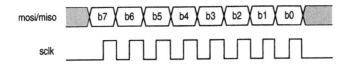

Figure 19.2   Basic timing diagram of SPI mode 0 data transfer.

byte-by-byte basis. In the beginning of the operation, both master and slave load
data into the registers. During the data transfer, data in both registers is shifted
to the left by one bit in each **sclk** cycle. After eight **sclk** cycles, eight data bits
are shifted and the master and slave have exchanged register values. The master
and slave then can processed the received data. This operation can be interpreted
that the master writes data to and reads data from the slave simultaneously, which
is known as *full-duplex* operation. In comparison, the I²C bus discussed in Section 18.2 supports only *half-duplex* operation since data can be transferred only in
one direction at a time.

   In addition to the **mosi**, **miso**, and **sclk** lines, a slave device may also have an
active-low chip select input, **ss** (for "slave select"). This can be used for the master
to select the desired slave device if there are multiple slave devices on the bus.

Basic timing   In SPI, the edges of the **sclk** signal are used for shifting and latching
a data bit. The *operation mode* defines the *polarity* and *phase* of **sclk** with respect
to the data bit. There are four modes. The master must know the mode of slave
devices in advance and generate proper polarity and phase accordingly. An SD card
uses *mode 0*, in which the base value of the clock is zero (i.e., polarity is 0) and
data are read at the rising edge and changed at the falling edge (i.e., phase is 0).
The conceptual timing diagram is shown in Figure 19.2. Basically, the data bits
are placed at the falling edge and retrieved at the rising edge of the **sclk** signal.
This arrangement eases the timing constraint since the shifting and retrieval are
done at opposite edges.

19.2.2   HDL implementation

The DE1 board uses an SPI interface to access the SD card and the FPGA chip
acts as a master. The data is grouped as an 8-bit packet and transferred. A custom
SPI controller can be constructed for this purpose. The design approach is similar
to that of the I²C controller in Section 18.2.2. The controller is basically an FSMD
that generates the **sclk** signal, shifts a data bit into an input buffer at the rising
edge of **sclk**, and shifts a data bit out from an output buffer at the falling edge
of **sclk**. The sketch of the control FSM is shown in Figure 19.3. The **sclk0** and
sclk1 states represent the low and high portions of the **sclk** signal. The FSM
circulates in these two states eight times to transfer eight bits of data. The HDL
code is shown in Listing 19.1.

Listing 19.1   SPI controller

```
library ieee;
use ieee.std_logic_1164.all;
use ieee.numeric_std.all;
4 entity spi is
```

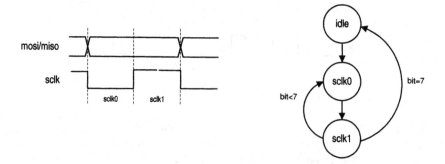

Figure 19.3   Sketch of SPI controller FSM.

```
    port (
       clk, reset: in  std_logic;
       din: in std_logic_vector(7 downto 0);
       dvsr: in std_logic_vector(7 downto 0);
9      wr_sd: in std_logic;
       dout: out std_logic_vector(7 downto 0);
       spi_clk: out std_logic;
       spi_miso: in std_logic;
       spi_mosi: out std_logic;
14     spi_done_tick, spi_idle: out std_logic
    );
  end spi;

  architecture arch of spi is
19    type statetype is (idle, sclk0, sclk1);
      signal state_reg, state_next: statetype;
      signal c_reg, c_next: unsigned(7 downto 0);
      signal spi_tick: std_logic;
      signal spi_clk_reg, spi_clk_next: std_logic;
24    signal bit_reg, bit_next: unsigned(2 downto 0);
      signal sin_reg, sin_next: std_logic_vector(7 downto 0);
      signal sout_reg, sout_next: std_logic_vector(7 downto 0);
  begin

29    — fsmd for transmitting one byte

      — registers
      process (clk, reset)
      begin
34       if reset='1' then
            state_reg <= idle;
            sin_reg <= (others=>'0');
            sout_reg <= (others=>'0');
            bit_reg <= (others=>'0');
39          c_reg <=(others=>'0');
            spi_clk_reg <= '0';
         elsif (clk'event and clk='1') then
            state_reg <= state_next;
            sin_reg <= sin_next;
44          sout_reg <= sout_next;
            bit_reg <= bit_next;
            c_reg <=c_next;
            spi_clk_reg <= spi_clk_next;
```

```
           end if;
49     end process;
       -- next-state logic
       process(state_reg,sin_reg,sout_reg,bit_reg,
               din,wr_sd,spi_miso)
       begin
54        state_next <= state_reg;
          spi_idle <= '0';
          spi_done_tick <='0';
          sin_next <= sin_reg;
          sout_next <= sout_reg;
59        bit_next <= bit_reg;
          c_next <= c_reg + 1;   -- timer runs continuously
          case state_reg is
             when idle =>
                spi_idle <= '1';
64              if wr_sd='1' then
                   sout_next <= din;
                   state_next <= sclk0;
                   bit_next <= (others=>'0');
                   c_next <= x"01";
69              end if;
             when sclk0 =>
                if c_reg=unsigned(dvsr) then  -- spi_clk 0-to-1
                   state_next <= sclk1;
                    sin_next <= sin_reg(6 downto 0) & spi_miso;
74                 c_next <= x"01";
                end if;
             when sclk1 =>
                if c_reg=unsigned(dvsr) then  -- spi_clk 1-to-0
                   if bit_reg=7 then
79                    spi_done_tick <='1';
                      state_next <= idle;
                   else
                      sout_next <= sout_reg(6 downto 0) & '0';
                      state_next <= sclk0;
84                    bit_next <= bit_reg + 1;
                      c_next <= x"01";
                   end if;
                end if;
          end case;
89     end process;
       -- lookahead output decoding
       spi_clk_next <= '1' when state_next=sclk1 else '0';
       -- output
       dout <= sin_reg;
94     spi_mosi <= sout_reg(7);
       spi_clk <= spi_clk_reg;
    end arch;
```

The din port is the 8-bit input data to be transferred and the dout port is the received data. The dvsr port specifies the clock divisor value, which in turn controls the frequency of sclk. The spi_clk, spi_miso, and spi_mosi ports are connected to the clock and data lines of the SPI bus. Note that the processing of the ss signal is in another module and not part of the SPI controller. In addition, the controller includes an input command, wr_sd, which initiates the data transfer, and two status signals, spi_idle and spi_done_tick.

There are several registers in the FSMD. The sin_reg and sout_reg registers are shift registers storing the input and output data, respectively. The input data from spi_miso is sampled and shifted into sin_reg at the rising edge of spi_clk, which occurs when the FSM transits from the sclk0 state to the sclk1 state. The MSB of sout_reg is connected to spi_mosi. The old bit is shifted out at the falling edge of spi_clk, which occurs when the FSM transits from the sclk1 state to the sclk0 state. The bit_reg register keeps track of the numbers of bits processed. The SPI clock signal is buffered via the spi_clk_reg to remove potential glitches.

The frequency of the sclk clock, which controls the data transfer rate of the SPI bus, may vary over a wide range. We use the c_reg register as a timer to keep track of the number of clock cycles spent in sclk0 and sclk1 states, which corresponds to half of the sclk period. It counts continuously and is cleared to 1 when the FSM exits the current state. To make the design more flexible, an external input, dvsr (for clock divisor), is used to set the upper boundary of the counter. If the value of dvsr is N, the FSM spends N system clock cycles in each state and the frequency of sclk becomes $\frac{f_{system}}{2*N}$. Since the system clock of the DE1 board is 50 MHz, sclk can run up to 25 MHz.

19.3   SPI CONTROLLER IP CORE DEVELOPMENT

19.3.1   Avalon interfaces

The wrapping circuit for the SPI controller should include an Avalon MM slave interface to interact with the host, a clock input interface for the system clock, and a conduit interface for the external signals.

19.3.2   Register map

A Nios II processor interacts with the SPI controller as follows:
- check (i.e., read) status signal.
- receive (i.e., read) 8-bit data via the SPI bus.
- set up (i.e., write) the ss (SD card chip select) signal.
- set up (i.e., write) the SPI clock divisor.
- transmit (i.e., write) 8-bit data via the SPI bus.

The registers, their address offsets, and fields are:
- Read addresses (data to cpu)
  - offset 0 (read data register)
    * bits 7 to 0: 8-bit data received from the SPI bus
    * bit 8: asserted (i.e., 1) when the SPI controller is idle (i.e., ready)
- Write addresses (data from cpu)
  - offset 1 (SD card chip select register)
    * bit 0: 1-bit SD card chip select
  - offset 2 (clock divisor register)
    * bits 7 to 0: 8-bit sclk clock divisor
  - offset 3 (write data register)
    * bits 7 to 0: 8-bit data to be transmitted via the SPI bus

19.3.3 Wrapped SPI controller

The HDL code of the wrapped SPI controller is shown in Listing 19.2.

Listing 19.2   Wrapped SPI controller

```vhdl
library ieee;
use ieee.std_logic_1164.all;
use ieee.numeric_std.all;
entity chu_avalon_sd is
   port (
      clk, reset: in  std_logic;
      -- external interface
      sd_cs, sd_clk, sd_di: out std_logic;
      sd_do: in std_logic;
      -- avalon interface
      sd_address: in std_logic_vector(1 downto 0);
      sd_chipselect: in std_logic;
      sd_write: in std_logic;
      sd_writedata: in std_logic_vector(31 downto 0);
      sd_read: in std_logic;
      sd_readdata: out std_logic_vector(31 downto 0)
   );
end chu_avalon_sd;

architecture arch of chu_avalon_sd is
   signal wr_en, wr_sd, wr_cs, wr_dvsr: std_logic;
   signal cs_reg: std_logic;
   signal dvsr_reg: std_logic_vector(7 downto 0);
   signal sd_out: std_logic_vector(7 downto 0);
   signal sd_ready: std_logic;

begin
   -- instantiate SPI unit
   spi_unit: entity work.spi(arch)
      port map(clk=>clk, reset=>reset,
               din=>sd_writedata(7 downto 0),
               dvsr=>dvsr_reg, dout=>sd_out, wr_sd=>wr_sd,
               spi_clk=>sd_clk, spi_mosi=>sd_di, spi_miso=>sd_do,
               spi_idle=>sd_ready, spi_done_tick=>open);
   -- registers
   process (clk, reset)
   begin
      if reset='1' then
         cs_reg <='0';
         dvsr_reg <= (others=>'1');
      elsif (clk'event and clk='1') then
         if (wr_cs='1') then
            cs_reg <=sd_writedata(0);
         end if;
         if (wr_dvsr='1') then
            dvsr_reg <=sd_writedata(7 downto 0);
         end if;
      end if;
   end process;
   -- write decoding
   wr_en <= '1' when sd_write='1' and sd_chipselect='1' else '0';
   wr_cs <= '1' when sd_address="01" and wr_en='1' else '0';
   wr_dvsr <= '1' when sd_address="10" and wr_en='1' else '0';
   wr_sd <= '1' when sd_address="11" and wr_en='1' else '0';
   -- read data
   sd_readdata <= x"00000" & "000" &  sd_ready & sd_out;
```

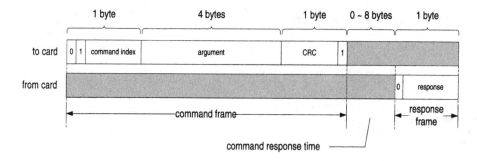

Figure 19.4 Basic timing diagram of an SD card command with R1 response.

```
    —— output
    sd_cs <= cs_reg;
59 end arch;
```

The code consists of two registers to store the `sclk` clock divisor and chip-select values and write decoding logic. Note that `mosi`, `miso`, `sclk`, and `ss` are renamed as `sd_di`, `sd_do`, `sd_clk`, and `sd_cs` to be consistent with the naming convention of the DE1 board.

19.3.4 SOPC component creation

After constructing the Avalon bus interface module, we can create a new SOPC component in Component Editor following the procedure outlined in Section 14.5.4. It then can be treated as a normal IP core and integrated into a Nios II system.

19.4 SD CARD PROTOCOL

On top of the SPI bus, an SD card communicates with an external device via its specific protocols. The communication and data transfer are performed via a series of commands. We provide a quick review of relevant commands, the initialization procedure, and data access procedure in this section. For clarity, only relevant commands and key procedures are examined. The complete command set and protocols can be found in the SD card documentation in the bibliographic section.

19.4.1 SD card command and response formats

In SPI mode, an SD card acts as a slave device and the SPI controller issues the commands. After receiving a command, the SD card responds with a response frame. For read and write operations, it continues with data token and bulk data.

The timing diagram of an SD card command and response is shown in Figure 19.4. The basic sequence is:

- A host issues a *command frame*.
- An SD card processes the command within a certain amount of time (known as *command response time*).
- An SD card responds with a *response frame*.

Although the SPI standard allows full-duplex operation, an SD card does not utilize this capability. The protocol is based on a half-duplex operation and the card either receives or transmits data but not both. The master and slave usually raise the data line to 1 when it is not used. This corresponds to transmit and receive the 0xff patterns. The unused portion is indicated by the gray regions of the timing diagram.

A command frame consists of six bytes. The first byte is command, which is started with 01, and then followed by the 6-bit command index. The next four bytes are the argument field that can accommodate up to 32 bits of information. The last byte consists of the 7-bit the CRC (cyclic redundancy check) code and a final stop bit 1. The use of the CRC code is optional in SPI mode and is disabled by default. We don't use this feature in our implementation. During the initialization process, CRC codes are required on two occasions and the values can be calculated and included manually. Even that the CRC feature is not used, the last byte is still required to compose the 6-byte command frame.

The command response time provides time for a card to process the command. During this interval, the master continues generating the `sclk` clock and the slave sends data bytes filled with all 1's (i.e., 0xff). The response time is between zero and eight bytes.

The SD card then responds with a response frame. There are several formats of response frames. The most common one is the *R1* response, which contains one byte. The MSB is always 0 and the seven LSBs indicate various conditions and error status:

- bit 7: always 0
- bit 6: parameter error
- bit 5: address error
- bit 4: erase sequence error
- bit 3: command CRC error
- bit 2: illegal command
- bit 1: erase reset
- bit 0: in idle state

The expected response from most commands should be 0x00, which indicates that no error occurs. In our implementation, we also use commands that respond with the *R3* and *R7* formats. These formats contain five bytes, in which the first byte is the same as R1 format and the remaining four bytes return certain status information.

The 6-bit command index field defines 64 (i.e., 2^6) commands, which are represented as CMDxx in SD card documentation, where "xx" is the index number. The SD card standard further extends the number of commands by defining an additional set of *application-specific commands*. An application-specific command is a sequence of *two commands*, in which a specific command follows the CMD55 command. These commands are represented as ACMDxx. For example, ACMD41 means the sequence of CMD55 and CMD41.

The commands used in our implementation are summarized in Table 19.1, which lists the command index, argument usage, and response type and explains its basic function. Note that one application specific command, ACMD41, is used for SD card initialization.

Table 19.1 Basic SD card commands

Command Index	Argument	Resp. type	Description
CMD0	[31:0]: stuff bits	R1	reset SD card to idle state
CMD8	[31:12]: reserved (0) [11:8]: voltage [7:0] check pattern	R7	send SD card interface condition
CMD17	[31:0]: address	R1	read a single block
CMD24	[31:0]: address	R1	write a single block
CMD55	[31:0]: stuff bits	R1	start an application command
CMD58	[31:0]: stuff bits	R3	read OCR register
ACMD41	[30]: HCS other: reserved (0)	R1	send HCS bit and initialize card

19.4.2 Initialization and identification process

An SD card must be properly set in SPI mode and initiated before data transfer. The basic steps of initialization and identification process are:

1. Power-on delay
2. SPI model selection
3. Interface condition check
4. SD card initialization
5. Card type check

Four types of cards can be used in an SD card slot, which are MMC card, standard capacity SD card version 1, standard capacity SD card version 2 and higher, and high capacity SDHC card, and their protocols are not identical. Since the former two are mainly for backward compatibility, our discussion just covers the latter two.

Although the SPI bus can run at a high data rate for data transfer, the `sclk` line should be set to a relative low rate, between 100 kHz and 400 kHz, during the initialization process.

Power-on delay After power-on or insertion, an SD card needs a small delay before it can accept a command. The host should hold the `mosi` and `ss` lines high and supply at least 74 cycles over the `sclk` line. After this, the SD card enters the SD mode and is in the *idle state*.

SPI mode selection By default, an SD card enters the SD mode after power-on. We must force it to switch to the SPI mode. This is done by holding `ss` low and sending the CMD0 (reset) command. The SD card should respond with an R1 response of 0x01, in which the bit 0 is asserted, an indication that the card is in the idle state. Note that the CMD0 command must be transmitted with a valid CRC value, which is 0x95.

Interface condition check For an SD card version 2 or higher, the host must issue the CMD8 command to verify that the SD card interface can operate within the host's supplied voltage range. The command include a 32-bit argument and its fields are:

- bits 31 to 12: reserved. The field should be set to 0.
- bits 11 to 8: supply voltage. The field should be set to 0x1 for a voltage range of 2.7 V to 3.6 V.
- bits 7 to 0: 8-bit check pattern. We use the 0xaa pattern.

The complete argument becomes 0x000001aa. A valid CRC value is also needed for this step, which is 0x87. The SD card should respond with a 5-byte R3 response. The first byte should still be 0x01 since the card is still in the idle state and the remaining four bytes should be 0x000001aa, which means the card can work within the voltage range and echoes the 0xaa check pattern. If the CMD8 command is rejected, the card is an MMC card or an older SD card.

SD card initialization After verifying the working voltage range, the master can issue the ACMD41 command (i.e., CMD55 and then CMD41) to send the host capacity support information and activate the card's initiation process. The command uses bit 30 of the argument to indicate whether the host supports high capacity SDHC cards. The other 31 bits are reserved and should be set to 0's. Thus, the argument should be either 0x00000000 or 0x40000000. The initialization process can take up to several hundred milliseconds. After completion, the card will exit the idle state and bit 0 of R1 response will be cleared to 0, which means the value of the R1 response changes from 0x01 to 0x00. The host should repeatedly issue the ACMD41 command and check the response until the idle state bit is cleared to 0.

Card type check In the last step, the host checks whether the card is a standard SD capacity card or a high capacity SDHC card. A card keeps this information in the *CCS (card capacity status) bit* of its internal *OCR (operation condition register) register*. The CCS bit is set to 1 in a high capacity SDHC card. The host can use the CMD58 command to retrieve the contents of the OCR register. The card responds with a 5-byte R3 response. The first byte is the R1 response and should be 0x00 and the remaining four bytes are the contents of the OCR register. Bit 30 is the CCS bit.

19.4.3 Data read and write process

After successful initialization, we can read data from or write data to an SD card. Unlike SRAM or DRAM memory, which usually accesses data by individual bytes or words, SD cards transfer data in blocks. For simplicity, our discussion assumes that the data is always accessed in properly aligned 512-byte blocks and only one block is transferred at a time. This is the default length for the standard capacity SD card and the mandatory length for the high capacity SDHC card. In our discussion, we also refer to a 512-byte block as a *sector*.

The host starts data transfer by sending a block read or write command. After the card responds with a valid R1 response, one or more data packets are transmitted by the host or the SD card. The packet consists of a 1-byte *data token* (0xfe), the actual 512-byte data block, and a 2-byte CRC-16 code. The use of CRC is optional in the SPI mode but the two bytes must be included in the data packet. To increase performance, the SPI clock rate can be increased after initialization. The maximal rate depends on the "speed class" of a card and the information can be retrieved from the card's internal register. In our setup, the limiting factor is the software driver and this issue is discussed in Section 19.8.

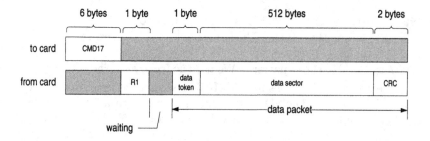

Figure 19.5   Basic timing diagram of a single block read operation.

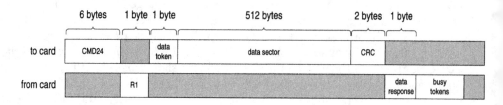

Figure 19.6   Basic timing diagram of a single block write operation.

Single block read process   The CMD17 command is used to read a single data block from an SD card. The 32-bit argument specifies the starting location of the data. The location is represented as the *byte address* for the standard capacity SD card and as the *sector address* for the high capacity SDHC card.

The basic timing diagram of a single block read operation is shown in Figure 19.5. The host issues the CMD17 command. The card acknowledges it with a valid R1 response (0x00) and then transmits the data packet. The card may require some time to process the request and thus the packet will not be transmitted immediately. The host should keep on generating the clock cycles and monitoring the input data stream for the data token (0xfe). After the token is detected, the 512-byte data can be retrieved. If there is an error, the card will send an 8-bit *data error token* in place of the data packet. The three MSBs of error token are 0's and the remaining five bits indicate the error conditions.

Single block write process   The CMD24 command is used to write a single data block to an SD card. The 32-bit argument specifies the starting location of the data, similar to that of the CMD17 command.

The basic timing diagram of a single block write operation is shown in Figure 19.6. The host first issues the CMD24 command. After receiving a valid R1 response (0x00), the host then transmits a data packet. The two CRC bytes can be any value but must be included. The card acknowledges the packet with an 8-bit *data response* token. If the data packet is accepted without errors, the five LSBs should be "00101". The card then starts the write operation and issues a continuous stream of 0x00 "busy" tokens (i.e., holds the output line low) when the device programming is in progress. Upon completion, the card returns the output line to high, which appears as 0xff data for the host.

19.5   SPI AND SD CARD DRIVER

There are two layers of drivers to access raw SD card data. The lower layer is the SPI driver consisting of routines to receive and transmit data over the SPI bus. The SD card driver is built on top of the SPI driver and consists of routines to initialize the card, read a sector from the card, and write a sector to the card.

19.5.1   SPI driver routines

The SPI controller has four addressable registers. To make the code clear, we define the address offsets as symbolic register names in the header file:

```
#define CHU_SD_RD_DATA_REG      0
#define CHU_SD_CS_REG           1
#define CHU_SD_DVSR_REG         2
#define CHU_SD_WR_DATA_REG      3
```

There are five basic functions:

- sd_spi_is_ready(): check whether the SPI controller is ready.
- sd_spi_wr_cs(): set the SD card's **ss** (chip select) signal.
- sd_spi_wr_dvsr(): write the SPI clock divisor value.
- sd_spi_wr_byte(): transmit a byte via the SPI bus.
- sd_spi_rd_byte(): retrieve a byte from the SPI bus.

These functions are shown in Listing 19.3.

Listing 19.3

```
int sd_spi_is_ready(alt_u32 sd_base)
{
  alt_u32 spi_reg;
  int spi_ready_bit;

  spi_reg = IORD(sd_base, CHU_SD_RD_DATA_REG);
  spi_ready_bit = (int) ((spi_reg & 0x00000100) >> 8);
  return spi_ready_bit;
}

void sd_spi_wr_cs(alt_u32 sd_base, int en_bit)
{
  IOWR(sd_base, CHU_SD_CS_REG, (alt_u32) en_bit);
}

void sd_spi_wr_dvsr(alt_u32 sd_base, alt_u8 dvsr)
{
  IOWR(sd_base, CHU_SD_DVSR_REG, (alt_u32) dvsr);
}

void sd_spi_wr_byte(alt_u32 sd_base, alt_u8 spi_data)
{
  /* wait until SPI is ready */
  while (!sd_spi_is_ready(sd_base)){};
  /* write a byte */
  IOWR(sd_base, CHU_SD_WR_DATA_REG, (alt_u32) spi_data);
}

alt_u8 sd_spi_rd_byte(alt_u32 sd_base)
{
```

```
    alt_u32 spi_reg;
    alt_u8 spi_data;

    /* write a dummy byte and shift in data as well */
    sd_spi_wr_byte(sd_base, 0xff);
    spi_reg = IORD(sd_base, CHU_SD_RD_DATA_REG);
    spi_data = (alt_u8) (spi_reg & 0x000000ff);
    return spi_data;
}
```

The **sd_spi_is_ready()** function extracts the idle bit from the data register. The **sd_spi_wr_cs()** function writes a bit to the SPI controller's 1-bit **cs_reg** register, which is connected to the SD card's **ss** pin. Note that **ss** is active low and thus writing 0 enables the card. The **sd_spi_wr_dvsr()** writes the SPI clock divisor value to the register. The role of the divisor is discussed in Section 19.2.2. If the value of **dvsr** is N, the frequency of **sclk** becomes $\frac{f_{system}}{2*N}$.

The SPI bus is a full-duplex system and the host writes data and reads data at the same time. However, the SD card protocol is a half-duplex protocol and a card transfers data only in one direction at a time. For clarity, we create two separate functions. The **sd_spi_wr_byte()** function performs the write operation and transmits a byte on the SPI bus. The received byte is ignored. The **sd_spi_rd_byte()** function performs the read operation and receives a byte from the SPI bus. The SPI controller actually performs a write operation at the same time. The stuffing data bytes (0xff) are transmitted by the host but are ignored by the SD card.

19.5.2   SD card driver routines

The SD card driver follows the protocol and consists of functions to send a command, initialize a card, and read and write a sector. To make the code clear, we define the commands, two special data tokens, and timeout periods as symbolic constants in the header file:

```
    #define SD_CMD0_RESET          0
    #define SD_CMD8_SEND_IF        8
    #define SD_CMD17_READ          17
    #define SD_CMD24_WRITE         24
    #define SD_CMD55_APP           55
    #define SD_CMD58_RD_OCR        58
    #define SD_ACMD41_INIT_SD      41
    /* SD card token */
    #define SD_TOKEN_DATA_START    0xfe
    #define SD_TOKEN_DATA_ACCEPT   0x05
    /* SD card timeout cycles */
    #define SD_INIT_TIME_OUT       900      // 0.50s w/ 200 kHz sclk
    #define SD_READ_TIME_OUT       312500   // 0.10s w/ 25 MHz sclk
    #define SD_WRITE_TIME_OUT      781250   // 0.25s w/ 25 MHz sclk
```

Sending command routine   The **sd_wr_cmd()** function sends a command to an SD card and retrieves the response. The code is shown in Listing 19.4.

Listing 19.4

```
alt_u8 sd_wr_cmd(alt_u32 sd_base, alt_u8 cmd, alt_u32 argu, alt_u32 *r3)
{
  int i;
  alt_u8 crc, rcode, byte;
  alt_u32 ocr=0;

  /* crc for CMD0 and CMD8 */
  if (cmd==SD_CMD0_RESET)
    crc=0x95;        // crc for CMD0(0)
  else
    crc =0x87;       // crc for CMD8(0x1aa)
  /* send command */
  sd_spi_wr_byte(sd_base, cmd|0x40);    //2 MSBs is 01
  /* send argument */
  sd_spi_wr_byte(sd_base, (alt_u8)(argu>>24));
  sd_spi_wr_byte(sd_base, (alt_u8)(argu>>16));
  sd_spi_wr_byte(sd_base, (alt_u8)(argu>>8));
  sd_spi_wr_byte(sd_base, (alt_u8)argu);
  /* send crc; only valid for CMD0 and CMD8 */
  sd_spi_wr_byte(sd_base, crc);
  /* wait for response, up to 8-byte delay */
  for(i=0; i<8; i++){
  /* read R1 response */
  rcode = sd_spi_rd_byte(sd_base);
  if (rcode != 0xff)
    break;
  }
  /* read 4 additional bytes for CMD8 and CMD58 response */
  if (cmd==SD_CMD8_SEND_IF || cmd==SD_CMD58_RD_OCR){
    for(i=0; i<4; i++){
      byte = sd_spi_rd_byte(sd_base);
      ocr = (ocr <<8 | byte);
    }
    // printf("rocde, ocr: 0x%02x, 0x%08x\n", rcode, ocr);
    *r3 = ocr;
  }
  return (rcode);
}
```

The function basically transmits a 6-byte command frame (1-byte command index, 4-byte argument, and 1-byte CRC) in sequence, waits for up to 8-byte delay, and retrieves the response. For simplicity, we only consider the subset discussed in Table 19.1. There are two caveats in the code. First, we choose not to use the 7-bit CRC code and ignore this field in general. However, the CMD0 and CMD8 commands in the initialization process require valid CRC codes. We manually calculate the CRC values for the two special cases and insert it into the command frame as needed. Second, while most commands respond with a 1-byte R1 response, the CMD8 and CMD58 commands respond with R7 and R3 responses, both consisting of four additional bytes. We use an additional for loop to read four more bytes as needed. The function returns the 1-byte R1 response. Its value is 0x00 most of the time, which means that there is no error and the card is not in the idle state.

Device initialization routine   The sd_init() function determines the type of card and initializes it in the SPI mode. The code is shown in Listing 19.5.

Listing 19.5

```
int sd_init(alt_u32 sd_base)
{
    int i;
    alt_u8 rcode;
    alt_u32 ocr;
    int hcs;

    /* set SPI sclk clock to 200 kHz */
    sd_spi_wr_dvsr(sd_base, 125);
    /* force the sd card to enter spi mode */
    sd_spi_wr_cs(sd_base, 1);                // set cs to 1
    for(i=0; i<10; i++){                     // generate 80 sd_clk cycles
        sd_spi_wr_byte(sd_base, 0xff);       // 8 cycles per write
    }
    sd_spi_wr_cs(sd_base, 0);                // set cs to 0 (enable)
    /* send reset command */
    rcode = sd_wr_cmd(sd_base, SD_CMD0_RESET, 0, &ocr);
    if (rcode != 0x01){                      // not entering idle state
        printf("CMD0 command fails: R1=0x%02x \n", rcode);
        return (-1);
    }
    /* send interface condition check command */
    rcode = sd_wr_cmd(sd_base, SD_CMD8_SEND_IF, 0x000001aa, &ocr);
    if (rcode!=0x01 || ocr!=0x000001aa){
        printf("CMD8 command fails: R1/data=0x%02x 0x%x\n", rcode, ocr);
        return (-1);
    }
    /* send sd card init command and wait for 0.3 sec */
    for(i=0; i<SD_INIT_TIME_OUT; i++){
        sd_wr_cmd(sd_base, SD_CMD55_APP, 0, &ocr);
        rcode = sd_wr_cmd(sd_base, SD_ACMD41_INIT_SD, 0x40000000, &ocr);
        if (rcode==0x00)                     // correct response received
            break;
    }
    if (rcode!=0x00){
        printf("ACMD41 command fails: R1=0x%02x\n", rcode);
        return (-1);
    }
    /* send read OCR register command */
    rcode = sd_wr_cmd(sd_base, SD_CMD58_RD_OCR, 0x000001aa, &ocr);
    if (rcode != 0x00){
        printf("CMD58 command fails: R1=0x%02x \n", rcode);
        return (-1);
    }
    /* extract hcs bit (bit 30) */
    hcs = (ocr & 0x40000000) >> 30;
    /* set SPI clock to 1 MHz */
    sd_spi_wr_dvsr(sd_base, 25);
    return(hcs);
}
```

The code follows the procedure discussed in Section 19.4.2. The host issues a sequence of commands and verifies the responses. The function returns -1 after receiving an incorrect response. Otherwise, it returns 0 for a standard capacity SD card and returns 1 for a high capacity SDHC card.

The ACMD41 command starts the SD card's internal initialization process and may require a fraction of a second to complete. The host must keep on issuing the command until a valid R1 response 0x00 (which indicates the card is no longer in idle state) is received or the timeout period is reached. We use a for loop for this task:

```
for(i=0; i<SD_INIT_TIME_OUT; i++){
  sd_wr_cmd(sd_base, SD_CMD55_APP, ...);
  rcode = sd_wr_cmd(sd_base, SD_ACMD41_INIT_SD, ...);
  if (rcode==0x00)      // correct response received
    break;
}
```

The SD_INIT_TIME_OUT term is a symbolic constant defined in the header file and used to impose the timeout period. We can estimate this value as follows. Sending CMD55 and ACMD41 commands and receiving two R1 responses require 14 bytes. Since the SPI sclk clock period is 5 us (i.e., $\frac{1}{200\ kHz}$), it takes about 0.56 ms (i.e., 14*8*5 us) to loop through one iteration. SD_INIT_TIME_OUT is set to 900, which provides a timeout period of 500 ms.

After successful initialization, we set the SPI sclk clock to 1 MHz for future data access. This value can be adjusted later to accommodate the actual system requirement.

Single sector reading routine   The sd_read_sector() function reads a 512-byte sector from an SD card and the code is shown in Listing 19.6.

Listing 19.6

```
int sd_read_sector(alt_u32 sd_base, int sdhc, alt_u32 sect, alt_u8 *buf)
{
  int i;
  alt_u8 rcode, token;
  alt_u32 addr, ocr;

  /* byte addr for SD card; sector # for SDHC card */
  if (sdhc==0)
    addr = (sect<<9);
  else
    addr = sect;
  /* send sd card read single block command */
  rcode = sd_wr_cmd(sd_base, SD_CMD17_READ, addr, &ocr);
  if (rcode != 0x00){
    printf("CMD17 command fails: R1=0x%02x \n", rcode);
    return (-1);
  }
  /* wait for data start token up to 100 ms */
  for(i=0; i<SD_READ_TIME_OUT; i++){
    token = sd_spi_rd_byte(sd_base);
    if (token==SD_TOKEN_DATA_START)    // correct response received
      break;
  }
  if (token!=SD_TOKEN_DATA_START){    //time-out
    printf("No data start token: last token=0x%02x\n", token);
    return (-1);
  }
  /* read one sector (512 bytes) */
  for (i=0; i<512; i++)
    buf[i] = sd_spi_rd_byte(sd_base);
  /* read and discard two crc bytes */
  sd_spi_rd_byte(sd_base);
  sd_spi_rd_byte(sd_base);
  return (0);
```

```
}
```

Note that the argument format of the CMD17 command is different for a standard capacity SD card and a high capacity SDHC card. The latter is the *sector number* and the former is the starting *byte address* of the sector. Since the `sect` argument of the `sd_read_sector()` function is the sector number, it must be converted to the starting byte address for a standard capacity SD card. The starting byte address of a 512-byte (i.e., 2^9 byte) sector can be obtained by concatenating the sector number with nine zeros, as in the `(sect<<9)` expression.

The remaining code follows the read procedure discussed in Section 19.4.3. The host first issues the CMD17 command. After the command is accepted, it polls the input continuously until the data start token (0xfe) is received and then reads the 512 bytes of data. The timeout period for the read process is 100 milliseconds. We use a loop structure similar to that of `sd_init()`:

```
for(i=0; i<SD_READ_TIME_OUT; i++){
  token = sd_spi_rd_byte(sd_base);
  if (token==SD_TOKEN_DATA_START)
    break;
}
```

The value of `SD_READ_TIME_OUT` depends on the SPI `sclk` rate. The loop body involves reading a byte from the SPI bus and thus needs eight `sclk` cycles, which takes $\frac{8}{f_{sclk}}$ seconds. It requires $0.1/\frac{8}{f_{sclk}}$ iterations to get a 100-ms timeout interval. For simplicity, we define `SD_READ_TIME_OUT` as a constant and set its value to 312500 to accommodate the highest possible 25-MHz `sclk` clock rate.

Single sector write routine   The `sd_write_sector()` function writes a 512-byte sector to an SD card and the code is shown in Listing 19.7.

Listing 19.7

```
int sd_write_sector(alt_u32 sd_base, int sdhc, alt_u32 sect, alt_u8 *buf)
{
  int i;
  alt_u8 rcode, token;
  alt_u32 addr, dummy;

  /* byte addr for SD card; sector # for SDHC card */
  if (sdhc==0)
    addr = (sect<<9);
  else
    addr = sect;
  /* send sd card write single block command */
  rcode = sd_wr_cmd(sd_base, SD_CMD24_WRITE, addr, &dummy);
  if (rcode != 0x00){            // error
    printf("CMD24 command fails: R1=0x%02x \n", rcode);
    return (-1);
  }
  /* initiate transfer by send data start token */
  sd_spi_wr_byte(sd_base, SD_TOKEN_DATA_START);
  /* send 512 bytes */
  for (i=0; i<512; i++){
    sd_spi_wr_byte(sd_base, buf[i]);
  }
  /* send two dummy crc bytes */
  sd_spi_wr_byte(sd_base, 0xff);
```

```
sd_spi_wr_byte(sd_base, 0xff);
/* wait for data acceptance token up to 0.25 sec */
for(i=0; i<SD_WRITE_TIME_OUT; i++){
  token = sd_spi_rd_byte(sd_base);
  token = token & 0x1f;                 // only 5 LSBs used
  if (token==SD_TOKEN_DATA_ACCEPT)      // correct response received
    break;
}
if (token!=SD_TOKEN_DATA_ACCEPT){
  printf("No data accept token: last token=0x%02x\n", token);
  return (-1);
}
/* wait for write completion */
for(i=0; i<SD_WRITE_TIME_OUT; i++){
  token = sd_spi_rd_byte(sd_base);
  if (token==0xff)                      // correct response received
    break;
}
if (token != 0xff){
  printf("Write completion timeout: last token=0x%02x\n", token);
  return (-1);
}
return (0);            // ok
}
```

The code follows the write procedure discussed in Section 19.4.3. The host first issues the CMD24 command. After the command is accepted, it sends the data packet, which consists of a 1-byte data start token, 512-byte data, and 2-byte stuffing CRC code. After receiving the data packet, the card acknowledges by sending the *data accept token* and then proceeds to perform the erasing and writing operation. It holds the output line low (which is read as 0x00 by the host) during the process. The timeout period for this process is 250 milliseconds. We use a loop structure similar to that in **sd_read_sector()** and set **RD_WRITE_TIME_OUT** to 781250.

19.6   FILE ACCESS

A computer file is a collection of related information. From an application program's point of view, a file is a linear contiguous storage space, in which data are accessed sequentially. A file system specifies how to map a logical file to a physical massive storage and organize multiple files in the storage. In a desktop computing environment, file management is part of an OS. It contains a collection of routines to create and delete directories, to create and delete files, and to read and write files. The routines are organized in layers. The lower layer routines deal with the access of physical I/O storage devices, similar to those discussed in Section 19.5.2. The upper layer routines deal with the aspects of the logical file and are incorporated into a special software module, such as the **stdio** library of C. A user application can use the generic file access functions, such as **fopen()**, and avoid the low-level details.

On the other hand, an embedded system sometimes does not include a full-fledged OS and an application usually does not need the full range of service. The file access is frequently handled in an ad hoc way, just implementing enough functionalities to match the application's need.

The most widely used file system for SD cards is the (file allocation table) file system. It was first developed by Microsoft in the late 1970s and continued to evolve to accommodate larger storage capacity and to incorporate advanced features. Two commonly used versions for flash cards are FAT16 and FAT32, which support storage sizes up to 4 GB and 2 TB, respectively. These file structures are used in portable devices and recognized by almost all current desktop computers.

In this section, we provide an overview of the FAT16 file structure and derive basic utility routines to open and read files from an SD card. File management is a complex task and a detailed discussion is beyond the scope of the book. For simplicity, we make the following assumptions about the file organization:

- The card must be preformatted in a FAT16 structure.
- The files must reside on the first partition.
- The files must be on the root directory (i.e., not in a subdirectory).
- The file names must be in 8.3 format, in which there are up to eight characters for the file name and up to three characters for the file extension. The "long file names" format should not be used.

The card and files must be prepared according to these constraints to be used with the utility routines. Note that the high capacity SDHC cards are formatted as FAT32 as the default. To be used with our implementation, they must be reformatted with FAT16. Because of the limitation of the FAT16 structure, only 4 GB of the storage capacity can be used.

19.6.1 Overview of FAT16 structure

The simplified layout of a single-partition "fixed-disk" FAT16 system is shown in Figure 19.7. It contains several key regions. The *boot record* (sometimes known as *partition boot record*) contains the basic information of the file system, such as the starting location and size of the file allocation table, root directory, and data region. *Root directory* maintains information for files, such as file names and sizes, and contains an entry for each individual file. The actual file data is stored in the *data region*. The region is organized as fixed-sized physical *clusters*. A file can occupy one or more clusters and this information is kept in a *file allocation table*.

As the storage capacity increases, a physical device can host multiple file systems, which are known as *partitions*. Each partition has its own boot record, file allocation table, etc. In such a system, the *MBR* (*master boot record*) uses a partition table to maintain the relevant information. The FAT16 system can support up to four partitions. When a device is formatted in a Microsoft operation system, a "fixed disk," such as a hard disk, is usually formatted with an MBR and a "removable disk," such as the old floppy diskette, is formatted without an MBR. An SD card can be formatted either way and thus may or may not include the MBR. In the latter case, the card normally contains a single partition.

For a massive storage device, the data is usually transferred in blocks. The FAT16 system uses a *512-byte sector* as the basic storage unit and the starting locations and sizes of the regions are represented in term of sectors (i.e., along the 512-byte address boundary). A cluster usually contains multiple sectors.

MBR and boot record   In computing, booting is the process to start the operating systems. In the DOS and early Windows, the MBR and boot record contain the basic file information and the starting-up code to load the main operating system.

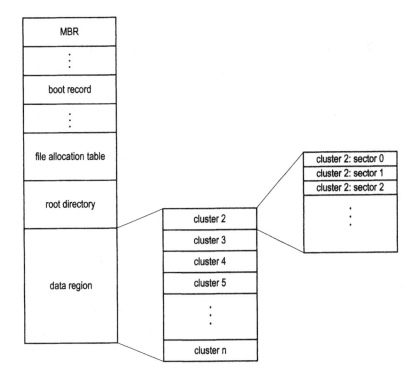

Figure 19.7   Simplified layout of a FAT16 file system.

For non-bootable storage, we only need to extract key FAT16 file system parameters from the MBR and boot record. The parameters are:

- Starting sector of the root directory
- Starting sector of the file allocation table
- Starting sector of the data region
- Number of file entries in the root directory
- Number of sectors per cluster
- Number of clusters in the data region

For an SD card formatted without the MBR, the boot record is located in sector 0. For an SD card formatted with the MBR, the MBR is located in sector 0 and the boot record is located in the first sector of the partition. The starting location of the first partition can be found in the MBR's partition table. In summary, obtaining FAT16 file system parameters from an SD card involves three basic steps:

1. Load the sector 0. If it is not the MBR (i.e., the sector is assumed to be the boot record), go to step 3.
2. Determine the first partition's location and load its first sector (i.e., the boot record).
3. Extract the relevant file system parameters from the boot record.

One way to distinguish between an MBR and a boot record is to examine the first byte of the sector. In a boot record, the first three bytes should be "jump code" and the first byte of the corresponding jump instructions is either 0xeb or

Table 19.2   MBR key fields

Offset	Size	Description	Value
0x1c2	1	file system type of the first partition	0x04, 0x06, or 0x0e
0x1c6	4	starting sector of the first partition	
0x1fe	2	MBR signature	0x55aa

Table 19.3   FAT16 boot record fields

Offset	Size	Description	Value
0x000	3	jump code	
0x003	8	optional manufacturer description	
0x00b *	2	bytes per sector (almost always 512)	512
0x00d *	1	sectors per cluster	
0x00e *	2	number of reserved sectors before file allocation table	1
0x010 *	1	copies of file allocation table	2
0x011 *	2	maximal number of file entries in root directory	
0x013 *	2	total number of sectors in partition (0 if > 32 MB)	0
0x015	1	media descriptor	
0x016 *	2	sectors per file allocation table	
0x018	2	sectors per track	
0x01a	2	number of heads	
0x01c	4	number of hidden sectors preceding the partition	0
0x020 *	4	total number of sectors in partition	
0x024	2	physical drive number (not used)	
0x026	1	extended boot record signature	
0x027	4	volume serial number	
0x02b	4	volume label	
0x036	8	file system identifier	
0x03c	448	OS bootstrap code	
0x1fe *	2	boot sector signature	0x55aa

0xe9. We assume the sector is a boot record if the first byte matches one of the patterns.

If the sector is an MBR, we need to retrieve information from the partition table. For our purposes, only three fields are needed. Their offsets, widths, purposes, and typical values are listed in Table 19.2. The boot record is located in the first sector of a partition. The file system type indicates the type of file system and its value should be 0x04, 0x06, or 0x0e for a FAT16 file system. The signature field occupies the last two bytes of the MBR and its value should be 0x55aa. This field can be used to verify that a card is formatted as a FAT file system.

Step 3 retrieves actual file system parameters from the boot record. The offsets, widths, purposes, and typical values of a boot record's fields are listed in Table 19.3. Many fields are used to support the OS booting process or existed for historical reasons. Only few key fields are needed for our purposes, as indicated by the * mark.

We first calculate the starting sectors of the root directory, file allocation table, and data region. Let the starting sectors of boot record, root directory, file allocation table, and data region be $br0$, $rdir0$, $fat0$, and $data0$, respectively. The $br0$ is obtained earlier from MBR and is 0 if MBR does not exist.

The OS sometimes needs additional sectors after the boot record and the reserved number is stored in offset 0x00e. The file allocation table starts after this and the first sector is

$$fat0 = br0 + \# \text{ of reserved sectors}$$

The size of the file allocation table (in terms of sectors) is specified in offset 0x016. Since it is the most critical part of the file system, OS sometimes maintains duplicate copies. The number of copies is specified in offset 0x010. The root directory region follows the file allocation tables and the first sector is

$$rdir0 = fat0 + (\# \text{ of copies of file allocation table}) * (\text{sectors per table})$$

The number of file entries in the root directory is specified in offset 0x011. Since each entry takes 32 bytes, the size of this region can be calculated accordingly. The data region follows the root directory and the first sector is

$$data0 = rdir0 + \frac{(\# \text{ of file entries}) * 32}{512}$$

Two other important parameters in the boot record are the number of sectors per cluster and the number of sectors in the partition. The former is specified in offset 0x00d. The latter is specified in offset 0x013 or 0x020. This field originally consisted of two bytes in offset 0x013 and can support storage capacity up to 32 MB (i.e., $2^{16} * 512$). A four-byte field in offset 0x020 was added later to accommodate larger storage. A 0 in offset 0x013 indicates that the capacity is greater than 32 MB and the information should be retrieved from offset 0x020. As in MBR, the boot record uses a signature of 0x55aa in the end of the sector. This can be used to verify the current sector is indeed a boot record.

File allocation table and data clusters   The file data is stored in the data region. The region is organized as a collection of *clusters*. The cluster is the basic unit used in allocating storage space and a file can be mapped to one or more clusters. The size of a cluster is fixed within a partition and its value is defined in the boot record. A cluster usually contains multiple sectors and can be as large as 32 KB.

A large file can take many clusters. Instead of allocating the file to one large consecutive chunk of storage space, the FAT16 file system uses a *linked list* to thread the clusters together. A linked list consists of a chain of records such that in each record there is a field that contains a reference (i.e., a link) to the next record in the chain. For example, assume that the file `myfile.c` is allocated to clusters 4, 5, 8, and 9. The conceptual linked list is shown in Figure 19.8. In the actual implementation, the data cluster resides in the data region and the linked list is realized separately by a lookup table, in which the index of the table represents the current record and its content is the corresponding link. The content of the previous example is shown in Figure 19.9. This table is referred to as the *file allocation table* and the FAT file system is named after this table.

In a FAT16 file structure, the width of the table entry (i.e., link) is 16 bits and thus the file system can accommodate up to 2^{16} clusters. The 16-bit value in the

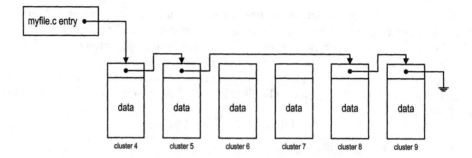

Figure 19.8 Conceptual file cluster chain.

	⋮
0x0004	0x0005
0x0005	0x0008
0x0006	0x0000
0x0007	0x0000
0x0008	0x0009
0x0009	0xffff
	⋮

Figure 19.9 Portion of file allocation table.

entry usually represents the number of the next cluster in the chain. However, several patterns are reserved for special use:

- 0x0000: free cluster.
- 0x0001: reserved.
- 0xfff0 to 0xfff6: reserved.
- 0xfff7: bad cluster.
- 0xfff8 to 0xffff: last cluster in file.

Since the first two cluster numbers are for special use, no physical cluster exists. The actual starting cluster in the data region is cluster 2, as shown in Figure 19.7.

Root directory The root directory contains individual file information on the root level. Each file is represented by a 32-byte entry, which consists of fields for the name, extension, attributes (archive, directory, hidden, read-only, system, and volume), the date and time of creation, the starting cluster of the file data, and the size of the file. The offsets and widths of these fields are summarized in Table 19.4. Our implementation requires the file name, file extension, starting cluster number, and file size fields. The file name and extension fields are used to identify a specific file. The file name field contains eight bytes and thus can support up to eight characters. The name starts with the first byte and the field is padded with space characters if a file name is fewer than eight characters in length. The format of the file extension field is similar except that it contains only three bytes. For example, the file **myfile.c** is represented as "MYFILE␣␣C␣␣" in these two fields.

Table 19.4   File entry fields

Offset	Size	Description
0x00	8	file name
0x08	3	file extension
0x0b	1	file attributes (directory, volume, read-only, etc.)
0x0c	10	reserved
0x16	2	time created or last updated
0x18	2	date created or last updated
0x1a	2	starting cluster number for file
0x1c	4	file size in bytes

The first byte of the file name is also used to indicate the status of the entry. Several special values are reserved for this purpose:

- 0x00: the entry is available and no subsequent entry is in use.
- 0x05: the first character of the filename is actually 0xe5.
- 0x2e: the entry is a special "dot" (i.e., "." or "..") directory entry.
- 0xe5: the entry has been deleted and is available.

In addition to a regular file, a file entry can also be used for the volume label, subdirectory, etc.

19.6.2   Read-only FAT16 file access driver routines

File management is a complex task and developing a complete suite of file access routines is beyond the scope of this book. In this subsection, we introduce the basic concept with a simple set of routines to read a file from an SD card. The major routines are:

- sd_file_mount(): initialize the FAT16 file system.
- sd_file_fopen(): open a file for read.
- sd_file_read_byte(): read a byte from an opened file.
- sd_file_fclose(): close a file.

The last three routines are somewhat like the fopen(), fgetc(), and fclose() functions used in Sections 17.9 and 18.7. Instead of relying on the host-based file system, these routines allow us to read image and audio files from an SD card directly. In our implementation, we make the data structures and procedure general to accommodate file write operations as well. A similar sd_file_write_byte() function can be developed if needed.

Auxiliary functions   We define two auxiliary functions to facilitate the file processing. The bget16() and bget32() functions obtain two-byte and four-byte data from a sector buffer, respectively. The code is shown in Listing 19.8.

Listing 19.8

```
alt_u16 bget16(alt_u8 *buf, int pos)
{
  alt_u16 b0, b1, r;

  b0 = (alt_u16) buf[pos];
```

```
    b1 = (alt_u16) buf[pos+1];
    r = (b1<<8) + b0;
    return(r);
}

alt_u32 bget32(alt_u8 *buf, int pos)
{
    alt_u32 b0, b1, b2, b3, r;

    b0 = (alt_u32) buf[pos];
    b1 = (alt_u32) buf[pos+1];
    b2 = (alt_u32) buf[pos+2];
    b3 = (alt_u32) buf[pos+3];
    r = (b3<<24) + (b2<<16) + (b1<<8) + b0;
    return(r);
}
```

The sd_file_conv_fname() function reads the input file name string and converts it to an 11-element character array whose format matches the name and extension fields of the root directory file entry. For example, the input sting "myfile.c" will be converted to "MYFILE␣␣C␣␣". The code is shown in Listing 19.9.

Listing 19.9

```
void sd_file_conv_fname(char *si, char *f83)
{
    int i, pos;
    char ch;

    /* blank by default */
    for (i=0; i<11; i++)
      f83[i] = ' ';
    /* file name */
    pos=0;
    for (i=0; i<9; i++){
      ch=toupper(si[pos]);
      pos++;
      if (ch=='\0')   // end of string
        return;
      if (ch=='.')
        break;
      f83[i]=ch;
    }
    /* file extension */
    for (i=8; i<11; i++){
      ch=toupper(si[pos]);
      pos++;
      if (ch=='\0')
        return;
      f83[i]=ch;
    }
}
```

The sd_file_mount() *function*   The sd_file_mount() function verifies the FAT16 file system structure of an SD card and extracts the relevant parameters. It essentially initializes (i.e., "mounts") the FAT16 file system. We use a special data structure to record the relevant parameters and it is defined as

```
typedef struct tag_fat
{
  alt_32 sd_base;      // base address of SPI controller
  int sdhc;            // type of SD card
  alt_u16 rdir0;       // starting sector of root directory
  alt_u16 fat0;        // starting sector of FAT table
  alt_u16 data0;       // starting sector of data region
  int cls_size;        // # sectors per cluster
  int rdir_size;       // max # file entries in root directory
  int data_size;       // max # clusters for data
} fat_para_type;
```

The basic procedure is:

1. Initialize the SD card.
2. Read sector 0.
3. Determine whether the sector is MBR. If not, skip the next three steps and go to step 7.
4. Verify the MBR signature (0x55aa) and verify that the FAT file system type is FAT16.
5. Extract the sector number of the first partition's boot record.
6. Read the boot record sector.
7. Verify the boot record signature (0x55aa).
8. Extract relevant FAT16 parameters.
9. Construct and return the FAT16 record.

The detailed code is shown in Listing 19.10.

Listing 19.10

```
int sd_file_mount(alt_u32 sd_base, fat_para_type *f)
{
  alt_u8 buf[512];    // buffer for a sector
  alt_u32 br0;        // boot record sector number
  alt_u8 sec_per_cls, fat_copy;
  alt_u16 reserved_sec, root_dir_size, sec_per_part_small, sec_per_fat;
  alt_u32 sec_per_part_large;
  int status, sdhc, sec_per_part;

  /* initialize SD card */
  sdhc=sd_init(sd_base);
  if (sdhc==-1){
    printf("SD card initialization failed \n");
    return (1);
  }
  f->sdhc = sdhc;
  f->sd_base = sd_base;
  /* fetch sector 0 */
  status=sd_read_sector(sd_base, sdhc, 0, buf);
  if (status!=0){
    printf("Sector 0 read failed\n");
    return (2);
  }
  br0=0;
  /* check whether the sector is the MBR */
  if (buf[0]!=0xeb && buf[0]!=0xe9) { /* not boot record */
    /* check MBR signature 0x55aa */
    if ((buf[510]!=0x55) || (buf[511]!=0xaa)){
      printf("MBR signature not match\n");
```

```
      return (3);
  }
  /* check FAT16 file system type at 0x1c2 */
  if ((buf[0x1c2]!=0x04) && (buf[0x1c2]!=0x06) && (buf[0x1c2]!=0x0e)){
    printf("FAT16 file type not match\n");
    return (4);
  }
  /* fetch boot record of first partition */
  br0 = bget32(buf, 0x1c6);
  status=sd_read_sector(sd_base, sdhc, br0, buf);
  if (status!=0){
    printf("Boot record read failed\n");
    return (5);
  }
} // end if
/* check boot record signature 0x55aa */
if (buf[510]!=0x55 || buf[511]!=0xaa){
  printf("Boot record signature not match\n");
  return (6);
}
/* extract key FAT16 parameters */
sec_per_cls = buf[0x0d];
reserved_sec = bget16(buf, 0x0e);
fat_copy = buf[0x10];
root_dir_size = bget16(buf, 0x11);
sec_per_part_small = bget16(buf, 0x13);
sec_per_fat = bget16(buf, 0x16);
sec_per_part_large = bget32(buf, 0x20);
/* construct the FAT16 record */
if (sec_per_part_small!=0)
  sec_per_part = (int) sec_per_part_small;
else
  sec_per_part = sec_per_part_large;
f->fat0 = br0 + reserved_sec;
f->rdir0 = f->fat0 + fat_copy * sec_per_fat;
f->data0 = f->rdir0 + root_dir_size*32/512;
f->cls_size = (int) sec_per_cls;
f->rdir_size = (int) root_dir_size;
f->data_size = (sec_per_part - f->data0)/sec_per_cls;
return(0);
}
```

If the operation is successful, the function returns a status code 0 and updates the FAT16 record pointed by f. Otherwise, it returns an error code and prints the error condition on the console.

The "file descriptor" data structure From the application point of view, a file is a linear contiguous storage space and the data is accessed sequentially. In the actual physical storage, the data are mapped to a chain of clusters and the data transfer is performed in blocks. To facilitate the implementation, we use a special data structure, *file descriptor*, to maintain basic file information, to keep track of the operation status, such as the current logical position and physical cluster, and to provide buffering space. The data structure is defined as

```
typedef struct tag_file_descriptor
{
  char name[11];        // 11-bit file name+extension
  int rdir_index;       // entry index in root directory table
  int size;             // size of file in byte
```

```
    alt_u16 cls0;          // starting cluster number
    alt_u16 clsi;          // current data cluster number
    int sect;              // current sector
    int idx;               // current index in sector
    int seek;              // current position in file
    char dbuf[512];        // data sector buffer
    int sf;                // current fat sector number
    char fbuf[512];        // fat sector buffer
    int open;              // file is properly opened
} file_dp_type;
```

The first four fields, name[11], rdir_index, size, and cls0, represent the 11-character array of the file name and extension, the entry number in the root directory, the file size in bytes, and the file's starting cluster number. The seek field keeps track of the current read position of the "logical file" viewed by the application. The physical position of the byte is stored in the clsi, sect, and idx fields, which are the current cluster number, current sector number, and the position within the sector. The open field is a status indicating whether the file is currently open.

The SD card data is retrieved one sector at a time and the individual byte is read afterward. Instead of loading the same sector repetitively, it is more efficient to store the 512 bytes of data in a buffer. The read process can continue until the current buffer is exhausted. The dbuf[512] field provides the needed buffer space. The file allocation table is also stored in the SD card and the cluster chain is retrieved in a similar fashion. A large file contains a large number of clusters and the table must be read repetitively. We can use another 512-byte buffer to store the data of the retrieved sector. This can reduce the access time significantly if the clusters are allocated in the adjacent area. The fbuf[512] field provides the file allocation table buffer space and the sf field records the corresponding sector.

The sd_file_fopen() *function*   Opening a file involves searching the matched file name on the root directory, retrieving the key parameters from the entry, and initializing the relevant fields of the file descriptor. The function is defined as

```
int sd_file_fopen(fat_para_type *fat, char *fname,
                  file_dp_type *fd)
```

The calling function must call the sd_file_mount() function in advance to obtain the FAT information and define a static variable for the file descriptor, which essentially allocates the needed buffer space. The pointers to the two data structures, along with the file name string, are passed to the sd_file_open() function. The key task of the function is to search a matched file in the root directory. The combined file name and extension fields of entries of the directory are compared sequentially until a match is found or until all the entries are exhausted. The starting sector of the root directory and the number of file entries can be obtained from the fat data structure. Since the size of an entry is 32 bytes, 16 entries can be stored in a 512-byte sector. A new sector must be loaded after every 16 entries. The function returns 0 if the file is found and returns an error code otherwise. The code is shown in Listing 19.11.

Listing 19.11

```
int sd_file_fopen(fat_para_type *fat, char *fname, file_dp_type *fd)
{
  int index, status, n;
  int sect, oft;
  alt_u8 *buf;    // pointer to a sector buffer
  char f83[11];   // file name in normalized 8.3 format

  sd_file_conv_fname(fname, f83);
  fd->open = 0;
  /* use file descriptor's data buffer as temp buffer */
  buf = fd->dbuf;
  /* search entire root directory */
  for (index=0; index<fat->rdir_size; index++){
    /* load a new sector after 16 file entries */
    if (index%16==0){
      sect = fat->rdir0 + index/16;
      status=sd_read_sector(fat->sd_base, fat->sdhc, sect, buf);
      if (status!=0){
        printf("SD read failed\n");
        return (2);
      }
    }
    /* offset in current sector */
    oft = 32 * (index%16);
    /* 0x00 in first char of file name means no more entries */
    if (buf[oft]==0x00){
      printf("File not found\n");
      return(1);   // file not found
    }
    /* compare normalized 8.3 name—extension */
    for (n=0; n<11; n++){
      if (f83[n] != buf[oft+n])
        break;
    }
    /* file entry matches; update file descriptor's fields */
    if (n==11){
      fd->rdir_index = index;
      fd->cls0 = bget16(buf, oft+0x1a);
      fd->size = (int) bget32(buf, oft+0x1c);
      for (n=0; n<12; n++)
        fd->name[n] = buf[oft+n];
      /* initialize file descriptor's counters */
      fd->clsi = fd->cls0;
      fd->sect = 0;
      fd->idx = 0;
      fd->seek = 0;
      break;   // success
    } // end if
  } // end for
  if (index==fat->rdir_size){ // file not found
    printf("File not found\n");
    return(1);
  }
  /* load file's 1st file allocation table sector to buffer */
  sect = fat->fat0 + fd->cls0/256;
  status=sd_read_sector(fat->sd_base, fat->sdhc, sect, fd->fbuf);
  if (status!=0){
    printf("File allocation table sector read failed\n");
    return (3);
  }
```

```
    fd->sf = sect;
    /* update open status */
    fd->open = 1;
    return(0);
}
```

Note that the code uses the **dbuf** field of the file descriptor as a temporary buffer to store the sector retrieved from the root directory and loads a new sector after every 16 entries. After a matched file is found, the function updates the relevant parameters and also loads the corresponding file allocation table sector (determined by the starting cluster) to the **fbuf** buffer.

The **sd_file_read_byte()** *function*   The **sd_file_read_byte()** function retrieves a byte from an opened file. It traverses the file sequentially. When it is called, it returns the current byte and advances the index one position. The file must be properly opened before being read.

The data is retrieved one sector at a time and is stored in the 512-byte **dbuf** buffer of the file descriptor structure. Most read operations obtain the byte for the buffer and involve no I/O operations. When the buffer is exhausted, a new sector will be loaded. It is usually the next sector. However, if the current sector is the last one in the cluster, we must follow the cluster chain in the file allocation table to obtain the new cluster and calculate the new sector number accordingly. The basic procedure is:

- Check whether the current position is the end of the file.
- If the **idx** field points to the first byte of the buffer, load the new sector to the data buffer.
- Read a byte from data buffer pointed by the **idx** field.
- Update the **idx** and **seek** fields.
- If the byte is the last byte in the buffer, reset the **idx** field to 0 and increment the **sect** field.
- If the sector is the last sector in the cluster, reset the **sect** field to 0 and follow the cluster chain to retrieve the next cluster. A new file allocation table sector will be loaded if necessary.

The function returns 0 if the operation is successful and returns an error code otherwise. The code is shown in Listing 19.12.

Listing 19.12

```
int sd_file_read_byte(fat_para_type *fat, file_dp_type *fd,alt_u8 *byte)
{
    int status;
    alt_u16 s;

    /* check end-of-file */
    if (fd->seek >= fd->size)
        return(1);
    /* must load a new sector at the beginning */
    if (fd->idx==0){
        s = fat->data0 + ((fd->clsi)-2)*(fat->cls_size) + fd->sect;
        status=sd_read_sector(fat->sd_base, fat->sdhc, s, fd->dbuf);
        if (status!=0){
            printf("Data sector read failed\n");
            return (2);
        }
```

```
}
/* read a byte */
*byte = fd->dbuf[fd->idx];
/* update file descriptor counter */
fd->seek++;
fd->idx++;
/* not reach end of buffer yet */
if (fd->idx!=512)
  return(0);  // success
/* reach the end of buffer */
fd->idx = 0;
fd->sect++;
/* not last sector in cluster */
if (fd->sect!=fat->cls_size)
  return(0);  // success
/* last sector in cluster; fetch next cluster from FAT table */
fd->sect = 0;
s = fat->fat0 + fd->clsi/256;  // sector containing the cluster
/* file allocation table sector not in buffer; load new sector */
if (s != fd->sf){
  status=sd_read_sector(fat->sd_base, fat->sdhc, s, fd->fbuf);
  if (status!=0){
    printf("File allocation table sector read failed\n");
    return (3);
  }
  fd->sf = s;
  // printf("Debug: sector %d in FAT\n", s);
}
fd->clsi = bget16(fd->fbuf, 2*(fd->clsi%256));
return(0);  // success
}
```

The sd_file_close() function   The sd_file_close() function closes the file. Since only the read function is implemented, this function only needs to reset the **open** field of the file descriptor to 0, as shown in Listing 19.13. It is included for completeness. If the write operation is supported, this function must write back the data stored in the two sector buffers.

Listing 19.13

```
void sd_file_fclose(file_dp_type *fd)
{
  fd->open = 0;
  return;
}
```

19.7 TESTING PROGRAM

The SPI controller can be instantiated and integrated to a Nios II system like a normal IP core. The system derived in Section 16.10.2 includes the SPI controller core and can be used for testing. We construct a simple program to demonstrate and verify file access routines and the SD card driver routines. The program consists of the following tests:

- Mount a FAT16 file system from an SD card and display the relevant parameters.

- Open a file in the root directory and display the relevant parameters.
- Read the opened file and display the first two sectors and the last two sectors of the file.
- Read and display data from the designated sector.
- Write sample data to the designated sector.
- Initialize an SD card.
- Set the frequency divisor of SPI's `sclk` clock.

The SD card must be formatted as a FAT16 file system. Because the write operation may damage the file system, it is a good idea to use a spare card for testing.

The first three tests are the steps to read a file and need to be performed in sequence. The other four tests exercise the low-level driver functionalities. The fourth test reads data from a single sector and displays it in a table. This test can be used as a tool to examine and study the structures of MBR, boot sector, root directory, etc. The fifth test writes data to a single sector. We use a for loop to generate data. For address offset i, eight LSBs of $s + i$ (where s is the sector number) is written to the location. Since the write operation permanently alters the data on an SD card, it must be used with care. Writing to a wrong sector can damage a file or even render the entire file system unusable. The sixth test initializes an SD card and it should be performed after a new card is inserted. The last test changes the rate of SPI's `sclk` clock.

The main program is shown in Listing 19.14.

Listing 19.14

```
int main(void)
{
  fat_para_type fat;
  file_dp_type fd;
  char fname[15], yes[3];
  int sw, btn;
  unsigned int num;
  int i, j, s, status, sdhc;
  alt_u32 sect;
  alt_u8 ch, dvsr;
  alt_u8 rbuf[512], wbuf[512];   // read and write buffers
  alt_u8 disp_msg[4]={sseg_conv_hex(5), sseg_conv_hex(13),
                      0xff, sseg_conv_hex(12)};

  sseg_disp_ptn(SSEG_BASE, disp_msg);      // show "Sd C" for display
  printf("SD card FAT16 file system test:\n");
  sdhc=sd_init(SDC_BASE);
  if (sdhc==-1)
    printf("SD card initialization failed \n");
  btn_clear(BTN_BASE);
  while (1){
    while (!btn_is_pressed(BTN_BASE)){ };       // wait for button
    btn=btn_read(BTN_BASE);                      // read button
    if (btn & 0x02){                             // key1 pressed
      sw=pio_read(SWITCH_BASE);                  // read switch
      // printf("key/sw: %d/%d\n", btn, sw);
    }
    btn_clear(BTN_BASE);
    switch (sw){
      case 0:  // mount FAT16 file system
        status=sd_file_mount(SDC_BASE, &fat);
        sdhc = fat.sdhc;
        if (status==0) {
```

```
      printf("FAT 16 mounted on SD card\n");
      printf("starting FAT table sector: %d\n", fat.fat0);
      printf("starting root dir sector:  %d\n", fat.rdir0);
      printf("starting data sector:      %d\n", fat.data0);
      printf("sectors per cluster:       %d\n", fat.cls_size);
      printf("clusters in data segment:  %d\n", fat.data_size);
      printf("file entries of root dir:  %d\n\n", fat.rdir_size);
    } else
  printf("Mount failed: status=%d\n\n", status);
  break;
  case 1:   // open a file
    printf("Enter file name in 8.3 format: ");
    scanf("%s", fname);
    status=sd_file_fopen(&fat, fname, &fd);
    if (status==0) {
      printf("\nFile open successful.\n");
      printf("file size (bytes): %d\n", fd.size);
      printf("root dir entry #:  %d\n", fd.rdir_index);
      sect = fat.rdir0 + fd.rdir_index/32;
      printf("entry sector #:    %d\n", (int)sect);
      printf("starting cluster#: %d\n", fd.cls0);
      sect = fat.data0 + (fd.cls0-2)*fat.cls_size;
      printf("starting sector #: %d\n\n", (int)sect);
    } else
      printf("\nFile open failed: status=%d\n\n", status);
    break;
  case 2:   // read file and list first 2 and last 2 sectors
    s = fd.size/512;
    if ((fd.size%512)!=0)   // fraction of a segment
      s++;
    printf("File size (bytes/sectors): %d/%d\n", fd.size, s);
    printf("Data dump for first 2 and last 2 sectors:\n\n");
    for (i=0; i<s; i++){
      for (j=0; j<512; j++){
        sd_file_read_byte(&fat, &fd, &ch);
          rbuf[j]=ch;
       }
      if (i<2 || i>s-3){
        printf("sector %d of file:\n", i);
        print_sector(rbuf);
      } else {
        printf(".");
      } // end if
    } // end for i
    break;
  case 3: // read and print a 512 byte sector
    printf("Enter sd card read sector number: ");
    scanf("%u", &num);
    sect = (alt_16) num;
    status=sd_read_sector(SDC_BASE, sdhc, sect, rbuf);
    if (status!=0)
      printf("read sd card failed\n\n");
    else{
      printf("\nsector %d(0x%x)\n", (int)sect, (int)sect);
      print_sector(rbuf);
    }
    break;
  case 4: // write a 512 byte sector
    printf("Low-level sector write may corrupt file system.\n");
    printf("Press Y to continue: ");
    scanf("%s", yes);
```

```
        if (yes[0]!='y' && yes[0]!='Y'){
          printf("Sector write abandoned.\n\n");
          break;
        }
        printf("Enter sd card write sector number: ");
        scanf("%u", &num);
        sect = (alt_16) num;
        for (i=0; i<512; i++)
          wbuf[i] = (alt_u8) (sect + i);
        status=sd_write_sector(SDC_BASE, sdhc, sect, wbuf);
        if (status!=0)
          printf("Write sd card failed\n\n");
        else
          printf("write sd card completed\n\n");
        break;
      case 5:  // set frequency divisor of SPI clock
        printf("Enter frequency divisor (1-255): ");
        scanf("%u", &num);
        dvsr = (alt_u8) num;
        sd_spi_wr_dvsr(SDC_BASE, dvsr);
        printf("SPI bus frequency set to %6.3f MHz.\n\n", 50.0/(2.0*dvsr));
        break;
      case 6:  // reinitialize SD card
        sdhc=sd_init(SDC_BASE);
        if (sdhc==-1)
          printf("SD card initialization failed.\n\n");
        else
          printf("SD card initialized (sdhc status=%d).\n\n", sdhc);
        break;
    } //end switch
  } // end while
}
```

The main program includes an auxiliary function, **print_sector()**, which displays 512-byte data retrieved from a sector. The code is shown in Listing 19.15.

Listing 19.15

```
void print_sector(alt_u8 *buf)
{
  int i, j;
  alt_u8 ch;

  for (j=0; j<32; j++){
    printf("0x%02x-  ", j);
    for (i=0; i <16; i++)
      printf("%02x ", buf[j*16+i]);
    printf("  ");
    for (i=0; i <16; i++){
      ch = buf[j*16+i];
      if (ch>127 || ch<32)    // non ascii
        ch = '.';
      printf("%c", ch);
    } // end for i
    printf("\n");
  } //end for j
}
```

It arranges the data as two tables, one shown as two-digit hexadecimal numbers and the other shown as ASCII characters (when printable). A sample screen is shown in Figure 19.10. It is the result after we perform a writing test on sector 2.

```
sector 2(0x2)
0x00- 02 03 04 05 06 07 08 09 0a 0b 0c 0d 0e 0f 10 11    ...............
0x01- 12 13 14 15 16 17 18 19 1a 1b 1c 1d 1e 1f 20 21    ............. !
0x02- 22 23 24 25 26 27 28 29 2a 2b 2c 2d 2e 2f 30 31    "#$%&'()*+,-./01
0x03- 32 33 34 35 36 37 38 39 3a 3b 3c 3d 3e 3f 40 41    23456789:;<=>?@A
0x04- 42 43 44 45 46 47 48 49 4a 4b 4c 4d 4e 4f 50 51    BCDEFGHIJKLMNOPQ
0x05- 52 53 54 55 56 57 58 59 5a 5b 5c 5d 5e 5f 60 61    RSTUVWXYZ[\]^_'a
0x06- 62 63 64 65 66 67 68 69 6a 6b 6c 6d 6e 6f 70 71    bcdefghijklmnopq
0x07- 72 73 74 75 76 77 78 79 7a 7b 7c 7d 7e 7f 80 81    rstuvwxyz{|}~-..
0x08- 82 83 84 85 86 87 88 89 8a 8b 8c 8d 8e 8f 90 91    ...............
0x09- 92 93 94 95 96 97 98 99 9a 9b 9c 9d 9e 9f a0 a1    ...............
0x0a- a2 a3 a4 a5 a6 a7 a8 a9 aa ab ac ad ae af b0 b1    ...............
0x0b- b2 b3 b4 b5 b6 b7 b8 b9 ba bb bc bd be bf c0 c1    ...............
0x0c- c2 c3 c4 c5 c6 c7 c8 c9 ca cb cc cd ce cf d0 d1    ...............
0x0d- d2 d3 d4 d5 d6 d7 d8 d9 da db dc dd de df e0 e1    ...............
0x0e- e2 e3 e4 e5 e6 e7 e8 e9 ea eb ec ed ee ef f0 f1    ...............
0x0f- f2 f3 f4 f5 f6 f7 f8 f9 fa fb fc fd fe ff 00 01    ...............
0x10- 02 03 04 05 06 07 08 09 0a 0b 0c 0d 0e 0f 10 11    ...............
0x11- 12 13 14 15 16 17 18 19 1a 1b 1c 1d 1e 1f 20 21    ............. !
0x12- 22 23 24 25 26 27 28 29 2a 2b 2c 2d 2e 2f 30 31    "#$%&'()*+,-./01
0x13- 32 33 34 35 36 37 38 39 3a 3b 3c 3d 3e 3f 40 41    23456789:;<=>?@A
0x14- 42 43 44 45 46 47 48 49 4a 4b 4c 4d 4e 4f 50 51    BCDEFGHIJKLMNOPQ
0x15- 52 53 54 55 56 57 58 59 5a 5b 5c 5d 5e 5f 60 61    RSTUVWXYZ[\]^_'a
0x16- 62 63 64 65 66 67 68 69 6a 6b 6c 6d 6e 6f 70 71    bcdefghijklmnopq
0x17- 72 73 74 75 76 77 78 79 7a 7b 7c 7d 7e 7f 80 81    rstuvwxyz{|}~-..
0x18- 82 83 84 85 86 87 88 89 8a 8b 8c 8d 8e 8f 90 91    ...............
0x19- 92 93 94 95 96 97 98 99 9a 9b 9c 9d 9e 9f a0 a1    ...............
0x1a- a2 a3 a4 a5 a6 a7 a8 a9 aa ab ac ad ae af b0 b1    ...............
0x1b- b2 b3 b4 b5 b6 b7 b8 b9 ba bb bc bd be bf c0 c1    ...............
0x1c- c2 c3 c4 c5 c6 c7 c8 c9 ca cb cc cd ce cf d0 d1    ...............
0x1d- d2 d3 d4 d5 d6 d7 d8 d9 da db dc dd de df e0 e1    ...............
0x1e- e2 e3 e4 e5 e6 e7 e8 e9 ea eb ec ed ee ef f0 f1    ...............
0x1f- f2 f3 f4 f5 f6 f7 f8 f9 fa fb fc fd fe ff 00 01    ...............
```

Figure 19.10   Sample screen of the `print_sector()` function.

19.8   PERFORMANCE OF SD CARD DATA TRANSFER

An SD card is mainly used as a massive storage. An important design criterion is to increase the data transfer rate between the card and the host. The rate depends on the card, the SPI controller, the processor, and software driver. SD cards are divided into different speed "classes" and their maximal transfer rates vary. Most cards can support a 25-MHz SPI clock rate.

In our configuration, although the SPI controller can generate an SPI clock rate up to 25 MHz (by setting the frequency divisor to 1), the processor and driver may not be able to keep up with the rate. Recall that the system clock period is 20 ns (i.e., 50 MHz). If the SPI clock period is s ns, transferring an 8-bit data takes $\frac{8*s}{20}$ system clocks. The processor must be able to process the data within this time slot. In an ideal setup, a fast Nios II processor (i.e., the Nios II/f configuration) can execute one instruction per system clock cycle in average and thus $\frac{8*s}{20}$ instructions can be executed in the given time slot. For an SPI clock rate of 25 MHz, we can use up to 16 instructions to process a byte of data. This appears possible since the processing usually involves only simple indexing and loop operations. However, this analysis is overly optimistic. An actual Nios II system usually involves a memory hierarchy of fast cache and slow DRAM and the execution may hold for many clock cycles for a cache miss. The system may also have various interrupts, which can preempt the normal execution any time and take at least several hundred cycles to complete. Thus, errors can occur from time to time. We can decrease the SPI clock rate to provide more time. For example, if the SPI clock rate is 1 MHz, 400 instructions can be used to process a byte of data. The large cushion can reduce the number of expected errors. However, errors can still occur if complex ISRs are used. To accommodate this type of errors, the SPI driver routines may need to be modified to repeat the data access several times before reporting failure. The situation becomes even worse when a slow Nios II processor (i.e., the Nios II/e configuration) is used.

To obtain reliable and robust high-speed transfer, we need to ensure that the driver resides in fast memory (such as Nios II's tightly coupled memory) and re-design the interrupt structure to facilitate the SPI data stream processing. This is by no means a simple task. A better alternative is to migrate certain needed functionalities to hardware. We can construct a custom SD card controller that can process an SD card command, read a single sector from a buffer, and write a single sector to a buffer. The processor now only needs to transfer 512 bytes between a buffer and its main memory and no tight timing constraint is imposed in this operation. If desired, we can further reduce the processor's load by adding an addition DMA (direct memory access) control circuit to automate the data transfer process.

19.9 BIBLIOGRAPHIC NOTES

The SPI bus is a de facto standard and thus there is no official documentation. More information can be found on the Wikipedia web site (searching by the keyword "serial peripheral interface bus"). The SD card standard is set by the SD Card Association and the simplified version, titled *SD Specifications Part 1: Physical Layer Simplified Specification Version 2.00*, can be found on its web site (www.sdcard.org). The FAT file system is somewhat involved and the article on the Wikipedia web site provides general information and links for the relevant documentation. Some open-source generic FAT system libraries for small embedded applications are available. One such implementation can be found at elm-chan.org.

19.10 SUGGESTED EXPERIMENTS

19.10.1 SD card data transfer performance test

We can determine the maximal data transfer rate of an SD card by gradually increasing the sclk clock rate of the SPI bus until the read or write errors occur. Derive the function and perform the test.

19.10.2 Robust SD card driver routines

As we discussed in Section 19.8, certain occasional events, such as interrupts and cache misses, may cause SD card access errors. One way to mitigate the problem is to repeat the read or write operation for an interval longer than the durations of these events. The routine reports errors after this interval. Modify the SD card driver routines in Section 19.5 to include these changes and verify the operation. Use the function in Experiment 19.10.1 to determine the maximal date transfer rate.

19.10.3   Dedicated processor for SD card access

As we discussed in Section 19.8, certain occasional events, such as interrupts and cache misses, may cause SD card access errors. One way to mitigate the problem is to use a dedicated Nios II system to remove timing uncertainties. We can configure the system by using FPGA's internal embedded memory modules for memory and disable all interrupt requests. Derive the new Nios II system and verify its operation. Use the function in Experiment 19.10.1 to determine the maximal date transfer rate.

19.10.4   Hardware-based SD card read and write operation

One main design goal of an SD card controller is to increase transfer rate. Instead of tweaking with software and a processor, one alternative is to migrate the functionalities from software to hardware. In other words, we can use dedicated hardware to implement the sd_read_sector() and sd_write_sector() functions. The controller should use an FSMD to realize the reading and writing procedure and include a 512-byte FIFO buffer to store the temporary data. The processor only needs to check the controller's status and read or write a sector of data accordingly. Design the new SD card controller, incorporate it into a Nios II system, and verify its operation. Use the function in Experiment 19.10.1 to determine the maximal date transfer rate.

19.10.5   SD card information retrieval

An SD card has several internal registers. One of them is the *CSD* (*card-specific data*) register, which maintains basic information about the card, such as the capacity, data transfer rate, etc. The contents of the CSD register can be retrieved by the CMD9 command. Consult the SD card standard and derive a function to determine the capacity and transfer rate of a card.

19.10.6   MMC card support

The sd_init() function in Section 19.5.2 assumes that the SD card is version 2.0 or later. Consult the SD card standard and the MMC card standards to extend the function to support earlier version SD cards and MMC cards.

19.10.7   Multiple sector read and write operation

The SD card protocol has commands to read and write multiple sectors. Consult the SD card standard and derive driver routines to read and write multiple sectors and verify their operations.

19.10.8   SD card driver routines with CRC checking

We can utilize the SD card's CRC feature to increase the reliability of SD card access. This can be done by issuing the CMD59 command to turn on this option and add proper codes to generate and check the CRC field. Modify the SD card driver routines in Section 19.5 to include the CRC option.

19.10.9   Digital music player

A digital music player reads .wav files from an SD card and plays the sound by sending the data to the audio codec. Design a user interface, derive the program, and verify its operation.

19.10.10   Digital picture frame

A digital picture frame reads .bmp files from an SD card and displays the images on the VGA monitor. Design a user interface, derive the program, and verify its operation.

19.10.11   Additional FAT functionalities

The routines in Section 19.6 only support basic file read operations. Many additional functionalities and enhancements can be added to access the FAT file system. These include:
- Support a subdirectory.
- Support long file names.
- Support both FAT16 and FAT32 file system structures.
- Delete a file from a directory.
- Create a new file in a directory.
- Read an arbitrary number of bytes from a file at a time.
- Write a byte to a file.
- Write an arbitrary number of bytes to a file at a time.
- Skip an arbitrary number of bytes.
- Format an SD card with the FAT16 structure.

Implement one or more of the features.

19.11   SUGGESTED PROJECTS

19.11.1   HAL API file access integration

Instead of using the ad hoc file access routines in Section 19.6, we can integrate the functionalities into the Altera HAL API and C library and use generic C file functions. Consult HAL documentation, develop a set of driver routines conforming to the specification, and integrate the SD card file access into the HAL development environment.

19.12 COMPLETE PROGRAM LISTING

Listing 19.16 chu_avalon_sd.h

```
/********************************************************************
 *
 * Module:  SPI, SD card and FAT16 driver header
 * File:    chu_avalon_sd.h
 * Purpose: Routines to initialize and read/write SD card and
 *          to read a file from FAT16 file system
 *
 ********************************************************************
 *  Register map
 ********************************************************************
 * Read (data to cpu):
 *   offset 0
 *     * bits 7-0: 8-bit read data
 *     * bit 8: SPI controller idle bit
 * Write (data from cpu):
 *   offset 1
 *     * bit:0 chip select (ss)
 *   offset 2
 *     * bits 7 to 0: sclk clock divisor
 *   offset 3
 *     * bits 7 to 0: 8-bit write data
 ********************************************************************/
/* file inclusion */
#include <alt_types.h>

/********************************************************************
 * data type definitions
 ********************************************************************/
/* data type for FAT parameters */
typedef struct tag_fat
{
  alt_32 sd_base;      // base address of SD card SPI controller
  int sdhc;            // type of SD card
  alt_u16 rdir0;       // starting sector of root directory
  alt_u16 fat0;        // starting sector of FAT table
  alt_u16 data0;       // starting sector of data region
  int cls_size;        // # sectors per cluster
  int rdir_size;       // max # file entries in root directory
  int data_size;       // max # clusters for data
} fat_para_type;

/* data type for file parameters */
typedef struct tag_file_descriptor
{
  char name[11];       // 11-bit file name+extension
  int rdir_index;      // entry index in root directory table
  int size;            // size of file in byte
  alt_u16 cls0;        // starting cluster number
  alt_u16 clsi;        // current data cluster number
  int sect;            // current sector
  int idx;             // current index in sector
  int seek;            // current position in file
  char dbuf[512];      // data sector buffer
  int sf;              // current fat sector number
  char fbuf[512];      // fat sector buffer
  int open;            // file is properly opened
} file_dp_type;
```

```
/*************************************************************************
* constant definitions
**************************************************************************/
/* SPI registers */
#define CHU_SD_RD_DATA_REG        0
#define CHU_SD_CS_REG             1
#define CHU_SD_DVSR_REG           2
#define CHU_SD_WR_DATA_REG        3

/* SD card commands */
#define SD_CMD0_RESET             0
#define SD_CMD1_INIT_MMC          1        // not used for SD
#define SD_CMD8_SEND_IF           8
#define SD_CMD17_READ            17
#define SD_CMD24_WRITE           24
#define SD_CMD55_APP             55
#define SD_CMD58_RD_OCR          58
#define SD_ACMD41_INIT_SD        41
/* SD card token */
#define SD_TOKEN_DATA_START      0xfe
#define SD_TOKEN_DATA_ACCEPT     0x05
/* SD card timeout cycles */
#define SD_INIT_TIME_OUT          900       // 0.50s w/ 200 kHz sclk
#define SD_READ_TIME_OUT        312500      // 0.10s w/ 25 MHz sclk
#define SD_WRITE_TIME_OUT       781250      // 0.25s w/ 25 MHz sclk


/*************************************************************************
* Function prototypes
**************************************************************************/
/* SPI functions */
int sd_spi_is_ready(alt_u32 sd_base);
void sd_spi_wr_cs(alt_u32 sd_base, int en_bit);
void sd_spi_wr_dvsr(alt_u32 sd_base, alt_u8 dvsr);
void sd_spi_wr_byte(alt_u32 sd_base, alt_u8 spi_data);
alt_u8 sd_spi_rd_byte(alt_u32 sd_base);

/* SD card functions */
alt_u8 sd_wr_cmd(alt_u32 sd_base, alt_u8 cmd, alt_u32 argument, alt_u32 *r3);
int sd_init(alt_u32 sd_base);
int sd_read_sector(alt_u32 sd_base, int sdhc, alt_u32 sect, alt_u8 *buf);
int sd_write_sector(alt_u32 sd_base, int sdhc, alt_u32 sect, alt_u8 *buf);

/* FAT16 functions */
int sd_file_mount(alt_u32 sd_base, fat_para_type *f);
int sd_file_fopen(fat_para_type *fat, char *fname, file_dp_type *fd);
void sd_file_conv_fname(char *si, char *so);
int sd_file_read_byte(fat_para_type *fat, file_dp_type *fd, alt_u8 *byte);
void sd_file_fclose(file_dp_type *fd);
```

Listing 19.17   chu_avalon_sd.c

```
/**********************************************************************
*
* Module:   SPI, SD card and FAT16 driver function prototypes
* File:     chu_avalon_sd.c
* Purpose:  Routines to initialize and read/write SD card and
*           to read a file from FAT16 file system
*
**********************************************************************/
/* include section */
#include <stdio.h>            // for printf()
#include <io.h>
#include <ctype.h>            // for toupper() function
#include "chu_avalon_sd.h"

/**********************************************************************
*  SPI related functions
**********************************************************************/
/**********************************************************************
* function: sd_spi_is_ready()
* purpose:  check ready signal of SPI controller
* argument:
*    sd_base: base address of SD card interface
* return: 1: ready; 0: not ready
**********************************************************************/
int sd_spi_is_ready(alt_u32 sd_base)
{
  alt_u32 spi_reg;
  int spi_ready_bit;

  spi_reg = IORD(sd_base, CHU_SD_RD_DATA_REG);
  spi_ready_bit = (int) ((spi_reg & 0x00000100) >> 8);
  return spi_ready_bit;
}


/**********************************************************************
* function: sd_spi_wr_cs()
* purpose:  enable or disable SD card's cs_n (ss) signal
* argument:
*    sd_base: base address of SD card interface
*    en_bit: 0 for enable; 1 for disable
* return:
**********************************************************************/
void sd_spi_wr_cs(alt_u32 sd_base, int en_bit)
{
  IOWR(sd_base, CHU_SD_CS_REG, (alt_u32) en_bit);
}
```

```
/****************************************************************************
 * function: sd_spi_wr_dvsr()
 * purpose:  write SPI clock divisor register
 * argument:
 *   sd_base: base address of SD card interface
 *   dvsr: dvsr value (between 1 and 2^8-1)
 * return:
 * note:
 *   - # system clocks in sclk period = 2*dvsr
 *   - sclk freq = system clock freq / (2*dvsr)
 ***************************************************************************/
void sd_spi_wr_dvsr(alt_u32 sd_base, alt_u8 dvsr)
{
  IOWR(sd_base, CHU_SD_DVSR_REG, (alt_u32) dvsr);
}


/****************************************************************************
 * function: sd_spi_wr_byte()
 * purpose:  write a byte to SPI bus
 * argument:
 *   sd_base: base address of SD card interface
 *   spi_data: 8-bit data
 * return:
 ***************************************************************************/
void sd_spi_wr_byte(alt_u32 sd_base, alt_u8 spi_data)
{
  /* wait until SPI is ready */
  while (!sd_spi_is_ready(sd_base)){};
  /* write a byte */
  IOWR(sd_base, CHU_SD_WR_DATA_REG, (alt_u32) spi_data);
}


/****************************************************************************
 * function: sd_spi_rd_byte()
 * purpose:  read a byte from SPI bus
 * argument:
 *   sd_base: base address of sd card interface
 * return: 8-bit data
 * note:
 *   - shift in/out are done simultaneously in SPI bus
 *   - read/write are the same except that the input data is retrieved
 *   - a dummy data is used for write
 ***************************************************************************/
alt_u8 sd_spi_rd_byte(alt_u32 sd_base)
{
  alt_u32 spi_reg;
  alt_u8 spi_data;

  /* write a dummy byte and shift in data as well */
  sd_spi_wr_byte(sd_base, 0xff);
  spi_reg = IORD(sd_base, CHU_SD_RD_DATA_REG);
  spi_data = (alt_u8) (spi_reg & 0x000000ff);
  return spi_data;
}
```

```
/***************************************************************************
 *  SD card related functions
 ***************************************************************************/
/***************************************************************************
 * function: sd_wr_cmd()
 * purpose:  send a command to sd card
 * argument:
 *   sd_base: base address of sd card interface
 *   cmd: 6-bit command
 *   argument: 32-bit argument
 *   r3: 32-bit additional data for R3 and R5 responses
 * return: response code (R1):
 *   - 0xff when no response
 *   - CMD0 (reset) command: 0x01 for no error
 *   - other commands: 0x00 for no error; other for error conditions
 * note:
 *   - 6 bytes send: command + 4-byte argument (MSB first) + crc
 *   - except for the CMD0/CMD8, crc is optional (not implemented)
 *   - command format: 01 + 6-bit value
 *   - sd card may need up to 8-byte transmission time to respond
 *   - sd card continues sending 1 when busy (or no data)
 *   - CMD8/CMD58 responds with 5 bytes
 ***************************************************************************/
alt_u8 sd_wr_cmd(alt_u32 sd_base, alt_u8 cmd, alt_u32 argument, alt_u32 *r3)
{
  int i;
  alt_u8 crc, rcode, byte;
  alt_u32 ocr=0;

  /* crc for CMD0 and CMD8 */
  if (cmd==SD_CMD0_RESET)
    crc=0x95;          // crc for CMD0(0)
  else
    crc =0x87;         // crc for CMD8(0x1aa)
  /* send command */
  sd_spi_wr_byte(sd_base, cmd|0x40);     // 2 MSBs is 01
  /* send argument */
  sd_spi_wr_byte(sd_base, (alt_u8)(argument>>24));
  sd_spi_wr_byte(sd_base, (alt_u8)(argument>>16));
  sd_spi_wr_byte(sd_base, (alt_u8)(argument>>8));
  sd_spi_wr_byte(sd_base, (alt_u8)argument);
  /* send crc; only valid for CMD0 and CMD8 */
  sd_spi_wr_byte(sd_base, crc);
  /* wait for response, up to 8-byte delay */
  for(i=0; i<8; i++){
  /* read R1 response */
  rcode = sd_spi_rd_byte(sd_base);
  if (rcode != 0xff)
    break;
  }
  /* read 4 additional bytes for CMD8 and CMD58 response */
  if (cmd==SD_CMD8_SEND_IF || cmd==SD_CMD58_RD_OCR){
    for(i=0; i<4; i++){
      byte = sd_spi_rd_byte(sd_base);
      ocr = (ocr <<8 | byte);
    }
    // printf("rocde, ocr: 0x%02x, 0x%08x\n", rcode, ocr);
    *r3 = ocr;
  }
  return (rcode);
}
```

```
/***************************************************************
 * function: sd_init()
 * purpose:  initialize sd card in SPI mode
 * argument:
 *   sd_base: base address of sd card interface
 * return:
 *   - status code:
 *      - -1: initialization fails
 *      -  0: initialization success with standard SD card
 *      - +1: initialization success with high capacity SDHC card
 * note:
 *   - initialize procedure
 *      1. send at least 74 sd_clk with sd card deselected (i.e., cs_n=1)
 *         to force the sd card to accept native command
 *      2. send CMD0 w/ cs asserted (cs_n=0) to force the card
 *         to enter SPI mode and reset the card to idle state
 *      3. send CMD55 and then CMD41 to initialize the card
 *   - different commands needed in step 2 to initialize MMC or SDHC card
 *   - initialization may take a fraction of a second
 ***************************************************************/
int sd_init(alt_u32 sd_base)
{
  int i;
  alt_u8 rcode;
  alt_u32 ocr;
  int hcs;

  /* set SPI sclk clock to 200 kHz */
  sd_spi_wr_dvsr(sd_base, 125);
  /* force the sd card to enter SPI mode */
  sd_spi_wr_cs(sd_base, 1);           // set cs to 1
  for(i=0; i<10; i++){                // generate 80 sclk cycles
    sd_spi_wr_byte(sd_base, 0xff);    // 8 cycles per write
  }
  sd_spi_wr_cs(sd_base, 0);           // set cs to 0 (enable)
  /* send reset command */
  rcode = sd_wr_cmd(sd_base, SD_CMD0_RESET, 0, &ocr);
  if (rcode != 0x01){                 // not entering idle state
    printf("CMD0 command fails: R1=0x%02x \n", rcode);
    return (-1);
  }
  /* send interface condition check command */
  rcode = sd_wr_cmd(sd_base, SD_CMD8_SEND_IF, 0x000001aa, &ocr);
  //printf("Debug CMD8 command: R1/data=0x%02x 0x%x\n", rcode, ocr);
  if ((rcode!=0x01 || ocr!=0x000001aa) && (rcode!=05)) {
    printf("CMD8 command fails: R1/data=0x%02x 0x%x\n", rcode, (int)ocr);
    return (-1);
  }
  /* send sd card init commend and wait for 0.3 sec */
  for(i=0; i<SD_INIT_TIME_OUT; i++){
    sd_wr_cmd(sd_base, SD_CMD55_APP, 0, &ocr);
    rcode = sd_wr_cmd(sd_base, SD_ACMD41_INIT_SD, 0x40000000, &ocr);
    if (rcode==0x00)                  // correct response received
      break;
  }
  if (rcode!=0x00){
    printf("ACMD41 command fails: R1=0x%02x\n", rcode);
    return (-1);
  }
```

```
    /* send read OCR register commend */
    rcode = sd_wr_cmd(sd_base, SD_CMD58_RD_OCR, 0x000001aa, &ocr);
    // printf("Debug CMD58 command: R1/data=0x%02x 0x%x\n", rcode, (int)ocr);
    if (rcode != 0x00){
        printf("CMD58 command fails: R1=0x%02x \n", rcode);
        return (-1);
    }
    /* extract hcs bit (bit 30) */
    hcs = (ocr & 0x40000000) >> 30;
    /* set SPI clock to 1 MHz */
    sd_spi_wr_dvsr(sd_base, 25);
    return(hcs);
}
```

```
/****************************************************************
* function: sd_read_sector()
* purpose:  retrieve one sector (512 2^9 bytes) from sd card to a buffer
* argument:
*   sd_base: base address of sd card interface
*   sdhc: 0 for SD card; 1 for SDHC card
*   sect: sector number
*   buf: pointer to buffer
* return: 0 for success; -1 otherwise; buf updated
* note:
*   - procedure:
*     1. send read-one-sector command
*     2. wait for sd to send data_start token
*     3. read 512 bytes
*     4. read 2 dummy crc bytes
*   - CMD17 argument
*     - standard capacity SD card: starting byte address (sect<<9)
*     - high capacity SDHC card: starting sector address (sect)
*   - 10 MHz sclk used to calculate # timeout cycles
*   - calling function must allocate 512 byte buffer space
****************************************************************/
int sd_read_sector(alt_u32 sd_base, int sdhc, alt_u32 sect, alt_u8 *buf)
{
  int i;
  alt_u8 rcode, token;
  alt_u32 addr, ocr;

  /* byte addr for SD card; sector # for SDHC card */
  if (sdhc==0)
    addr = (sect<<9);
  else
    addr = sect;
  /* send sd card read single block commend */
  rcode = sd_wr_cmd(sd_base, SD_CMD17_READ, addr, &ocr);
  if (rcode != 0x00){
    printf("CMD17 command fails: R1=0x%02x \n", rcode);
    return (-1);
  }
  /* wait for data start token up to 0.1 sec */
  for(i=0; i<SD_READ_TIME_OUT; i++){
    token = sd_spi_rd_byte(sd_base);
    if (token==SD_TOKEN_DATA_START)   // correct response received
      break;
  }
  if (token!=SD_TOKEN_DATA_START){   //time-out
    printf("No data start token: last token=0x%02x\n", token);
    return (-1);
  }
  /* read one sector (512 bytes) */
  for (i=0; i<512; i++)
    buf[i] = sd_spi_rd_byte(sd_base);
  /* read and discard two crc bytes */
  sd_spi_rd_byte(sd_base);
  sd_spi_rd_byte(sd_base);
  return (0);
}
```

```
/**************************************************************************
 * function: sd_write_sector()
 * purpose:  write one sector (512 (2^9) bytes) from a buffer to sd card
 * argument:
 *   sd_base: base address of sd card interface
 *   sdhc: 0 for SD card; 1 for SDHC card
 *   sect: sector number
 *   buf: pointer to buffer
 * return: 0 for success; -1 otherwise
 * note:
 *   - procedure:
 *     1. send write-one-sector command
 *     2. send data_start token
 *     3. send 512 bytes
 *     4. send 2 dummy crc bytes
 *     5  check data accept token
 *     5. wait for completion
 *   - CMD24 argument
 *     - standard capacity SD card: starting byte address (sect<<9)
 *     - high capacity SDHC card: starting sector address (sect)
 *   - 10 MHz sclk used to calculate # timeout cycles
 **************************************************************************/
int sd_write_sector(alt_u32 sd_base, int sdhc, alt_u32 sect, alt_u8 *buf)
{
  int i;
  alt_u8 rcode, token;
  alt_u32 addr, dummy;

  /* byte addr for SD card; sector # for SDHC card */
  if (sdhc==0)
    addr = (sect<<9);
  else
    addr = sect;
  /* send sd card write single block commend */
  rcode = sd_wr_cmd(sd_base, SD_CMD24_WRITE, addr, &dummy);
  if (rcode != 0x00){           // error
    printf("CMD24 command fails: R1=0x%02x \n", rcode);
    return (-1);
  }
  /* initiate transfer by send data start token */
  sd_spi_wr_byte(sd_base, SD_TOKEN_DATA_START);
  /* send 512 bytes */
  for (i=0; i<512; i++){
    sd_spi_wr_byte(sd_base, buf[i]);
  }
  /* send two dummy crc bytes */
  sd_spi_wr_byte(sd_base, 0xff);
  sd_spi_wr_byte(sd_base, 0xff);
  /* wait for data acceptance token up to 0.25 sec */
  for(i=0; i<SD_WRITE_TIME_OUT; i++){
    token = sd_spi_rd_byte(sd_base);
    token = token & 0x1f;                 // only 5 LSBs used
    if (token==SD_TOKEN_DATA_ACCEPT)      // correct response received
      break;
  }
  if (token!=SD_TOKEN_DATA_ACCEPT){
    printf("No data accept token: last token=0x%02x\n", token);
    return (-1);
  }
```

```
  /* wait for write completion */
  for(i=0; i<SD_WRITE_TIME_OUT; i++){
    token = sd_spi_rd_byte(sd_base);
    if (token==0xff)                    // correct response received
      break;
  }
  if (token != 0xff){
    printf("Write completion timeout: last token=0x%02x\n", token);
    return (-1);
  }
  return (0);            // ok
}
```

```
/************************************************************************
 *  FAT16 related functions
 ************************************************************************/
/************************************************************************
 * function: bget16()
 * purpose:  get 16 bits from a buffer
 * argument:
 *    buf: pointer to buffer
 *    pos: position in buffer
 * return: a half word in alt_u16 data type
 * note: "little endian" byte ordering
 ************************************************************************/
alt_u16 bget16(alt_u8 *buf, int pos)
{
   alt_u16 b0, b1, r;

   b0 = (alt_u16) buf[pos];
   b1 = (alt_u16) buf[pos+1];
   r = (b1<<8) + b0;
   return(r);
}


/************************************************************************
 * function: bget32()
 * purpose:  get 32 bits (a word) from a buffer
 * argument:
 *    buf: pointer to buffer
 *    pos: position in buffer
 * return: a word in alt_u32 data type
 * note: "little endian" byte ordering
 ************************************************************************/
alt_u32 bget32(alt_u8 *buf, int pos)
{
   alt_u32 b0, b1, b2, b3, r;

   b0 = (alt_u32) buf[pos];
   b1 = (alt_u32) buf[pos+1];
   b2 = (alt_u32) buf[pos+2];
   b3 = (alt_u32) buf[pos+3];
   r = (b3<<24) + (b2<<16) + (b1<<8) + b0;
   return(r);
}
```

```
/*************************************************************************
* function: sd_file_conv_fname()
* purpose:  convert a string to normalized FAT16 8.3 name-extension entry
* argument:
*   si: pointer to file name string input (e.g., "myfile.c")
*   f83: pointer to 11-char array of normalize 8.3 file name-extension
* return:
*   updated f83
* note:
*   - normalized 8.3 entry:
*       - name: 8 char, padding with \0 if needed
*       - extension: 3 char, padding with \0 if needed
*       - all uppercase
*   - e.g., convert "myfile.c" to "MYFILE   C  "
*************************************************************************/
void sd_file_conv_fname(char *si, char *f83)
{
    int i, pos;
    char ch;

    /* blank */
    for (i=0; i<11; i++)
      f83[i] = ' ';
    /* file name */
    pos=0;
    for (i=0; i<9; i++){
      ch=toupper(si[pos]);
      pos++;
      if (ch=='\0')  // end of string
        return;
      if (ch=='.')
        break;
      f83[i]=ch;
     }
    /* file extension */
    for (i=8; i<11; i++){
      ch=toupper(si[pos]);
      pos++;
      if (ch=='\0')
        return;
      f83[i]=ch;
    }
}
```

```
/***************************************************************************
 * function: sd_file_mount()
 * purpose:  extract key parameters from a FAT16 file system
 * argument:
 *   sd_base: base address of SD card interface
 *   f: pointer to the FAT parameter record
 * return:
 *   0: successful; f updated
 *   1: SD card initialization failed
 *   2: SD card read MBR failed
 *   3: MBR signature not match
 *   4: FAT16 type not match
 *   5: SD card read boot record failed
 *   6: Boot record signature not match
 * note:
 *   - file system must be FAT16 and in first partition
 *   - procedure
 *       1. initialize SD card
 *       1. read sector 0
 *       2. determine whether the sector is MBR. If not, go to step 6.
 *       3. verify MBR signature and FAT file system type
 *       4. extract sector number for the first partition's boot record
 *       5. read boot record
 *       6. verify boot record signature
 *       7. extract relevant FAT16 parameters
 *       8. construct and return the FAT parameter record
 ***************************************************************************/
int sd_file_mount(alt_u32 sd_base, fat_para_type *f)
{
  alt_u8 buf[512];      // buffer for a sector
  alt_u32 br0;          // boot record sector number
  alt_u8 sec_per_cls, fat_copy;
  alt_u16 reserved_sec, root_dir_size, sec_per_part_small, sec_per_fat;
  alt_u32 sec_per_part_large;
  int status, sdhc, sec_per_part;

  /* initialize SD card */
  sdhc=sd_init(sd_base);
  if (sdhc==-1){
    printf("SD card initialization failed \n");
    return (1);
  }
  f->sdhc = sdhc;
  f->sd_base = sd_base;
  /* fetch sector 0 */
  status=sd_read_sector(sd_base, sdhc, 0, buf);
  if (status!=0){
    printf("Sector 0 read failed\n");
    return (2);
  }
  br0=0;
  /* check whether the sector is the MBR */
  if (buf[0]!=0xeb && buf[0]!=0xe9) { /* not boot record */
    /* check MBR signature 0x55aa */
    if ((buf[510]!=0x55) || (buf[511]!=0xaa)){
      printf("MBR signature not match\n");
      return (3);
    }
```

```
  /* check FAT16 file system type at 0x1c2 */
  if ((buf[0x1c2]!=0x04) && (buf[0x1c2]!=0x06) && (buf[0x1c2]!=0x0e)){
    printf("FAT16 file type not match\n");
    return (4);
  }
  /* fetch boot record of first partition */
  br0 = bget32(buf, 0x1c6);
  status=sd_read_sector(sd_base, sdhc, br0, buf);
  if (status!=0){
    printf("Boot record read failed\n");
    return (5);
  }
} // end if
/* check boot record signature 0x55aa */
if (buf[510]!=0x55 || buf[511]!=0xaa){
  printf("Boot record signature not match\n");
  return (6);
}
/* extract key FAT16 parameters */
sec_per_cls = buf[0x0d];
reserved_sec = bget16(buf, 0x0e);
fat_copy = buf[0x10];
root_dir_size = bget16(buf, 0x11);
sec_per_part_small = bget16(buf, 0x13);
sec_per_fat = bget16(buf, 0x16);
sec_per_part_large = bget32(buf, 0x20);
/* construct the FAT16 record */
if (sec_per_part_small!=0)
  sec_per_part = (int) sec_per_part_small;
else
  sec_per_part = sec_per_part_large;
f->fat0 = br0 + reserved_sec;
f->rdir0 = f->fat0 + fat_copy * sec_per_fat;
f->data0 = f->rdir0 + root_dir_size*32/512;
f->cls_size = (int) sec_per_cls;
f->rdir_size = (int) root_dir_size;
f->data_size = (sec_per_part - f->data0)/sec_per_cls;
return(0);
}
```

```
/******************************************************************************
 * function: sd_file_fopen()
 * purpose:  open a FAT16 file and store relevant info in file descriptor
 * argument:
 *   fat: pointer to the FAT parameter record
 *   fname: pointer to file name string (such as "myfile.c")
 *   fd: pointer to file descriptor
 * return:
 *   0: successful; fd updated
 *   1: file not found
 *   2: SD card read data sector failed
 *   3: SD card read FAT sector failed
 * note:
 ******************************************************************************/
int sd_file_fopen(fat_para_type *fat, char *fname, file_dp_type *fd)
{
  int index, status, n;
  int sect, oft;
  alt_u8 *buf;      // pointer to a sector buffer
  char f83[11];     // file name in normalized 8.3 format

  sd_file_conv_fname(fname, f83);
  fd->open = 0;
  /* use file descriptor's data buffer as temp buffer */
  buf = fd->dbuf;
  /* search entire root directory */
  for (index=0; index<fat->rdir_size; index++){
    /* load a new sector after 16 file entries */
    if (index%16==0){
      sect = fat->rdir0 + index/16;
      status=sd_read_sector(fat->sd_base, fat->sdhc, sect, buf);
      if (status!=0){
        printf("SD read failed\n");
        return (2);
      }
    }
    /* offset in current sector */
    oft = 32 * (index%16);
    /* 0x00 in first char of file name means no more entries */
    if (buf[oft]==0x00){
      printf("File not found\n");
      return(1);  // file not found
    }
    /* compare normalized 8.3 name-extension */
    for (n=0; n<11; n++){
      if (f83[n] != buf[oft+n])
        break;
    }
    /* file entry matches; update file descriptor's fields */
    if (n==11){
      fd->rdir_index = index;
      fd->cls0 = bget16(buf, oft+0x1a);
      fd->size = (int) bget32(buf, oft+0x1c);
      for (n=0; n<12; n++)
        fd->name[n] = buf[oft+n];
      /* initialize file descriptor's counters */
      fd->clsi = fd->cls0;
      fd->sect = 0;
      fd->idx = 0;
      fd->seek = 0;
      break;  // success
```

```
      } // end if
    } // end for
    if (index==fat->rdir_size){ // file not found
      printf("File not found\n");
      return(1);
    }
    /* load file's 1st file allocation table sector to buffer */
    sect = fat->fat0 + fd->cls0/256;
    status=sd_read_sector(fat->sd_base, fat->sdhc, sect, fd->fbuf);
    if (status!=0){
      printf("File allocation table sector read failed\n");
      return (3);
    }
    fd->sf = sect;
    /* update open status */
    fd->open = 1;
    return(0);
}


/***********************************************************************
* function:  sd_file_fclose()
* purpose:   close a FAT16 file
* argument:
*    fd: pointer to file descriptor
* return:
* note: included for completeness
***********************************************************************/
void sd_file_fclose(file_dp_type *fd)
{
  fd->open = 0;
  return;
}
```

```
/*****************************************************************************
 * function: sd_file_read_byte()
 * purpose:  get 32 bits (a word) from a buffer
 * argument:
 *   fat: pointer to the FAT parameter record
 *   fd: pointer to file descriptor
 *   byte: pointer to the returned byte
 * return:
 *   0: successful; fd updated; byte updated
 *   1: end of file
 *   2: SD card read data sector failed
 *   3: SD card read FAT sector failed
 * note:
 *   - file must be opened before read
 *****************************************************************************/
int sd_file_read_byte(fat_para_type *fat, file_dp_type *fd,alt_u8 *byte)
{
  int status;
  alt_u16 s;

  /* check end-of-file */
  if (fd->seek >= fd->size)
    return(1);
  /* must load a new sector at the beginning */
  if (fd->idx==0){
    s = fat->data0 + ((fd->clsi)-2)*(fat->cls_size) + fd->sect;
    status=sd_read_sector(fat->sd_base, fat->sdhc, s, fd->dbuf);
    if (status!=0){
      printf("Data sector read failed\n");
      return (2);
    }
  }
  /* read a byte */
  *byte = fd->dbuf[fd->idx];
  /* update file descriptor counter */
  fd->seek++;
  fd->idx++;
  /* not reach end of buffer yet */
  if (fd->idx!=512)
    return(0);  // success
  /* reach the end of buffer */
  fd->idx = 0;
  fd->sect++;
  /* not last sector in cluster */
  if (fd->sect!=fat->cls_size)
    return(0);  // success
  /* last sector in cluster; fetch next cluster from FAT table */
  fd->sect = 0;
  s = fat->fat0 + fd->clsi/256;  // sector containing the cluster
  /* file allocation table sector not in buffer; load new sector */
  if (s != fd->sf){
    status=sd_read_sector(fat->sd_base, fat->sdhc, s, fd->fbuf);
    if (status!=0){
      printf("File allocation table sector read failed\n");
      return (3);
    }
    fd->sf = s;
  }
  fd->clsi = bget16(fd->fbuf, 2*(fd->clsi%256));
  return(0);  // success
}
```

Listing 19.18   chu_main_sd_test.c

```
/************************************************************************
 *
 * Module:   SD card and FAT test
 * File:     chu_main_sd_test.c
 * Purpose: Test SD card access
 * IP core base addresses:
 *    - SWITCH_BASE: slide switch
 *    - BTN_BASE: pushbutton
 *    - SSEG_BASE: 7-segment LED display
 *    - SDC_BASE: SD card controller
 *
 ************************************************************************/
/* file inclusion */
#include <stdio.h>
#include "system.h"
#include "chu_avalon_gpio.h"
#include "chu_avalon_sd.h"


/************************************************************************
 * function: print_sector()
 * purpose: print a 512-byte sector in hex/ascii table
 * argument:
 *    buf: pointer to the 512-byte buffer
 * return:
 * note:
 ************************************************************************/
/* print a 512-byte sector in hex and ASCII format */
void print_sector(alt_u8 *buf)
{
  int i, j;
  alt_u8 ch;

  for (j=0; j<32; j++){
    printf("0x%02x-  ", j);
    for (i=0; i <16; i++)
      printf("%02x ", buf[j*16+i]);
    printf("  ");
    for (i=0; i <16; i++){
      ch = buf[j*16+i];
      if (ch>127 || ch<32)    // non ascii
        ch = '.';
      printf("%c", ch);
    } // end for i
    printf("\n");
  } //end for j
  printf("\n");
}
```

```
/**************************************************************************
* function: main()
* purpose: test SD card and FAT16
* note:
**************************************************************************/
int main(void)
{
  fat_para_type fat;
  file_dp_type fd;
  char fname[15], yes[3];
  int sw, btn;
  unsigned int num;
  int i, j, s, status, sdhc;
  alt_u32 sect;
  alt_u8 ch, dvsr;
  alt_u8 rbuf[512], wbuf[512];    // read and write buffers
  alt_u8 disp_msg[4]={sseg_conv_hex(5), sseg_conv_hex(13),
                      0xff, sseg_conv_hex(12)};

  sseg_disp_ptn(SSEG_BASE, disp_msg);      // show "Sd C" for display
  printf("SD card FAT16 file system test:\n");
  sdhc=sd_init(SDC_BASE);
  if (sdhc==-1)
    printf("SD card initialization failed \n");
  btn_clear(BTN_BASE);
  while (1){
    while (!btn_is_pressed(BTN_BASE)){ };       // wait for button
    btn=btn_read(BTN_BASE);                      // read button
    if (btn & 0x02){                             // key1 pressed
      sw=pio_read(SWITCH_BASE);                  // read switch
      // printf("key/sw: %d/%d\n", btn, sw);
    }
    btn_clear(BTN_BASE);
    switch (sw){
      case 0:   // mount FAT16 file system
        status=sd_file_mount(SDC_BASE, &fat);
        sdhc = fat.sdhc;
        if (status==0) {
          printf("FAT 16 mounted on SD card\n");
          printf("starting FAT table sector: %d\n", fat.fat0);
          printf("starting root dir sector: %d\n", fat.rdir0);
          printf("starting data sector:     %d\n", fat.data0);
          printf("sectors per cluster:      %d\n", fat.cls_size);
          printf("clusters in data segment: %d\n", fat.data_size);
          printf("file entries of root dir: %d\n\n", fat.rdir_size);
        } else
          printf("Mount failed: status=%d\n\n", status);
        break;
      case 1:   // open a file
        printf("Enter file name in 8.3 format: ");
        scanf("%s", fname);
        status=sd_file_fopen(&fat, fname, &fd);
        if (status==0) {
          printf("\nFile open successful.\n");
          printf("file size (bytes): %d\n", fd.size);
          printf("root dir entry #:  %d\n", fd.rdir_index);
          sect = fat.rdir0 + fd.rdir_index/32;
          printf("entry sector #:    %d\n", (int)sect);
          printf("starting cluster#: %d\n", fd.cls0);
          sect = fat.data0 + (fd.cls0-2)*fat.cls_size;
          printf("starting sector #: %d\n\n", (int)sect);
```

```
   } else
     printf("\nFile open failed: status=%d\n\n", status);
break;
case 2:   // read file and list first 2 and last 2 sectors
  s = fd.size/512;
  if ((fd.size%512)!=0)   // fraction of a segment
    s++;
  printf("File size (bytes/sectors): %d/%d\n", fd.size, s);
  printf("Data dump for first 2 and last 2 sectors:\n\n");
  for (i=0; i<s; i++){
    for (j=0; j<512; j++){
      sd_file_read_byte(&fat, &fd, &ch);
        rbuf[j]=ch;
     }
    if (i<2 || i>s-3){
      printf("sector %d of file:\n", i);
      print_sector(rbuf);
    } else {
      printf(".");
    } // end if
  } // end for i
  break;
case 3: // read and print a 512 byte sector
  printf("Enter sd card read sector number: ");
  scanf("%u", &num);
  sect = (alt_16) num;
  status=sd_read_sector(SDC_BASE, sdhc, sect, rbuf);
  if (status!=0)
    printf("read sd card failed\n\n");
  else{
    printf("\nsector %d(0x%x)\n", (int)sect, (int)sect);
    print_sector(rbuf);
  }
  break;
case 4: // write a 512 byte sector
  printf("Low-level sector write may corrupt file system.\n");
  printf("Press Y to continue: ");
  scanf("%s", yes);
  if (yes[0]!='y' && yes[0]!='Y'){
    printf("Sector write abandoned.\n\n");
    break;
  }
  printf("Enter sd card write sector number: ");
  scanf("%u", &num);
  sect = (alt_16) num;
  for (i=0; i<512; i++)
    wbuf[i] =  (alt_u8) (sect + i);
  status=sd_write_sector(SDC_BASE, sdhc, sect, wbuf);
  if (status!=0)
    printf("Write sd card failed\n\n");
  else
    printf("write sd card completed\n\n");
  break;
case 5:   // set frequency divisor of SPI clock
  printf("Enter frequency divisor (1-255): ");
  scanf("%u", &num);
  dvsr = (alt_u8) num;
  sd_spi_wr_dvsr(SDC_BASE, dvsr);
  printf("SPI bus frequency set to %6.3f MHz.\n\n", 50.0/(2.0*dvsr));
  break;
```

```
      case 6:   // reinitialize SD card
        sdhc=sd_init(SDC_BASE);
        if (sdhc==-1)
          printf("SD card initialization failed.\n\n");
        else
          printf("SD card initialized (sdhc status=%d).\n\n", sdhc);
        break;
    } //end switch
  } // end while
}
```

PART IV

HARDWARE ACCELERATOR
CASE STUDIES

CHAPTER 20

GCD ACCELERATOR

The GCD (greatest common divisor) of two non-zero integers is the largest positive integer that divides the numbers without a remainder. The *binary GCD algorithm* is a method to obtain the GCD. In this chapter, we implement the algorithm using a software routine and a hardware accelerator and compare their performances.

20.1 INTRODUCTION

The $\gcd(a, b)$ function returns the GCD of two positive integers, a and b. It is the largest positive integer that divides the numbers without a remainder. For example, $\gcd(1,12)$ is 1 and $\gcd(24,15)$ is 3. The gcd() function has the following property:

$$\gcd(a, b) = \begin{cases} a & \text{if } a = b \\ \gcd(a - b, b) & \text{if } a > b \\ \gcd(a, b - a) & \text{if } a < b \end{cases}$$

We can apply this equation repetitively to reduce the values of two operands and eventually obtain the GCD. However, the convergence rate of this scheme can be really slow when one of the numbers is small. For an N-bit input, computing $\gcd(1, 2^N - 1)$ requires $2^N - 1$ iterations. One way to improve the algorithm is to take advantage of the binary number system. For a binary number, we can tell whether it is odd or even by checking the LSB. Based on the LSBs of two inputs, several simplification rules can be applied in the derivation of the gcd() function:

Embedded SOPC Design with Nios II Processor and VHDL Examples. By Pong P. Chu **619**
Copyright © 2011 John Wiley & Sons, Inc.

- If both a and b are even, $\gcd(a,b) = 2\gcd(\frac{a}{2}, \frac{b}{2})$.
- If a is odd and b is even, $\gcd(a,b) = \gcd(a, \frac{b}{2})$.
- If a is even and b is odd, $\gcd(a,b) = \gcd(\frac{a}{2}, b)$.

The previous equation can be extended:

$$
\gcd(a,b) = \begin{cases}
a & \text{if } a = b \\
2\gcd(\frac{a}{2}, \frac{b}{2}) & \text{if } a \neq b \text{ and } a, b \text{ even} \\
\gcd(a, \frac{b}{2}) & \text{if } a \neq b \text{ and } a \text{ odd}, b \text{ even} \\
\gcd(\frac{a}{2}, b) & \text{if } a \neq b \text{ and } a \text{ even}, b \text{ odd} \\
\gcd(a - b, b) & \text{if } a > b \text{ and } a, b \text{ odd} \\
\gcd(a, b - a) & \text{if } a < b \text{ and } a, b \text{ odd}
\end{cases}
$$

Note that this equation uses only subtraction and divided-by-2 operations. Since the divided-by-2 operation can be realized by shifting the dividend to the right by one position, no expensive general-purpose division operation is needed. This is known as the *binary GCD algorithm*. In the following sections, we implement this algorithm in software and with a custom hardware accelerator and compare their performance.

20.2 SOFTWARE IMPLEMENTATION

The previous GCD equation can be converted to C code by repetitively examining the conditions of two operands and applying the corresponding reduction rule until the terminating condition (i.e., the first rule) is met. These rules generally reduce the values of the operands. However, in the second rule, which is

$$
\gcd(a,b) = 2\gcd(\frac{a}{2}, \frac{b}{2}) \quad \text{if } a \neq b \text{ and } a, b \text{ even}
$$

a common factor (i.e., 2) is extracted and special treatment is needed. In addition to normal reduction, we use a variable, n, to keep track of the number of occurrences of this condition. The final GCD value can be restored by multiplying the initial result by 2^n. The pseudo-code is

```
while (1){
  if (a==b)
    break;
  if (even(a)){       // a even
    a = a>>1;
    if (even(b)){     // b even
      b = b>>1;
      n++;
    }
  } else {            // a odd
    if (even(b))      // b even
      b = b>>1;
    else {            // b odd
      if (a > b)
        a = a - b;
      else
        b = b - a;
```

```
        } // end else (b odd)
      } // end else (a odd)
20   } // end while
     a = a<<n;
```

Note that multiplying by 2^n corresponds to shifting the initial result left n positions, as in the expression in the last statement.

20.3 HARDWARE IMPLEMENTATION

20.3.1 ASMD chart

The algorithm used in Section 20.2 can be converted to an ASMD chart and then realized by an FSMD in hardware. The resulting ASMD chart is shown in Figure 20.1.

It is helpful to identify the key operations and their corresponding functional units in the data path. We assume that the inputs are 32-bit unsigned integers. There are three arithmetic operations, including two subtractions, a-b and b-a, and one increment operation, n++. The former infers two 32-bit subtractors and the latter infers a small 5-bit incrementor. There are four shift operations. However, three of them shift a constant amount and thus do not introduce a physical logic component. Only the a<<n expression infers a 32-bit barrel shifter.

Since the two subtractions and barrel shifting operation are performed in parallel, putting them in the same state will not prolong the clock period. Thus we group all operations to a single op state.

20.3.2 HDL implementation

After constructing the ASMD chart, we can derive the HDL code accordingly. The code is shown in Listing 20.1.

Listing 20.1   GCD engine

```
library ieee;
use ieee.std_logic_1164.all;
use ieee.numeric_std.all;
4 entity gcd_engine is
    port(
        clk, reset: in std_logic;
        start: in std_logic;
        a_in, b_in: in std_logic_vector(31 downto 0);
9       gcd_done_tick, ready: out std_logic;
        r: out std_logic_vector(31 downto 0)
    );
  end gcd_engine;

14 architecture arch of gcd_engine is
    type state_type is (idle, op);
    signal state_reg, state_next: state_type;
    signal a_reg, a_next, b_reg, b_next: unsigned(31 downto 0);
    signal n_reg, n_next: unsigned(4 downto 0);
19 begin
    -- state & data registers
    process(clk,reset)
    begin
```

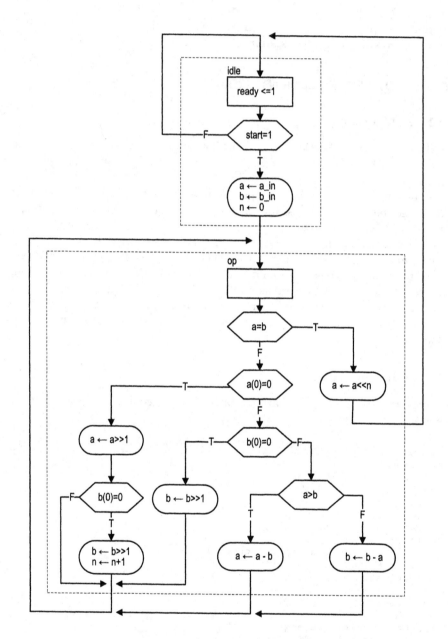

Figure 20.1 GCD ASMD chart.

```
         if reset='1' then
24           state_reg <= idle;
             a_reg <= (others=>'0');
             b_reg <= (others=>'0');
             n_reg <= (others=>'0');
         elsif (clk'event and clk='1') then
29           state_reg <= state_next;
             a_reg <= a_next;
             b_reg <= b_next;
             n_reg  <= n_next;
         end if;
34   end process;
     -- next-state logic & data path functional units/routing
     process(state_reg,a_reg,b_reg,n_reg,start,a_in,b_in,n_next)
     begin
         a_next <= a_reg;
39       b_next <= b_reg;
         n_next <= n_reg;
         state_next <= state_reg;
         gcd_done_tick<='0';
         case state_reg is
44           when idle =>
                 if start='1' then
                     a_next <= unsigned(a_in);
                     b_next <= unsigned(b_in);
                     n_next <= (others=>'0');
49                   state_next <= op;
                  end if;
             when op =>
                 if (a_reg=b_reg) then
                     state_next <= idle;
54                   gcd_done_tick <= '1';
                     a_next <= shift_left(a_reg, to_integer(n_reg));
                 else
                     if (a_reg(0)='0') then -- a_reg even
                         a_next <= '0' & a_reg(31 downto 1);
59                       if (b_reg(0)='0') then  -- both even
                             b_next <= '0' & b_reg(31 downto 1);
                             n_next <= n_reg + 1;
                         end if;
                     else -- a_reg odd
64                       if (b_reg(0)='0') then  -- b_reg even
                             b_next <= '0' & b_reg(31 downto 1);
                         else -- both a_reg and b_reg odd
                             if (a_reg > b_reg) then
                                 a_next <= a_reg - b_reg;
69                           else
                                 b_next <= b_reg - a_reg;
                             end if;
                         end if;
                     end if;
74               end if;
         end case;
     end process;
     --output
     ready <= '1' when state_reg=idle else '0';
79   r <= std_logic_vector(a_reg);
 end arch;
```

20.4   TIME MEASUREMENT

To compare the performance of the software routine and hardware accelerator, we need to measure the time spent on the computation. This can be done by using HAL's time stamp mechanism or a custom counter.

20.4.1   HAL time stamp driver

The Altera HAL framework provides a high-resolution timing function using a *time stamp* driver. The driver utilizes a monotonically increasing counter driven by the system clock. Its content can be retrieved any time and thus a "time stamp" can be obtained at that particular point. The two main functions are:

- `alt_timestamp_start()`: initiate the time stamp counter and clear its content to 0.
- `alt_timestamp()`: read the current time stamp.

To use the HAL time stamp driver, it must be properly configured in the BSP setting. We must first instantiate a timer core and then designate the instance as the stamp timer. For the Nios II system in Section 16.10.2, we create two timer modules. The module named `stamp_timer` is used for this purpose. It is a 32-bit timer designated as a stamp timer in the `timestamp_timer` field of BSP, as shown in Figure 16.10. After setting up the BSP, we can then include the header file, `sys/alt_timestamp.h`, and use the related functions.

By taking two time stamps at two designated points and find their difference, we can determine the number of elapsed clock cycles. Although the elapsed time is represented in terms of clock cycles, the resolution is not as accurate as it appears because of the overhead associated with the execution of `alt_timestamp()`. One way to reduce the error is to estimate the overhead and then subtract it from the final difference. The basic code outline is:

```
/* estimate the overhead of alt_timestamp() */
t0 = alt_timestamp();
t1 = alt_timestamp();
overhead = t1 - t0;
/* start actual measurement */
t0 = alt_timestamp();
   . . .
activities to be measured
   . . .
t1 = alt_timestamp();
elapsed_cycles = t1 - t0 - overhead;
```

20.4.2   Custom hardware counter

For the GCD hardware accelerator, some input combinations may take only a few clock cycles to complete and the potential error introduced by the `alt_timestamp()` method is too large in comparison. To overcome the problem, we create a companion auxiliary counter in the GCD hardware accelerator. The counter starts when the GCD operation is initiated and stops when the operation is done. We then read the counter and thus obtain the exact elapsed clock cycles.

20.5 GCD ACCELERATOR IP CORE DEVELOPMENT

20.5.1 Avalon interfaces

We can add a wrapping circuit and the auxiliary counter to the GCD system and create an SOPC component. It includes an Avalon MM slave interface to interact with the host and a clock input interface for the system clock.

20.5.2 Register map

A Nios II processor interacts with the GCD accelerator by passing the two input operands and retrieving the calculated GCD result. In addition, it initiates the operation by writing dummy data to a specific address, checks the GCD accelerator status for completion, and reads the value from the measurement counter. The registers, their address offsets, and fields are:
- Write addresses (data from cpu)
  - offset 0 (a register)
    * bits 31 to 0: 32-bit a input
  - offset 1 (b register)
    * bits 31 to 0: 32-bit b input
  - offset 2
    * Dummy data used to generate an enable pulse to start GCD operation
- Read addresses (data to cpu)
  - offset 3 (status register)
    * bit 31: 1-bit ready signal asserted (i.e., 1) when the GCD accelerator is idle
    * bits 15 to 0: 16-bit measurement counter value
  - offset 4 (r register)
    * bits 31 to 0: 32-bit GCD result

20.5.3 Wrapped GCD accelerator

The wrapped GCD accelerator contains two registers to store the two input operands and write decoding logic to generate enable signals. In addition, it contains an auxiliary counter that keeps track of the number of clock cycles spent on GCD computation. The counter starts when the GCD operation is initiated and stops when the operation is done. The HDL code is shown in Listing 20.2.

Listing 20.2   GCD engine Avalon interface

```
library ieee;
use ieee.std_logic_1164.all;
use ieee.numeric_std.all;
entity chu_avalon_gcd is
    port (
        clk, reset: in  std_logic;
        --- avalon interface
        gcd_address: in std_logic_vector(2 downto 0);
        gcd_chipselect: in std_logic;
```

```vhdl
        gcd_write: in std_logic;
        gcd_writedata: in std_logic_vector(31 downto 0);
        gcd_readdata: out std_logic_vector(31 downto 0)
    );
14 end chu_avalon_gcd;

   architecture arch of chu_avalon_gcd is
      signal gcd_start, gcd_done_tick: std_logic;
      signal a_in_reg, b_in_reg: std_logic_vector(31 downto 0);
19    signal r_out: std_logic_vector(31 downto 0);
      signal gcd_ready: std_logic;
      signal wr_en, wr_a, wr_b: std_logic;
      type state_type is (idle, count);
      signal state_reg, state_next: state_type;
24    signal c_reg, c_next: unsigned(15 downto 0);

   begin

   --============================================
   -- instantiation
29 --============================================
      gcd_unit: entity work.gcd_engine(arch)
         port map
            (clk=>clk, reset=>reset, start=>gcd_start,
             a_in=>a_in_reg, b_in=>b_in_reg, r=>r_out,
34           gcd_done_tick=>gcd_done_tick, ready=>gcd_ready);

   --============================================
   -- registers, decoding, and multiplexing
   --============================================
39 -- register
      process (clk, reset)
      begin
         if reset='1' then
            a_in_reg <= (others=>'0');
44          b_in_reg <= (others=>'0');
         elsif (clk'event and clk='1') then
            if wr_a='1' then
               a_in_reg <= gcd_writedata;
            end if;
49          if wr_b='1' then
               b_in_reg <= gcd_writedata;
            end if;
         end if;
      end process;
54    -- write encoding
      wr_en <= '1' when gcd_write='1' and gcd_chipselect='1' else '0';
      wr_a <= '1' when gcd_address="000" and wr_en='1' else '0';
      wr_b <= '1' when gcd_address="001" and wr_en='1' else '0';
      gcd_start <= '1' when gcd_address="010" and wr_en='1' else '0';
59    -- read multiplexing
      gcd_readdata <= r_out when gcd_address="100" else
                      gcd_ready & "000" & x"000" &
                      std_logic_vector(c_reg);

   --============================================
64 -- auxiliary counter to measure the cycles in execution
   --============================================
      process (clk, reset)
      begin
         if reset='1' then
69          state_reg <= idle;
            c_reg <= (others=>'0');
```

```
              elsif (clk'event and clk='1') then
                 state_reg <= state_next;
                 c_reg <= c_next;
74            end if;
        end process;
        — next-state logic & data path functional units
        process(state_reg,c_reg,gcd_start,gcd_done_tick)
        begin
79            c_next <= c_reg;
              state_next <= state_reg;
              case state_reg is
                 when idle =>
                    if gcd_start='1' then
84                      c_next <= x"0001";
                       state_next <= count;
                    end if;
                 when count =>
                    if (gcd_done_tick='1') then
89                      state_next <= idle;
                    else
                       c_next <= c_reg + 1;
                    end if;
              end case;
94       end process;
     end arch;
```

Base on this top-level HDL file, we can create a new SOPC component in Component Editor and integrate it into a Nios II system.

20.6 TESTING PROGRAM

The system derived in Section 16.10.2 includes the GCD accelerator core (named g_engine) and can be used for testing. Furthermore, it also contains a timer module for the time stamp operation and the BSP is configured to use this timer, as discussed in Section 20.4. Since the purpose of the core is to demonstrate the use of hardware accelerator, it is unlikely to be used in other applications. We do not derive a separate device driver library and simply integrate the I/O access in the testing routines.

20.6.1 GCD routines

The software GCD routine is shown in Listing 20.3.

Listing 20.3

```
#define even(x) (((x) & 1)==0)      // check whether x is an even number

alt_u32 calc_gcd_soft(alt_u32 a, alt_u32 b, int *ticks)
{
   int n=0;
   alt_32 t0, t1, overhead;

   alt_timestamp_start();          // initialize time_stamp timer to 0
   /* determine the overhead of alt_timestamp() */
   t0 = alt_timestamp();
   t1 = alt_timestamp();
   overhead = t1 - t0;
```

```
/* start actual measurement */
t0 = alt_timestamp();
  while (1){
   if (a==b)
     break;
   if (even(a)){        // a even
     a = a>>1;
     if (even(b)){      // b even
       b = b>>1;
       n++;
     }
   } else {             // a odd
     if (even(b))        // b even
       b = b>>1;
     else {             // b odd
       if (a > b)
         a = a - b;
       else
         b = b - a;
     } // end else (b odd)
   } // end else (a odd)
 } // end while
 a = a<<n;
 t1 = alt_timestamp();
 *ticks = (int) (t1-t0-overhead);
 return(a);
}
```

The routine follows the algorithm in Section 20.2 but with the added time measurement functionality.

The hardware GCD routine is shown in Listing 20.4.

Listing 20.4

```
#define GCD_A_REG         0
#define GCD_B_REG         1
#define GCD_START_REG     2
#define GCD_STATUS_REG    3
#define GCD_R_REG         4
#define GCD_CNT_FLD       0x0000ffff
#define GCD_READY_BIT     0x80000000

alt_u32 calc_gcd_hard(alt_u32 gcd_base, alt_u32 a, alt_u32 b, int *clks)
{
  alt_u32 status, result;

  IOWR(gcd_base, GCD_A_REG, a);       // write a
  IOWR(gcd_base, GCD_B_REG, b);       // write b
  IOWR(gcd_base, GCD_START_REG, 0);   // write dummy data to start operation
  while(1){
    /* loop until GCD data is ready */
    status=IORD(gcd_base, GCD_STATUS_REG);
    if (status & GCD_READY_BIT)
      break;
  }
  *clks = (int) (status & GCD_CNT_FLD);
  result=IORD(gcd_base, GCD_R_REG);
  return(result);
}
```

The routine simply loads the two inputs and starts the operation and then retrieves the result and the elapsed clock cycles. The custom auxiliary timer, not the Altera core, is used to keep track of the computation time.

20.6.2 Main program

The main program is shown in Listing 20.5.

Listing 20.5

```
int main(void)
{
  alt_u32 a, b, rs, rh;
  int i;
  int s=0, c=0;
  int samples=10000;
  int s_ticks, clks;

  printf("GCD test starts:\n\n");
  srand(100);                       // set the random number seed
  for (i=0; i<samples; i++){        // obtain gcd for 10,000 random samples
    a = rand() + 1;                 // +1 to avoid 0
    b = rand() + 1;
    rs = calc_gcd_soft(a, b, &s_ticks);
    rh = calc_gcd_hard(G_ENGINE_BASE, a, b, &clks);
    if (rs!=rh)                     // sanity check
      printf("Inconsistency: gcd(%d,%d)=%d/%d\n", a, b, rs, rh);
    s = s + s_ticks;
    c = c + clks;
  }
  printf("average clocks (soft/hard): %d/%d\n", s/samples, c/samples);
  printf("hardware acclerator speedup: %d\n", s/c);
}
```

It performs the GCD computations of 10,000 randomly selected pairs with both software and hardware accelerator-based routines and reports and summarizes the results.

20.7 PERFORMANCE COMPARISON

Benchmarking is a complicated task since the performance depends on a wide variety of factors, such as hardware configuration, compiling option, etc. Recall the testing system used in 20.6.2 is constructed with a fast processor core (i.e., Nios II/f). During the processing, the default setting is used in Quartus II for hardware synthesis and the default setting is used in Nios IDE for compiling the C routines. The statistics from the previous section shows that the average times to perform one GCD computation is about 2,000 clock cycles for the software routine and about 60 clock cycles for the hardware accelerator. Thus, the hardware accelerator speeds up the computation about 30 times. Because the GCD circuit is much simpler than the Nios II processor, it can run at a much faster rate. The Quartus II timing analysis shows that the maximal operating frequency of the accelerator is about twice that of the Nios II system. It is possible to use a PLL core to generate a separate clock for the GCD circuit. The approach can double the performance and speed up computation more than 60 times.

Note that the GCD function is not a "computation-intensive" algorithm since it contains mainly comparison, branching, and looping but not many arithmetic operations. There is not much "inherent parallelism" in the algorithm. The speedup is due to several factors. First, execution in a processor involves fetching instruction, decoding, and fetching and storing data. The GCD accelerator is controlled by a simple FSM and thus avoids this type of overhead. Second, the hardware accelerator can convert the sequential conditional branches into a routing network, as discussed in Sections 4.2 and 4.4, and thus performs comparison and branching concurrently. Finally, the hardware accelerator can implement and combine many simple computations, such as testing for an even condition and shifting one position, together. In the GCD hardware accelerator implementation, all computations are lumped into a single op state without severe penalty to performance (i.e., significantly prolonging the clock cycle).

In a custom hardware system, it is frequently possible to increase the system's performance with additional hardware resource. For example, we can construct a special combinational circuit that counts the trailing zeros and shifts them out. If the input a has m trailing zeros, the circuit outputs the count, m, and the shifted result, which corresponds to $\frac{a}{2^m}$. The corresponding new GCD equation becomes

$$
\gcd(a,b) = \begin{cases}
a & \text{if } a = b \\
2^n \gcd\left(\frac{a}{2^m}, \frac{b}{2^n}\right) & \text{if } a \neq b;\ a,\ b \text{ have } m,\ n \text{ trailing 0's};\ m > n \\
2^m \gcd\left(\frac{a}{2^m}, \frac{b}{2^n}\right) & \text{if } a \neq b;\ a,\ b \text{ have } m,\ n \text{ trailing 0's};\ m \leq n \\
\gcd(a - b, b) & \text{if } a > b;\ a,\ b \text{ odd} \\
\gcd(a, b - a) & \text{if } a < b;\ a,\ b \text{ odd}
\end{cases}
$$

With this circuit, the GCD accelerator can process multiple trailing zeros in one step and thus further improve the performance.

20.8 BIBLIOGRAPHIC NOTES

A more detailed discussion of binary GCD algorithm and Euclid's algorithm can be found on the Wikipedia web site. *Nios II Software Developers Handbook* provides the specification and use of the Altera HAL time stamp driver.

20.9 SUGGESTED EXPERIMENTS

20.9.1 Performance with other processor configuration

Resynthesize the Nios II system with the economic configuration (i.e., Nios II/e) and the standard configuration (i.e., Nios II/s) and compare their performances with the fast configuration.

20.9.2 GCD accelerator with minimal size

The three major components in the GCD accelerator's data path are two subtractors and one combinational barrel shifter. We can use a sequential shifter (which shifts one bit in each clock cycle) and share the subtractor (i.e., to reduce two subtractors to one subtractor). Modify the GCD accelerator design, resynthesize the Nios II system, and compare its performance.

20.9.3   GCD accelerator with trailing zero circuit

Design the trailing zero counting and shifting circuit discussed in Section 20.7 and integrate the circuit to the GCD accelerator. Resynthesize the Nios II system and compare its performance.

20.9.4   GCD accelerator with 64-bit data

Modify the GCD software routine and hardware accelerator for 64-bit input operands and compare their performance.

20.9.5   GCD accelerator with 128-bit data

Modify the GCD software routine and hardware accelerator for 128-bit input operands and compare their performance.

20.9.6   GCD by Euclid's algorithm

The Euclid algorithm is an alternative method to obtain GCD and involves division (actually modulo operation). For simplicity, assume that $\gcd(x, 0) = x$. Euclid's algorithm can be written as

$$\gcd(a, b) = \begin{cases} a & \text{if } b = 0 \\ \gcd(b, a \bmod b) & \text{otherwise} \end{cases}$$

Derive a software routine and a hardware accelerator based on this algorithm and compare their performance with the binary GCD algorithm. The hardware accelerator needs to incorporate a custom circuit to perform the modulo operation.

20.10   COMPLETE PROGRAM LISTING

Listing 20.6   chu_main_gcd_test.c

```
/***********************************************************************
 *
 * Module:   GCD function prototypes and main
 * File:     chu_main_gcd_test.c
 * Purpose:  software- and hardware-accelerator-based GCD routines
 * IP core base addresses:
 *    - G_ENGINE_BASE: GCD engine
 *
 ***********************************************************************
 *  Register map
 ***********************************************************************
 * Write (data from cpu):
 *    offset 0
 *      * bits 31-0: 32-bit a
 *    offset 1
 *      * bits 31-0: 32-bit b
 *    offset 2
 *      * dummy data to start operation
 * Read (data to cpu):
 *    offset 3
 *      * bit 32: 1-bit ready signal
 *      * bits 15 - 0: 16-bit timer count
 *    offset 4
 *      * bits 31-0: 32-bit GCD result
 ***********************************************************************/
/* file inclusion */
/* General C library */
#include <stdio.h>
#include <stdlib.h>    // to use rand()
/* Altera-specific library */
#include <io.h>
#include <alt_types.h>
#include <sys/alt_timestamp.h>
#include "system.h"

/* address and field definition */
#define GCD_A_REG          0
#define GCD_B_REG          1
#define GCD_START_REG      2
#define GCD_STATUS_REG     3
#define GCD_R_REG          4
#define GCD_CNT_FLD        0x0000ffff
#define GCD_READY_BIT      0x80000000

/* macro definition */
#define even(x) (((x) & 1)==0)      // check whether x is an even number
```

```
/*************************************************************************
 * function: calc_gcd_soft()
 * purpose:   calculate GCD by binary GCD algorithm
 * argument:
 *   a, b: 2 operands of gcd function
 *   *ticks: # clock cycles calculated by alt_timestamp() function
 * return:
 *   - gcd(a,b)
 *   - *ticks updated
 * note:
 *   - call/return/execution of alt_timestamp() introduces extra overhead
 *   - use stop_stamp_timer()/start_stamp_timer() to pause the timer to
 *     get more accurate result
 *************************************************************************/
alt_u32 calc_gcd_soft(alt_u32 a, alt_u32 b, int *ticks)
{
  int n=0;
  alt_32 t0, t1, overhead;

  alt_timestamp_start();        // initialize time_stamp timer to 0
  /* determine the overhead of alt_timestamp() */
  t0 = alt_timestamp();
  t1 = alt_timestamp();
  overhead = t1 - t0;
  /* start actual measurement */
  t0 = alt_timestamp();
  while (1){
    if (a==b)
      break;
    if (even(a)){        // a even
      a = a>>1;
      if (even(b)){      // b even
        b = b>>1;
        n++;
      }
    } else {             // a odd
      if (even(b))       // b even
        b = b>>1;
      else {             // b odd
        if (a > b)
          a = a - b;
        else
          b = b - a;
      } // end else (b odd)
    } // end else (a odd)
  } // end while
  a = a<<n;
  t1 = alt_timestamp();
  *ticks = (int) (t1-t0-overhead);
  return(a);
}
```

```
/****************************************************************************
* function: calc_gcd_hard()
* purpose:   calculate GCD via hardware accelerator
* argument:
*   gcd_base: base address of GCD engine
*   a, b: 2 operands of gcd function
*   *clks:  # clock cycles from hardware accelerator
* return:
*   - gcd(a,b)
*   - *clks updated
* note:
****************************************************************************/
alt_u32 calc_gcd_hard(alt_u32 gcd_base, alt_u32 a, alt_u32 b, int *clks)
{
  alt_u32 status, result;

  IOWR(gcd_base, GCD_A_REG, a);      // write a
  IOWR(gcd_base, GCD_B_REG, b);      // write b
  IOWR(gcd_base, GCD_START_REG, 0); // write dummy data to start operation
  while(1){
  /* loop until GCD data is ready */
    status=IORD(gcd_base, GCD_STATUS_REG);
    if (status & GCD_READY_BIT)
      break;
  }
  *clks = (int) (status & GCD_CNT_FLD);
  result=IORD(gcd_base, GCD_R_REG);
  return(result);
}
```

```c
/************************************************************************
 * function:  main()
 * purpose:   calculate gcd and collect statistics for randomly selected
 *            numbers by software routine and hardware accelerator
 * note:
 ************************************************************************/
int main(void)
{
  alt_u32 a, b, rs, rh;
  int i;
  int s=0, c=0;
  int samples=10000;
  int s_ticks, clks;

  printf("GCD test:\n\n");
  srand(100);                     // set the random number seed
  for (i=0; i<samples; i++){      // obtain gcd for 10,000 random samples
    a = rand() + 1;               // +1 to avoid 0
    b = rand() + 1;
    rs = calc_gcd_soft(a, b, &s_ticks);
    rh = calc_gcd_hard(G_ENGINE_BASE, a, b, &clks);
    if (rs!=rh)                   // sanity check
      printf("Inconsistency: gcd(%d,%d)=%d/%d\n", a, b, rs, rh);
    s = s + s_ticks;
    c = c + clks;
  }
  printf("average clocks (soft/hard): %d/%d\n", s/samples, c/samples);
  printf("hardware accelerator speedup: %d\n", s/c);
}
```

CHAPTER 21

MANDELBROT SET FRACTAL ACCELERATOR

The *Mandelbrot set* can be used to generate interesting fractal images. The algorithm to determine whether a point is within the set is computation intensive and thus drawing the entire screen requires a significant amount of time. In this chapter, we design a system that can zoom in on a specific area of the set and display the fractal on a VGA screen. We implement the same algorithm using a software routine and a hardware accelerator and compare their performances.

21.1 INTRODUCTION

A *fractal* is a fragmented geometric shape that can be split into smaller parts, each of which has a similar (but not necessarily identical) appearance to the full shape. The boundary of the *Mandelbrot set* forms a fractal. A sample zoom sequence of the Mandelbrot set is shown in Figure 21.1. The initial image of the Mandelbrot set is shown at the top left. A small region is selected and magnified (i.e., "zoomed in") to form the second image at the top right. The process is repeated two more times and the images are shown on the bottom. Note that the set has an elaborate boundary and the boundary does not simplify at any given magnification.

Despite the complexity of the image, the fractal is usually governed by simple mathematical equations, as in the Mandelbrot set. Studying the mathematical properties of fractals is beyond the scope of this book. We simply treat these equations as an algorithm to generate fractal images and discuss the computation procedure in the following subsections.

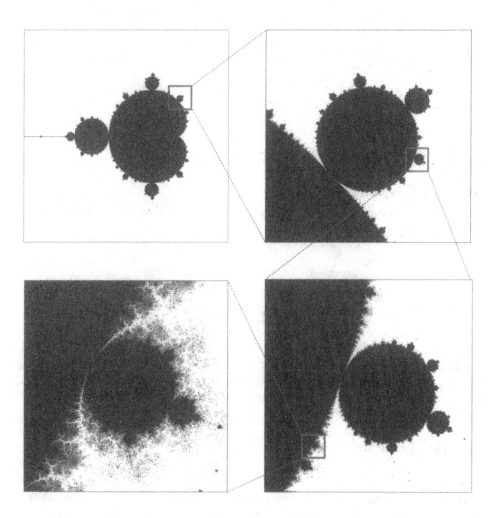

Figure 21.1 Sample zoom sequence of the Mandelbrot set.

21.1.1   Overview of the Mandelbrot set

The Mandelbrot set, named after mathematician Benot B. Mandelbrot, is a set of complex numbers that has a convoluted fractal boundary when plotted. Mathematically, it is defined as follows. Consider a sequence defined in the complex plane:

$$z_n = \begin{cases} 0 & \text{if } n = 0 \\ z_{n-1}^2 + c & \text{otherwise} \end{cases}$$

For a given complex number c, c is in the Mandelbrot set if the absolute value of z_n is bounded as n goes to infinity.

Instead of complex numbers, we use a pair of real numbers in our discussion. The previous definition can be rewritten as follows. Consider a two-dimensional sequence:

$$(x_n, y_n) = \begin{cases} (0,0) & \text{if } n = 0 \\ (x_{n-1}^2 - y_{n-1}^2 + c_x, 2 * x_{n-1} * y_{n-1} + c_y) & \text{otherwise} \end{cases}$$

For a given pair (c_x, c_y), the pair is in the Mandelbrot set if $x_n^2 + y_n^2$ is bounded as n goes to infinity. For example, if (c_x, c_y) is $(1, 0)$, the sequence becomes $(0,0)$, $(1,0)$, $(2,0)$, $(2,0)$, $(5,0)$, $(26,0)$, ..., which tends to infinity, and thus the point does not belong to the Mandelbrot set. On the other hand, if (c_x, c_y) is $(-1,0)$, the sequence becomes $(0,0)$, $(-1,0)$, $(0,0)$, $(-1,0)$, ..., which is clearly bounded, and thus the point belongs to the Mandelbrot set. The Mandelbrot set fractal is generated in the (c_x, c_y) plane (i.e., c_x as the x axis and c_y as the y axis) by plotting each point in the image plane. The points are colored black if they are in the set and white if they are not in the set. The resulting image is shown in Figure 21.2.

21.1.2   Determination of a Mandelbrot set point

The previous bound sequence generated by $(-1, 0)$ is extremely simple and a special case. Mathematically, there is no formula to determine whether a point is in the set and we would have to perform an infinite number of iterations. A special property can partially help us ease the computation. It states that the sequence will diverge and grow infinitely large once $x_n^2 + y_n^2$ ever gets to be greater than 4.0. This is sometimes known as the *escape condition*. For example, let (c_x, c_y) be $(0.2, 0.7)$. The sequence becomes $(0.00, 0.00)$, $(0.20, 0.70)$, $(-0.25, 0.98)$, $(-0.70, 0.21)$, $(0.64, 0.41)$, $(0.45, 1.22)$, $(-1.10, 1.80)$, and $(-1.82, -3.23)$, which reaches the escape condition in the seventh iteration, and thus the point is not in the set.

There is no simple test to ensure that a point is in the set and to determine whether or when a sequence diverges. In practice, we usually set an upper limit for the number of iterations and declare that a point is in the set if the iteration sequence reaches the limit without diverging. A computer-generated Mandelbrot set fractal is thus at best an approximation of the actual set. For example, if a point diverges after 150 iterations, it is considered to be in the set if the upper limit is set to 100 and considered not to be in the set if the upper limit is set to 200. Some points in the boundary regions may take many iterations before they diverge and we must set the upper limit to a larger number to obtain a more detailed image.

We can derive an algorithm to perform the iteration sequence for a given point. Assume that the value of the point is (`cx`,`cy`) and the maximal number of iteration is `max_itr`. The pseudo-code of this function is

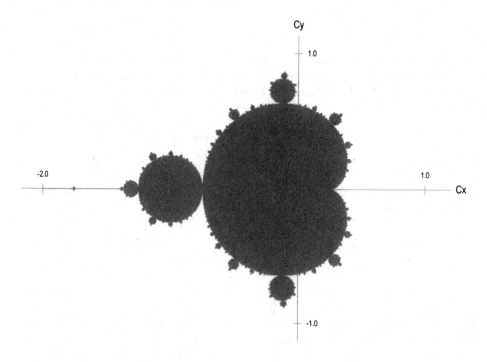

Figure 21.2   Mandelbrot set.

```
calc_frac_point(cx, cy, max_itr){
   x = cx;
   y = cy;
   itr=0;
   do {
      xx = x * x;
      yy = y * y;
      xy = x * y;
      y =  x2 - y2 + cx;
      x = xy*2 + cy;
      itr++;
   } while ((xx+yy)<4.0 && itr<max_itr);
   return(itr);
}
```

If the sequence does not diverge, the while loop ends when **max_itr** iterations are reached and the function returns **max_itr**. In this case, we declare that (**cx,cy**) is in the set.

The **calc_frac_point()** function is computationally intensive. It requires three multiplications in each iteration and may need up to 3***max_itr** multiplications if the sequence does not diverge.

21.1.3   Coloring scheme

For our purposes, the Mandelbrot set is used as a mechanism to generate visual fractal patterns. For a more interesting image, we can "color" the region just outside the Mandelbrot set. For a point outside the set, `calc_frac_point()` returns a value that is smaller than `max_itr`. We can assign a color according to the value and the resulting image can reveal the delicate characteristic in the set's boundary region.

Assume the value of `max_itr` is M and `calc_frac_point()` returns n and the value of color i is C_i. In the original Mandelbrot set fractal of Figure 21.2, the escaped points are assigned to white and the non-escaped points are assigned to black. The color map function is

$$color = \begin{cases} C_{black} & \text{if } n = M \\ C_{white} & \text{otherwise} \end{cases}$$

A more interesting "checkerboard" pattern can be obtained by assigning the black and white colors alternatively:

$$color = \begin{cases} C_{black} & \text{if } n = M \\ C_{white} & \text{if } n < M \text{ and } n \text{ is even} \\ C_{black} & \text{if } n < M \text{ and } n \text{ is odd} \end{cases}$$

A graphic display can show different intensity levels of a color. For example, the intensity level can range from 0 to 15 for a four-bit representation. If we assume that C_{black} and C_{white} represent two extreme values of the range of a monochrome display, we can use different levels of shades to show the "contour" around the Mandelbrot set's boundary by assigning the intensity levels according to the number of iterations:

$$color = \begin{cases} C_{black} & \text{if } n = M \\ (1 - a) * C_{white} + a * C_{black} & \text{if } n < M \text{ and } a = \frac{n}{M} \end{cases}$$

We can also reverse the intensity level around the boundary to make the fractal look like "lightning." The color function becomes:

$$color = \begin{cases} C_{black} & \text{if } n = M \\ a * C_{white} + (1 - a) * C_{black} & \text{if } n < M \text{ and } a = \frac{n}{M} \end{cases}$$

More striking fractal images can be obtained with a color graphic display. A simple way to do it is using a modulo function. Assume that the display supports K colors. The color function becomes

$$color = \begin{cases} C_{black} & \text{if } n = M \\ C_p & \text{if } n < M \text{ and } p = n\%K \end{cases}$$

A representative colored fractal image is shown in Figure 21.3. The previous "intensity" schemes can also be combined to generate more interesting images.

21.1.4   Generation of a fractal image

A fractal image can be obtained by selecting and magnifying a specific area in the c_x-c_y plane, as demonstrated in Figure 21.1. To generate an image in a graphic display, we need to map the corresponding (c_x, c_y) values to the pixel location and determine the escape iteration and color accordingly. Assume that the size of the display is H_{pix} pixels by V_{pix} pixels and the bottom left corner and the top right corner of the selected area are (c_{x0}, c_{y0}) and (c_{xn}, c_{yn}). The horizontal and vertical

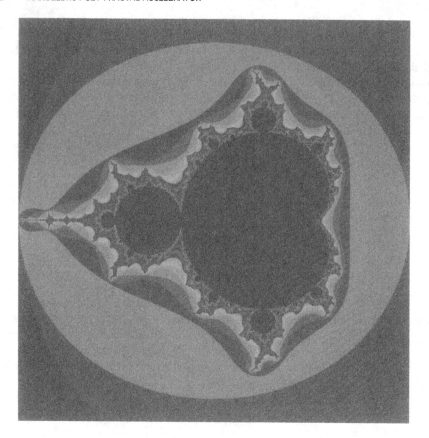

Figure 21.3   Colored Mandelbrot set.

distances between successive pixels are $\frac{c_{xn}-c_{x0}}{H_{pix}}$ and $\frac{c_{yn}-c_{y0}}{V_{pix}}$. If the shape of the graphic display and the selected area are similar, the two distances are identical. It is defined as *delta* in our discussion. The pseudo code of generating a fractal image is

```
cy = cy0;
for(j=0; j<HPIX; j++){
  cx = cx0;
  for(i=0; i<VPIX; i++){
    itr = calc_frac_point(cx, cy, max_itr);
    color = assign_color(itr);
    display_pix(i, j, color);
    cx = cx + delta;
  }
  cy = cy + delta;
}
```

21.2 FIXED-POINT ARITHMETIC

The Mandelbrot set deals with the points in a complex plane, which are represented as a pair of real numbers, and thus a proper data type should be selected to accommodate the needed computation. Ideally, the *floating-point* data type should be used since it provides a large range and good precision. However, its complexity can severely degrade the software performance and complicate the design of a dedicated hardware accelerator. An alternative is to use the *fixed-point* data type, which requires only simple modification over the integer data type.

The fixed-point data type is essentially an integer that is scaled by a specific factor. For explanation purposes, let us first consider the fixed-point data type in the decimal number system. For example, we assume a fixed-point data type with five decimal digits and a scaling factor of 10^{-3}. The value 12.345 can be represented as 12345 (i.e., $12345 * 10^{-3}$). The scaling factor corresponds to the exponent of the floating-point data type. Unlike floating-point data types, the scaling factor is the same for all entities of the same type and does not change during the computation. This particular data type can be interpreted as five-digit integer type with an implicit decimal point after the second most significant digit. Thus, the value 12345 becomes 12.345. One way to represent a fixed-point data type is the $\mathbf{Q}m.f$ notation, in which m represents the number of integer digits and f represents the number of fractional digits. The previous data type can be represented as the $\mathbf{Q}2.3$ format.

To add or subtract two numbers in the same fixed-point type, it is sufficient to add or subtract the underlying integers and keep their common scaling factor. The result is in the same type, as long as no overflow occurs. In other words, if the two numbers are in $\mathbf{Q}m.f$ format, the sum will also be in $\mathbf{Q}m.f$ format. The addition and subtraction thus are actually identical to those in the integer data type except that we assume that there is an implicit decimal point after the mth digits.

To multiply two fixed-point numbers, it suffices to multiply the two underlying integers. Unlike the addition operation, the resulting scaling factor changes. It becomes *the product of two scaling factors of the two numbers*. For example, assume that 12345 and 00025 are in $\mathbf{Q}2.3$ format, which are interpreted as $12345 * 10^{-3}$ and $00025 * 10^{-3}$ (i.e., 12.345 and 0.025). The five-digit integer multiplication leads to a ten-digit product 0000308625, which is interpreted as $308625 * 10^{-6}$ (i.e., 0.308625). In other words, if the two numbers are in $\mathbf{Q}m.f$ format, the product will be in $\mathbf{Q}2m.2f$ format. For fixed-point operation, it is desirable to have the product in the same $\mathbf{Q}m.f$ format. This can be done by trimming the m most significant digits and f least significant digits. In the previous example, the trimmed product becomes 00308, which is interpreted as $308 * 10^{-3}$ (i.e., 0.308). When using a fixed-point data type, we usually know the range of operation in advance and select a format to ensure that no overflow occurs after multiplication i.e., the m most significant digits in the product are always 0's. Thus, we only need to performed the shift-right operation to remove the f least significant digits.

The main advantage of the fixed-point data type is in its implementation. We can use the same integer arithmetic operations, plus shift operation, for the real numbers. On the other hand, the fixed-point data type operations tend to lose precision even if all operations are within the range. For example, the previous fixed-point multiplication can be rewritten as $12345 * 10^{-3} * 25 * 10^{-3} = 308 * 10^{-3}$, in which the product only has three digits of accuracy. If the floating-point data type is used, the multiplication can be expressed as $12345*10^{-3} * 25*10^{-3} = 30862*10^{-5}$, in which the product maintains five digits of accuracy. If the computation involves a larger number of iterations, the inaccuracy may be accumulated and may lead to a much larger error.

In a digital system, the fixed-point data type is based on the binary system. The basic properties are similar to those in the decimal system except that binary bits are used in the places of decimal digits. In the remaining chapter, both software and hardware implementation are based on the binary representation.

21.3 SOFTWARE IMPLEMENTATION OF CALC_FRAC_POINT()

The C language has no inherent support for the fixed-point data type and thus we must manually manipulate the proper integer operation according to the interpretation of a specific fixed-point format. Since the native data width of the Nios II processor is 32 bits, we choose this as the total width of the fixed-point data. Recall that the area of interest of the Mandelbrot set resides within the $(-2,+2)$ range of both axes and the iteration stops when the escape condition, $x_n^2 + y_n^2 > 4.0$, is reached. The largest value of intermediate calculation, which is likely to be $x_n^2 + y_n^2$, should be smaller than 8.0. Based on this observation, we select **Q**4.28 for our implementation, in which the four MSBs are interpreted as the integer portion and 28 LSBs are interpreted as the faction portion. In this format, the most negative value is -8.0 (i.e., $1000 \cdots 00$) and the most positive value is close to $+8.0$ (i.e., $0111 \cdots 11$, which is $2^2 + 2^1 + 2^0 + 2^{-1} + 2^{-2} + 2^{-3} + \cdots + 2^{-28}$).

To emphasize the new interpretation of the fixed-point representation, we define a new 32-bit data type for **Q**4.28 format and a new 64-bit data type for **Q**8.56 format:

```
typedef alt_32 fixed428;
typedef alt_64 fixed856;
```

The pseudo code of `calc_frac_point()` of Section 21.1.1 can be modified to accommodate the new format and the code is shown in Listing 21.1.

Listing 21.1

```
int calc_frac_point_soft(fixed428 cx, fixed428 cy, alt_u16 max_itr)
{
    fixed428 x, y, xx, yy, xy2;
    fixed856 xx_raw, yy_raw, xy_raw;
    int itr;

    x = cx;
    y = cy;
    itr=0;
    do {
        /* Q4.28 multiplications */
        xx_raw = (fixed856)(x) * (fixed856)(x);
        xx = (fixed428)(xx_raw >> 28);
        yy_raw = (fixed856)(y) * (fixed856)(y);
        yy = (fixed428)(yy_raw >> 28);
        xy_raw = (fixed856)(x) * (fixed856)(y);
        xy2 = (fixed428)(xy_raw >> 27);  // 2* is same as <<1
        /* iteration equation */
        x = xx - yy + cx;
        y = xy2 + cy;
```

```
     itr++;;
   } while (((xx+yy)<0x40000000) && (itr<max_itr));
   return(itr);
}
```

The first part of the loop performs multiplications in **Q**4.28 format. The two operands are extended to 64-bit **Q**8.56 format and multiplied as two 64-bit integers. The result is then converted back to **Q**4.28 format by shifting the product to the right by 28 positions. The *2 operation in calculation of $x * y * 2$ corresponds to shift $x * y$ to the left by one position. It is merged with the shift-right operation and becomes the xy_raw>>27 expression.

21.4 HARDWARE IMPLEMENTATION OF CALC_FRAC_POINT()

21.4.1 ASMD chart

The calc_frac_point_soft() function in Section 21.3 can be converted to an ASMD chart and then realized by an FSMD in hardware. The resulting ASMD chart is shown in Figure 21.4. The main computation is done in the op state and it corresponds to the C statements within the while loop. The FSM returns to the idle state when the escape condition is met or the loop reaches the maximal number of iterations.

21.4.2 HDL implementation

After constructing the ASMD chart, we can derive the HDL code accordingly. The code is shown in Listing 21.2. For clarity, we isolate the fixed-point multiplication portion and code them in a separate segment.

Listing 21.2 Fractal engine

```
1 library ieee;
  use ieee.std_logic_1164.all;
  use ieee.numeric_std.all;
  entity frac_engine is
     generic(
6       W: integer := 32;   -- width (# bits) of Qm.f format
        M: integer := 4     -- # of bits in m
        );
     port(
        clk, reset: in std_logic;
11       frac_start: in std_logic;
        cx, cy: in std_logic_vector(W-1 downto 0);
        max_it: in std_logic_vector(15 downto 0);
        iter: out std_logic_vector(15 downto 0);
        frac_ready, frac_done_tick: out std_logic
16       );
  end frac_engine;

  architecture arch of frac_engine is
     constant F: integer := W - M; -- # of bits in fraction
21    type state_type is (idle, op);
     signal state_reg, state_next: state_type;
     signal it_reg, it_next: unsigned(15 downto 0);
     signal xx_raw, yy_raw, xy_raw: signed(2*W-1 downto 0);
```

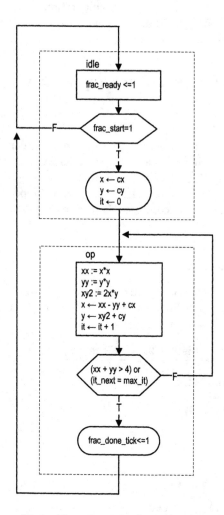

Figure 21.4   Fractal point calculation ASMD chart.

```vhdl
        signal xx, yy, xy2: signed(W-1 downto 0);
26      signal x_reg, x_next, y_reg, y_next: signed(W-1 downto 0);
        signal cx_reg, cx_next: signed(W-1 downto 0);
        signal cy_reg, cy_next: signed(W-1 downto 0);
        signal escape: std_logic;
   begin
31      -- FSMD state & data registers
        process(clk, reset)
        begin
           if reset='1' then
              state_reg <= idle;
36            it_reg  <= (others=>'0');
              x_reg   <= (others=>'0');
              y_reg   <= (others=>'0');
              cx_reg  <= (others=>'0');
              cy_reg  <= (others=>'0');
41         elsif (clk'event and clk='1') then
              state_reg <= state_next;
              it_reg  <= it_next;
              x_reg   <= x_next;
              y_reg   <= y_next;
46            cx_reg  <= cx_next;
              cy_reg  <= cy_next;
           end if;
        end process;

51      -- fixed-point multiplications
        xx_raw <= x_reg*x_reg;                           -- in Q2m.2f
        xx <= xx_raw((2*W-1)-M downto F);                -- back to Qm.f
        yy_raw <= y_reg*y_reg;                           -- in Q2m.2f
        yy <= yy_raw((2*W-1)-M downto F);                -- back to Qm.f
56      xy_raw  <= x_reg*y_reg;                          -- xy in Q2m.2f
        xy2 <= xy_raw((2*W-1)-M-1 downto F-1);  -- 2xy in Qm.f
        -- escape condition
        escape <= '1' when (xx + yy > x"40000000") else '0';

61      -- FSMD next-state logic
        process(state_reg,it_reg,it_next,x_reg,y_reg,cx_reg,cy_reg,
                max_it,cx,cy,xx,yy,xy2,frac_start,escape)
        begin
           state_next <= state_reg;
66         it_next <= it_reg;
           x_next <= x_reg;
           y_next <= y_reg;
           cx_next <= cx_reg;
           cy_next <= cy_reg;
71         frac_ready <= '0';
           frac_done_tick <= '0';
           case state_reg is
              when idle =>
                 frac_ready <= '1';
76               if frac_start='1' then
                    x_next <= signed(cx);
                    y_next <= signed(cy);
                    cx_next <= signed(cx);
                    cy_next <= signed(cy);
81                  it_next <= (0=>'0',others=>'0');
                    state_next <= op;
                 end if;
              when op =>
                 x_next <= xx - yy + cx_reg;
```

```
86            y_next <= xy2 + cy_reg;
              it_next <= it_reg +  1;
              if (escape='1' or it_next=unsigned(max_it)) then
                 state_next <= idle;
                 frac_done_tick <= '1';
91            end if;
        end case;
     end process;
     iter <= std_logic_vector(it_reg);
  end arch;
```

21.5   MANDELBROT SET FRACTAL ACCELERATOR IP CORE DEVELOPMENT

21.5.1   Avalon interface

We can add a wrapping circuit for the Mandelbrot set fractal accelerator and create an SOPC component. It includes an Avalon MM slave interface to interact with the host and a clock input interface for system clock.

21.5.2   Register map

A Nios II processor interacts with the fractal accelerator by passing the coordinates of (c_x, c_y) and the maximal number of iterations and retrieving the calculated result. In addition, it initiates the operation by writing dummy data to a specific address and checks the fractal accelerator status for completion. The registers, their address offsets, and fields are:
- Write addresses (data from cpu)
  - offset 0 (cx register)
    * bits 31 to 0: 32-bit c_x input
  - offset 1 (cy register)
    * bits 31 to 0: 32-bit c_y input
  - offset 2 (max_it register)
    * bits 15 to 0: 16-bit maximal iteration value
  - offset 3
    * dummy data write to start operation
- Read addresses (data to cpu)
  - offset 0 (itr_out register)
    * bit 16: 1-bit fractal accelerator ready signal
    * bits 15 to 0: 16-bit iteration value

21.5.3   Wrapped Mandelbrot set fractal accelerator

The wrapped Mandelbrot set fractal accelerator contains three registers to store the input values of c_x, c_y, and maximal iterations and write decoding logic to generate enable signals. The HDL code is shown in Listing 21.3.

Listing 21.3 Fractal engine Avalon interface

```vhdl
library ieee;
use ieee.std_logic_1164.all;
use ieee.numeric_std.all;
entity chu_avalon_frac is
   port (
      clk, reset: in  std_logic;
      frac_address: in std_logic_vector(1 downto 0);
      frac_write: in std_logic;
      frac_chipselect: in std_logic;
      frac_writedata: in std_logic_vector(31 downto 0);
      frac_readdata: out std_logic_vector(31 downto 0)
   );
end chu_avalon_frac;

architecture arch of chu_avalon_frac is
   signal frac_start: std_logic;
   signal cx_reg, cy_reg: std_logic_vector(31 downto 0);
   signal max_it_reg: std_logic_vector(15 downto 0);
   signal iter_out: std_logic_vector(15 downto 0);
   signal frac_ready: std_logic;
   signal wr_en, wr_cx, wr_cy, wr_max: std_logic;
begin
   -- instantiation
   frac_unit: entity work.frac_engine(arch)
      port map(clk=>clk, reset=>reset,
               frac_start=>frac_start,
               cx=>cx_reg, cy=>cy_reg,
               max_it=>max_it_reg,
               iter=>iter_out,
               frac_ready=>frac_ready,
               frac_done_tick=>open);
   -- registers
   process (clk, reset)
   begin
      if reset='1' then
         cx_reg <= (others=>'0');
         cy_reg <= (others=>'0');
         max_it_reg <= (others=>'0');
      elsif (clk'event and clk='1') then
         if wr_cx='1' then
            cx_reg <= frac_writedata;
         end if;
         if wr_cy='1' then
            cy_reg <= frac_writedata;
         end if;
         if wr_max='1' then
            max_it_reg <= frac_writedata(15 downto 0);
         end if;
      end if;
   end process;
   -- write encoding
   wr_en <= '1' when frac_write='1' and frac_chipselect='1' else '0';
   wr_cx <= '1' when frac_address="00" and wr_en='1' else '0';
   wr_cy <= '1' when frac_address="01" and wr_en='1' else '0';
   wr_max <= '1' when frac_address="10" and wr_en='1' else '0';
   frac_start <= '1' when frac_address="11" and wr_en='1' else '0';
   -- read
   frac_readdata <= x"000" & "000" & frac_ready & iter_out;
end arch;
```

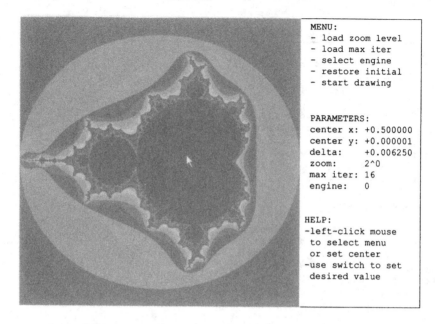

```
MENU:
- load zoom level
- load max iter
- select engine
- restore initial
- start drawing


PARAMETERS:
center x: +0.500000
center y: +0.000001
delta:    +0.006250
zoom:     2^0
max iter: 16
engine:   0


HELP:
-left-click mouse
 to select menu
 or set center
-use switch to set
 desired value
```

Figure 21.5   Sketch of the fractal graphic user interface.

Base on this top-level HDL file, we can create a new SOPC component in Component Editor and integrate it into a Nios II system.

21.6   TESTING PROGRAM

21.6.1   Fractal graphic user interface

The system derived in Section 16.10.2 includes the fractal hardware accelerator core (named f_engine) and can be used for testing. The fractal testing program verifies the operation of the hardware accelerator and compares its performance with the software routine. We create a simple graphic user interface on a VGA display and use the mouse and switches to control the fractal generations. The sketch of the interface is shown in Figure 21.5. The 640-pixel-by-480-pixel VGA display is divided into two areas. The left portion is a 480-by-480 square for the fractal graphic and the right portion is a 240-by-480 rectangle for textual information. The text consists of an action menu (top), relevant Mandelbrot set fractal parameters (middle), and simple help information (bottom).

We need the following information to generate a fractal graphic:

- The coordinates of the fractal's center, $(c_{x(center)}, c_{y(center)})$.
- The zoom factor, m, which represents the magnification over the initial fractal.
- The maximal number of iterations.
- The "engine" used for the computation, which can be the software routine or the hardware accelerator.

The cx0, cy0, and delta values, which are the bottom left coordinates and the distance between two successive points, of the pseudo code in Section 21.1.4 can be derived from $c_{x(center)}$, $c_{y(center)}$, and m. We create a fractal on the VGA display that covers the entire Mandelbrot set during initialization. The ranges of the c_x and c_y axes of the initial fractal are between $[-2.0, 1.0]$ and $[-1.5, 1.5]$ and the resulting graph is similar to that in Figure 21.2. The distance between two points (i.e., delta) is $\frac{3.0}{480}$. To magnify a region in the fractal graph by a factor of m corresponds to reduce the distance between two points by a factor of m. The corresponding delta becomes $\frac{3.0}{480*m}$. Once delta is determined, cx0 and cy0 can be derived, which are $(c_{x(center)} - 240 * \frac{3.0}{480*m})$ and $(c_{y(center)} - 240 * \frac{3.0}{480*m})$, respectively.

The interface uses a mouse and switches to set parameters and to control the fractal drawing. The basic procedure is:

- Move the mouse pointer to the graphic area and click the left mouse button to select the center coordinates of the new fractal.
- Use the slide switches to set the desired zoom level and then click the left mouse button on the menu's "load zoom level" region. If the binary value of the slide switches is sw, the magnification factor (m) is set to 2^{sw}. The maximal value of m is limited to 2^{20}.
- Use the slide switches to set the desired maximal number of iterations and then click the left mouse button on the menu's "load max iter" region. If the binary value of the slide switches is sw, the maximal number of iterations is set to $10 * sw$. The maximal value of iterations is limited to 4095.
- Use the switch to set the desired "computation engine" and then click the left mouse button on the menu's "select engine" region. If the slide switch 0 (labeled sw0 on the DE1 board) is 0, the software routine is used. If it is 1, the hardware accelerator is used.
- Check the parameters in the middle part and, if they are correct, click the left mouse button on the menu's "start drawing" region to initiate the operation.

There is an additional "restore initial" item on the menu. Clicking the button on this region restores the initial parameter setting.

Since the fractal engine is mainly used to demonstrate the use of the hardware accelerator, it is unlikely to be used in other applications. We do not derive a separate device driver library and simply integrate the I/O access in the testing program. The remaining section discusses the testing and user interface routines.

21.6.2 Fractal hardware accelerator engine control routine

Using fractal hardware accelerator to perform the calc_frac_point() function requires a simple routine to write and read its I/O registers. The code is show in Listing 21.4.

Listing 21.4

```
#define CX_REG              0
#define CY_REG              1
#define MAX_ITR_REG         2
#define FRAC_START_REG      3
#define ITR_DATA_REG        0
#define ITR_FIELD           0x0000ffff
#define FRAC_READY_BIT      0x00010000

int calc_frac_point_hard(alt_u32 frac_base, fixed428 cx, fixed428 cy,
                         alt_u16 max_itr)
{
```

```
alt_u32 data;
int itr;

IOWR(frac_base, CX_REG, (alt_u32) cx);
IOWR(frac_base, CY_REG, (alt_u32) cy);
IOWR(frac_base, MAX_ITR_REG, (alt_u32) max_itr);
IOWR(frac_base, FRAC_START_REG, 0); // write dummy to start operation
while(1){
  data=IORD(frac_base, ITR_DATA_REG);
  if (data & FRAC_READY_BIT)            // check 17th bit for ready signal
    break;
}
itr = data & ITR_FIELD;                 // get 16-bit result
return(itr);
}
```

21.6.3 Fractal drawing routine

The fractal drawing routine is based on the pseudo code in Section 21.1.4. The **engine** argument specifies whether the software routine or the hardware accelerator is selected to perform the computation.

The DE1 board supports 12-bit color depth. However, due to the memory constraint, the data width of our VGA controller is limited to 8 bits and the colors are arranged in the form of $r_3r_2r_1g_3g_2g_1b_3b_2$, as discussed in Section 17.4. Thus only 2^8 colors can be selected for the fractal image. The routine uses the following color function:

$$color = \begin{cases} C_{black} & \text{if } n = M \\ C_p & \text{if } n < M \text{ and } p = n\%2^8 \end{cases}$$

Note that the $n\%2^8$ operation corresponds to extract the 8 LSBs of n. We also rearrange the color representation as $b_3b_2r_3r_2r_1g_3g_2g_1$ to make smoother color transitions. The code is shown in Listing 21.5.

Listing 21.5

```
void draw_fractal(alt_u32 vga_grf_base, alt_u32 frac_base, int engine,
          fixed428 cx0, fixed428 cy0, fixed428 delta, int max_itr)
{
  int i, j;
  int itr;
  alt_u8 color;
  fixed428 cx, cy;

  cy = cy0;
  for(j=0; j<FPIX; j++){
    cx = cx0;
    for(i=0; i<FPIX; i++){
      /* calculate one point */
      if (engine==0)
        itr = calc_frac_point_soft(cx, cy, max_itr);
      else
        itr = calc_frac_point_hard(frac_base, cx, cy, max_itr);
      /* coloring scheme */
      if (itr==max_itr)
        color = 0x00;  // no escape; black
      else
        color = (alt_u8)((0xd0 & itr)>>6) + (alt_u8)(itr<<2) ;
```

```
      /* write a pixel on VGA */
      vgag_wr_pix(vga_grf_base, i, 479-j, color);
      cx = cx + delta;
    } // end for i
    cy = cy + delta;
  } // end for j
}
```

21.6.4   Text panel display routines

The right text panel can be constructed with the bit-mapped text driver routines of Section 17.6.4. We divide the text generation operation into two parts and create two functions. The disp_msg() function generates the fixed portion of the text and the disp_param() function processes the parameter values, which are updated according to mouse activities. The code is shown in Listing 21.6.

Listing 21.6

```
#define f2f(x) (x)*(1.0/(1<<28))   // convert Q4.28 to floating
. . .
void disp_msg(alt_u32 vga_grf_base)
{
  char para_msg[]=
    "PARAMETERS        \n"
    "center x:         \n"
    "center y:         \n"
    "delta    :        \n"
    "zoom     :        \n"
    "max iter:         \n"
    "engine   :        \n";
  char menu_msg[]=
    "MENU:             \n"
    "- load zoom level \n"
    "- load max iter   \n"
    "- select engine   \n"
    "- restore initial \n"
    "- start drawing   \n";
  char help_msg[]=
    "HELP:             \n"
    "-left-click mouse \n"
    " to select menu   \n"
    " or set center    \n"
    "-use switch to set \n"
    " desired value    \n";

  vgag_wr_bit_str(vga_grf_base, 485, 0, menu_msg, 0xff, 1);
  vgag_wr_bit_str(vga_grf_base, 485, 16*10, para_msg, 0xff, 1);
  vgag_wr_bit_str(vga_grf_base, 485, 16*20, help_msg, 0xff, 1);
}

void disp_param(alt_u32 vga_grf_base, fixed428 cx_c, fixed428 cy_c,
                fixed428 delta, int zoom, int max_itr, int engine)
{
  char s[]="123456789";
  char erase[]= "\x7f\x7f\x7f\x7f\x7f\x7f\x7f\x7f\x7f";
  int x_offset = 485+8*10;
  int y0 = 11;
  int i;
```

```
/* clear old field */
for (i=0; i<6; i++)
  vgag_wr_bit_str(vga_grf_base, x_offset, 16*(y0+i), erase, 0x00, 1);
/* draw new values */
sprintf(s, "%+2.6f", f2f(cx_c));
vgag_wr_bit_str(vga_grf_base, x_offset, 16*y0, s, 0xff, 1);
sprintf(s, "%+2.6f", f2f(cy_c));
vgag_wr_bit_str(vga_grf_base, x_offset, 16*(y0+1), s, 0xff, 1);
sprintf(s, "%+2.6f", f2f(delta));
vgag_wr_bit_str(vga_grf_base, x_offset, 16*(y0+2), s, 0xff, 1);
sprintf(s, "2^%d", zoom);
vgag_wr_bit_str(vga_grf_base, x_offset, 16*(y0+3), s, 0xff, 1);
sprintf(s, "%d", max_itr);
vgag_wr_bit_str(vga_grf_base, x_offset, 16*(y0+4), s, 0xff, 1);
sprintf(s, "%d", engine);
vgag_wr_bit_str(vga_grf_base, x_offset, 16*(y0+5), s, 0xff, 1);
}
```

The `disp_msg()` function simply writes three text strings to specific locations. The `disp_param()` function first erases the old values with the special `0x7f` character and then forms new strings using the `sprintf()` function and writes the strings to the designated locations.

21.6.5   Mouse processing routine

The fractal user interface is controlled by a mouse. The `proc_mouse()` function monitors the mouse activities and updates the parameters as needed. It exits when the "start drawing" item of the menu is selected. The code is shown in Listing 21.7.

Listing 21.7

```
void proc_mouse(alt_u32 vga_grf_base, alt_u32 ps2_base, alt_u32 sw_base,
                fixed428 *cx0, fixed428 *cy0, fixed428 *delta,
                int *max_itr, int *engine)
{
  alt_u32 sw;
  /* fractal related variables */
  static int zoom = 0;
  static int mitr = 30;
  static fixed428 cx_c = 0xf8000000;   // center of cx axis
  static fixed428 cy_c = 0x00000000;   // center of cy axis
  fixed428 cx_c_new, cy_c_new, delta_old, delta_new;
  /* mouse pointer related variables */
  static int px = 240;     // x coordinate of mouse
  static int py = 240;     // y coordinate of mouse
  int act, px_new, py_new;
  mouse_mv_type mv;
  /* mouse pointer pixel buffer */
  static alt_u8 bdata[20*12];
  bmp_type below={12, 20, bdata};

  delta_old = (WD/480) >> zoom;   //delta of current graph
  delta_new = delta_old;          // for panel display
  cx_c_new = cx_c;
  cy_c_new = cy_c;

  /*  draw initial pointer */
  ps2_flush_fifo(ps2_base);
  vrag_init_mouse_ptr(vga_grf_base,ps2_base,px,py,&MOUSE_BMP,&below);
  /* processing mouse activities */
```

```
while (1){
  act = vrag_move_mouse_ptr(vga_grf_base, ps2_base,
      px, py, &below, &px_new, &py_new, &MOUSE_BMP, &mv);
  if (act){
    px = px_new;
    py = py_new;
    if(mv.lbtn){ /* left button pressed */
      sw=IORD_ALTERA_AVALON_PIO_DATA(sw_base);
      /* update parameters according to mouse pointer position */
      if (px<480){                   // in drawing area,
        /* set new center of drawing */
        cx_c_new = cx_c + delta_old*(px-240);
        cy_c_new = cy_c + delta_old*(240-py);
        printf("center:(%+2.6f, %+2.6f) \n",
               f2f(cx_c_new), f2f(cy_c_new));
      }
      if (px>480 && 16<=py && py<32){   // in "load zoom" area
        if (sw>20)
          sw = 20;
        zoom = sw;
        delta_new = (WD/480)>>zoom;
        printf("zoom factor: %d (2^%d)\n", 1<<zoom, zoom);
      }
      if (px>480 && 32<=py && py<48){   // in "load max_itr" area
        mitr = (sw & 0x000003ff)*10;
        if (mitr>4095)
          mitr=4095;
        printf("max iteration: %d\n", mitr);
      }
      if (px>480 && 48<=py && py<64){   // in "select engine" area
        *engine = sw & 0x00000001;
        printf("engine #: %d\n", *engine);
      }
      if (px>480 && 64<=py && py<80){   // in "restore initial" area
        *engine = 0;
        zoom = 0;
        delta_new = (WD/480) >> zoom;
        mitr = 30;
        cx_c_new = 0xf8000000;   // -0.5 in Q4.28
        cy_c_new = 0x00000000;   //  0.0 in Q4.28
      }
      if (px>480 && 80<=py && py<96){   // in "start drawing" area
        printf("start drawing\n\n");
        break;
      }
      disp_param(vga_grf_base, cx_c_new, cy_c_new, delta_new,
                 zoom, mitr, *engine);
    } // end if (mv.lbtn)
  } // end if (act)
} // end while
/* update fractal parameters */
cx_c = cx_c_new;
cy_c = cy_c_new;
*delta = delta_new;
*max_itr = mitr;
*cx0 = cx_c_new - delta_new*240;
*cy0 = cy_c_new - delta_new*(479-240);
// restore image below mouse
vgag_wr_bitmap(vga_grf_base, px, py, &below, 0);
}
```

The main part of the function is the infinite while loop and a collection of static variables that keep track of various parameters. The loop monitors the mouse activities and continuously updates the mouse pointer's locations. When the left button is pressed, it determines the region that the pointer resides and acts accordingly. If the pointer is in the left graphic area, the center of the new fractal is calculated. If it is in the action menu region, the corresponding parameter is updated. The loop exits when the "start drawing" item of the menu is selected. The function then calculates the relevant parameters and passes them to the calling function.

21.6.6 Main program

The code of the main program is shown in Listing 21.8. It first initializes various components and draws the basic top-level fractal and then enters an infinite loop. Within the loop, proc_mouse() monitors the mouse activities and obtains the relevant parameters. When a user select the "start drawing" command, the function exits and draw_fractal() draws the complete fractal on the VGA screen.

Listing 21.8

```
int main(void)
{
  fixed428 cx0 = 0xe0000000;          // -2 in Q4.28
  fixed428 cy0 = 0xe8000000;          // -1.5 in Q4.28
  fixed428 delta = WD/FPIX;
  int max_itr = 30;
  int engine = 0;
  alt_u8 frac_msg[4]={sseg_conv_hex(15), 0x2f, 0x27, sseg_conv_hex(1)};

  sseg_disp_ptn(SSEG_BASE, frac_msg);     // show "Frcl" for Fractal
  printf("Fractal test \n\n");
  vga_clr_screen(VRAM_BASE,0);
  disp_msg(VRAM_BASE);
  disp_param(VRAM_BASE, cx0+240*delta, cy0+240*delta, delta,
             0, max_itr, engine);
  draw_fractal(VRAM_BASE, 0, engine, cx0, cy0, delta, max_itr);
  mouse_init(PS2_BASE);
  while (1){
    proc_mouse(VRAM_BASE, PS2_BASE, SWITCH_BASE,
               &cx0, &cy0, &delta, &max_itr, &engine);
    draw_fractal(VRAM_BASE, F_ENGINE_BASE, engine,
                 cx0, cy0, delta, max_itr);
  }
}
```

21.7 DISCUSSION

The fractal drawing system is a representative embedded SoPC application. The core of the algorithm is computation intensive. A custom hardware accelerator can be used to perform the main computation and a relatively slow processor can be used to handle the remaining housekeeping tasks, such as the user interface and video control.

The design in this chapter uses a hardware engine to determine the escape iterations. The engine consists of hardware multipliers and adders to perform the main computation within the loop body of the calc_frac_point() routine:

```
xx = x * x;
yy = y * y;
xy = x * y;
y  =  x2 - y2 + cx;
x  =  xy*2 + cy;
itr++;
```

The operation is performed in parallel and completed within a single clock cycle. By selecting a hardware or software engine in the testing program, we can visually observe the effect of acceleration. The speedup is due to concurrent computation and the elimination of overhead associated with processor execution, as discussed in Section 20.7.

It is very difficult to further improve the performance of the `calc_frac_point()` routine. However, since the fractal drawing is required to calculate the escaped iterations for tens of thousands pixels, this application exhibits a significant amount of "inherent parallelism." Multiple engines can be instantiated to work on multiple pixels concurrently. The number of engines is limited by the number of embedded multipliers within an FPGA device. Large modern FPGA devices contain hundreds of multipliers and thus can further speed the operation by several orders of magnitude.

For simplicity and clarity, a fixed-point format is used in our computation. This format has less precision when the value is small and the error can be accumulated through iterations. The resulting fractal image may show some "random" color dots and this effect becomes more noticeable when the range of the plot is small. The floating-point format can provide a much better result and leads to "smoother" images. Constructing a floating-point adder and multiplier is much more difficult and requires more resources. However, with proper design technique, such as pipelining, we can still obtain a hardware accelerator with good performance. On the other hand, since the Nios II processor does not have a floating-point unit, the floating-point addition and multiplication will be implemented by software functions. This will introduce a large overhead and further degrade the performance of a software-based drawing routine.

21.8 BIBLIOGRAPHIC NOTES

A general overview and discussion of the Mandelbrot set can be found on the Wikipedia web site. The new VHDL 2008 standard includes a fixed-point package, which provides more detailed information about this format and the implementation of relevant fixed-point operations.

21.9 SUGGESTED EXPERIMENTS

21.9.1 Hardware accelerator with one multiplier

Three 32-bit combinational multipliers are needed to perform three multiplications in the hardware accelerator in Section 21.4.1. An alternative is to use one multiplier and distribute the three multiplications over three clock cycles (i.e., three states in FSMD). Modify the FSMD, resynthesize the design, and verify its operation.

21.9.2 Hardware accelerator with modified escape condition

The main computation of Mandelbrot set iteration involves $x^2 - y^2$ and $x * y$, which are realized by three multiplications, $x * x$, $y * y$, and $x * y$, in our design. An alternative to reduce the number of multiplications is to calculate the $x^2 - y^2$ expression with $(x + y) * (x - y)$, which requires a single multiplication. To accommodate this scheme, the escape condition can be changed from $x^2 + y^2 < 4$ to $|x| + |y| < 4$. Modify the FSMD with the new algorithm, resynthesize the design, and verify its operation.

21.9.3 Hardware accelerator with Q4.12 format

To reduce the size of the combinational multiplier, we can use **Q4.12** format, which only requires 16-bit combinational multipliers. Redesign the hardware accelerator using this format, resynthesize the circuit, and compare the fractal image with the original **Q4.28** format. Examine the synthesis report to compare the number of embedded multipliers used in the two formats.

21.9.4 Hardware accelerator with multiple fractal engines

Follow the discussion in Section 21.7 and instantiate multiple "fractal engines" to calculate the escape iterations of multiple pixels. Since the EPC2C20 device on a DE1 board only has a limited number of multipliers, the **Q4.12** format should be used for the engine. Examine the synthesis report to determine the usage of embedded multipliers and instantiate as many fractal engines as possible. Redesign the hardware accelerator and the corresponding software routines. resynthesize the system and observe its performance.

21.9.5 "Burning-ship" fractal

A different fractal image can be obtained by modifying the original Mandelbrot iteration equation:

$$(x_n, y_n) = \begin{cases} (0,0) & \text{if } n = 0 \\ (x_{n-1}^2 - y_{n-1}^2 + c_x, 2 * |x_{n-1} * y_{n-1}| + c_y) & \text{otherwise} \end{cases}$$

The image looks like a ship on fire and thus is known as a "burning-ship" fractal. Modify the hardware accelerator and software routines for the new fractal and verify the operation.

21.9.6 Enhanced testing program

Several additional features can be added to the testing program:
- Add a new software "engine" that uses a floating-point data type to implement the `calc_frac_point()` routine.
- Add a status item that shows the speedup of the hardware accelerator over the software routine.
- Include multiple coloring schemes and add this to the menu selection.
- Modify the mouse interface to allow the user to press and the drag the mouse pointer to select the new drawing area.

Implement one or more of the features.

21.10 SUGGESTED PROJECTS

21.10.1 Floating-point hardware accelerator

IEEE 754 Standard defines the format for single-precision (32 bits) and double-precession (64 bits) floating-point representations. Design a floating-point adder and a floating-point multiplier that conform to the standard and redesign the fractal hardware accelerator using the floating-point format.

21.10.2 General fractal drawing platform

A free online book, titled *Strange Attractors: Creating Patterns in Chaos* by Julien C. Sprott, describes a method of generating fractal patterns from a class of iteration equations. Develop an embedded system to generate and display the images. The system should include a hardware accelerator to support the key computation needed and a keyboard and VGA user interface for entering commands and displaying the resulting image.

21.11 COMPLETE PROGRAM LISTING

Listing 21.9 chu_main_frac_test.c

```
/*********************************************************************
 *
 * Module:  Fractal function prototypes and main
 * File:    chu_main_frac_test.c
 * Purpose: software- and hardware-accelerator-based fractal routines
 * IP core base addresses:
 *   - SSEG_BASE: 7-segment LED display
 *   - VRAM_BASE: video SRAM
 *   - PS2_BASE: PS2 controller
 *   - SWITCH_BASE: slide switch
 *   - F_ENGINE_BASE: fractal hardware accelerator engine
 *
 *********************************************************************
 *  Register map
 *********************************************************************
 * Write (data from cpu):
 *   offset 0
 *     * bits 31-0: cx
 *   offset 1
 *     * bits 31-0: cy
 *   offset 2
 *     * bits 15-0: maximal iterations
 *   offset 3
 *     * dummy data to start operation
 * Read (data to cpu):
 *   offset 0
 *     * bit 16: 1-bit ready signal
 *     * bits 15 - 0: 16-bit iteration value
 *********************************************************************/
/* file inclusion */
/* General C library */
#include <stdio.h>
#include <unistd.h>
/* Altera-specific library */
#include "system.h"
#include "chu_avalon_gpio.h"
#include "chu_avalon_vga.h"
#include "chu_avalon_ps2.h"

/* data type definition */
typedef alt_32 fixed428;          // define 32-bit Q4.28 fixed data type
typedef alt_64 fixed856;          // define 64-bit Q8.56 fixed data type

/* macro definition */
#define f2f(x) (x)*(1.0/(1<<28))   // convert Q4.28 to floating

/* address and field definition */
#define CX_REG            0
#define CY_REG            1
#define MAX_ITR_REG       2
#define FRAC_START_REG    3
#define ITR_DATA_REG      0
#define ITR_FIELD         0x0000ffff
#define FRAC_READY_BIT    0x00010000
```

```
/* constant definition */
#define WD      (0x30000000)    // width of initial fractal: +3 in Q4.28
#define FPIX    480             // # pixels in each axis in fractal

/* 12-row-by-20-column 8-bit-color mouse pointer bitmap array */
alt_u8 MOUSE_DATA[]=
   {
   0x00, 0x00, 0x00, 0x00, 0x00, 0x00, 0x00, 0x00, 0x00, 0x00, 0x00, 0x00,
   0x00, 0xff, 0x00, 0x00, 0x00, 0x00, 0x00, 0x00, 0x00, 0x00, 0x00, 0x00,
   0x00, 0xff, 0x6d, 0x00, 0x00, 0x00, 0x00, 0x00, 0x00, 0x00, 0x00, 0x00,
   0x00, 0xff, 0x92, 0x6d, 0x00, 0x00, 0x00, 0x00, 0x00, 0x00, 0x00, 0x00,
   0x00, 0xff, 0x92, 0x92, 0x6d, 0x00, 0x00, 0x00, 0x00, 0x00, 0x00, 0x00,
   0x00, 0xff, 0x92, 0x92, 0x92, 0x6d, 0x00, 0x00, 0x00, 0x00, 0x00, 0x00,
   0x00, 0xff, 0x92, 0x92, 0x92, 0x92, 0x6d, 0x00, 0x00, 0x00, 0x00, 0x00,
   0x00, 0xff, 0x92, 0x92, 0x92, 0x92, 0x92, 0x6d, 0x00, 0x00, 0x00, 0x00,
   0x00, 0xff, 0x92, 0x92, 0x92, 0x92, 0x92, 0x92, 0x6d, 0x00, 0x00, 0x00,
   0x00, 0xff, 0x92, 0x92, 0x92, 0x92, 0x92, 0x92, 0x92, 0x6d, 0x00, 0x00,
   0x00, 0xff, 0x92, 0x92, 0x92, 0x92, 0x92, 0x92, 0x92, 0x92, 0x6d, 0x00,
   0x00, 0xff, 0x92, 0x92, 0x92, 0x92, 0x6d, 0x6d, 0x6d, 0x6d, 0x6d, 0x6d,
   0x00, 0xff, 0x92, 0x92, 0x6d, 0x92, 0x92, 0x6d, 0x00, 0x00, 0x00, 0x00,
   0x00, 0xff, 0x92, 0x92, 0x00, 0x6d, 0x92, 0x92, 0x6d, 0x00, 0x00, 0x00,
   0x00, 0xff, 0x92, 0x00, 0x00, 0x6d, 0x92, 0x92, 0x6d, 0x00, 0x00, 0x00,
   0x00, 0xff, 0x00, 0x00, 0x00, 0x00, 0x6d, 0x92, 0x92, 0x6d, 0x00, 0x00,
   0x00, 0x00, 0x00, 0x00, 0x00, 0x00, 0x6d, 0x92, 0x92, 0x6d, 0x00, 0x00,
   0x00, 0x00, 0x00, 0x00, 0x00, 0x00, 0x00, 0x6d, 0x92, 0x92, 0x6d, 0x00,
   0x00, 0x00, 0x00, 0x00, 0x00, 0x00, 0x00, 0x6d, 0x92, 0x92, 0x6d, 0x00,
   0x00, 0x00, 0x00, 0x00, 0x00, 0x00, 0x00, 0x00, 0x6d, 0x6d, 0x00, 0x00
   };
/* 2-row-by-20-column 8-bit-color mouse pointer bitmap data structure */
bmp_type MOUSE_BMP={
   12,              // width
   20,              // height
   MOUSE_DATA       // bitmap array
};
```

```
/**************************************************************************
* function: calc_frac_point_soft()
* purpose:  calculate one fractal point with Q4.28 fixed-point data type
* argument:
*   cx, cy: cx and cy values in Q4.28
*   max_itr: max number of iterations
* return: # iteration when escape condition reached
* note: return max_itr if sequence not diverge
**************************************************************************/
int calc_frac_point_soft(fixed428 cx, fixed428 cy, alt_u16 max_itr)
{
  fixed428 x, y, xx, yy, xy2;
  fixed856 xx_raw, yy_raw, xy_raw;
  int itr;

  x = cx;
  y = cy;
  itr = 0;
  do {
    /* Q4.28 multiplications */
    xx_raw = (fixed856)(x) * (fixed856)(x);
    xx = (fixed428)(xx_raw >> 28);
    yy_raw = (fixed856)(y) * (fixed856)(y);
    yy = (fixed428)(yy_raw >> 28);
    xy_raw = (fixed856)(x) * (fixed856)(y);
    xy2 = (fixed428)(xy_raw >> 27);   // 2* is same as <<1
    /* iteration equation */
    x = xx - yy + cx;
    y = xy2 + cy;
    itr++;;
  } while (((xx+yy)<0x40000000) && (itr<max_itr));
  return(itr);
}


/**************************************************************************
* function: calc_frac_point_hard()
* purpose:  calculate one fractal point with hardware accelerator
* argument:
*   frac_base: base address of fractal engine
*   cx, cy: cx and cy values in Q4.28
*   max_itr: max number of iterations
* return: # iteration when escape condition reached
* note: return max_itr if sequence not diverge
**************************************************************************/
int calc_frac_point_hard(alt_u32 frac_base, fixed428 cx, fixed428 cy,
                          alt_u16 max_itr)
{ alt_u32 data;
  int itr;

  IOWR(frac_base, CX_REG, (alt_u32) cx);
  IOWR(frac_base, CY_REG, (alt_u32) cy);
  IOWR(frac_base, MAX_ITR_REG, (alt_u32) max_itr);
  IOWR(frac_base, FRAC_START_REG, 0); // write dummy to start operation
  while(1){
    data=IORD(frac_base, ITR_DATA_REG);
    if (data & FRAC_READY_BIT)            // check 17th bit for ready signal
      break;
  }
  itr = data & ITR_FIELD;                 // get 16-bit result
  return(itr);
}
```

```
/****************************************************************
 * function: draw_fractal()
 * purpose: draw fractal graphic
 * argument:
 *   vga_base: base address of VGA video ram
 *   frac_base: base address of fractal engine
 *   engine: engine to do computation (0 for software; 1 for accelerator)
 *   cx0: initial cx value in Q4.28 (left bottom point in cx-cy plane)
 *   cy0: initial cy value in Q4.28 (left bottom point in cx-cy plane)
 *   delta: value between two successive points in fractal in Q4.28
 *   max_itr: max number of iterations
 * return:
 * note:
 ****************************************************************/
void draw_fractal(alt_u32 vga_base, alt_u32 frac_base, int engine,
                  fixed428 cx0, fixed428 cy0, fixed428 delta, int max_itr)
{
  int i, j;
  int itr;
  alt_u8 color;
  fixed428 cx, cy;

  cy = cy0;
  for(j=0; j<FPIX; j++){
    cx = cx0;
    for(i=0; i<FPIX; i++){
      /* calculate one point */
      if (engine==0)
        itr = calc_frac_point_soft(cx, cy, max_itr);
      else
        itr = calc_frac_point_hard(frac_base, cx, cy, max_itr);
      /* coloring scheme */
      if (itr==max_itr)
        color = 0x00;   // no escape; black
      else
        color = (alt_u8)((0xd0 & itr)>>6) + (alt_u8)(itr<<2) ;
      /* write a pixel on VGA */
      vga_wr_pix(vga_base, i, 479-j, color);
      cx = cx + delta;
    } // end for i
    cy = cy + delta;
  } // end for j
}
```

```c
/****************************************************************************
* function: disp_msg()
* purpose:  display message on right panel of VGA
* argument:
*   vga_base: base address of VGA video ram
* return:
* note:
****************************************************************************/
void disp_msg(alt_u32 vga_base)
{
  char help_msg[]=
    "HELP:             \n"
    "-left-click mouse  \n"
    " to select menu    \n"
    " or set center     \n"
    "-use switch to set \n"
    " desired value     \n";
  char para_msg[]=
    "PARAMETERS         \n"
    "center x:          \n"
    "center y:          \n"
    "delta   :          \n"
    "zoom    :          \n"
    "max iter:          \n"
    "engine  :          \n";
  char menu_msg[]=
    "MENU:              \n"
    "- load zoom level  \n"
    "- load max iter    \n"
    "- select engine    \n"
    "- restore initial  \n"
    "- start drawing    \n";

  vga_wr_bit_str(vga_base, 485, 0, menu_msg, 0xff, 1);
  vga_wr_bit_str(vga_base, 485, 16*10, para_msg, 0xff, 1);
  vga_wr_bit_str(vga_base, 485, 16*20, help_msg, 0xff, 1);
}
```

```
/**************************************************************************
* function: disp_param()
* purpose:  display parameter values on right panel of VGA
* argument:
*   vga_base: base address of VGA video ram
*   cx_c: center cx value in Q4.28 (center of cx-cy plane)
*   cy_c: center cy value in Q4.28 (center of cx-cy plane)
*   delta: value between two successive points in fractal in Q4.28
*   zoom: zoom factor
*   max_itr: max number of iterations
*   engine: engine used (0 for software; 1 for accelerator)
* return:
* note:
**************************************************************************/
void disp_param(alt_u32 vga_base, fixed428 cx_c, fixed428 cy_c,
                fixed428 delta, int zoom, int max_itr, int engine)
{
  char s[]="123456789";
  char erase[]= "\x7f\x7f\x7f\x7f\x7f\x7f\x7f\x7f\x7f";
  int x_offset = 485+8*10;
  int y0 = 11;
  int i;

  /* clear old field */
  for (i=0; i<6; i++)
    vga_wr_bit_str(vga_base, x_offset, 16*(y0+i), erase, 0x00, 1);
  /* draw new values */
  sprintf(s, "%+2.6f", f2f(cx_c));
  vga_wr_bit_str(vga_base, x_offset, 16*y0, s, 0xff, 1);
  sprintf(s, "%+2.6f", f2f(cy_c));
  vga_wr_bit_str(vga_base, x_offset, 16*(y0+1), s, 0xff, 1);
  sprintf(s, "%+2.6f", f2f(delta));
  vga_wr_bit_str(vga_base, x_offset, 16*(y0+2), s, 0xff, 1);
  sprintf(s, "2^%d", zoom);
  vga_wr_bit_str(vga_base, x_offset, 16*(y0+3), s, 0xff, 1);
  sprintf(s, "%d", max_itr);
  vga_wr_bit_str(vga_base, x_offset, 16*(y0+4), s, 0xff, 1);
  sprintf(s, "%d", engine);
  vga_wr_bit_str(vga_base, x_offset, 16*(y0+5), s, 0xff, 1);
}
```

```
/******************************************************************
 * function:  vga_init_mouse_ptr()
 * purpose:   initialize the mouse pointer by writing pointer bitmap
 *            at (x,y) and saving the underlying pixels in below buffer
 * argument:
 *   vga_base: base address of VGA video ram
 *   ps2_base: base address of PS2 device
 *   x: x-axis coordinate, 10 LSBs used
 *   y: y-axis coordinate, 9 LSBs used
 *   mouse: pointer to mouse pointer bitmap
 *   below: pointer to the buffer storing pixels below mouse pointer
 * return:
 *   underlying pixels at (x,y) are stored into bellow buffer
 * note:
 *   - the calling function must allocate memory for "below" buffer
 ******************************************************************/
void vga_init_mouse_ptr(alt_u32 vga_base,  alt_u32 ps2_base,
                        int x, int y, bmp_type *mouse, bmp_type *below)
{
  /* read hidden pixels */
  vga_rd_bitmap(vga_base, x, y, below);
  /* draw initial pointer */
  vga_wr_bitmap(vga_base, x, y, mouse, 1);
}
```

```
/*************************************************************************
 * function: vga_move_mouse_ptr()
 * purpose:  move the mouse pointer by writing pointer bitmap
 *           to a new coordinate
 * argument:
 *   vga_base: base address of VGA video ram
 *   ps2_base: base address of PS2 device
 *   xold: current x-axis coordinate, 10 LSBs used
 *   yold: current y-axis coordinate, 9 LSBs used
 *   below: pointer to the buffer storing pixels below mouse pointer
 *   xnew: current x-axis coordinate, 10 LSBs used
 *   ynew: current y-axis coordinate, 9 LSBs used
 *   mouse: pointer to mouse pointer bitmap
 *   mv: pointer to mouse activity data
 * return:
 *   - 1 if mouse has activities; 0 otherwise
 *   - new mouse pointer coordinates: xnew, ynew
 *   - mouse activity data: mv
 *   - underlying pixels at (xnew,ynew) are stored into below buffer
 * note:
 *   - the calling function must allocate memory for "below" buffer
 *************************************************************************/
int vga_move_mouse_ptr(alt_u32 vga_base,  alt_u32 ps2_base,
        int xold, int yold, bmp_type *below, int *xnew, int *ynew,
        bmp_type *mouse, mouse_mv_type *mv)
{
  if (mouse_get_activity(ps2_base, mv)==0)   // no movement
    return(0);
  /* calculate new mouse pointer position */
  *xnew = xold + mv->xmov;
  if (*xnew > (639 - mouse->width))
    *xnew = 639- mouse->width;
  if (*xnew<0)
    *xnew=0;
  *ynew = yold - mv->ymov;                    // VGA y-axis goes downward
  if (*ynew>(479 - mouse->height))
    *ynew = 479 - mouse->height;
  if (*ynew<0)
    *ynew=0;
  /* draw the updated mouse pointer, restore "under" area */
  vga_move_bitmap(vga_base, xold, yold, below, *xnew, *ynew, mouse);
  return(1);
}
```

```
/****************************************************************************
 * function: proc_mouse()
 * purpose:  process mouse activities
 * argument:
 *   vga_base: base address of VGA video ram
 *   ps2_base: base address of PS2 mouse
 *   sw_base: base address of switches
 *   btn_base: base address of push buttons
 *   *cx0: initial cx value in Q4.28 (left bottom point in cx-cy plane)
 *   *cy0: initial cy value in Q4.28 (left bottom point in cx-cy plane)
 *   *delta: value between two successive points in fractal in Q4.28
 *   *max_itr: max number of iterations
 *   *engine: engine used (0 for software; 1 for accelerator)
 * return:
 *   update values of cx0, cy0, delta, max_itr, engine
 * note:
 *   - function exits when clicking on "start drawing"
 ****************************************************************************/
void proc_mouse(alt_u32 vga_base, alt_u32 ps2_base, alt_u32 sw_base,
                fixed428 *cx0, fixed428 *cy0, fixed428 *delta,
                int *max_itr, int *engine)
{
  alt_u32 sw;
  /* fractal related variables */
  static int zoom = 0;
  static int mitr = 30;
  static fixed428 cx_c = 0xf8000000;   // center of cx axis
  static fixed428 cy_c = 0x00000000;   // center of cy axis
  fixed428 cx_c_new, cy_c_new, delta_old, delta_new;
  /* mouse pointer related variables */
  static int px = 240;   // x coordinate of mouse
  static int py = 240;   // y coordinate of mouse
  int act, px_new, py_new;
  mouse_mv_type mv;
  /* mouse pointer pixel buffer */
  static alt_u8 bdata[20*12];
  bmp_type below={12, 20, bdata};

  delta_old = (WD/480) >> zoom;   //delta of current graph
  delta_new = delta_old;          // for panel display
  cx_c_new = cx_c;
  cy_c_new = cy_c;

  /* draw initial pointer */
  ps2_flush_fifo(ps2_base);
  vga_init_mouse_ptr(vga_base, ps2_base, px, py,&MOUSE_BMP, &below);

  while (1){
    act = vga_move_mouse_ptr(vga_base, ps2_base, px, py, &below,
                             &px_new, &py_new, &MOUSE_BMP, &mv);
    if (act){
      px = px_new;
      py = py_new;
      if(mv.lbtn){ /* left button pressed */
      sw = pio_read(sw_base);
        /* update parameters according to mouse pointer position */
        if (px<480){                           // in drawing area,
          /* set new center of drawing */
          cx_c_new = cx_c + delta_old*(px-240);
          cy_c_new = cy_c + delta_old*(240-py);
          printf("center:(%+2.6f, %+2.6f) \n",
```

```
                    f2f(cx_c_new), f2f(cy_c_new));
        }
        if (px>480 && 16<=py && py<32){      // in "load zoom" area
          if (sw>20)
            sw = 20;
          zoom = sw;
          delta_new = (WD/480)>>zoom;
          printf("zoom factor: %d (2^%d)\n", 1<<zoom, zoom);
        }
        if (px>480 && 32<=py && py<48){      // in "load max_itr" area
          mitr = (sw & 0x000003ff)*10;
          if (mitr>4095)
            mitr=4095;
          printf("max iteration: %d\n", mitr);
        }
        if (px>480 && 48<=py && py<64){      // in "select engine" area
          *engine = sw & 0x00000001;
          printf("engine #: %d\n", *engine);
        }
        if (px>480 && 64<=py && py<80){      // in "restore initial" area
          *engine = 0;
          zoom = 0;
          delta_new = (WD/480) >> zoom;
          mitr = 30;
          cx_c_new = 0xf8000000;   // -0.5 in Q4.28
          cy_c_new = 0x00000000;   //  0.0 in Q4.28
        }
        if (px>480 && 80<=py && py<96){      // in "start drawing" area
          printf("start drawing\n\n");
          break;
        }
        disp_param(vga_base, cx_c_new, cy_c_new, delta_new,
                   zoom, mitr, *engine);
      } // end if (mv.lbtn)
    } // end if (act)
  } // end while
  /* update fractal parameters */
  cx_c = cx_c_new;
  cy_c = cy_c_new;
  *delta = delta_new;
  *max_itr = mitr;
  *cx0 = cx_c_new - delta_new*240;
  *cy0 = cy_c_new - delta_new*(479-240);
  // restore image below mouse
  vga_wr_bitmap(vga_base, px, py, &below, 0);
}
```

```c
/*****************************************************************************
* function: main()
* purpose:  Draw fractal on VGA screen with mouse/switch interface
* note:
*****************************************************************************/
int main(void)
{
  fixed428 cx0 = 0xe0000000;           // -2 in Q4.28
  fixed428 cy0 = 0xe8000000;           // -1.5 in Q4.28
  fixed428 delta = WD/FPIX;
  int max_itr = 30;
  int engine = 0;
  alt_u8 frac_msg[4]={sseg_conv_hex(15), 0x2f, 0x27, sseg_conv_hex(1)};

  sseg_disp_ptn(SSEG_BASE, frac_msg);     // show "Frcl" for Fractal
  printf("Fractal test \n\n");
  vga_clr_screen(VRAM_BASE,0);
  disp_msg(VRAM_BASE);
  disp_param(VRAM_BASE, cx0+240*delta, cy0+240*delta, delta,
             0, max_itr, engine);
  draw_fractal(VRAM_BASE, 0, engine, cx0, cy0, delta, max_itr);
  mouse_init(PS2_BASE);
  while (1){
    proc_mouse(VRAM_BASE, PS2_BASE, SWITCH_BASE,
               &cx0, &cy0, &delta, &max_itr, &engine);
    draw_fractal(VRAM_BASE, F_ENGINE_BASE, engine,
                 cx0, cy0, delta, max_itr);
  }
}
```

CHAPTER 22

DIRECT DIGITAL FREQUENCY SYNTHESIS

DDFS (*direct digital frequency synthesis*) is a scheme that uses digital circuits and a DAC to generate tunable analog waveforms from a single fixed clock source. In this chapter, we implement this scheme and treat this circuit as a specialized Nios II peripheral. Since only the audio-frequency DACs (within the WM8731 codec device) are available on a DE1 board, we use the DDFS circuit as a sound and music synthesizer.

22.1 INTRODUCTION

Many communication-related applications need to generate a waveform of specific frequency and phase. DDFS is a method of producing a frequency- and phase-tunable digital or analog waveforms from a single fixed clock source. The data points of the waveform are first generated in digital format and then converted to analog format by using a DAC and a low-pass filter. Since the operations within a DDFS circuit are primarily digital, it can offer fast switching between output frequencies, fine frequency resolution, and operation over a wide range of frequencies.

22.2 DESIGN AND IMPLEMENTATION

The DDFS scheme can be used to generate a variety of waveforms. In this section, we examine the synthesis and implementation of three types of waveforms:

Embedded SOPC Design with Nios II Processor and VHDL Examples. By Pong P. Chu **671**

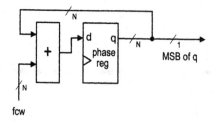

Figure 22.1   Block diagram for synthesizing digital waveform.

- Digital waveform, which is a square wave with constant amplitude.
- Unmodulated analog waveform, typically a sinusoidal waveform.
- Modulated analog waveform, in which the phase, frequency, and amplitude of the output signal can be controlled by another signal.

Once understanding the basic implementation, we can derive HDL codes accordingly.

22.2.1   Direct synthesis of a digital waveform

To synthesize a digital waveform, the DDFS scheme requires a register, which is known as the *phase register* or *phase accumulator*, and an adder, as shown in Figure 22.1. The output of the circuit is the MSB of the register. It is a square wave with the designated frequency and the duty cycle is close to 50%. The input is the `fcw` (for *frequency control word*) signal, which controls the frequency of the output signal. The value of `fcw` is added to the phase register in every clock cycle.

To explain the operation of this scheme, let us first define the relevant parameters:

- N: width (i.e., number of bits) of the register and adder.
- f_{sys} and T_{sys}: frequency and period of the system clock.
- f_{out} and T_{out}: frequency and period of the output signal.
- M: the value of `fcw`.

This system works as follows. For the phase register, its value starts from 0 and gradually increments to $2^N - 1$ and then wraps around. If we observe the MSB of the register in the process, it starts as 0, changes to 1 when the phase register reaches halfway of $2^N - 1$, and then returns to 0 and repeats when the phase register wraps around. The duration of incrementing from 0 to $2^N - 1$ can be considered as one period of the MSB (i.e., T_{out}). Since M is added to the phase register each time, it requires $\frac{2^N}{M}$ additions to complete one circulation and the corresponding duration is $\frac{2^N}{M} * T_{sys}$; i.e.,

$$T_{out} = \frac{1}{M} * 2^N * T_{sys}$$

The equation can be rewritten in terms of frequencies:

$$f_{out} = M * \frac{f_{sys}}{2^N}$$

The $\frac{f_{sys}}{2^N}$ term can be considered as the "resolution" of a DDFS system. As N increases, finer frequency can be obtained accordingly. The typical width of N is

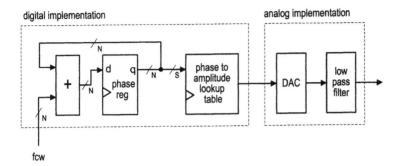

Figure 22.2 Block diagram for synthesizing analog waveform.

between 24 and 48 bits. Clearly, we can set M to a proper value to obtain the desired frequency:

$$M = \frac{f_{out}}{f_{sys}} * 2^N$$

The value of M must be rounded to a whole integer. The rounding error may introduce a small variation in T_{out} and the effect is known as *jitter*.

22.2.2 Direct synthesis of an unmodulated analog waveform

Assume the N bits of the phase register are $p_{N-1}p_{N-2}\cdots p_0$. The digital waveform uses the MSB, p_{N-1}, as the output. If we ignore the small jitter, it is a square wave with one half 0 and one half 1, each lasting $\frac{T_{out}}{2}$. The p_{N-1} bit essentially divides T_{out} into two equal regions. The second MSB, p_{N-2}, switches twice faster than p_{N-1}. If we consider the p_{N-1} and p_{N-2} bits together, they divide T_{out} into four equal regions. We can continue the process and divide T_{out} to smaller and smaller regions. The regions are commonly referred to as *phases* of the period and this is the reason that the register is known as a *phase register*.

We can generate an unmodulated analog waveform by mapping phases to digitized amplitude points and then converting the value to analog format by a DAC. The conceptual diagram is shown in Figure 22.2. The phase-to-amplitude lookup table performs the mapping and can be implemented by a ROM or RAM. The DAC converts digitized amplitude value to an analog value and the low-pass filter removes unwanted high-frequency signals. The "shape" of an analog waveform is determined by the values loaded to the lookup table and thus the DDFS scheme can generate any type of analog waveform. The sinusoidal waveform is used in most applications.

It is neither practical and nor necessary to use all N bits for the lookup table. We usually use 8 to 10 MSBs from the N-bit phase register output. It is labeled S in the diagram. For example, to use an 8-bit (i.e., 2^8 entries) lookup table to implement the sinusoidal function, we can divide one period into 256 equally spaced points, obtain the corresponding values, and load them into the lookup table. During the DDFS operation, the lookup table is swept every T_{out} seconds and the corresponding output waveform is $\sin(2\pi f_{out}t)$. A larger S value increases the size of the lookup table but puts less constraint on the low-pass filter.

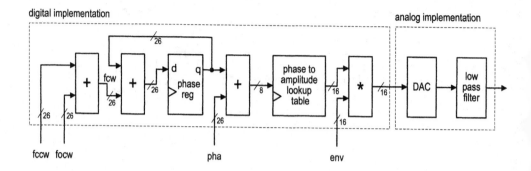

Figure 22.3   Block diagram for synthesizing modulated analog waveform.

22.2.3   Direct synthesis of a modulated analog waveform

DDFS is used widely in communication applications. A typical communication system involves two types of signals: a low-frequency "message" signal, such as an audio signal, and a high-frequency "carrier" signal to convey the message, such as the radio-frequency sinusoidal signal. *Modulation* is the process to modify the carrier signal in accordance with the message signal. The carrier signal is usually a sinusoidal waveform and the message can be used to adjust its amplitude, frequency, phase, or a combination of them. Assume that the carrier signal is $\sin(2\pi ft)$. The modulated signals become the following:

- Amplitude modulation: $A(t) * \sin(2\pi ft)$.
- Frequency modulation: $\sin(2\pi(f + \Delta f(t))t)$.
- Phase modulation: $\sin(2\pi ft + \Delta p(t))$.

The $A(t)$, $\Delta f(t)$, and $\Delta p(t)$ terms are slow time-varying signals that embed the message.

A DDFS system can incorporate the desired modulation scheme by inserting an additional adder or multiplier in its path. We can actually construct an extended DDFS system that supports all three modulation schemes. Instead of $\sin(2\pi ft)$, the extended system generates $A(t) * \sin(2\pi(f + \Delta f(t))t + \Delta p(t))$. The expanded diagram is shown in Figure 22.3. We assume that the $A(t)$, $\Delta f(t)$, and $\Delta p(t)$ terms are pre-processed and converted to proper digitized format:

- fccw: frequency control word to generate the carrier frequency, f.
- focw: frequency control word to generate the offset frequency, $\Delta f(t)$.
- pha: phase value corresponding to the desired phase offset, $\Delta p(t)$.
- env (for envelope): digitized value of $A(t)$.

The bus widths are the ones used in our implementation in Section 22.2.4.

22.2.4   HDL implementation

We can construct the digital portion of a DFFS circuit following the diagram in Figure 22.3. To accommodate the resource available on a DE1 board, we use the following parameters in the design:

- f_{sys}: 50 MHz, which is the frequency of the onboard clock.
- N: 26 bits.

- Width of the lookup table address: 8 bits (i.e., 2^8 entries).
- Number of bits per lookup table entry: 16 bits (in 2's complement signed format).

We select N to be 26 bits to obtain about 1-Hz resolution (i.e., $\frac{50*10^6}{2^{26}}$) and use the 16-bit signed format to match the codec DAC configuration.

The size of the lookup table is 2^8-by-16 (i.e., 4K bits). It is relatively small and can be implemented by using FPGA's internal embedded memory block. The embedded memory block is versatile and flexible. It can be configured as dual-port memory and initialized with specific values when the FPGA is programmed. We implement the lookup table as follows:

- Initialize the memory block with the sinusoidal waveform lookup data.
- Use one port for DDFS to read the amplitude value.
- Use one port for external system to write a different amplitude mapping, in case a different waveform is needed.

The corresponding HDL code is shown in Listing 22.1.

Listing 22.1 Sinusoidal lookup table

```vhdl
library ieee;
use ieee.std_logic_1164.all;
use ieee.numeric_std.all;
entity altera_ram_lut is
   generic(
      ADDR_WIDTH: integer:=8;
      DATA_WIDTH:integer:=16
   );
   port(
      clk: in std_logic;
      we: in std_logic;
      addr_w: in std_logic_vector(ADDR_WIDTH-1 downto 0);
      addr_r: in std_logic_vector(ADDR_WIDTH-1 downto 0);
      din: in std_logic_vector(DATA_WIDTH-1 downto 0);
      dout: out std_logic_vector(DATA_WIDTH-1 downto 0)
   );
end altera_ram_lut;

architecture beh_arch of altera_ram_lut is
   type ram_type is array (0 to 2**ADDR_WIDTH-1)
        of std_logic_vector (DATA_WIDTH-1 downto 0);
   -- sinusoidal LUT
   -- for symmetry, 0x8000 (i.e., -1) is replaced by 0x8001
   constant SIN_LUT: ram_type:=(   -- 2^8-by-16
      x"0000", x"0324", x"0648", x"096B", x"0C8C", x"0FAB", x"12C8",
      x"15E2", x"18F9", x"1C0C", x"1F1A", x"2224", x"2528", x"2827",
      x"2B1F", x"2E11", x"30FC", x"33DF", x"36BA", x"398D", x"3C57",
      x"3F17", x"41CE", x"447B", x"471D", x"49B4", x"4C40", x"4EC0",
      x"5134", x"539B", x"55F6", x"5843", x"5A82", x"5CB4", x"5ED7",
      x"60EC", x"62F2", x"64E9", x"66D0", x"68A7", x"6A6E", x"6C24",
      x"6DCA", x"6F5F", x"70E3", x"7255", x"73B6", x"7505", x"7642",
      x"776C", x"7885", x"798A", x"7A7D", x"7B5D", x"7C2A", x"7CE4",
      x"7D8A", x"7E1E", x"7E9D", x"7F0A", x"7F62", x"7FA7", x"7FD9",
      x"7FF6", x"7FFF", x"7FF6", x"7FD9", x"7FA7", x"7F62", x"7F0A",
      x"7E9D", x"7E1E", x"7D8A", x"7CE4", x"7C2A", x"7B5D", x"7A7D",
      x"798A", x"7885", x"776C", x"7642", x"7505", x"73B6", x"7255",
      x"70E3", x"6F5F", x"6DCA", x"6C24", x"6A6E", x"68A7", x"66D0",
      x"64E9", x"62F2", x"60EC", x"5ED7", x"5CB4", x"5A82", x"5843",
      x"55F6", x"539B", x"5134", x"4EC0", x"4C40", x"49B4", x"471D",
      x"447B", x"41CE", x"3F17", x"3C57", x"398D", x"36BA", x"33DF",
```

```
41        x"30FC", x"2E11", x"2B1F", x"2827", x"2528", x"2224", x"1F1A",
          x"1C0C", x"18F9", x"15E2", x"12C8", x"0FAB", x"0C8C", x"096B",
          x"0648", x"0324", x"0000", x"FCDC", x"F9B8", x"F695", x"F374",
          x"F055", x"ED38", x"EA1E", x"E707", x"E3F4", x"E0E6", x"DDDC",
          x"DAD8", x"D7D9", x"D4E1", x"D1EF", x"CF04", x"CC21", x"C946",
46        x"C673", x"C3A9", x"C0E9", x"BE32", x"BB85", x"B8E3", x"B64C",
          x"B3C0", x"B140", x"AECC", x"AC65", x"AA0A", x"A7BD", x"A57E",
          x"A34C", x"A129", x"9F14", x"9D0E", x"9B17", x"9930", x"9759",
          x"9592", x"93DC", x"9236", x"90A1", x"8F1D", x"8DAB", x"8C4A",
          x"8AFB", x"89BE", x"8894", x"877B", x"8676", x"8583", x"84A3",
51        x"83D6", x"831C", x"8276", x"81E2", x"8163", x"80F6", x"809E",
          x"8059", x"8027", x"800A", x"8001", x"800A", x"8027", x"8059",
          x"809E", x"80F6", x"8163", x"81E2", x"8276", x"831C", x"83D6",
          x"84A3", x"8583", x"8676", x"877B", x"8894", x"89BE", x"8AFB",
          x"8C4A", x"8DAB", x"8F1D", x"90A1", x"9236", x"93DC", x"9592",
56        x"9759", x"9930", x"9B17", x"9D0E", x"9F14", x"A129", x"A34C",
          x"A57E", x"A7BD", x"AA0A", x"AC65", x"AECC", x"B140", x"B3C0",
          x"B64C", x"B8E3", x"BB85", x"BE32", x"C0E9", x"C3A9", x"C673",
          x"C946", x"CC21", x"CF04", x"D1EF", x"D4E1", x"D7D9", x"DAD8",
          x"DDDC", x"E0E6", x"E3F4", x"E707", x"EA1E", x"ED38", x"F055",
61        x"F374", x"F695", x"F9B8", x"FCDC");
      signal ram: ram_type:=SIN_LUT;
   begin
      process(clk)
      begin
66       if (clk'event and clk = '1') then
            if (we = '1') then
               ram(to_integer(unsigned(addr_w))) <= din;
            end if;
            dout <= ram(to_integer(unsigned(addr_r)));
71       end if;
      end process;
   end beh_arch;
```

Note that the code is only valid for an FPGA device with this type of embedded memory block.

After constructing the lookup table, we can derive the top-level DDFS system code accordingly, as shown in Listing 22.2.

Listing 22.2   DDFS system

```
  library ieee;
2 use ieee.std_logic_1164.all;
  use ieee.numeric_std.all;
  entity ddfs is
     generic(PW: integer:=26); — width of phase accumulator
     port (
7       clk, reset: in  std_logic;
        fccw: in std_logic_vector(PW-1 downto 0);
        focw: in std_logic_vector(PW-1 downto 0);
        pha: in std_logic_vector(PW-1 downto 0);
        env: in std_logic_vector(15 downto 0);
12      — p2a ram interface
        p2a_we: in std_logic;
        p2a_waddr: in std_logic_vector(7 downto 0);
        p2a_din: in std_logic_vector(15 downto 0);
        p2a_aout: out std_logic_vector(15 downto 0);
17      p2a_pout: out std_logic_vector(PW-1 downto 0)
        );
  end ddfs;
```

```
   architecture arch of ddfs is
22     signal fcw, p_reg, p_next, pcw: unsigned(PW-1 downto 0);
       signal p2a_raddr: std_logic_vector(7 downto 0);
       signal amp: std_logic_vector(15 downto 0);
       signal modu: signed(31 downto 0);

27 begin
       -- instantiate sin ROM
       p2a_ram: entity work.altera_ram_lut
       port map(
         clk=>clk, we=> p2a_we, addr_w=>p2a_waddr, din=>p2a_din,
32       addr_r=>p2a_raddr, dout=>amp);
       -- phase register
       process (clk, reset)
       begin
         if reset='1' then
37           p_reg <= (others=>'0');
         elsif (clk'event and clk='1') then
             p_reg <= p_next;
         end if;
       end process;
42     -- frequency modulation
       fcw <= unsigned(fccw) + unsigned(focw);
       -- phase accumulation
       p_next <= p_reg + fcw;
       -- phase modulation
47     pcw <= p_reg + unsigned(pha);
       -- phase to amplitude mapping address
       p2a_raddr <= std_logic_vector(pcw(PW-1 downto PW-8));
       -- amplitude modulation
       -- Q16.0 * Q1.15 => modu is Q17.15
52     -- the -1 is not used and MSB of modu is always 0
       modu <= signed(env) * signed(amp);   -- modulated output
       p2a_aout <= std_logic_vector(modu(30 downto 15));
       p2a_pout <= std_logic_vector(p_reg);
   end arch;
```

The code follows the diagram in Figure 22.3. Since there are combinational multiplication modules in a Cyclone II device, the * operator is used in the code directly.

Note that both the lookup table output (i.e., **amp**) and **env** are 16 bits wide. After multiplication, we must trim the 32-bit multiplication result back to 16 bits. This issue can be solved by representing the signals in fixed-point format, as discussed in Section 21.2. We assume that **amp** is in $Q16.0$ format and **env** is in $Q1.15$ format (i.e., between -1 and $+1$). The multiplication result, **modu**, is in $Q17.15$ format. We can select the proper portion of the **modu** signal and trim it back to $Q16.0$ format (i.e., 16-bit signed integer) to match the input format of the codec DAC.

22.3 DDFS IP CORE DEVELOPMENT

22.3.1 Avalon interface

We can add a wrapping circuit for the Mandelbrot set fractal accelerator and create an SOPC component. It includes an Avalon MM slave interface to interact with the host, a clock input interface for the system clock, and a conduit interface for the DDFS circuit's output signal.

22.3.2   Register map

A Nios II processor configures the DDFS circuit by specifying the values of four key parameters, fccw, focw, pha and env, and by loading data to the lookup table when needed. The wrapping circuit contains four registers to store the values. Since the output of the DDFS circuit is too fast for a processor to handle, it is not connected to the Avalon MM interface and thus no read operation is needed. The registers, their address offsets, and fields are:

- Write addresses (data from cpu)
  - offset 0x000 (fccw register)
    * bits 25 to 0: 26-bit carrier frequency control word
  - offset 0x001 (focw register)
    * bits 25 to 0: 26-bit offset frequency control word
  - offset 0x002 (phase register)
    * bits 25 to 0: 26-bit phase offset word
  - offset 0x003 (envelope register)
    * bits 15 to 0: 16-bit amplitude modulation envelope
  - offset 0x100 to 0x1ff (phase-to-amplitude lookup table entries)
    * bits 15 to 0: 16-bit lookup table data

22.3.3   Wrapped DDFS circuit

The wrapped DDFS circuit contains four registers for the relevant DDFS parameters and write decoding logic to generate enable signals. The HDL code is shown in Listing 22.3.

Listing 22.3   DDFS circuit Avalon interface

```
library ieee;
use ieee.std_logic_1164.all;
use ieee.numeric_std.all;
4 entity chu_avalon_ddfs is
   port (
      clk, reset: in  std_logic;
      -- external interface
      ddfs_data_out: out std_logic_vector(15 downto 0);
9     -- avalon interface
      ddfs_address: in std_logic_vector(8 downto 0);
      ddfs_chipselect: in std_logic;
      ddfs_write: in std_logic;
      ddfs_writedata: in std_logic_vector(31 downto 0)
14    );
   end chu_avalon_ddfs;

   architecture arch of chu_avalon_ddfs is
      signal fccw_reg, focw_reg: std_logic_vector(25 downto 0);
19     signal pha_reg: std_logic_vector(25 downto 0);
      signal env_reg: std_logic_vector(15 downto 0);
      signal wr_en, wr_p2a_ram: std_logic;
      signal wr_fccw, wr_focw, wr_pha, wr_env: std_logic;

24 begin
      -- instantiation
      ddfs_unit: entity work.ddfs
```

```
      generic map(PW=>26)
      port map(
29       clk=>clk, reset=>reset,
         fccw=>fccw_reg, focw=>focw_reg,
         pha=>pha_reg, env=>env_reg,
         -- p2a ram interface
         p2a_we=>wr_p2a_ram,
34       p2a_waddr=>ddfs_address(7 downto 0),
         p2a_din=>ddfs_writedata(15 downto 0),
         -- ddfs output
         p2a_aout=>ddfs_data_out,
         p2a_pout=>open);
39    -- registers
      process (clk, reset)
      begin
         if reset='1' then
            fccw_reg <= (others=>'0');
44          focw_reg <= (others=>'0');
            pha_reg  <= (others=>'0');
            env_reg  <=  x"7fff";  -- almost 1.00
         elsif (clk'event and clk='1') then
            if wr_fccw='1' then
49             fccw_reg <= ddfs_writedata(25 downto 0);
            end if;
            if wr_focw='1' then
               focw_reg <= ddfs_writedata(25 downto 0);
            end if;
54          if wr_pha='1' then
               pha_reg <= ddfs_writedata(25 downto 0);
            end if;
            if wr_env='1' then
               env_reg <= ddfs_writedata(15 downto 0);
59          end if;
         end if;
      end process;
      -- write decoding
      wr_en <= '1' when ddfs_write='1' and ddfs_chipselect='1' else '0';
64    wr_fccw <= '1' when ddfs_address="000000000" and wr_en='1' else '0';
      wr_focw <= '1' when ddfs_address="000000001" and wr_en='1' else '0';
      wr_pha  <= '1' when ddfs_address="000000010" and wr_en='1' else '0';
      wr_env  <= '1' when ddfs_address="000000011" and wr_en='1' else '0';
      wr_p2a_ram <= '1' when  ddfs_address(8)='1' and wr_en='1' else '0';
69 end arch;
```

Based on this top-level HDL file, we can create a new SOPC component in Component Editor and integrate the DDFS circuit into a Nios II system.

22.3.4 Codec DAC integration

The last stage of a DDFS circuit consists of a DAC and analog low-pass filter. The only DAC available on the DE1 board is the DACs inside the audio codec device. When the audio IP core is developed, additional multiplexing logic is included to route the external data streams to the DACs, as discussed in Section 18.4.4. To integrate the DACs, we can instantiated a DDFS module and an audio module and connect the DDFS circuit output to the DAC input of the audio codec controller in the top-level HDL file. This scheme is used in the comprehensive Nios II testing system in Section 16.10.2. The corresponding code segment of Listing 16.14 is

```
    signal ddfs_data: std_logic_vector(15 downto 0);
  signal dac_load_tick: std_logic;
  . . .
    -- within Nios II port mapping
    ddfs_data_out_from_the_d_engine=>ddfs_data,
  . . .
    codec_dac_data_in_to_the_audio=>(ddfs_data & ddfs_data),
    codec_dac_wr_to_the_audio=>dac_load_tick,
    codec_sample_tick_from_the_audio=>dac_load_tick,
  . . .
```

The connection is done via the port mapping of the instantiated Nios II module, in which the DDFS circuit's output is connected to ddfs_data and then duplicated (for both the left and right channels) and passed to the audio controller's DACs.

Note that the data rate of the DDFS circuit is much higher than the sampling rate of the audio codec DAC. Our configuration does not attempt to synchronize the operation. The DDFS circuit generates the data continuously and the audio controller only retrieves the data when the sampling tick, which is connected to dac_load_tick, is asserted. An external board with a fast DAC device is needed to fully utilize the capability of the implemented DDFS system.

22.4 DDFS DRIVER

The driver consists of routines to initialize the DDFS circuit and to configure its parameters. To make the code clear, we define the address offsets as symbolic register names and define two constants for the phase register width and system frequency in the header file:

```
#define CHU_DDFS_FCCW_REG  0
#define CHU_DDFS_FOCW_REG  1
#define CHU_DDFS_PHA_REG   2
#define CHU_DDFS_ENV_REG   3
#define DDFS_PW            26        // phase register width
#define DDFS_F_SYS         50000000 // sys frequency (50 MHz)
```

22.4.1 Configuration routines

We use four routines to configure the DDFS circuit's four register, as shown in Listing 22.4.

Listing 22.4

```
#define power2n(n) (1<<(n))   //2^n is same as shifting 1 to left n bits

void ddfs_wr_carrier_freq(alt_u32 ddfs_base, alt_u32 freq)
{
  float tmp;
  alt_u32 fccw;

  tmp = ((float) power2n(DDFS_PW)) / ((float) DDFS_F_SYS);
  fccw = (alt_u32) (freq * tmp);
  IOWR(ddfs_base, CHU_DDFS_FCCW_REG, fccw);
}
```

```
void ddfs_wr_offset_freq(alt_u32 ddfs_base, alt_u32 freq)
{
  float tmp;
  alt_u32 focw;

  tmp = ((float) power2n(DDFS_PW)) / ((float) DDFS_F_SYS);
  focw = (alt_u32) (freq * tmp);
  IOWR(ddfs_base, CHU_DDFS_FOCW_REG, focw);
}

void ddfs_wr_pha(alt_u32 ddfs_base, int offset)
{
  alt_u32 pcw;

  pcw =  offset*DDFS_F_SYS/360 ;
  IOWR(ddfs_base, CHU_DDFS_PHA_REG, pcw);
}

void ddfs_wr_env(alt_u32 ddfs_base, alt_16 env)
{
  IOWR(ddfs_base, CHU_DDFS_ENV_REG, env);
}
```

The ddfs_wr_carrier_freq() and ddfs_wr_offset_freq() functions calculate and write carrier and offset frequency control words, respectively. The formula, $M = \frac{f_{out}}{f_{sys}} * 2^N$, is used in the calculation.

The ddfs_wr_pha() calculates the phase control word and writes the register. The input phase offset is represented in term of degrees. Since 2^N steps in the phase registers represents one period, which is 360 degrees, the amount to be added to the phase register is $\frac{offset}{360} * 2^N$.

22.4.2 Initialization routine

The ddfs_init() function initializes audio codec, connects the DDFS circuit's output to DAC, and configures the DFFS registers. It is shown in Listing 22.5.

Listing 22.5

```
void ddfs_init(alt_u32 audio_base, alt_u32 ddfs_base)
{
  audio_init(audio_base);
  audio_wr_src_sel(audio_base, 1, 0);    // DAC connected to external bus
  ddfs_wr_carrier_freq(ddfs_base, 440);  // mid-A frequency
  ddfs_wr_offset_freq(ddfs_base, 0);
  ddfs_wr_pha(ddfs_base, 0);
  ddfs_wr_env(ddfs_base, 0x7fff);         // close to 1.0
}
```

22.5 TESTING

The main application of DDFS is in communication systems, usually involving the generation and modulation of high-frequency signals. However, since the FPGA prototyping board contains only an audio DAC, our testing program is oriented to the generation and synthesis of sound and music. The system derived in Section 16.10.2 includes a DDFS module (named d_engine) and can be used for testing.

Table 22.1   Note frequencies of nine octaves

	oct 0	oct 1	oct 2	oct 3	oct 4	oct 5	oct 6	oct 7	oct 8
C	16.4	32.7	65.4	130.8	261.6	523.3	1046.5	2093.0	4186.0
C♯	17.3	34.7	69.3	138.6	277.2	554.4	1108.7	2217.5	4434.9
D	18.4	36.7	73.4	146.8	293.7	587.3	1174.7	2349.3	4698.6
D♯	19.5	38.9	77.8	155.6	311.1	622.3	1244.5	2489.0	4978.0
E	20.6	41.2	82.4	164.8	329.6	659.3	1318.5	2637.0	5274.0
F	21.8	43.7	87.3	174.6	349.2	698.5	1396.9	2793.8	5587.7
F♯	23.1	46.3	92.5	185.0	370.0	740.0	1480.0	2960.0	5919.9
G	24.5	49.0	98.0	196.0	392.0	784.0	1568.0	3136.0	6271.9
G♯	26.0	51.9	103.8	207.7	415.3	830.6	1661.2	3322.4	6644.9
A	27.5	55.0	110.0	220.0	440.0	880.0	1760.0	3520.0	7040.0
A♯	29.1	58.3	116.5	233.1	466.2	932.3	1864.7	3729.3	7458.6
B	30.9	61.7	123.5	246.9	493.9	987.8	1975.5	3951.1	7902.1

22.5.1   Overview of music notes and synthesis

Music notes   In music, a *note* means a specific frequency. There are 12 notes in an *octave*, represented by C, C♯, D, D♯, E, F, F♯, G, G♯, A, A♯, and B. The frequencies from octave 0 to octave 8 are summarized in Table 22.1

There is a simple relationship between two successive notes. If the frequencies of two successive notes are f_i and f_{i+1}, then

$$f_{i+1} = 2^{\frac{1}{12}} * f_i$$

The notes are standardized around the A note of octave 4 (A4), which is 440.00 Hz. The frequencies of other notes are then derived accordingly.

The previous equation indicates that a frequency is doubled after one octave, i.e.,

$$f_{i+12} = (2^{\frac{1}{12}})^{12} * f_i = 2 * f_i$$

If the frequency of a note in octave 0 is f_0, the frequency in octave i becomes $2^i * f_0$.

Music synthesis and ADSR envelope   A music instrument creates a direct acoustic sound. A music synthesizer imitates an instrument by producing electronic signals and playing them through a speaker. The unmodulated DDFS circuit can be used to generate the basic tone. Additional schemes, such as adding harmonic components, performing special frequency modulation, and applying an ADSR envelope, can be combined to produce more interesting effects. We discuss the last scheme in more detail.

When a real musical instrument produces a note, the loudness changes over time. It rises quickly from zero and then decays over time. One scheme to model the variation is to multiply the constant tone by a loudness *ADSR envelope*, which contains the *attack, decay, sustain*, and *release* segments. The ADSR envelope for a piano-like sound is shown in Figure 22.4. When a key is pressed, the loudness quickly rises to the maximum (attack segment), then falls fast (decay segment) to a rather constant level (sustain segment), which continues until the key is released.

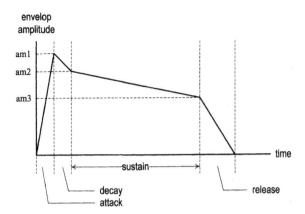

Figure 22.4   ADSR envelope for piano.

The sound then quickly fades away (release segment). We can imitate the sound of different instruments by adjusting the levels and lengths of various segments.

In communication terms discussed in Section 22.2.3, the original tone can be considered as the carrier signal and the ADSR envelope can be considered as the message. The ADSR scheme essentially performs the amplitude modulation over the original tone.

22.5.2 Testing program

We construct a simple program to verify the DDFS operation, and demonstrate the frequency and amplitude modulations. The program consists of the following tests:

- Initialize the DDFS circuit.
- Produce a single tone.
- Generate a dual-tone siren sound.
- Generate a continuously sweeping siren sound to demonstrate the frequency modulation.
- Play the musical notes in six octaves.
- Play the musical notes with the piano-like ADSR envelope to demonstrate the amplitude modulation.

The main program is shown in Listing 22.6.

Listing 22.6

```
int main(void)
{
  int sw, btn, freq, i;
  alt_u8 disp_msg[4]={SSEG_HEX_TABLE[13], SSEG_HEX_TABLE[13],
                      SSEG_HEX_TABLE[15], SSEG_HEX_TABLE[5]};

  sseg_disp_ptn(SSEG_BASE, disp_msg);          // show "ddFS" for display
  ddfs_init(ADUIO_BASE, D_ENGINE_BASE);
  btn_clear(BTN_BASE);
  printf("DDFS test \n\n");

  while (1){
    while (!btn_is_pressed(BTN_BASE)){ };        // wait for button
```

```
      btn=IORD_ALTERA_AVALON_PIO_EDGE_CAP(BTN_BASE);  // read button
      if (btn & 0x02){                                // key1 pressed
        sw=IORD_ALTERA_AVALON_PIO_DATA(SWITCH_BASE);  // read switch
        printf("key/sw: %d/%d\n", btn, sw);
      }
      btn_clear(BTN_BASE);
      switch (sw){
        case 0:
          ddfs_init(ADUIO_BASE, D_ENGINE_BASE);
          break;
        case 1:   // set ddfs frequency
          printf("enter frequency:");
          scanf("%d", &freq);
          ddfs_wr_carrier_freq(D_ENGINE_BASE, freq);
          break;
        case 2:   // play siren 1
          ddfs_play_siren1(D_ENGINE_BASE);
          break;
        case 3:   // play siren 2
          ddfs_play_siren2(D_ENGINE_BASE);
          break;
        case 4:   // play 6 octaves without envelope
          ddfs_play_oct(D_ENGINE_BASE);
          break;
        case 5:   // play 6 octaves with piano adsr envelope
          ddfs_play_piano(D_ENGINE_BASE);
          break;
      } //end switch
   } // end while
}
```

Siren generation routines The two siren generation functions demonstrate the frequency modulation and the code is shown in Listing 22.7

Listing 22.7

```
void ddfs_play_siren1(alt_u32 ddfs_base)
{
  int i;

  ddfs_wr_carrier_freq(ddfs_base, 600);       // 600 Hz carrier
  for (i=0; i<10; i++){                        // 10 cycles
    ddfs_wr_offset_freq(ddfs_base, 0);         // 0 Hz offset
    usleep(300*1000);                          // 300 ms
    ddfs_wr_offset_freq(ddfs_base, 300);       // 300 Hz offset
    usleep(300*1000);                          // 300 ms
  }
  ddfs_wr_carrier_freq(ddfs_base, 0);
  ddfs_wr_offset_freq(ddfs_base, 0);
}

void ddfs_play_siren2(alt_u32 ddfs_base)
{
  int i, j;

  ddfs_wr_carrier_freq(ddfs_base, 800);
  for (i=0; i<10; i++){                          // 10 cycles
    for (j=0; j<30; j++){                        // sweep 30 steps (750 ms)
      ddfs_wr_offset_freq(ddfs_base, j*40);      // 40 Hz increment
      usleep(25*1000);                           // 25 ms
    } // end j loop
```

```
  } // end i loop
  ddfs_wr_carrier_freq(ddfs_base, 0);
  ddfs_wr_offset_freq(ddfs_base, 0);
}
```

The `ddfs_play_siren1()` function generates a dual-tone (600 Hz and 900 Hz) siren sound, alternating every 300 ms. The `ddfs_play_siren2()` function generates a "sweeping" siren sound that increases its frequency gradually from 800 Hz to 2000 Hz within a 750-ms interval.

Note generation routines The frequency of a specific note can be obtained by applying the $f_{i+1} = 2^{\frac{1}{12}} * f_i$ formula. Since the direct calculation involves expensive floating-point operations, we use an alternative lookup table scheme. This scheme is based on the observation that if the frequency of a note in octave 0 is f_0, the frequency in octave i becomes $2^i * f_0$. We can store the precalculated frequencies of octave 0 in a table and obtain the frequencies in other octaves by multiplying the octave-0 frequency with 2^i. The multiplication operation corresponds to shift f_0 to left i positions. The code is shown in Listing 22.8.

Listing 22.8

```
void ddfs_set_note(alt_u32 ddfs_base, int oct, int ni)
{
  // frequency table for octave 0
  // octave n: f*2^n
  const float NOTES[]={
    16.3516,   //  0 C
    17.3239,   //  1 C#
    18.3541,   //  2 D
    19.4454,   //  3 D#
    20.6017,   //  4 E
    21.8268,   //  5 F
    23.1247,   //  6 F#
    24.4997,   //  7 G
    25.9565,   //  8 G#
    27.5000,   //  9 A
    29.1352,   // 10 A#
    30.8677    // 11 B
  };
  alt_u32 freq;

  ddfs_wr_offset_freq(ddfs_base, 0);
  freq = NOTES[ni] * (1<<oct);            // NOTES[]*(2^oct)
  ddfs_wr_carrier_freq(ddfs_base, freq);
}
```

The `ddfs_play_oct()` function plays the notes from octave 3 to octave 6, each lasting 500 ms. The code is shown in Listing 22.9.

Listing 22.9

```
void ddfs_play_oct(alt_u32 ddfs_base)
{
  int oct, ni;

  for (oct=3; oct<7; oct++){
    for (ni=0; ni<12; ni++){
      ddfs_set_note(ddfs_base, oct, ni);
      usleep(500*1000);
```

```
    } // end ni loop
  } // end oct loop
  ddfs_wr_carrier_freq(ddfs_base, 0);
}
```

Piano ADSR envelope routine The `ddfs_play_piano()` function modulates the original sinusoidal wave amplitude with an ADSR envelope. We assume that the envelope register is updated in every millisecond and calculate the envelope increment or decrement according to the ASDR parameters. For example, the decrement in the decay region is $\frac{am1-am2}{decay\ time}$. The code is shown in Listing 22.10. As in `ddfs_play_oct()`, it plays the notes from octave 3 to octave 6.

Listing 22.10

```
void ddfs_play_piano(alt_u32 ddfs_base)
{
  int oct, ni;
  int i, estep;
  int atime, dtime, stime, rtime;
  alt_16 env, am1, am2, am3;

  atime=10;
  dtime=50;
  stime=400;
  rtime=40;
  am1=0x7f00;
  am2=am1*0.8;
  am3=am2*0.5;

  for (oct=3; oct<7; oct++){
    for (ni=0; ni<12; ni++){
      ddfs_set_note(ddfs_base, oct, ni);  // set note frequency
      env = 0;
      /* attack region */
      estep = am1/atime;
      for (i=0; i<atime; i++){
        env = env + estep;
        usleep(1000);
        ddfs_wr_env(ddfs_base, env);
      }
      /* decay region */
      estep = (am1-am2)/dtime;
      for (i=0; i<dtime; i++){
        env = env - estep;
        usleep(1000);
        ddfs_wr_env(ddfs_base, env);
      }
      /* sustain region */
      estep = (am2-am3)/stime;
      for (i=0; i<stime; i++){
        env = env - estep;
        usleep(1000);
        ddfs_wr_env(ddfs_base, env);
      }
      /* release region */
      estep = am3/rtime;
      for (i=0; i<rtime; i++){
        env = env - estep;
        usleep(1000);
        ddfs_wr_env(ddfs_base, env);
```

```
        }
      } // end ni loop
    } // end oct loop
    ddfs_wr_carrier_freq(ddfs_base, 0);
    ddfs_wr_env(ddfs_base, 0x7fff);
}
```

22.6 BIBLIOGRAPHIC NOTES

The DDFS circuit is a key component is today's communication systems. *Direct Digital Synthesizers: Theory, Design and Applications* by J. Vankka provides detailed coverage of this subject. Basic concepts behind modulation and ADSR envelope can be found on the Wikipedia web site. *The Theory and Technique of Electronic Music* by M. Puckette discusses various techniques and their mathematical foundations of music synthesis. The book is available online and can be found by searching the title.

22.7 SUGGESTED EXPERIMENTS

22.7.1 Quadrature phase carrier generation

Many communication schemes require an additional 90 degree out-of-phase signal, known as the *quadrature* component. In other words, the $\sin(2\pi ft)$ and $\cos(2\pi ft)$ waveforms must be generated at the same time. Expand the DDFS circuit to generate both signals at the same time. Note that the FPGA's embedded memory block supports dual-port operation and thus two lookup operations can be done by using the same memory module.

22.7.2 Reduced-size phase-to-amplitude lookup table

The size of the lookup table can grow large when high-resolution output is needed. However, it can be reduced to one quarter of the original size by taking advantage of the symmetry of the sinusoidal function. We only need to include data points in the first quadrant (i.e., between 0 and $\frac{\pi}{2}$) and derive the rest data points using the following equations:

$$\sin(x) = \sin(\pi - x) \quad \text{if } \tfrac{\pi}{2} < x \leq \pi$$
$$\sin(x) = -\sin(x - \pi) \quad \text{if } \pi < x < 2\pi$$

Design the new DDFS circuit using this approach, derive the HDL code, and verify its operation.

22.7.3 Synthetic music player

A piece of music is represented by a sequence of notes and their durations. We can implement a synthetic music player using the DDFS circuit. Derive software, convert a simple song to this format and store them in an array, and verify its operation.

22.7.4   Keyboard piano

A PS2 keyboard can be used as piano keyboard to play synthesized music. Select 12 keys for the 12 notes and use digits 0 to 8 to select the desired octave. Note that the duration of the sustain segment is not fixed. It lasts until the amplitude decays to zero or when the key is released. Derive software and verify its operation.

22.7.5   Keyboard recorder

We can combine Experiments 22.7.3 and 22.7.4 and add additional recording functionality. Use keys R and T to start and stop the recording session. During a recording session, the notes and their durations are stored to an array. Use key P to play back the stored information.

22.7.6   Hardware envelope generator

We can implement the envelope generation function in hardware. The circuit should include registers storing the relevant amplitude and segment duration information and an FSMD to generate the envelope points. Note that the duration of the sustain segment is not fixed. It lasts until the amplitude decays to zero or when the key is released. Modify the DDFS IP core to incorporate the new functionality, resynthesize the Nios II system, create the necessary software driver, and verify its operation.

22.7.7   Additive harmonic synthesis

A harmonic is signal whose frequency is an integer multiple of the fundamental frequency. For example, if the fundamental frequency is f, its harmonics are $2f$, $3f$, $4f$, $\cdots$. One scheme to generate synthesized music is to add attenuated harmonics to the original signal. Expand the DDFS IP core to allow the addition of three harmonics. The integer multiple and attenuation level of each harmonic can be controlled individually. Resynthesize the Nios II system, create the necessary software driver, and verify its operation.

22.7.8   Sample-based synthesis

The waveform produced by a real music instrument is usually very complex rather than a simple sinusoidal function. One scheme to create better sound is to record a sample from a real instrument and store the waveform in the phase-to-amplitude lookup table. We can use the microphone input to record a note of an instrument (e.g., a harmonica), extract data points from one cycle, and store the results to the lookup table. Derive the software and verify its operation.

22.8   SUGGESTED PROJECTS

22.8.1   Sound generator

A sound generator is an IC chip designed to produce sound and were used widely in early computers and arcade games. A representative device is Yamaha YM2149.

Search its data sheet on the web, study its schematics and specifications, and design a similar system using HDL and FPGA.

22.8.2 Function generator

A function generator is an instrument that can produce various waveform patterns (such as sinusoidal, square, triangular, ramp, etc.) over a range of frequencies and amplitudes. It is used to test the response of circuits to common input signals. Research the basic operation and layout of a functional generator. Use the DDFS circuit to generate the desired waveform and use the VGA monitor and keyboard to create a virtual control panel. Note that because of the low sampling rate of an audio DAC, the range of the frequencies is limited between 20 and 20,000 Hz.

22.8.3 Full-fledged electric synthesizer

An electric synthesizer produces a variety of sounds. It can mimic other musical instruments and generate special effects. Several previous experiments implement parts of the functionalities. Research the functionalities, operation, and layout of an actual music keyboard. Expand the DDFS circuit to support a full-fledged electric synthesizer, and use the VGA monitor and keyboard to create a virtual control panel.

22.9 COMPLETE PROGRAM LISTING

Listing 22.11 chu_avalon_ddfs.h

```
/*****************************************************************
 *
 * Module:   ddfs header
 * File:     chu_avalon_ddfs.h
 * Purpose:  Routines to configure ddfs
 *
 *****************************************************************
 *  Register map
 *****************************************************************
 * Write (data from cpu):
 *   offset 0x000
 *     * bits 25-0: 26-bit freq (carrier) control word
 *   offset 0x001
 *     * bits 25-0: 26-bit freq (offset) control word
 *   offset 0x002
 *     * bits 25-0: 26-bit phase offset
 *   offset 0x003
 *     * bits 15-0: 16-bit envelope
 *   offset 0x100 to 0xff
 *     * bits 15-0: 16-bit phase-to-amplitude LUT data
 *
 *****************************************************************/
/* file inclusion */
#include <alt_types.h>

/*****************************************************************
 * constant definitions
 *****************************************************************/
#define CHU_DDFS_FCCW_REG    0
#define CHU_DDFS_FOCW_REG    1
#define CHU_DDFS_PHA_REG     2
#define CHU_DDFS_ENV_REG     3

#define DDFS_PW     26         // width (#bits) of ddfs phase accumulator
#define DDFS_F_SYS  50000000   // ddfs system reference frequency 50 MHz

/*****************************************************************
 * Function prototypes
 *****************************************************************/
void ddfs_wr_carrier_freq(alt_u32 ddfs_base, alt_u32 freq);
void ddfs_wr_offset_freq(alt_u32 ddfs_base, alt_u32 freq);
void ddfs_wr_pha(alt_u32 ddfs_base, int offset);
void ddfs_wr_env(alt_u32 ddfs_base, alt_16 env);
void ddfs_init(alt_u32 audio_base, alt_u32 ddfs_base);
```

Listing 22.12   chu_avalon_ddfs.c

```
/********************************************************************
 *
 * Module:  DDFS driver function prototypes
 * File:    chu_avalon_ddfs.c
 * Purpose: Routines to configure ddfs
 *
 ********************************************************************
/* file inclusion */
#include <io.h>
#include "chu_avalon_ddfs.h"
#include "chu_avalon_audio.h"

#define power2n(n) (1<<(n))    //2^n is same as shifting 1 to left n bits

/********************************************************************
 * function: ddfs_wr_carrier_freq()
 * purpose:  set freq (carrier) control word of dffs
 *           fccw = freq*2^DDFS_PW/DDFS_F_SYS
 * argument:
 *   ddfs_base: base address of ddfs
 *   freq: carrier frequency
 * return:
 ********************************************************************/
void ddfs_wr_carrier_freq(alt_u32 ddfs_base, alt_u32 freq)
{
  float tmp;
  alt_u32 fccw;

  tmp = ((float) power2n(DDFS_PW)) / ((float) DDFS_F_SYS);
  fccw = (alt_u32) (freq * tmp);
  IOWR(ddfs_base, CHU_DDFS_FCCW_REG, fccw);
}

/********************************************************************
 * function: ddfs_wr_offset_freq()
 * purpose:  set freq (offset) control word of dffs
 *           focw = freq*2^DDFS_PW/DDFS_F_SYS
 * argument:
 *   ddfs_base: base address of ddfs
 *   freq: offset frequency
 * return:
 ********************************************************************/
void ddfs_wr_offset_freq(alt_u32 ddfs_base, alt_u32 freq)
{
  float tmp;
  alt_u32 focw;

  tmp = ((float) power2n(DDFS_PW)) / ((float) DDFS_F_SYS);
  focw = (alt_u32) (freq * tmp);
  IOWR(ddfs_base, CHU_DDFS_FOCW_REG, focw);
}
```

```
/*************************************************************************
 * function:  ddfs_wr_pha()
 * purpose:   set phase offset
 *        offset = (deg/360) *2^DDFS_PW
 * argument:
 *    ddfs_base: base address of ddfs
 *    offset: phase offset in degrees
 * return:
 * note:
 *************************************************************************/
void ddfs_wr_pha(alt_u32 ddfs_base, int offset)
{
  alt_u32 pcw;

  pcw =  offset*DDFS_F_SYS/360 ;
  IOWR(ddfs_base, CHU_DDFS_PHA_REG, pcw);
}

/*************************************************************************
 * function:  ddfs_wr_env()
 * purpose:   set envelope
 * argument:
 *    ddfs_base: base address of ddfs
 *    env: envelope
 * return:
 * note:
 *************************************************************************/
void ddfs_wr_env(alt_u32 ddfs_base, alt_16 env)
{
  IOWR(ddfs_base, CHU_DDFS_ENV_REG, env);
}

/*************************************************************************
 * function:  ddfs_init()
 * purpose:   initialize ddfs
 * argument:
 *    ddfs_base: ase address of ddfs
 * note: use audio_init() to initialize codec and then
 *       route dac data to external bus
 *************************************************************************/
void ddfs_init(alt_u32 audio_base, alt_u32 ddfs_base)
{
  audio_init(audio_base);
  audio_wr_src_sel(audio_base, 1, 0);     // dac to external bus
  ddfs_wr_carrier_freq(ddfs_base, 262);   // mid C frequency
  ddfs_wr_offset_freq(ddfs_base, 0);
  ddfs_wr_pha(ddfs_base, 0);
  ddfs_wr_env(ddfs_base, 0x7fff);         // close to 1.0
}
```

Listing 22.13   chu_main_ddfs_test.c

```
/***************************************************************************
* Module:   ddfs test
* File:     chu_main_ddfs_test.c
* Purpose:  Test audio frequency DDFS
* IP core base addresses:
*    - SWITCH_BASE: switch
*    - BTN_BASE: pushbutton
*    - SSEG_BASE: 7-segment LED
*    - D_ENGINE_BASE: ddfs
*    - AUDIO_BASE: audio codec
***************************************************************************/
/* file inclusion */
#include <stdio.h>
#include <unistd.h>
#include "system.h"
#include "chu_avalon_gpio.h"
#include "chu_avalon_ddfs.h"


/***************************************************************************
* function: ddfs_play_siren1()
* purpose:  play two-frequency (600/900 Hz) siren for few seconds
* argument:
*    ddfs_base: base address of ddfs
* return:
***************************************************************************/
void ddfs_play_siren1(alt_u32 ddfs_base)
{
  int i;

  ddfs_wr_carrier_freq(ddfs_base, 600);       // 600 Hz carrier
  for (i=0; i<10; i++){                        // 10 cycles
    ddfs_wr_offset_freq(ddfs_base, 0);         // 0 Hz offset
    usleep(300*1000);                          // 300 ms
    ddfs_wr_offset_freq(ddfs_base, 300);       // 300 Hz offset
    usleep(300*1000);                          // 300 ms
  }
  ddfs_wr_carrier_freq(ddfs_base, 0);
  ddfs_wr_offset_freq(ddfs_base, 0);
}


/***************************************************************************
* function: ddfs_play_siren2()
* purpose:  play sweeping-frequncy (800-2000 Hz) siren for few seconds
* argument:
*    ddfs_base: base address of ddfs
***************************************************************************/
void ddfs_play_siren2(alt_u32 ddfs_base)
{
  int i, j;

  ddfs_wr_carrier_freq(ddfs_base, 800);
  for (i=0; i<10; i++){                             // 10 cycles
    for (j=0; j<30; j++){                           // sweep 30 steps
      ddfs_wr_offset_freq(ddfs_base, j*40);         // 40 Hz increment
      usleep(25*1000);                              // 25 ms
    } // end j loop
  } // end i loop
  ddfs_wr_carrier_freq(ddfs_base, 0);
  ddfs_wr_offset_freq(ddfs_base, 0);
}
```

```
/**************************************************************************
 * function: ddfs_set_note()
 * purpose:  set ddfs frequency for a music note
 * argument:
 *   ddfs_base: base address of ddfs
 *   oct: octave from 0 to 9 (16 to 16 kHz; "middle" is 4)
 *   ni: note index 0 to 11 (C, C#, D, D#, ..., to B)
 * return:
 * note:
 *   - 1st octave frequencies listed in a lookup table
 *   - corresponding note freq in octave n: f*2^n
 *   - listening to low- and high- octave freq needs good audio amp
 **************************************************************************/
void ddfs_set_note(alt_u32 ddfs_base, int oct, int ni)
{
   // frequency table for octave 0
   // octave n: f*2^n
   const float NOTES[]={
      16.3516,    //  0 C
      17.3239,    //  1 C#
      18.3541,    //  2 D
      19.4454,    //  3 D#
      20.6017,    //  4 E
      21.8268,    //  5 F
      23.1247,    //  6 F#
      24.4997,    //  7 G
      25.9565,    //  8 G#
      27.5000,    //  9 A
      29.1352,    // 10 A#
      30.8677     // 11 B
   };
   alt_u32 freq;

   ddfs_wr_offset_freq(ddfs_base, 0);
   freq = NOTES[ni] * (1<<oct);           // NOTES[]*(2^oct)
   ddfs_wr_carrier_freq(ddfs_base, freq);
}

/**************************************************************************
 * function: ddfs_play_oct()
 * purpose:  play notes in octaves 3 to 6; 500 ms each
 * argument:
 *   ddfs_base: base address of ddfs
 * return:
 * note:
 **************************************************************************/
void ddfs_play_oct(alt_u32 ddfs_base)
{
   int oct, ni;

   for (oct=3; oct<7; oct++){
     for (ni=0; ni<12; ni++){
       ddfs_set_note(ddfs_base, oct, ni);
       usleep(500*1000);
     } // end ni loop
   } // end oct loop
   ddfs_wr_carrier_freq(ddfs_base, 0);
}
```

```
/*********************************************************************
 * function:  ddfs_play_piano()
 * purpose:   play notes in octaves 3 to 6 with piano adsr envelope
 * argument:
 *    ddfs_base: base address of ddfs
 * return:
 * note:
 *********************************************************************/
void ddfs_play_piano(alt_u32 ddfs_base)
{
  int oct, ni;
  int i, estep;
  int atime, dtime, stime, rtime;
  alt_16 env, am1, am2, am3;

  atime=10;
  dtime=50;
  stime=400;
  rtime=40;
  am1=0x7f00;
  am2=am1*0.8;
  am3=am2*0.5;

  for (oct=3; oct<7; oct++){
    for (ni=0; ni<12; ni++){
      ddfs_set_note(ddfs_base, oct, ni);   // set note frequency
      env = 0;
      /* attack region */
      estep = am1/atime;
      for (i=0; i<atime; i++){
        env = env + estep;
        usleep(1000);
        ddfs_wr_env(ddfs_base, env);
      }
      /* decay region */
      estep = (am1-am2)/dtime;
      for (i=0; i<dtime; i++){
        env = env - estep;
        usleep(1000);
        ddfs_wr_env(ddfs_base, env);
      }
      /* sustain region */
      estep = (am2-am3)/stime;
      for (i=0; i<stime; i++){
        env = env - estep;
        usleep(1000);
        ddfs_wr_env(ddfs_base, env);
      }
      /* release region */
      estep = am3/rtime;
      for (i=0; i<rtime; i++){
        env = env - estep;
        usleep(1000);
        ddfs_wr_env(ddfs_base, env);
      }
    } // end ni loop
  } // end oct loop
  ddfs_wr_carrier_freq(ddfs_base, 0);
  ddfs_wr_env(ddfs_base, 0x7fff);
}
```

```
/***************************************************************************
* function: main()
* purpose:  test audio frequency ddfs
* note:
***************************************************************************/
int main(void)
{
  int sw, btn, freq;
  alt_u8 disp_msg[4]={sseg_conv_hex(13), sseg_conv_hex(13),
                      sseg_conv_hex(15), sseg_conv_hex(5)};

  sseg_disp_ptn(SSEG_BASE, disp_msg);        // show "ddFS" for display
  ddfs_init(AUDIO_BASE, D_ENGINE_BASE);
  btn_clear(BTN_BASE);
  printf("DDFS test \n\n");

  while (1){
    while (!btn_is_pressed(BTN_BASE)){ };       // wait for button
    btn=btn_read(BTN_BASE);                      // read button
    if (btn & 0x02){                             // key1 pressed
      sw=pio_read(SWITCH_BASE);                  // read switch
      printf("key/sw: %d/%d\n", btn, sw);
    }
    btn_clear(BTN_BASE);
    switch (sw){
      case 0:
        ddfs_init(AUDIO_BASE, D_ENGINE_BASE);
        break;
      case 1:  // set ddfs frequency
        printf("enter frequency:");
        scanf("%d", &freq);
        ddfs_wr_carrier_freq(D_ENGINE_BASE, freq);
        break;
      case 2:  // play siren 1
        ddfs_play_siren1(D_ENGINE_BASE);
        break;
      case 3:  // play siren 2
        ddfs_play_siren2(D_ENGINE_BASE);
        break;
      case 4:  // play 6 octaves without envelope
        ddfs_play_oct(D_ENGINE_BASE);
        break;
      case 5:  // play 6 octaves with piano adsr envelope
        ddfs_play_piano(D_ENGINE_BASE);
        break;
    } //end switch
  } // end while
}
```

REFERENCES

1. Altera, *Cyclone II Device Handbook*, Altera Co.

2. Altera, *DE1 Development and Education Board User Manual*, Altera Co.

3. Altera, *DE2 Development and Education Board User Manual*, Altera Co.

4. Altera, *Embedded Peripherals IP User Guide*, Altera Co.

5. Altera, *Nios II Processor Reference Handbook*, Altera Co.

6. Altera, *Nios II Software Developers Handbook*, Altera Co.

7. Altera, *Quartus II Handbook*, Altera Co.

8. Altera, *SOPC Builder User Guide*, Altera Co.

9. P. J. Ashenden, *The Designer's Guide to VHDL*, 3rd ed., Morgan Kaufmann, 2008.

10. M. Barr, *Programming Embedded Systems in C and C++*, 2nd ed., O'Reilly Media, 2006.

11. L. Bening and H. D. Foster, *Principles of Verifiable RTL Design*, 2nd ed., Springer-Verlag, 2001.

12. J. Bergeron, *Writing Testbenches: Functional Verification of HDL Models*, Springer-Verlag, 2003.

13. A. Chapweske, "PS/2 Mouse/Keyboard Protocol," http://www.computer-engineering.org.

14. A. Chapweske, "PS/2 Keyboard Interface," http://www.computer-engineering.org.

15. A. Chapweske, "PS/2 Mouse Interface," http://www.computer-engineering.org.

16. P. P. Chu, *RTL Hardware Design Using VHDL: Coding for Efficiency, Portability, and Scalability*, Wiley-IEEE Press, 2006.

17. P. P. Chu, *FPGA Prototyping by VHDL Examples: Xilinx Spartan-3 version*, Wiley-IEEE Press, 2008.

18. P. P. Chu, *FPGA Prototyping by Verilog Examples: Xilinx Spartan-3 version*, Wiley-IEEE Press, 2008.

19. M. D. Ciletti, *Advanced Digital Design with the Verilog HDL*, 2nd ed., Prentice Hall, 2010.

20. M. D. Ciletti, *Starter's Guide to Verilog 2001*, Prentice Hall, 2003.

21. A. Feldman, *Designing Arcade Computer Game Graphics*, Wordware Publishing, 2000.

22. D. D. Gajski, *Principles of Digital Design*, Prentice Hall, 1997.

23. J. O. Hamblen et al., *Rapid Prototyping of Digital Systems: Quartus II Edition*, Springer, 2005.

24. IEEE, *IEEE Standard for Verilog Hardware Description Language (IEEE Std 1364-2005)*, Institute of Electrical and Electronics Engineers, 2006.

25. IEEE, *IEEE Standard VHDL Language Reference Manual (IEEE Std 1076-2008)*, Institute of Electrical and Electronics Engineers, 2009.

26. Integrated Silicon Solution, "Data Sheet of IS61LV25616AL SRAM," Integrated Silicon Solution, Inc.

27. Integrated Circuit Solution, "Data Sheet of IS42S16400 SDRAM," Integrated Circuit Solution, Inc.

28. B. Jacob et al., *Memory Systems: Cache, DRAM, Disk*, Morgan Kaufmann, 2007.

29. L. Di Jasio,*Programming 32-bit Microcontrollers in C: Exploring the PIC32*, Newnes, 2008

30. R. H. Katz and G. Borriello, *Contemporary Logic Design*, 2nd ed., Prentice Hall, 2004.

31. M. Keating and P. Bricaud, *Methodology Manual for System-on-a-Chip Designs*, 3rd ed., Springer-Verlag, 2002.

32. B. W. Kernighan and D. M. Ritchie, *C Programming Language*, 2nd ed., Prentice Hall, 1988.

33. J. J. Labrosse, *Embedded Systems Building Blocks*, 2nd ed., CMP, 1999.

34. J. J. Labrosse, *MicroC/OS II: The Real Time Kernel*, Newnes, 2002.

35. J. J. Labrosse, *uC/OS-III, The Real-Time Kernel, or a High Performance, Scalable, ROMable, Preemptive, Multitasking Kernel for Microprocessors, Microcontrollers and DSPs*, Micrium Press, 2009.

36. C. M. Maxfield, *The Design Warrior's Guide to FPGAs*, Newnes, 2004.

37. Mentor Graphics, *ModelSim Tutorial*, Mentor Graphics Corporation.

38. NXP Semiconductor, *I^2C-Bus Specification and User Manual*, NXP Semiconductor.

39. J. Nurmi, *Processor Design: System-on-Chip Computing for ASICs and FPGAs*, Springer, 2007.

40. S. Palnitkar, *Verilog HDL*, 2nd ed., Prentice Hall, 2003.

41. D. A. Patterson and J. L. Hennessy, *Computer Organization and Design: The Hardware/Software Interface*, 4th ed., Morgan Kaufmann, 2008.

42. M. Puckette, *The Theory and Technique of Electronic Music*, World Scientific Publishing, 2007.

43. J. M. Rabaey, *Digital Integrated Circuits*, 2nd ed., Prentice Hall, 2002.

44. P. R. Schaumont, *A Practical Introduction to Hardware/Software Codesign*, Springer, 2010.

45. P. Shirley and S. Marschner, *Fundamentals of Computer Graphics, 2nd ed.*, A K Peters, 2009.

46. J. Vankka, *Direct Digital Synthesizers: Theory, Design and Applications*, Springer, 2001.

47. J. F. Wakerly, *Digital Design: Principles and Practices*, Prentice Hall, 2002.

48. F. Vahid and T. D. Givargis, *Embedded System Design: A Unified Hardware/Software Introduction*, Wiley, 2001.

49. W. Wolf, *FPGA-Based System Design*, Prentice Hall, 2004.

50. W. Wolf, *Computers as Components: Principles of Embedded Computing System Design*, 2nd ed., Morgan Kaufmann, 2008.

51. Wolfson Microelectronics, *Data Sheet of WM8731*, Wolfson Microelectronics PLC.

INDEX

Printed in the United States
By Bookmasters